HYDROGEN DEFICIENT STARS AND RELATED OBJECTS

ASTROPHYSICS AND SPACE SCIENCE LIBRARY

A SERIES OF BOOKS ON THE RECENT DEVELOPMENTS
OF SPACE SCIENCE AND OF GENERAL GEOPHYSICS AND ASTROPHYSICS
PUBLISHED IN CONNECTION WITH THE JOURNAL
SPACE SCIENCE REVIEWS

Editorial Board

R.L.F. BOYD, *University College, London, England*

W. B. BURTON, *Sterrewacht, Leiden, The Netherlands*

L. GOLDBERG, *Kitt Peak National Observatory, Tucson, Ariz., U.S.A.*

C. DE JAGER, *University of Utrecht, The Netherlands*

J. KLECZEK, *Czechoslovak Academy of Sciences, Ondřejov, Czechoslovakia*

Z. KOPAL, *University of Manchester, England*

R. LÜST, *European Space Agency, Paris, France*

L. I. SEDOV, *Academy of Sciences of the U.S.S.R., Moscow, U.S.S.R.*

Z. ŠVESTKA, *Laboratory for Space Research, Utrecht, The Netherlands*

VOLUME 128
PROCEEDINGS

HYDROGEN DEFICIENT STARS AND RELATED OBJECTS

PROCEEDINGS OF THE 87TH COLLOQUIUM OF THE
INTERNATIONAL ASTRONOMICAL UNION
HELD AT MYSORE, INDIA, 10-15 NOVEMBER 1985

Edited by

KURT HUNGER

and

DETLEF SCHÖNBERNER

*Institute for Theoretical Physics and Observatory,
University of Kiel, F.R.G.*

and

N. KAMESWARA RAO

Indian Institute of Astrophysics, Bangalore, India

D. REIDEL PUBLISHING COMPANY

A MEMBER OF THE KLUWER ACADEMIC PUBLISHERS GROUP

DORDRECHT / BOSTON / LANCASTER / TOKYO

Library of Congress Cataloging in Publication Data

International Astronomical Union. Colloquium (87th : 1985 : Mysore, India)
Hydrogen deficient stars and related objects.

(Astrophysics and space science library; v. 128)
1. A stars—Congresses.   2. B stars—Congresses.   3. Cool stars—
Congresses.   4. White dwarfs—Congresses.   5. Cosmochemistry—
Congresses.   I. Hunger, Kurt.   II. Schönberner, Detlef.   III. Kameswara
Rao, N.   IV. Title.   V. Series.
QB843.A12I57   1985   523.8   86–17857
ISBN 90–277–2326–5

Published by D. Reidel Publishing Company,
P.O. Box 17, 3300 AA Dordrecht, Holland.

Sold and distributed in the U.S.A. and Canada
by Kluwer Academic Publishers,
101 Philip Drive, Assinippi Park, Norwell, MA 02061, U.S.A.

In all other countries, sold and distributed
by Kluwer Academic Publishers Group,
P.O. Box 322, 3300 AH Dordrecht, Holland.

All Rights Reserved
© 1986 by D. Reidel Publishing Company, Dordrecht, Holland
No part of the material protected by this copyright notice may be reproduced or
utilized in any form or by any means, electronic or mechanical
including photocopying, recording or by any information storage and
retrieval system, without written permission from the copyright owner

Printed in The Netherlands

TABLE OF CONTENTS

PREFACE   ix
EDITORIAL NOTE   xi
LIST OF PARTICIPANTS   xv

I. INTRODUCTION

W.P. BIDELMAN: Introductory comments   3

II. BASIC DATA

J.S. DRILLING: Basic data on hydrogen-deficient stars (Review)   9
J.S. DRILLING, U. HEBER: Radial velocities of extreme helium stars
              and of hot sdO stars   23

III HOT EXTREME HELIUM STARS

U. HEBER: Spectroscopic analyses of hot extreme helium stars
          (Review)   33
A.U. LANDOLT: Photometric properties of the extreme helium stars
          (Review)   51
U. HEBER, G. JONAS, J.S. DRILLING: High resolution spectroscopy
          of six new extreme helium stars   67
U. HEBER: Emission lines in high resolution spectra of extreme
          helium stars   73
C.S. JEFFERY: The peculiar spectrum of the extreme helium star
          BD $-9°4395$ 81
A.E. LYNAS-GRAY, D. KILKENNY, I. SKILLEN, C.S. JEFFERY: Non-radial
          pulsations in the extreme helium star HD 160641   87
C.S. JEFFERY, P.W. HILL, K. MORRISON: The period of the extreme
          helium star BD $+1°4381$ 95
C.S. JEFFERY, U. HEBER, P.W. HILL: A preliminary analysis of the
          pulsating extreme helium star V 652 Her (BD $+13°3224$)   101
P.W. HILL, C.S. JEFFERY: The radial velocity curve of V 652
          Her (BD $+13°3224$)   109
A.E. LYNAS-GRAY, D. KILKENNY: The light curve of the pulsating
          extreme helium star BD $+13°3224$: further evidence of a
          decline in the period decrease rate   117

IV COOL HYDROGEN DEFICIENT STARS

D.L. LAMBERT: The chemical composition of cool stars: II-the hydrogen deficient stars (Review)	127
M.W. FEAST: The RCB stars and their circumstellar material (Review)	151
A.V. RAVEENDRAN, N. KAMESWARA RAO, M.R. DESHPANDE, U.C. JOSHI, A.K. KULSHRESTHA: Polarimetric observations of hydrogen deficient stars	167
A.E. ROSENBUSH: Distribution of light minima of R Coronae Borealis type stars	173
S. GIRIDHAR, N. KAMESWARA RAO: Abundance analysis of R CrB variable UW Cen	177
N. KAMESWARA RAO, R. VASUNDHARA, B.N. ASHOKA: Spectrophotometric observations of R CrB during 1972, 74 minima	185
A.V. RAVEENDRAN, B.N. ASHOKA, N. KAMESWARA RAO: Photometric and radial velocity variations of R CrB near maximum light	191
R. SURENDIRANATH, K.E. RANGARAJAN, N. KAMESWARA RAO: Preliminary analysis of the broad He I emission lines in R CrB	199
K. NANDY, N. KAMESWARA RAO, D.H. MORGAN: 3.0 to 3.5 micron spectrum of V 348 Sgr and R CrB	203
J.W. MENZIES: RY Sgr: Can the time of the next deep minimum be predicted?	207
W.A. LAWSON: RY Sgr: Pulsation related phenomenon	211
D. SCHÖNBERNER, U. HEBER: Anomalous UV-extinction and the effective temperature of V 348 Sgr	217
D. SCHÖNBERNER: On the mass and luminosity of V 348 Sgr	221
D.H. MORGAN, K. NANDY, N. KAMESWARA RAO: The Large Magellanic Cloud R CrB star - HV 12842	225

V HYDROGEN DEFICIENT BINARIES

M.J. PLAVEC: Hydrogen-poor binary stars (Review)	231
K. MORRISON, J.S. DRILLING, U. HEBER, P.W. HILL, C.S. JEFFERY: Photometric and spectroscopic variability of the hydrogen-deficient binary CPD -58°2721	245
P. NAGAR, K.D. ABHYANKAR: Hydrogen deficiency in Algol secondaries	251

VI INTERMEDIATE HELIUM STARS

K. HUNGER: Intermediate helium stars: Atmospheric parameters, oblique rotators and shells (Review)	261
P.K. BARKER: Magnetic fields and winds of the intermediate helium stars (Review)	277
A.P. ODELL, S.A. VOELS: Helium-rich stellar atmosphere models for B stars	297
A.P. ODELL: Analysis of the helium strong star HD 37017	301
G. LANGHANS, U. HEBER: SB 939 - a new intermediate helium star at high galactic latitudes	309
J.M. MATTHEWS, R.W. SLAWSON, W.H. WEHLAU: Spectral variations of the rapidly oscillating Ap star HD 60435	313

VII RELATED OBJECTS

R.H. MÉNDEZ, C.H. MIGUEL, U. HEBER, R.P. KUDRITZKI: Helium rich subdwarf O stars and central stars of planetary nebulae (Review) — 323

U. HEBER, J.S. DRILLING, D. HUSFELD: UV- and visual spectroscopy of nine extremely helium rich subluminous O-stars — 345

D. HUSFELD, U. HEBER, J.S. DRILLING: NLTE-analysis of three extremely helium-rich O-type subdwarfs — 353

S.R. POTTASCH, A. MAMPASO, A. MANCHADO, J. MENZIES: Hydrogen deficient planetary nebulae: preliminary results — 359

J. LIEBERT: The origin and evolution of helium-rich white dwarfs (Review) — 367

J. LIEBERT, F. WESEMAEL, C.J. HANSEN, G. FONTAINE, H.L. SHIPMAN, E.M. SION, D.E. WINGET, R.F. GREEN: Temperatures for hot and pulsating helium-rich (DB) white dwarfs obtained with the IUE observatory — 387

I. BUES: Line band profiles in the spectra of cool magnetic helium-rich white dwarfs — 391

K.R.N. KUTTY, T.M.K. MARAR, V.N. PADMINI, S. SEETHA, K. KASTURIRANGAN, U.R. RAO, J.C. BHATTACHARYYA, S. MOLIN, K. JAYAKUMAR: Detection of an extremely active state of AM Canum Venaticorum — 397

VIII IRAS - RESULTS

H.J. WALKER: IRAS results for hydrogen deficient stars (Review) — 407

IX THEORY

H. SAIO: Pulsations of hydrogen deficient stars (Review) — 425

Y.A. FADEYEV: Theory of dust formation in R Coronae Borealis stars (Review) — 441

G. MICHAUD: Diffusion and He overabundances: hydrodynamical implications (Review) — 453

D. SCHÖNBERNER: Evolutionary status and origin of extremely hydrogen-deficient stars (Review) — 471

A. TUTUKOV: On the origin of helium rich stars — 483

P.W. HILL: Summary — 489

X APPENDIX

J.S. DRILLING, P.W. HILL: Appendix A: A catalogue of hydrogen-deficient stars — 499

PREFACE

The first helium star was discovered in 1942, the first scientific meeting on the subject, however, took place in 1985. The meeting was hence long overdue for, in the meantime, a substantial amount of material had been accumulated by a rather small, but active scientific community. Hence, it appeared necessary to review the field in order to define the subject, assess its present status and discuss future developments.

Hydrogen deficiency is a widespread phenomenon, occurring in a large variety of stellar and nonstellar objects. It can be readily detected in B stars as these exhibit both hydrogen and helium lines, if the elements are present in appreciable amounts. It becomes less manifest in cool stars, where the temperature is too low to excite helium and where one has to devise indirect methods for proving hydrogen deficiency. Clearly, it was not possible to discuss the whole complex of hydrogen deficiency, i.e. in both stars and diffuse matter, but rather to concentrate on the issue of helium stars.

The scope of the meeting was further determined by the intention to bring together predominantly those scientists who work in the actual field of hydrogen-deficient stars, as it was vital in this first meeting on the subject to set the right accents. To outline this in some detail: the helium stars are divided into two distinct classes, those with hydrogen down by a factor 1000, and those with equal amounts by number of hydrogen and helium. The former we call "extreme helium stars", the others "intermediate helium stars". These two groups represent two totally distinct groups with respect to age, mass and evolution. The extreme helium stars appear to be old, evolved stars with masses of the order of unity, while the intermediate helium stars in most cases appear to belong to rather young or intermediate populations, with masses of the order of 3 solar masses or even main sequence masses. While in the extreme helium stars the helium enrichment of the photospheres appears to be genuine, that in the intermediate helium stars may be the result of diffusion. At least, this subgroup of intermediate helium stars, which has near main sequence star masses, is intimately related to the Ap-stars. However, as we do not want to reiterate the Ap-star physics, a topic that has been dealt with abundantly in the past, we made a cut in the program. We also made a cut at the hot end of the H.-R. diagram for similar reasons: we left out the WR stars, although they are definitively hydrogen-deficient objects. However, their physics differs widely from that of our helium stars, and meetings on WR stars have also been quite frequent in the past. A slight concession was made, however, towards the white dwarfs as some of these stars are no doubt genetically linked to our helium stars.

The central and most startling problem in the field of helium stars, something which has puzzled us from the very beginning, is how extreme helium stars are formed and how a star of one solar mass may get rid of all its original hydrogen. A few rivalling hypotheses are known but up to now none of them are convincing.

The aim of the meeting was to bring us closer to the answer and discuss paths along which a solution to the above problem can be found, both theoretically and, probably more so, by new methods of observation. To this end, the item "joint discussion" was included in the program, the discussion centering on the point as to whether the Hubble Space Telescope can be used for our key problem. As a result, a number of international collaborative programs have been started during the meeting, comprising further instruments such as IRAS, ESO, CASPEC and, possibly, SEST.

The colloquium was organized by a scientific organizing committee consisting of: J.S. Drilling, M.V. Feast, G.H. Herbig, P.W. Hill, I.M. Kopylov, M. Peimbert, N. Kameswara Rao, D. Schönberner, A.V. Tutukov and K. Hunger (Chairman), and a local organizing committee consisting of: K.R. Anantharamaiah, R.C. Kapoor, P.V. Kulkarni, D.C.V. Mallik, T.M.K. Marar, V.R. Venugopal and N. Kameswara Rao (Chairman). The colloquium was jointly sponsored by the presidents of the IAU commissions 27, 29, 34 and 35. The meetings were held at the famous Lalitha Mahal Palace in Mysore.

The participants are very much indebted to the following supporting organizations: International Astronomical Union; Indian Institute of Astrophysics; Indian National Science Academy; C.Z. Instruments India Pvt. Ltd.; Central Food Technological Research Institute, Mysore; Indian Tourism Development Corporation Jaycees, Mysore; Karnataka Tourism Development Corporation; Tata Consulting Engineers; Vikrant Tyres; Walchandnagar Industries Ltd..

Institut für Theoretische Physik   K. Hunger
und Sternwarte, Kiel
April, 1986

EDITORIAL NOTE

Due to technical problems, the transcripts of the discussion recordings were incomplete and occasionally damaged. The editors tried their best to correct for this. It cannot be excluded, though, that in a few cases our printed version does not fully reflect what the speaker intended to state. The editors apologize for this.

Even more regrettable is that for a major part of the contributions the discussions have been lost entirely. This may lead to the impression that no discussions took place, whereas the opposite was the case: there was not a single contribution without discussion. The editors apologize to the authors concerned.

The editors are happy that John Drilling and Phil Hill agreed to compile a list of objects which is reproduced in the Annex. In view of the many newly discovered helium stars, the reports of which are scattered in literature, such a list appears especially important and will add to the value of this volume.

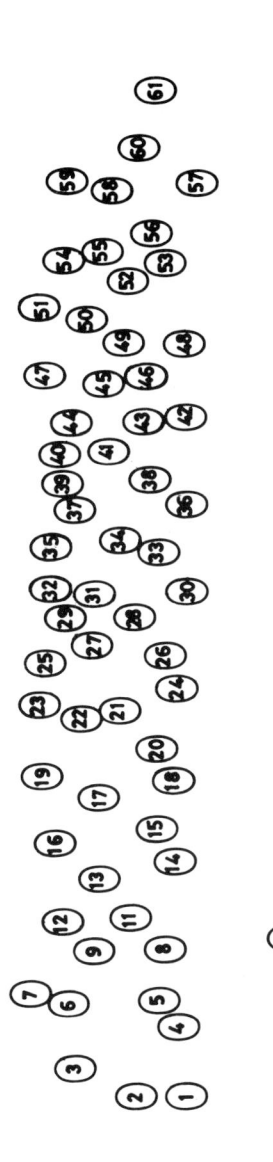

1. Babu,G.S.D.
2. Balakrishnan,A.P.
3. Lawson,W.A.
4. Balasubramaniam,K.S.
5. Tyagaraj,D.
6. Saio,H.
7. Wing,R.F.
8. Kapoor,R.C.
9. Liebert,J.W.
10. Kalyani Rao
11. Lyubimkov,L.S.
12. Mendez,R.H.
13. Drilling,J.S.
14. Hunger,K.
15. Wehlau,A.
16. Praveen Nagar
17. Wehlau,W.H.
18. Barker,P.K.
19. Rautela,B.S.
20. Kilambi,G.C.
21. Garrison,R.F.
22. Lynas-Gray,A.E.
23. Odell,A.P.
24. Deshpande (Jr.)
25. Feast,M.W.
26. Abhyankar,K.D.
27. Schönberner,D.
28. Bues,I.
29. Deshpande,M.R.
30. Bhattacharyya,J.C.
31. Rangarajan,K.E.
32. Mrs.Deshpande
33. Shylaja,B.S.
34. Mohan Rao,D.
35. Lambert,D.L.
36. Walker,H.J.
37. Gurm,H.S.
38. Badalia,J.K.
39. Jyotsna,V.
40. Vasundhara,R.
41. Ashoka,B.N.
42. Bhatt,H.C.
43. Mallik,D.C.V.
44. Raveendran,A.V.
45. Joshi,S.C.
46. Vardya,M.S.
47. Tutukov,A.V.
48. Rao,N.K.
49. Marar,T.M.K.
50. Kutty,N.
51. Pugach,A.F.
52. Rao,P.V.
53. Pottasch,S.R.
54. Mrs.Rao,A.R.
55. Rao,A.R.
56. Heber,U.
57. Seal,P.
58. Syl Reis
59. Vyas,M.L.
60. Jeffery,C.S.
61. Surendiranath,R.

LIST OF PARTICIPANTS

Abhyankar K.D., Osmania University, Hyderabad

Ashoka B.N., Indian Institute of Astrophysics, Bangalore

Babu G.S.D., Indian Institute of Astrophysics, Bangalore

Badalia J.K., Punjabi University, Patiala

Bagare S.P., Indian Institute of Astrophysics, Bangalore

Bhatt H.C., Indian Institute of Astrophysics, Bangalore

Bhattacharyya J.C., Indian Institute of Astrophysics, Bangalore

Balasubramaniam K.S., Indian Institute of Astrophysics, Bangalore

Barker P.K., University Of Western Ontario, Ontario

Bues I., Astronomisches Institut der Universität Erlangen, Bamberg

Drilling J.S., Louisiana State University, Baton Rouge

Deshpande M.R., Physical Research Laboratory, Ahmedabad

Feast M.W., South African Astronomical Observatory, Cape

Garrison R.F., David Dunlop Observatory, Ontario

Gurm H.S., Punjabi University, Patiala

Heber P.W., Institut für Theoretische Physik und Sternwarte der Universität, Kiel

Hill P.W., University Observatory, St. Andrews

Hunger K., Institut für Theoretische Physik und Sternwarte der Universität, Kiel

Jeffery C.S., University Observaory, St. Andrews

Joshi S.C., U.P. State Observatory, Nainital

Jyotsna V., Indian Institute of Astrophysics, Bangalore

Kameswara Rao N., Indian Institute of Astrophysics, Bangalore

Kapoor R.C., Indian Institute of Astrophysics, Bangalore

Kilambi G.C., Osmania University, Hyderabad

Kutty K.R.N., ISRO Satellite Centre, Bangalore

Lambert D.L., University of Texas, Austin

Lawson W.A., University of Canterbury, Christchurch

Liebert J.W., Steward Observatory, Tucson

Lynas-Gray A.E., University College, London

Lyubimkov L.S., Crimean Astrophysical Observatory, Crimea

Mallik D.C.V., Indian Institute of Astrophysics, Bangalore

Marar T.M.K., ISRO Satellite Centre, Bangalore

Mendez R.H., Instituto de Astronomia Fisica del Espacio, Buenos Aires

Michaud G.J., Université de Montréal, Montréal

Mohan Rao D., Indian Institute of Astrophysics, Bangalore

Odell A.P., Institute for Astronomy, University of Vienna

Pottasch S.R., Kapteyn Laboratorium, The Netherlands

Praveen Nagar, Osmania University, Hyderabad

Pugach A.F., Main Astronomical Observatory, Ukranian

Ramadurai S., Indian Institute of Science, Bangalore

Rangarajan K.E., Indian Institute of Astrophysics, Bangalore

Rao A.R., Tata Institute of Fundamental Research, Bombay

Raveendran A.V., Indian Institute of Astrophysics, Bangalore

Rautela B.S., U.P.State Observatory, Nainital

Saio D., University of Tokyo, Tokyo

Schönberner D., Institut für Theoretische Physik und Sternwarte der Universität, Kiel

LIST OF PARTICIPANTS

Seal P., Indian Institute of Astrophysics, Bangalore

Shylaja B.S., Indian Institute of Astrophysics, Bangalore

Surendiranath R., Indian Institute of Astrophysics, Bangalore

Tapde S.C., Indian Institute of Astrophysics, Bangalore

Tutukov A.V., Astronomical Council of the Academy of Sciences, Moscow

Vardya M.S., Tata Institute of Fundamental Research, Bombay

Vasundhara R., Indian Institute of Astrophysics, Bangalore

Venugopal V.R., Radio Astronomy Centre, Ootacamund

Vivekananda Rao P., Osmania University, Hyderabad

Vyas M.L., Osmania University, Hyderabad

Walker H.J., Queen Mary College, London

Wehlau W.H., University of Western Ontario, Ontario

Wing R.F., Ohio State University, Columbus

HIGHLIGHTS OF THE DISCUSSIONS

It is very easy to make a guess; it is very difficult to prove something.
                                                                    Tutukov

The unwritten rule whenever a talk on magnetic stars is given: never to ask basic questions.
                                                                    Liebert

There are no real spectral features, but some spectroscopists never give up.
                                                                    Walker

I. INTRODUCTION

INTRODUCTORY COMMENTS

        William P. Bidelman
        Warner & Swasey Observatory
        Case Western Reserve University
        Cleveland, OH 44106   USA

    The chairman has kindly encouraged, not to say entreated, me to write a few remarks concerning the subject of your symposium, which I was unfortunately unable to attend. This I am happy to do: the hydrogen-deficient stars are dear to my heart and even though I haven't contributed anything to the subject for several years, it is certainly nice to be remembered. From an outsider, then, a few thoughts.

    To quote from Miss Payne, in her classical study of 1925: The uniformity of composition of stellar atmospheres appears to be an established fact." Certainly for the time that statement was beyond reproach. Yet even then the seeds of hydrogen deficiency had already been sown. Mrs. Fleming, in noting the presence of bright Hβ in υ Sagittarii, in 1891, further states that its spectrum "is remarkable, since the hydrogen lines are very faint and of the same intensity as the additional dark lines." Further, Ludendorff, in a paper written on Aug. 16, 1906, discovered the complete absence of the Hγ line in R Coronae Borealis (a similar situation with respect to Hβ and Hδ being confirmed by Frost). And by a remarkable coincidence, a Harvard objective-prism plate taken the very same day was described by Miss Cannon as showing very little absorption at the G band. Both HD 30353 and RY Sagittarii are stated in the Henry Draper Catalogue to show a spectral resemblance to R CrB. And finally, the non-typical weakness of the G band of the carbon star HD 182040 was pointed out by Rufus as early as 1923.

    There was considerable reluctance to accept the possibility of a deficiency of hydrogen in stellar atmospheres: in 1923 Joy and Humason noted that the hydrogen lines were "greatly weakened by partial emission" in the spectrum of R CrB. Plaskett's 1927 study of υ Sgr suggested that the simultaneous appearance of helium and metallic lines in its spectrum might be "due to a supernormal abundance of helium or to the star being an exaggerated form of pseudo-cepheid or giant." The latter point of view was adopted by Miss Payne in her 1930 monograph. It was only with Berman's study of R CrB in 1935 and Struve and Sherman's and Greenstein's work on υ Sgr in 1940 that astronomers were forced to the conclusion that, somehow, a very substantial amount of hydrogen had been lost in a few exceptional stars.

Since that time a considerable number of luminous hydrogen-deficient stars--a recent list being contained in the writer's paper in IAU Symposium No. 83 (from which the star CoD -37°9248 should be deleted)-- have been studied. A few are single-line spectroscopic binaries, but the majority, including the numerous stars of the R CrB type, do not appear to be so.

In recent years the hydrogen-deficient character of a considerable number of subluminous O and B stars has been recognized, as has also the existence of the H-poor white dwarfs. Further, the hydrogen-deficient nature of at least the carbon Wolf-Rayet stars seems to have been finally established. This then completes the roster of the generally-recognized hydrogen-deficient objects. I shall mention some other candidates a bit later.

On the theoretical side the first consideration of hydrogen deficiency that I know of is due to Russell, who in 1933 wrote: "Suppose that some stars contain a considerably less overwhelming excess of hydrogen than the average. If the difference extends to the interior, as well as the surface, these stars will be brighter than the mass-luminosity relation indicates; they will be, or tend to be, supergiants...owing to the low density, the (Balmer) lines will be sharper and may appear fainter than in normal stars. The lines of other elements will be stronger than usual...especially the enhanced lines of the metals...Lines of high excitation, ordinarily absent, may appear." Unfortunately, he went too far, as we all tend to do, by adding that "these predictions of theory amount almost to a description of the spectrum of Alpha Cygni, and the suggestion that this and perhaps other c-stars are deficient in hydrogen appears plausible. To attribute the deficiency to partial exhaustion of hydrogen by processes of atomic synthesis in these very luminous stars is tempting."

Scanning the early literature on the hydrogen-deficient stars does not provide any enlightenment as to the cause of the phenomenon. For the binaries among them the suggestion---by Louis Henyey in the mid-50's to the best of my recollection---that they might result from a tidal-stripping process a la β Lyrae seems quite plausible now, but the large number of presumably single stars in the group would appear to indicate that the stars can lose their hydrogen-rich regions on their own. You will no doubt hear more of both explnations during the next few days.

Finally, a few suggestions for further work. First, I believe that a study of additional stars for which the evidence of hydrogen-deficiency is not quite so obvious would be well worthwhile. Are the so-called helium-rich stars actually also to some extent hydrogen-poor? What about the not-so-typical Ap star HR 6870 whose spectrum shows high-excitation lines of Ti III and Cl II? How about 3 Puppis, an early A-type spectroscopic binary with the same period as υ Sgr whose spectrum shows strong emission lines of [O I] and Ca II and which has enormous infrared excess? What about <u>any</u> stars that have suffered significant mass loss? Second, I have always been greatly disappointed by the lack of complete data on the light variations of υ Sgr and HD 30353. There must be <u>some</u> significance to this observable, perhaps very important, perhaps not.

INTRODUCTORY COMMENTS

Thus I conclude my opening remarks. I hope that they have not been too content-deficient! I am sure that by the end of this conference you will all have more than enough new observations to make and theories to concoct. May the force---or even better all four (or more) be with you!

REFERENCES

Berman, L. 1935, Astrophys. J. **81**, 369.
Bidelman, W. P. 1979, in IAU Symposium no. 83, Mass Loss and Evolution of O-type Stars, P. S. Conti and C. W. H. de Loore, eds., p. 305.
Cannon, A. J. 1912, Harvard Ann. **56**, p. 107.
Fleming, W. P. 1891, Astron. Nach. **126**, 165.
Greenstein, J. L. 1940, Astrophys. J., **91**, 438.
Joy, A. H. and Humason, M. L. 1923, Publs. Astron. Soc. Pacific **35**, 327.
Ludendorff, H. 1906, Astron. Nach. **173**, 3.
Payne, C. H. 1925, Stellar Atmospheres (Cambridge, Mass.: the Observatory), p. 189.
Payne, C. H. 1930, The Stars of High Luminosity (New York: McGraw-Hill), p. 146.
Plaskett, J. S. 1927, Publ. Dom. Astrophys. Obs. **4**, p. 1.
Rufus, W. C. 1923, Pub. Obs. Michigan **3**, p. 260.
Russell, H. N. 1933, Astrophys. J. **78**, 296.
Struve, O., and Sherman, F. 1940, Astrophys. J. **91**, 428.

II.   BASIC DATA

BASIC DATA ON HYDROGEN-DEFICIENT STARS

J. S. Drilling
Department of Physics and Astronomy
Louisiana State University
Baton Rouge, USA

ABSTRACT. The current state of our knowledge on the distribution and motions of various types of hydrogen-deficient stars, and of their positions in the H-R diagram, is reviewed. It is concluded that the extreme helium stars (with the exception of the H-deficient binaries) and cool hydrogen-deficient stars belong to the population of the Galactic nuclear bulge, whereas the intermediate helium stars are young stars of Population I. The helium-rich sdO stars appear to be a local sample which is predominantly Population I.

1. INTRODUCTION.

This review will cover the basic data (classification, surveys, distribution, motions, effective temperatures, and absolute magnitudes) of hot extreme helium stars (including the hydrogen-deficient binaries), cool hydrogen-deficient stars, intermediate helium stars, and helium-rich subdwarf O stars. The positions of these objects in the Hertzsprung-Russell diagram, as well as they can be determined at present, are given in Fig. 1. The helium-rich white dwarfs (which lie below the objects shown in Fig. 1), and helium-rich central stars of planetary nebulae (which lie to the left of most of the objects shown in Fig. 1) will be discussed elsewhere in this volume.

2. HOT EXTREME HELIUM STARS

The hot extreme helium stars are characterized by strong lines of HeI and weak or absent Balmer lines at a spectral resolution of 2 Å (Hunger 1975). This definition also includes some of the helium-rich sdO stars which, however, also show evidence of higher surface gravities. At the much lower resolution of the Case-Hamburg OB star surveys (see Stephenson and Sanduleak 1971), the spectra have a nearly featureless appearance, and are classified as OB+. It has thus been possible to obtain a complete sample of extreme helium stars down to photographic magnitude 12 by observing at a resolution of 2 Å or better all of the OB+ stars in the Case-Hamburg surveys (which cover the entire Milky Way) and their extension to b = ±30° for ℓ = ±60° (Drilling 1980). This sample, along with all of the other known extreme helium stars, is given in Table A1 (in the appendix) along with equatorial coordinates for 2000.0, galactic coordinates, V magnitudes, B-V colors, and heliocentric radial velocities (when known).

In Fig. 2, the distribution of these objects on the sky is shown. As mentioned above, the sample is complete down to photographic magnitude 12 within the area of the sky covered by the Case-Hamburg surveys and their extension by Drilling to b = ±30° for ℓ = ±60° (indicated by the solid lines in Figure 2). If one compares this figure with Fig. II-1 of Pottasch (1984), it is seen that the distribution of extreme helium stars is very similar to that of planetary nebulae with angular diameters less than 12", indicating that most of the extreme helium stars belong to the stellar population of the Galactic nuclear bulge (Whitford 1985).

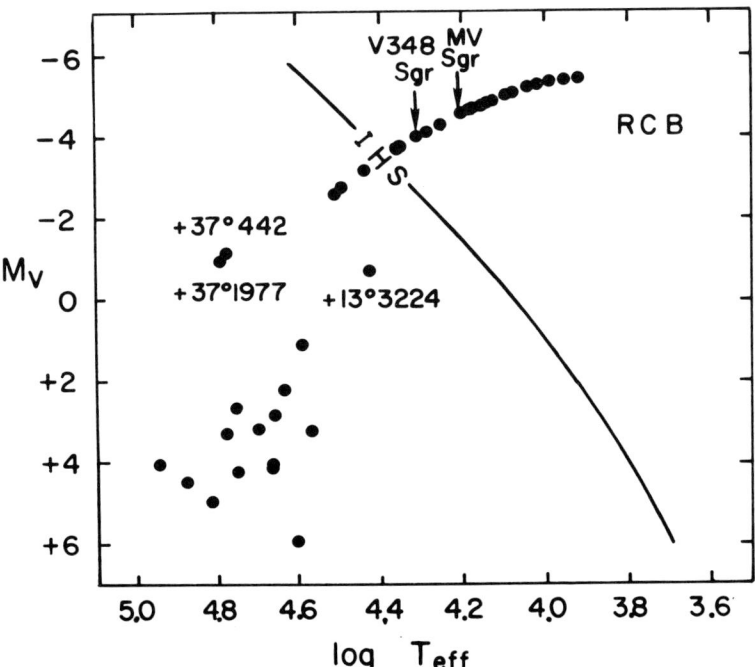

Figure 1. Hertzsprung-Russell diagram showing the positions of the cool hydrogen-deficient stars (RCB), intermediate helium stars (IHS), extreme helium stars (dots in upper half of figure), and helium-rich sdO stars (dots in lower half of figure). The position of the main sequence (solid curve) is also shown.

This conclusion is reinforced by Fig. 3 which shows the radial velocities given in Table A1 plotted against galactic longitude. The radial velocities have been corrected for the basic solar motion in all cases. The solid curves are the relations given by Pottasch (1984) for the expected radial velocities of objects at three different distances from the sun which are traveling in circular orbits about the galactic center. Comparison with Fig. II-5 of Pottasch (1984) shows that the velocity distribution of the extreme helium stars is very similar to that of planetary nebulae whose angular diameters are less than 20" and which lie within 10° of the galactic plane. These objects have a velocity dispersion of 140 km/sec in the direction of the galactic center, which again indicates that they belong to the bulge population, and could therefore not have had initial masses larger than one solar mass.

Three of the objects included in Table A1 and Figs. 2 and 3, MV Sgr, V348 Sgr, and DY Cen, show R CrB-type light variations (see next section). Near maximum light, MV Sgr fits the spectroscopic definition given above for the extreme helium stars, but shows in addition a number of weak, narrow emission lines (Herbig 1964). V348 Sgr has a predominantly emission-line spectrum at maximum light, with strong HeI and CII and little or no (stellar) contribution from the Balmer lines (Herbig 1958, Houziaux 1968, Dahari and Osterbrock 1984). Low-resolution IUE spectra of V348 Sgr in the wavelength region 1200-3000 Å are very similar to those of HD 124448 and HDE 225642 (Heber et al. 1984). The spectrum of DY Cen, for which

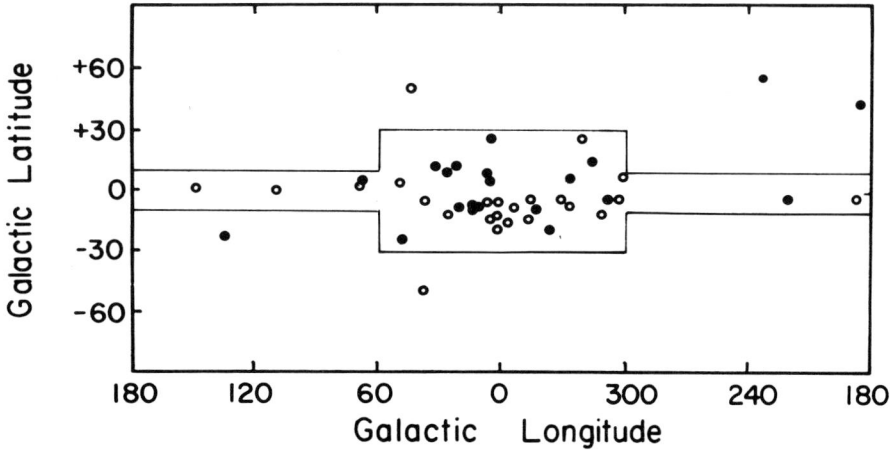

Figure 2. The distribution of extreme helium stars (filled circles) and cool hydrogen-deficient stars (open circles) on the sky. The boundaries of the Case-Hamburg OB star survey and its extension are shown by the solid lines.

Kilkenny and Whittet (1984) find T(eff) = 10,000°K, shows Hδ ≈ HeI 4026 and strong carbon features (Hill 1986).

The four H-deficient binaries (υ Sgr, KS Per, HDE 320156 and CPD -58°2721) have been included in Table A1, but are not plotted in Figs. 2 and 3. These stars fit the spectroscopic definition given above, but differ from the other extreme helium stars in the following ways: (a) the visible spectra indicate that they are single-lined spectroscopic binaries (the companions have been detected in the UV in two cases), (b) the N/C atmospheric abundance ratios are considerably higher, (c) the spectra show strong Hα emission, (d) in two cases there is a large infrared excess, (e) the system radial velocities are close to those expected for circular orbits about the galactic center, and (f) the distances from the galactic plane are less than 200 pc (see Drilling 1980; Drilling and Schönberner 1982; Schönberner and Drilling 1983; Schönberner and Drilling 1984; Rao and Venugopal 1985; Drilling, Heber, and Jefferey 1985; Jeffery and Drilling 1986). These stars, unlike the other extreme helium stars, appear to be relatively young stars of Population I, and therefore must have had larger initial masses than the other extreme helium stars to be seen at present in a similar evolutionary state.

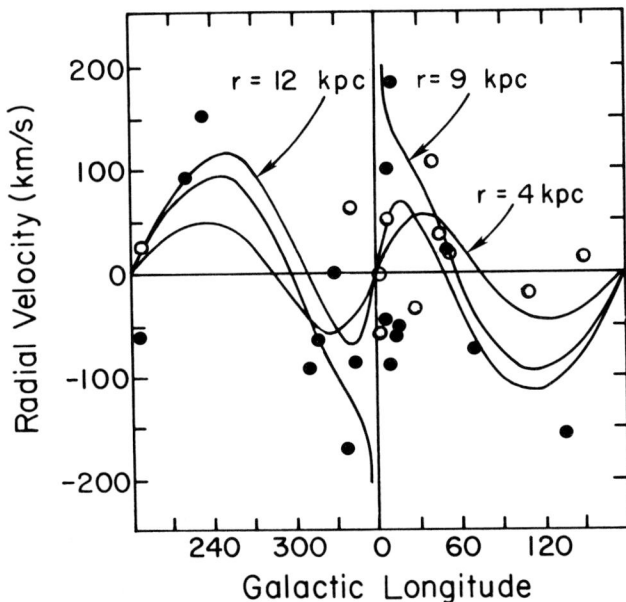

Figure 3. Radial velocity (corrected for basic solar motion) vs. galactic longitude for extreme helium stars (filled circles) and cool hydrogen-deficient stars (open circles). The solid curves are the expected loci of objects at three different distances from the sun which are traveling in circular orbits about the galactic center (Pottasch 1984).

The peculiar object V652 Her (BD+13°3224) has also been included in Table A1 but is not plotted in Figs. 2 and 3. V652 Her fits the spectroscopic definition given above, but appears to be less luminous and less helium-rich than the other extreme helium stars and, unlike the other objects listed in Table A1, is a regular pulsating variable with a period of 0.108 days (Hill et al. 1981). Like the hydrogen-deficient binaries, V652 Her appears to have a much higher N/C abundance ratio than the other objects listed in Table A1. The effective temperature and absolute magnitude used to plot V652 Her in Fig. 1 are those given by Hill et al. (1981).

Unless otherwise noted, the effective temperatures used to plot the objects listed in Table A1 in Fig. 1 are those determined by Drilling et al. (1984) or by Schönberner et al. (1982) from the observed continuous energy distribution between 1200 and 34,000 Å. In all but two cases, the strength of the 2200 Å interstellar feature and the Seaton (1979) reddening law were used to correct for the effects of interstellar absorption, and helium-rich model atmospheres used to estimate both the angular diameters and the flux shortward of 1200 Å. In the cases of the H-deficient binaries υ Sgr and KS Per, the effective temperatures of the primary components were determined by fitting the model atmospheres directly to the dereddened continua. Similar methods were used by Darius, Gidding, and Wilson (1979) to determine the effective temperatures of BD+37°442 and BD+37°1977 from low-resolution IUE spectra. In six of the above cases effective temperatures have also been determined by the fine analysis of high-resolution line spectra (Wolff, Pilachowski, and Wolstencroft 1974; Schönberner and Wolf 1974; Kaufmann and Schönberner 1977; Schönberner 1978; Walker and Schönberner 1981; Heber 1983) and the agreement is quite satisfactory. The effective temperature of HDE 320156 has also been determined by the fine analysis of high-resolution line spectra (Schönberner and Drilling 1984), and Heber et al. (1984) have concluded that T(eff) = 15,000 - 16,000°K for V348 Sgr from the similarity of its low-resolution IUE spectrum to those of HD 124448 and HDE 225642. Drilling (1986) has estimated the effective temperatures of LSS 99, LSS 3184, LSS 4357, LSIV+6°2, and LSS 5121 from the similarity of their low resolution optical spectra to those of other objects listed in Table A1.

In plotting the objects listed in Table A1 in Fig. 1, it has been assumed that all of them except V652 Her have the Schönberner luminosity, log (L/L_0) = 4.1. This luminosity is consistent with the positions in the log T(eff) - log g plane of the 6 extreme helium stars and 3 R CrB stars whose positions are known from the fine analysis of high-resolution spectra (see Walker and Schönberner 1981), and with the post-AGB evolutionary tracks computed by Schönberner (1977). Bolometric corrections were determined from the effective temperatures using Planck's Law, and do not differ significantly from those given by the helium-rich model atmospheres used by Schönberner et al. (1982) and Drilling et al. (1984). The absolute magnitudes obtained in this way are not significantly different from those obtained for υ Sgr (M_v = -4.8 ± 1.0; Rao and Venugopal 1985) and CPD-58°2721 (M_v = -5.0 ± 1.0, Drilling 1986) from the distribution of interstellar reddening, polarization, and interstellar line strength for nearby stars. They are also consistent with M_v = -5.4 for BE 202, which lies in Large Magellanic Cloud, if it is assumed that this star is similar to the hydrogen-deficient binaries (Bohannon 1981). Finally, they are consistent with the mean absolute magnitude of R CrB stars in the LMC, -4.4 ± 0.6 (Feast 1972).

3. COOL HYDROGEN-DEFICIENT STARS

The R CrB variables are stars which undergo irregular decreases in light of as much as 9 magnitudes in a few weeks. The relatively rapid decrease is followed by a slower return to maximum light, where the star spends most of its time. There are, in fact, stars which are spectroscopically identical to the cool R CrB stars which have never been observed to vary in light, the so-called non-variable hydrogen-deficient carbon (HdC) stars (Bidelman 1953, Warner 1967). At a spectroscopic resolution of 2 Å, both the cool R CrB stars and non-variable HdC stars show strong carbon features (except for CH, which is very weak) and very weak Balmer lines. All stars which fit the spectroscopic definition given above are listed in Table A2, along with equatorial coordinates for 2000.0, galactic coordinates, V magnitudes, B-V colors, and heliocentric radial velocities (when known). According to Warner (1967), HD 148839 is not as hydrogen-deficient as the other stars listed in Table A2, and may represent a type intermediate to the cool hydrogen-deficient stars and normal carbon stars.

The distribution on the sky of the objects listed in Table A2 is shown in Fig. 2, along with that of the extreme helium stars. All but 5 of these objects are R CrB stars, which are relatively easy to discover because of the large light variations. The sample is, however, effected by the incompleteness of the variable star surveys. The non-variable HdC stars are much harder to pick out because spectra of 2 Å resolution or better are required to identify them. For this reason, Warner (1967) has estimated that their space densities may actually be on the order of 10 times greater than those of the R CrB stars. Taking these selection effects into account, we conclude that the cool hydrogen-deficient stars also have a space distribution which is similar to that of the distant planetary nebulae, i.e. that most of them belong to the population of the Galactic nuclear bulge. The known radial velocities of cool hydrogen-deficient stars (corrected for the basic solar motion) are plotted against galactic longitude in Fig. 3, along with those of the extreme helium stars. Again, the radial velocities reinforce the conclusion that we are dealing with the bulge population, but the case is not as strong as it is for the extreme helium stars.

The effective temperatures used to define the locus of the cool hydrogen-deficient stars in Fig. 1 are those given by Kilkenny and Whittet (1984), which were determined by fitting Planck's Law to the continuous energy distributions of 10 cool RCrB stars as determined from UBVRIJHKLMNQ photometry. Color excesses were determined using published maps of the color excess and assuming that all of the cool hydrogen-deficient stars lie beyond the absorbing layer. The observed colors were then corrected for interstellar reddening using van de Hulst's curve No. 15 (Johnson 1968). Schönberner (1977) has determined effective temperatures for three cool RCrB stars from the fine analysis of high-resolution spectra. He finds an effective temperature for RY Sgr, the one star in common to the two studies, which is 900°K higher than that found by Kilkenny and Whittet. The absolute magnitudes used to define the locus of the cool hydrogen-deficient stars in Fig. 1 are those given by Feast (1972) for the three R CrB stars which lie in the Large Magellanic Cloud.

4. INTERMEDIATE HELIUM STARS

The spectra of intermediate helium stars are similar to those of normal stars of MK spectral type B2V, but have abnormally high HeI/H line strengths (Walborn 1983). Because the Balmer lines are stronger than they are in the case of the extreme helium stars, they can be seen at the resolution of the OB star surveys, where the intermediate helium stars are usually classified as OB or OB$^-$. A much higher resolution (2 Å) is, however, needed to resolve the helium lines and separate these objects from normal stars with spectral types near B2V. For this reason, the sample of intermediate helium stars is only complete down to the limiting magnitude of the HD catalog (B $\sim$ 8.5), for which most of the early-type stars have been observed at higher resolution. The sample of intermediate helium stars is complete to B = 10 for the southern sky because of the surveys of MacConnell, Frye, and Bidelman (1970) and Garrison, Hiltner, and Schildt (1977). The problem of completeness is complicated by the fact that the helium spectra are sometimes variable, in one case going from normal to pronouncedly helium-rich and back in less than 10 days (Bond and Levato 1976).

All of the known intermediate helium stars have been listed by Walborn (1983), and this list is repeated in Table A3 along with the 2000.0 equatorial coordinates, galactic coordinates, V magnitudes, B-V colors, and heliocentric radial velocities (when known). According to Walborn (1983) and Hunger (1975), HD 144941 is intermediate to the extreme and intermediate helium stars according to the spectroscopic definitions. Walborn also lists additional objects which he has not confirmed to be helium-rich or which he believes to be related to the intermediate helium stars.

Fig. 4 shows the distribution of known intermediate helium stars on the sky. If we take into account the selection effects mentioned above, it is seen that the brighter objects show a preference for Gould's belt, whereas the fainter objects tend to concentrate towards the galactic plane, but not the galactic center. In Fig. 5, the radial velocities listed in Table A3 are plotted against galactic longitude after correction for the basic solar motion. These stars are seen to have the radial velocities expected for objects moving in circular orbits about the galactic center if they are located within 4 kpc of the sun, which is the case if these stars have the absolute magnitudes of normal B2V stars. That the absolute magnitudes and effective temperatures of these stars are similar to those of normal B2V stars is indicated by the apparent magnitudes and colors of the six stars which are members of associations (Walborn 1983; Drilling 1981) and is consistent with the fine analysis of high-resolution spectra for most of the stars (see Hunger 1975 and in this volume). We conclude that the intermediate helium stars are predominantly young, massive stars of population I. A possible exception is the star SB 939, which is discussed by Langhans, Heber, and Hunger elsewhere in this volume.

5. HELIUM-RICH SUBLUMINOUS O STARS

Like the extreme helium stars, helium-rich sdO stars with $n(H)/n(He) \lesssim 1$ show strong lines of HeI (and/or HeII for the hotter stars) and very weak or absent Balmer lines at a spectral resolution of 2 Å (Greenstein and Sargent 1974; Hunger 1975; Hunger et al. 1981;

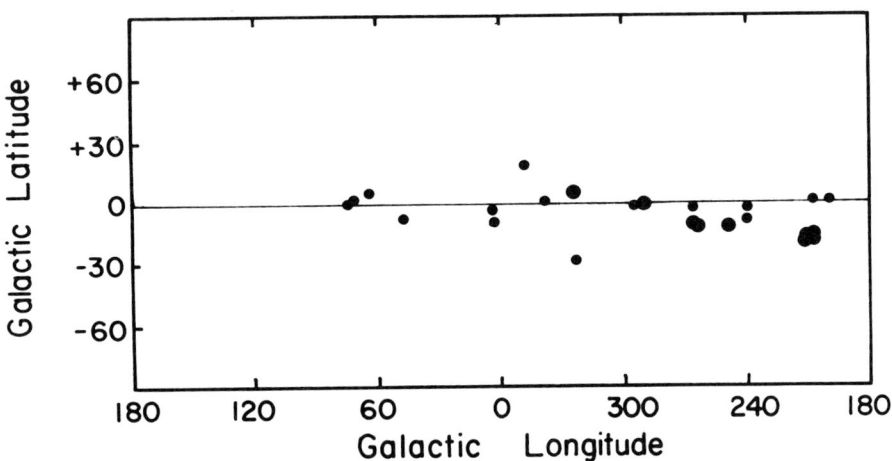

Figure 4. The distribution of intermediate helium stars on the sky. Large dots represent stars with V < 7.5, and small dots, stars with V ≥ 7.5.

Schönberner and Drilling 1984). For this reason, the sample is complete to photographic magnitude 12 for the area of the sky covered by the Case-Hamburg OB star surveys and their extension. Unlike the extreme helium stars, however, the spectra of these stars show evidence of surface gravities greater than log g = 4 (Stark broadened H and He lines and strong HeII 4686). Included in Table A4 are all stars classified by Drilling (1983,1986) as belonging to this category plus all additional helium-rich sdO stars known to this writer which are brighter than B = 14.5.

The distribution of these stars on the sky is shown in Fig. 6. The selection effects mentioned above are very important because very few of the objects brighter than B = 12 lie within the region of the sky covered by the Case-Hamburg surveys or their extension (shown by the solid boundaries in Fig. 6). The excess of faint objects in the Southern Milky Way is undoubtedly due to the superior seeing at Cerro Tololo as compared to that at Hamburg and Cleveland. The clustering of faint objects near the South Galactic Pole is due to the survey of Slettebak and Brundage (1971), which allowed all OB stars brighter than B = 14 to be identified within an 824 square degree region surrounding the South Galactic Pole. Taking all of this into account, one may conclude that we are dealing here with a more-or-less local

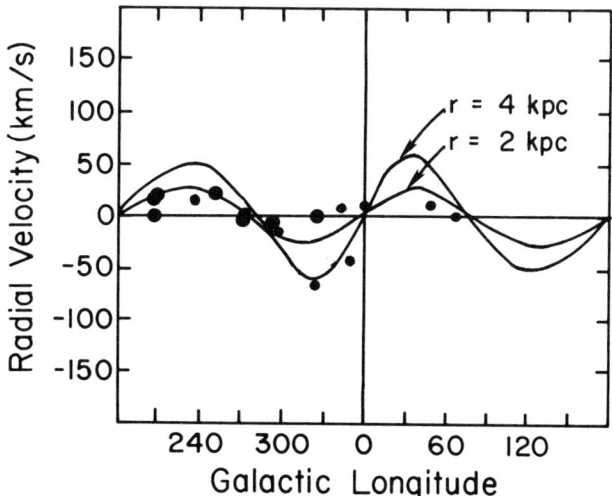

Figure 5. Radial velocity (corrected for basic solar motion) vs. galactic longitude for intermediate helium stars. Large dots represent stars with V < 7.5, and small dots, stars with V > 7.5. The solid curves are the expected loci of objects at two different distances from the sun which are traveling in circular orbits about the galactic center (Pottasch 1984).

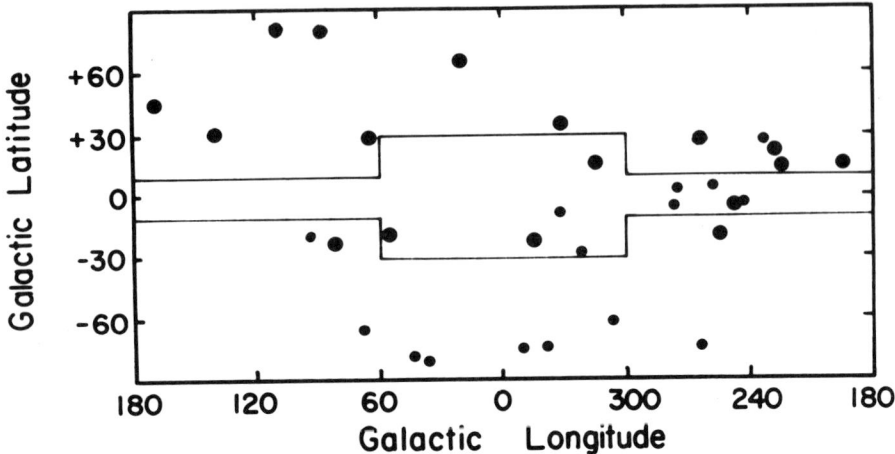

Figure 6. The distribution of helium-rich sdO stars on the sky. Large dots represent stars with V < 12.0, and small dots, stars with V ⩾ 12.0. The boundaries of the Case-Hamburg OB star survey and its extension are shown by the solid lines.

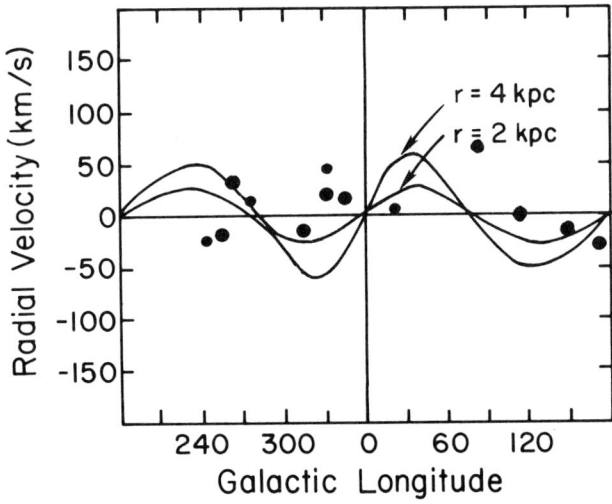

Figure 7. Radial velocity (corrected for basic solar motion) vs. galactic longitude for helium-rich sdO stars. Large dots represent stars with V < 12.0, and small dots, stars with V ⩾ 12.0. The solid curves are the expected loci of objects at two different distances from the sun which are traveling in circular orbits about the galactic center (Pottasch 1984).

sample, as one would expect from the absolute magnitudes indicated in Fig. 1. This impression is reinforced by Fig. 7, where the radial velocities given in Table A4 are plotted against galactic longitude after correction for the basic solar motion. Here, the radial velocities tend to cluster around zero to all longitudes, as one would expect for a local sample of objects which more or less share the sun's orbit about the galactic center (i.e. Population I). BD+39°3226, which appears to be a high-velocity star (Dworetsky, Whitelock, and Carnochan 1982), is not plotted in Fig. 7.

The effective temperatures and absolute magnitudes plotted in Fig. 1 are those given by Schönberner and Drilling (1984). For the four hottest objects plotted (LSE 153, LSE 259, LSE 263, and LSIV+10°9), the effective temperatures were determined from the slope of the ultraviolet continuum. The strength of the 2200 Å feature was used along with the Seaton (1979) reddening law and published maps of the color excess in the solar neighborhood to correct for the effects of interstellar reddening and to estimate the distances of these objects. The effective temperatures and surface gravities of the other stars have been determined from the fine analysis of high resolution spectra (Hunger et al. 1981), and the absolute magnitudes were estimated using Schönberner's (1981) bolometric corrections and assuming a mass of 0.5 M_o.

Partial support of this work by the National Science Foundation (Grant Nos. AST-8018766 and INT-8219240) and by the National Aeronautics and Space Administration (Grant No. NAG5-71) is gratefully acknowledged.

REFERENCES

Bidelman, W. P. 1953, *Ap. J.*, **117**, 25.
Bohannon, B. 1981, *Bull. American Astron. Soc.* **13**, 526.
Bond, H. E., and Levato, H. 1976, *Publ. Astron. Soc. Pacific* **88**, 905.
Dahari, O., and Osterbrock, D. E. 1984, *Ap. J.* **277**, 648.
Darius, J., Gidding, J. R., and Wilson, R. 1979, *The First Year of IUE*, p. 36.
Drilling, J. S. 1980, *Ap. J.* **242**, L43.
Drilling, J. S. 1981, *Ap. J.* **250**, 701.
Drilling, J. S. 1983, *Ap. J.* **270**, L13.
Drilling, J. S. 1986, in preparation.
Drilling, J. S., Heber, U., and Jefferey, C. S. 1985, IAU Circular No. 4086.
Drilling, J. S., and Schönberner D. 1982, *Astron. Astrophys.* **113**, L22.
Drilling, J. S., Schönberner, D., Heber, U., and Lynas-Gray, A. E. 1984, *Ap. J.* **278**, 224.
Dworetsky, M. M., Whitelock, P. A., and Carnochan, D. J. 1982, *Mon. Not. R. Astron. Soc.* **201**, 901.

Feast, M. W. 1972, Mon. Not. R. Astr. Soc. **158**, 11P.
Garrison, R. F., Hiltner, W. A., and Schild, R. E. 1977, Ap. J. Suppl. **35**, 111.
Greenstein, J. L., and Sargent, A. I. 1974, Ap. J. Suppl. **28**, 157.
Heber, U. 1983, Astron. Astrophys. **118**, 39.
Heber, U., Heck, A., Houziaux, L., Manfroid, J., and Schönberner, D. 1984, Proc. Fourth European IUE Conference (ESA SP-218), 367.
Herbig, G. H. 1958, Ap. J. **127**, 312.
Herbig, G. H. 1964, Ap. J. **140**, 1317.
Hill, P. W. 1986, private communication.
Hill, P. W., Kilkenny, D., Schönberner, D., and Walker, H. J. 1981, Mon. Not. R. Astr. Soc. **197**, 81.
Houziaux, L. 1968, Bull. Astron. Inst. Czech. **19**, 265.
Hunger, K. 1975, Problems in Stellar Atmospheres and Envelopes, eds. B. Baschek, W. H. Kegel, and G. Traving (New York: Springer-Verlag), p. 57.
Hunger, K., Gruschinske, J., Kudritzki, R. P., and Simon, K. P. 1981, Astron. Astrophys. **95**, 244.
Jeffery, C. S., and Drilling, J. S. 1986, in preparation.
Johnson, H. L. 1968, Nebulae and Interstellar Matter, eds. B. M. Middlehurst and L. H. Allor (University of Chicago Press), p. 167.
Kaufmann, J. P., and Schönberner, D. 1977, Astron. Astrophys. **57**, 169.
Kilkenny, D., and Whittet, D. C. B. 1984, Mon. Not. R. Astr. Soc. **208**, 25.
MacConnell, D. J., Frye, R. L., and Bidelman, W. P. 1970, Publ. Astron. Soc. Pacific **82**, 730.
Pottasch, S. R. 1984, Planetary Nebulae: A Study of Late Stages of Stellar Evolution (Dordrecht, Boston, Lancaster: D. Reidel).
Rao, N. K., and Venugopal, V. R. 1985, J. Astrophys. Astron. **6**, 101.
Schönberner, D. 1975, Astron. Astrophys. **44**, 383.
Schönberner, D. 1977, Astron. Astrophys. **57**, 437.
Schönberner, D. 1978, Mitt. Astron. Gesellschaft **43**, 266.
Schönberner, D. 1981, Astron. Astrophys. **103**, 119.
Schönberner, D., and Drilling, J. S. 1983, Ap. J. **268**, 225.
Schönberner, D., and Drilling, J. S. 1984, Ap. J. **276**, 229.
Schönberner, D., Drilling, J. S., Lynas-Gray, A. E., and Heber, U. 1982, Advances in Ultraviolet Astronomy: Four Years of IUE Research (NASA CP-2238), 593.
Schönberner, D., and Wolf, R. E. A. 1974, Astron. Astrophys. **37**, 87.
Seaton, M. J. 1979, Mon. Not. R. Astr. Soc. **187**, 73P.
Slettebak, A., and Brundage, R. K. 1971, Astron. J. **76**, 338.
Stephenson, C. B., and Sanduleak, N. 1971, Publ. Warner and Swasey Obs. **1**, 1.
Walborn, N. R. 1983, Ap. J. **268**, 195.
Walker, H. J., and Schönberner, D. 1981, Astron. Astrophys. **97**, 291.
Warner, B. 1967, Mon. Not. R. Astr. Soc. **137**, 119.
Whitford, A. E. 1985, Publ. Astron. Soc. Pacific **97**, 205.
Woolf, S. C., Pilachowski, C. A., and Wolstencroft, R. D. 1974, Ap. J. **194**, L83.

DISCUSSION

FEAST: This is a comment rather than a question. I think DY Cen has a spectrum definitely like a hot star. I don't know what the temperature is, but I would say more like an A-type star.

N.K. RAO: I have been observing this star at CTIO. Herbig has identified several C II lines in absorption and in emission.

FEAST: In addition to W Men there are 2 other R CrB stars in the LMC. W Men and HV 12842 have $M_v \approx -5$, and HV 5637 (which has strong C_2 bands) has $M_v \approx -3$. I am a little worried that you want to call the R CrB stars population II. I would prefer to call them old population I. You are talking about very old metal-deficient objects.

DRILLING: I did not want to imply that they are metal-deficient, but they are old and have a Pop. II distribution.

FEAST: U Aqr has a very high radial velocity. But I think that is the only one.

N.K. RAO: May I add that Prof. Herbig is presently looking into the radial velocities of R CrB stars from both hemispheres. According to him there are only a few stars which have very high velocities, in addition to U Aqr and VZ Sgr, which I remember have about $+200$ kms^{-1}; both are genuine members of the class. The rest of them are within about 30 kms^{-1}. So they are not as a group of high radial velocity.

DRILLING: I did seem to me that the velocity dispersion for the cooler stars was less than for the extreme helium stars and so perhaps I am wrong to throw them all into the same bag.

LIEBERT: How does the reddening affect the completeness of the more luminous groups - R Cor Bor and Extreme Helium Stars - for these magnitude-limited surveys? (It seems to me from the evidence presented that they may at least include old disk/Pop. I stars as well as Pop. II stars and there may be a considerable selection effect against finding them in the plane.)

DRILLING: The stars are very luminous. For the solar neighbourhood, up to 2 or 4 kpc, I would think the sample would be complete.

LIEBERT: What are the scale heights, based on the statistics that you have, for the extreme helium stars and the cool hydrogen-deficient stars?

DRILLING: I haven't attempted to derive the scale heights because of the small numbers of stars involved.

TUTUKOV: Can you estimate the total numbers of helium rich stars of different types in our galaxy?

DRILLING: Again, I think that would be very difficult to do because of the fact that if they are of population II, they drop off with the cube of the distance as you go away from the galactic center. Incompleteness sets in after about 2 to 4 kpcs from the sun and so I think it's very difficult to determine the variation with distance from the galactic center and hence to determine the total number in the galaxy.

POTTASCH: Can you make a guess?

DRILLING: Brian Warner has made an estimate for the cool hydrogen-deficient stars.

GARRISON: How does the absolute magnitude determined from the 2200 feature compare with the stark broadened profile estimates for the sdO stars?

DRILLING: Heber has been working on this. I believe the result so far is that the absolute magnitudes derived from surface gravities (assuming a mass of 0.5 s.m.) are more luminous than the ones that have been estimated from the 2200 feature.

HEBER: Yes. The absolute magnitudes derived from the line profiles of sdO stars indicate that the stars are in general more luminous than those inferred from the reddening laws. Unfortunately, these discrepancies are very large in a few cases (up to 5 magnitudes).

DRILLING: I have compared low resolution spectra of these stars with those of BD+37°1977, and the line broadening is much greater. So they are certainly less luminous than the extreme helium stars.

N.K.RAO: I remember Dr. Feast measured the cool R CrB stars in the LMC. I thought he came up with the result that the cool R CrB stars are 1 magnitude fainter than the F type stars.

FEAST: I can tell you this afternoon. (Laughter)

RADIAL VELOCITIES OF EXTREME HELIUM STARS AND OF HOT sdO STARS

John S. Drilling[1] and U. Heber[2]
[1]Department of Physics and Astronomy
Louisiana State University
Baton Rouge, LA 70803-4001, U.S.A.
[2]Institut für Theoretische Physik und Sternwarte
der Universität Kiel
Olshausenstr. 40, 2300 Kiel, F.R.G.

ABSTRACT. Radial velocities have been determined for nearly all the extreme helium stars and for the very hot sdO stars discovered by Drilling from high-resolution spectra obtained at ESO and at CTIO.

1. INTRODUCTION

A systematic survey (see Drilling, these proceedings) of stars classified as OB+ in the Case-Hamburg surveys, and their extension by Drilling to b = $\pm$ 30° for l = $\pm$ 60°, led to the discovery of ten new extreme helium stars and twelve very hot sdO stars (Drilling, 1983). High-resolution spectra for nearly all of these stars were obtained at ESO and CTIO. Presented here are radial velocities measured from these spectra.

2. OBSERVATIONS AND RADIAL VELOCITY MEASUREMENTS

Radial velocities of 14 extreme helium stars and 12 hot subluminous O stars were obtained using the ESO-Cassegrain Echelle spectrograph (CASPEC) attached to the 3.6 m telescope at La Silla. These spectra, which covered the spectral range from 3900 Å to 4800 Å, were reduced and wavelength calibrated, as described by Heber, Jonas and Drilling (these proceedings). We have estimated the errors in the calibration to be less than 0.03 Å.

For the extreme helium stars about 20 unblended, sharp stellar lines were used to determine their radial velocities. The standard deviation of the mean proved to be very low ($\pm$ 1 km/s) in almost all cases. The hot sdO stars displayed only a few (mostly broad) lines in their spectra and, therefore, their radial velocities were less accurately determined. The resulting radial velocities, corrected for the earth's orbital motion, are given in Table I along with those of intervening interstellar Ca II, obtained from measurements of H and K lines (if visible in the spectrum).

High-resolution (0.15 Å) spectra of extreme helium stars were also obtained with the 4 m telescope, Echelle spectrograph and Singer image-tube camera at CTIO for those stars marked with asterisks in Table Ia). Because these plates were wavelength calibrated by ThA comparison spectra taken on other plates, radial velocities could not be measured accurately. However, differences between the stellar and interstellar radial velocities could be measured with an accuracy comparable to that obtained from the CASPEC spectra. In only one case, that of CPD−58°2721 (Drilling, Heber and Jeffery, 1985), was there a measurable change in the stellar radial velocity with respect to the interstellar H and K lines. In this case, the radial velocity increased by 140 km/s between 30th May, 1983, and 9th April, 1985, indicating that CPD−58°2721 is a hydrogen-deficient binary similar to υ Sgr and KS Per (see also Morrison et al., these proceedings).

For some of the sdO stars (those marked with a plus sign in Table Ib), radial velocities were obtained with the 4.0 m telescope, Echelle spectrograph and SIT Vidicon detector at CTIO. The accuracy of the latter is somewhat lower (m.e. $\approx$ 8 km/s) than that of the CASPEC data. In only one case, that of LSS 2018 (Drilling, 1985), was there a significant difference in the radial velocities obtained with the two instruments. This star, which is the nucleus of a planetary nebula, has been found to be a double-lined spectroscopic binary with a period of 8.6 hours and a system velocity of −25 km/sec.

3. DISCUSSION

The stars listed in Table I (except CPD−58°2721) are plotted in Figure 1. In this figure, the radial velocities given in Table I (in the case of LSS 2018, the system velocity) are plotted against galactic longitude after correction for basic solar motion. The extreme helium stars are displayed in the top panel, the sdO stars in the lower panel.

3.1. The extreme helium stars

In Figure 1 stars which lie close to the galactic plane ($|b| < 10°$) are indicated by filled circles, the others by open circles.

The extreme helium stars are luminous stars (log $L/L_\odot \approx 4.1$) which are a great distance from the sun (Heber and Schönberner, 1981). Therefore, predicted radial velocities for circular orbits and for three distances from the sun (4, 9 and 12 kpc, respectively) are also plotted in Figure 1 (solid lines, adapted from Pottasch, 1984). The large dispersion in radial velocity for the galactic plane extreme helium stars, particularly for those which lie in the direction of the galactic center, indicates that they do not have circular orbits and, hence, belong to an old population.

TABLE I: HELIOCENTRIC RADIAL VELOCITIES

Star name	Stellar (km/s)	Interstellar (km/s)	Date (UT)
a) Extreme helium stars			
LSS 99	+109 ± 1		8 Apr 1985
BD+10°2179*	+158 ± 1	+5 ± 3	8 Apr 1985
CPD-58°2721	+55 ± 1	+2 ± 1	9 Apr 1985
LSS 3184	-89 ± 1	-26 ± 2	8 Apr 1985
HD 124448*	-65 ± 1	+5 ± 1	8 Apr 1985
CoD-48°10153*	-4 ± 1		9 Apr 1985
CoD-48°10153*	-4 ± 1		9 Apr 1985
BD-9°4395	-56 ± 1	-13 ± 6	8 Apr 1985
V2076 Oph*	+77 ± 1	-5 ± 1	3 Apr 1984
V2076 Oph*	+79 ± 1	-3 ± 1	9 Apr 1985
LSE 78*	-92 ± 1	-13 ± 4	8 Apr 1985
LSE 78*	-90 ± 1		4 Oct 1984
LSS 4357	-99 ± 1	+10 ± 4	12 May 1985
HD 168476*	-172 ± 1		9 Apr 1985
HD 168476*	-171 ± 1		9 Apr 1985
LSS 5121	-62 ± 3	+10 ± 3	9 Apr 1985
LSII+33°5*	-88 ± 1	-8 ± 2	12 May 1985
BD+1°4381*	+12 ± 1	+26 ± 4	12 May 1985
b) Subdwarf-O-stars			
LSS 982	+126 ± 4		6 Oct 1984
LSS 1274	+24 ± 1	+10 ± 11	9 Apr 1985
LSS 1362	+7 ± 5		4 Apr 1984
LSS 1362	-1 ± 9	0 ± 7	8 Apr 1985
LSS 1362	+6 ± 10		9 Apr 1985
LSS 2018+	+6 ± 5	+2 ± 1	3 Apr 1984
LSE 21	-14 ± 5	-3	7 Oct 1985
LSE 44+	-50 ± 4	-15 ± 11	3 Apr 1984
LSE 153+	-17 ± 4	-19 ± 16	3 Apr 1984
LSE 125	-14 ± 5	-17 ± 2	3 Apr 1984
LSIV-12°1+	-179 ± 2	-12 ± 11	3 Apr 1984
LSE 259+	+43 ± 4	-16 ± 16	3 Apr 1984
LSE 234	-30 ± 5	-22 ± 20	4 Apr 1984
LSE 263+	+13 ± 7	-17 ± 10	3 Apr 1984

* = stars also observed at CTIO (Singer image-tube camera), see text
\+ = stars also observed at CTIO (SIT Vidicon), see text

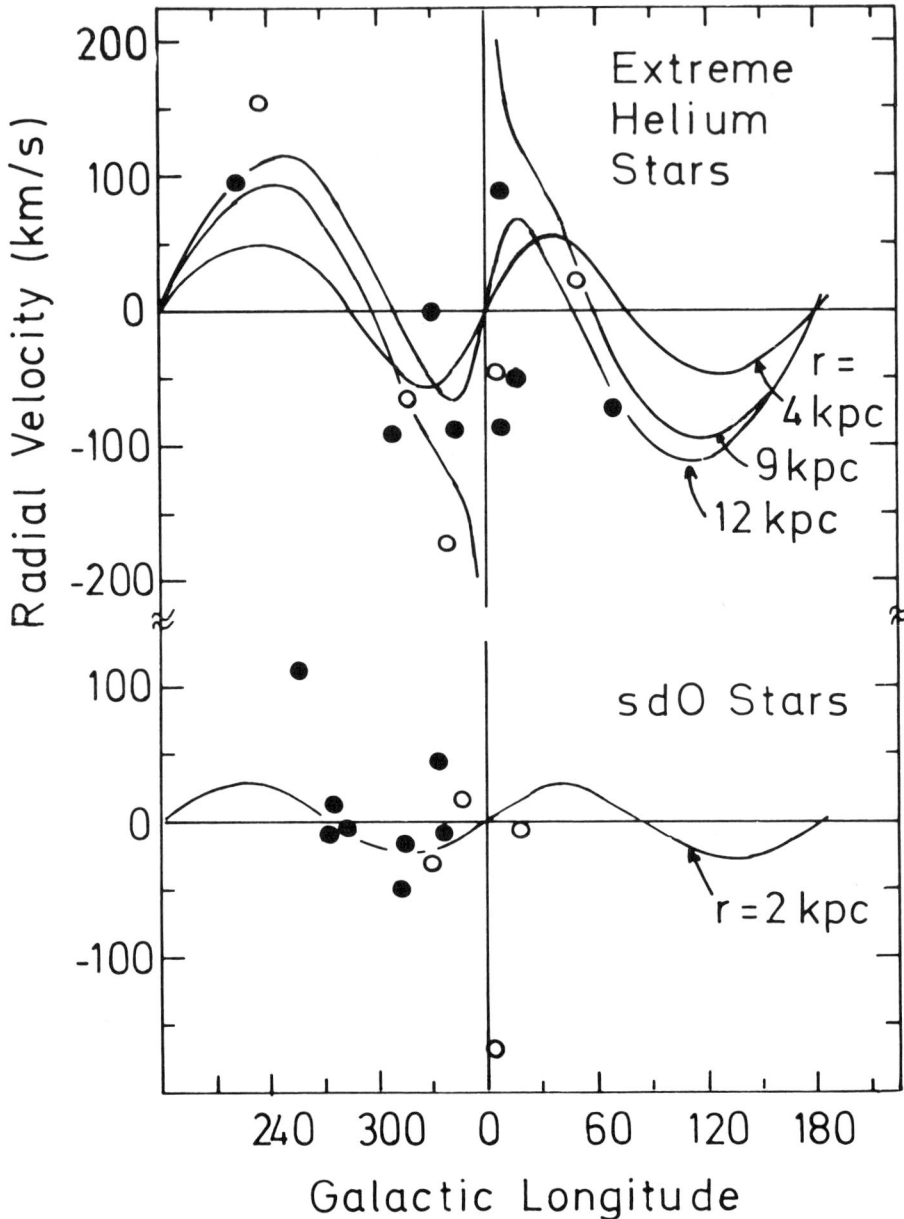

Fig. 1 Radial velocities (corrected for basic solar motion) of extreme helium stars (top) and sdO stars (bottom). Stars which lie close to the galactic plane ($|b| < 10°$ in the top panel and $|b| < 20°$ in the lower panel) are marked as filled circles. The other stars are marked by open circles. The solid curves are the predicted radial velocities for circular orbits.

3.2. The hot sdO stars

In Figure 1 stars which lie close to the galactic plane ($|b| < 20°$) are indicated by filled circles (lower panel), the others by open circles. Since these sdOs are at a smaller distance from the sun than the extreme helium stars, predicted radial velocities are only plotted for a circular orbit and a distance of 2 kpc from the sun (solid line). Note that 2 kpc is a firm upper limit for the distances of these stars (see Schönberner and Drilling, 1984). Since the sdO stars are probably related to the central stars of planetary nebulae (Mendez et al., these proceedings), we can compare their radial velocity distribution to that of the CPNs. In fact, three stars from our ensemble are surrounded by planetary nebulae. Pottasch (1984; Fig. II-6) plotted the radial velocities of nearby PNs against galactic longitude. Comparison with Figure II-6 by Pottasch (1984) shows that the dispersion in radial velocity of the sdOs is similar to that of the "local" planetary nebulae. The very large radial velocities for two of the sdO stars (LS IV-12°001 and LSS 982) indicate that they may belong to Population II.

This research was supported in part by NSF Grants AST 8018766, AST 8514574 and INT 8219240 and by NASA Grant NAG 5-71.

REFERENCES

Drilling, J.S.: 1983, Astrophys. J. **270**, L13
Drilling, J.S.: 1985, Astrophys. J. **294**, L107
Drilling, J.S., Heber, U., Jeffery, C.S.: 1985, IAU Circular No. 4086
Heber, U., Schönberner, D.: 1981, Astron. Astrophys. **102**, 73
Pottasch, S.R.: 1984, Planetary Nebulae: A Study of Late Stages of Stellar Evolution (Dordrecht, Boston, Lancaster: D. Reidel), p. 26
Schönberner, D., Drilling, J.S.: 1984, Astrophys. J. **278**, 702

DISCUSSION

HUNGER: You obtained radial velocities with an accuracy of ± 1 km s^{-1} for stars which are in the range of 10-12 magnitude, which is something new.

MENDEZ: How many spectrograms are used?

DRILLING: Usually we got a second spectrogram which was uncalibrated in the sense that the comparison spectrum is not on the same plate. We have been able to judge whether the star is variable by using interstellar lines.

MENDEZ: Can you say most of them do not show radial velocity variations?

DRILLING: That's correct. Only one star showed radial velocity variations greater than 10 km^{-1}.

SCHÖNBERNER: This is a question to Prof. Pottasch. Do the planetaries close to the galactic bulge belong to population II?

POTTASCH: The time of evolution to the planetary nebular stage depends on the mass of the object so that some could be young and some could be old. Their spatial distribution indicates a strong concentration towards the galactic center and at the same time a concentration towards the galactic plane. I would interpret this as meaning that the planetaries in the galactic bulge are older, population II, objects which have predominantly evolved from low mass stars. This is consistent with their radial velocities (at least within 10° of the galactic center) which do not follow circular orbits at all. The planetaries in the galactic plane on the other hand, seem to be predominantly younger objects which have evolved from higher mass stars. The evidences for this are not only the radial velocities which conform more to circular orbits, but the fact that all the planetaries with high nitrogen and helium abundances (the high mass progenitors) lie in the galactic plane but not in the galactic bulge.

TUTUKOV: Did you check for variability of different types from your previous paper and this one

DRILLING: The only one of this sample for which there is evidence of binarity is LSS 2018, a binary planetary nebula nucleus.

WALKER: What was the time between the individual spectra?

DRILLING: The time between observations was typically 2 years.

LIEBERT: We have observed 19 stars from the Palomar Green survey at the Steward Observatory and at Kitt Peak. They were found to show composite spectra, generally with a hot subdwarf primary and a cool main-sequence secondary star (Ferguson et al., Ap. J. **287**, 1984). Approximately half of all the PG subdwarfs could be in undetected binaries.

HEBER: We have carried out an analysis of SdB and SdO stars (Heber, A+A 155, 33, 1986) for a survey at the south galactic pole (Slettebak and Brundage). This survey is smaller than the Palomar Green survey. But, nevertheless, it contains 20 subdwarfs. We carried out model atmosphere analyses for all of them, based on IDS and IUE spectra. We found that 20% have binary companions of the F or G type. This results is in good agreement with the conclusion of Ferguson et al..

BUES: I would like to comment on the photometric determination of
binaries within a sample of blue stars. For our survey of white dwarf
suspects, we use combined color diagrams (U–B)/(R–I) and (U–V)/(R–I).
Hot single white dwarfs have positions just at the black-body line,
whereas binaries have a red excess of 0.2 to 0.5 magnitudes. From
this experience I would guess that the fraction of binaries in the
Drilling's sample could increase up to 30 percent.

III HOT EXTREME HELIUM STARS

SPECTROSCOPIC ANALYSES OF HOT EXTREME HELIUM STARS

U. Heber
Institut für theoretische Physik und Sternwarte
Olshausenstr. 40
D 2300 Kiel, F.R.G.

ABSTRACT. Spectroscopic fine analyses of hot extreme helium stars are reviewed. The chemical composition is discussed in detail and conclusions as to the nuclear history of the atmospheric material are drawn. Evidence of inhomogeneities among the hot extreme helium stars is presented and it is concluded that the extreme helium stars can be found in the halo population (BD+10°2179) as well as in the disc population. Their mass loss rates are of the same order of magnitude (10^{-9} $M_\odot$/yr) as those of normal stars. Four helium rich sdO stars are identified as possible descendants of the B-type extreme helium stars.

1. INTRODUCTION

The hot extreme helium stars represent the most obvious cases for large atmospheric hydrogen depletion and helium enrichment among all classes of hydrogen deficient stars. Unlike in the case of cool hydrogen deficient stars, it is immediately evident from their spectra that helium replaces hydrogen as the most abundant element and that hydrogen, indeed, is a trace element only.

Spectroscopically, the class of extreme helium stars is defined by the following criteria:
i) They are giant stars with spectral types ranging from early A to late O.
ii) The helium absorption line spectrum is unusually strong.
iii) Hydrogen lines are very weak or absent.
iv) The carbon lines are strong.

Drilling (these proceedings) lists 17 stars which meet all criteria and, thus, are true class members.

Some other hot helium stars should be mentioned which, however, do not meet all the above criteria. In view of binary star evolutionary scenarios (e.g. Iben and Tutukov, 1985) it is important to point out that the four hydrogen deficient binaries known today (υ Sgr, KS Per, LSS 1922 and LSS 4300) can not be considered as extreme helium stars because they are carbon weak lined. The presumedly single helium stars BD+13°3224 and HD 144941 are also not class members for the same reason.

Since the properties of the hydrogen deficient binaries are reviewed by Plavec (these proceedings) only the latter will be discussed here. Some helium and carbon rich sdO stars (despite their somewhat larger gravities) appear to be closely related to the extreme helium stars as are two hot RCrB stars (see the review of Rao in these proceedings).

In the absence of reliable distance determinations for hot extreme helium stars, the best way to discuss their properties and evolutionary status is to find their position in the (g, T_{eff})- diagram. Clues to the origin of the hydrogen deficiency may come from their chemical composition which may also help to identify their progenitors and descendants. Detailed spectroscopic analyses are required to derive these quantities.

Quantitative spectroscopy of extreme helium stars began with the curve of growth analysis of HD 160641 (Aller, 1953) and was continued by the coarse analyses of BD+10°2179 (Klemola, 1961, Hill, 1965), HD 124448 and HD 168476 (Hill, 1965). The first modern quantitative spectral analysis using LTE model atmospheres was carried out for BD+10°2179 (Hunger and Klinglesmith, 1969). These early analyses have been discussed in previous reviews (Hunger ,1975, Scholz, 1972, Dinger, 1970, Hack, 1967, Underhill, 1966) to which the reader is referred. Here I will review only the more recent developments.

Since 1975, the observation techniques have improved considerably. In the mid seventies, high quality photographic spectra became available which provided a firm basis for four fine analyses. The IUE satellite opened a new spectral range - the ultraviolet - for the analysis. Very recently, the advent of efficient linear detectors combined with Echelle spectrographs (e.g. ESO - CASPEC) allowed all known extreme helium stars - even the faintest ones - to be observed at high spectral resolution and S/N.

The review begins with a brief outline of the model atmospheres and the method of analyses (section 2). Then I will summarize the results of four fine analyses of photographic spectra and discuss the conclusions as to their nuclear history (section 3) . The effective temperatures as derived from photometry and ultraviolet fluxes will be discussed in section 4 and the mass loss rates in section 5. The properties of some closely related helium stars are summarized in section 6. I conclude the review with an outlook on what can be expected from the new CASPEC spectra.

2. MODEL ATMOSPHERES AND METHOD OF ANALYSIS

The stellar spectra are analyzed by means of model atmosphere techniques. This method is usually referred to as a fine analysis and has been widely applied to many classes of stars. Because of the extreme helium stars' peculiar chemical composition, the model atmospheres have to be adapted adequately. The atmospheric structure of a normal B star is determined esentially by two parameters: the effective temperature and the gravity. In the case of an extreme helium B-star an additional atmospheric parameter - the C/He ratio - enters into the analysis. Because neutral helium is a poor absorber, carbon dominates the opacity at many wavelengths and its abundance therefore influences the atmospheric structure. It is worthwhile to present here some details of the model atmospheres in use and compare them to normal composition model atmospheres.

2.1. Model atmospheres

Model atmospheres are calculated assuming LTE, plane parallel geometry and hydrostatic equilibrium. The continous opacities are (almost) the same as used by Kurucz (1979) except for carbon. All excited levels for which atomic data are available (Peach, 1970) have to be included in the opacity calculations. The effect of carbon continous opacities is largest in the UV. Several carbon absorption edges show up in the model flux distribution and cause a shallower UV-flux gradient than normal composition models have.

Line blanketing is taken into account by including the detailed bound-bound opacity of the strongest UV metal lines as well as of helium lines. About 800 line transitions are taken into account as Voigt profiles. This approach allows to treat the line blanketing of the strongest lines properly but fails to include the effect of millions of weak lines. The following differences to normal composition model fluxes can be noted: Carbon resonance lines are the strongest absorption features. In the absence of a Lyman absorption edge, the emergent flux below 900 Å is considerably larger than for normal composition (see Figure 1). Since many ions of relevance have resonance lines below 900 Å, their line blocking is of greater importance for an extreme helium star than for a normal composition B star model.

2.2 Determination of atmospheric parameters

The three atmospheric parameters (T_{eff}, g, C/He) have to be determined simultaneously. The model fitting proceeds as follows: In the first step a reasonable C/He is assumed and kept fixed throughout the calculations. Then the gravity as a function of T_{eff} can be derived by matching the line profiles of the He I lines, which are Stark broadened and, therefore, sensitive to electron density. The ionization equilibria of various elements can be used to determine T_{eff} as a function of gravity. The intersection of these fit curves in the (g, T_{eff})- diagram defines a first estimate of the atmospheric parameters. Now the carbon abundance has to be determined and the fitting procedure has to be repeated for a fine adjustment of the atmospheric parameters. A fit diagram for BD+10°2179 is displayed in Figure 2 for illustration.

3. FINE ANALYSES OF VISUAL AND ULTRAVIOLET SPECTRA

Only four stars have so far been analyzed using model atmospheres: HD 124448 (Schönberner and Wolf, 1974), BD-9°4395 (Kaufmann and Schönberner, 1977), HD 168476 (Walker and Schönberner, 1981) and BD+10°2179 (Hunger and Klinglesmith, 1969, Heber, 1983). These analyses

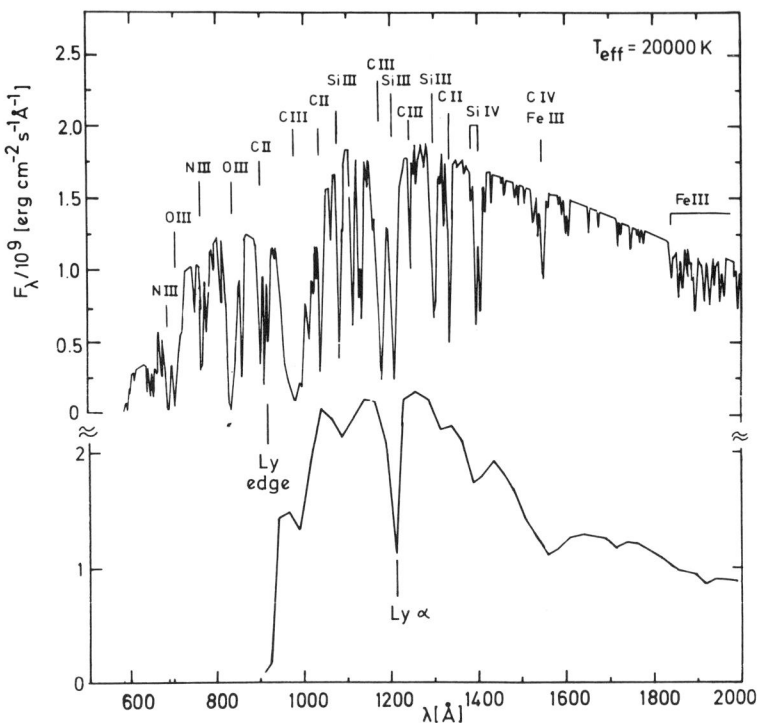

Figure 1: UV-model flux distribution for a He-C-atmosphere (He/C = 0.01, top) and for a normal composition (Kurucz, 1979; bottom). Both models have T_{eff} = 20000 K and log g =2.5. The He-C-model-flux distribution is sampled at 3Å whereas the other is sampled at 25Å.

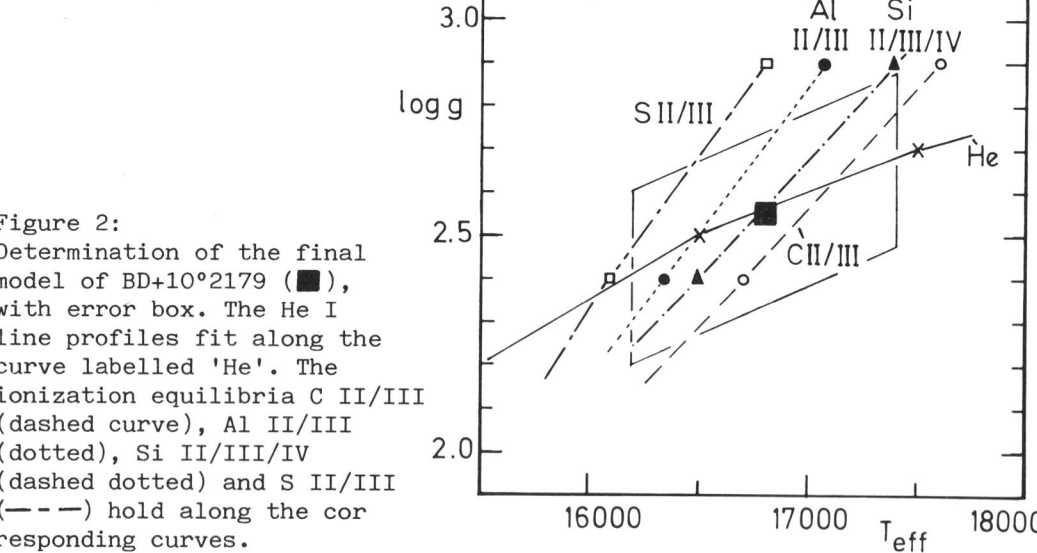

Figure 2: Determination of the final model of BD+10°2179 (■), with error box. The He I line profiles fit along the curve labelled 'He'. The ionization equilibria C II/III (dashed curve), Al II/III (dotted), Si II/III/IV (dashed dotted) and S II/III (— - —) hold along the corresponding curves.

were based on photographic spectra from which abundances of ten elements in all four stars were determined. Besides the light elements H, He, C, N and O, abundances for intermediate Z elements (Mg, Al, SI, P, S) were derived. For elements heavier than sulphur the analyses are rather incomplete, the reason being the weakness of their visual spectral lines. HD 168476 is considerably cooler than the other stars analyzed and, therefore, displays a rather large number of spectral lines of iron group elements (Ti to Ni) in the visual spectrum. In the hotter extreme helium stars the abundances of these elements can be derived from UV high resolution spectra. Other light elements of interest (e.g. boron) also become accessible to the analysis. However, an abundance analysis of ultraviolet spectra has so far been carried out for BD+10°2179 only (Heber, 1983). In this case the number of elements analyzed was almost doubled. However, there is a major drawback – the crowding of the lines in the UV – which renders an analysis difficult or even impossible. Extensive spectrum synthesis is required to derive the abundances.

3.1 Effective temperatures and gravities

The positions of the four analyzed stars in the $(g, \log T_{eff})$ plane as derived from the fine analyses are shown in Figure 3. Since T_{eff} and gravity allow the luminosity over mass ratio L/M to be determined, lines of constant L/M are also shown in Figure 3. The four stars have $\log L/M = 4.2$ (solar units), in the mean. Because the individual errors are smaller than the star to star scatter (see Figure 3) it has to be concluded that differences in L/M are real, HD 168476 having the largest L/M ($\log L/M = 4.6$, solar units) and BD+10°2179 the lowest L/M ($\log L/M = 3.7$, solar units). These L/M ratios are important observational constraints to be met by any evolutionary scenario. Additional constraints arise from the observed atmospheric composition.

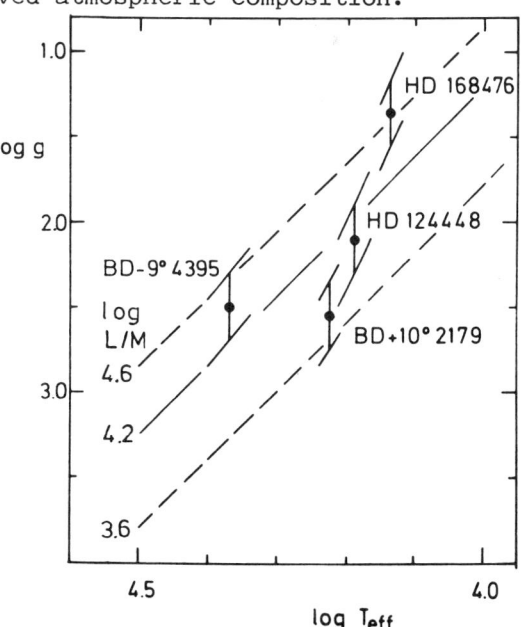

Figure 3: Position of the extreme helium stars in the $\log g - \log T_{eff}$ plane. Dashed: lines of constant L/M, labelled with $\log L/M$ in solar units.

3.2 The abundance patterns

The elemental abundances derived by means of the fine analyses are summarized in Table I. The primary data are the abundances relative to helium. In order to compare them to normal composition stars, especially to the sun, they have to be normalized. Abundances normalized to $\log \Sigma \mu_i n_i = 12.15$ (μ_i being the atomic weight number and n_i the number fraction of element i) are given in Table I. This normalization is appropriate when we assume that all hydrogen has been burnt to helium (see below). Abundances relative to the sun are plotted in Figure 4. Before we discuss the implications of the observations with respect to the stars' nuclear history and their population caracteristics let us discuss the observational results for individual elements in some detail.

3.2.1 **Hydrogen**. Balmer lines are detected in BD-9°4395 and BD+10°2179 while they are absent in HD 168476 and HD 124448. Hence only upper limits to the hydrogen abundance were derived for the latter. Hydrogen is found to be a trace element and its abundance varies considerably form star to star.

3.2.2 **Boron**. Boron has been analyzed from its UV resonance line (B II) only for BD+10°2179. Boron is deficient by more than 1 dex.

3.2.3 **Carbon, nitrogen and oxygen**. Carbon is overabundant with respect to the sun by a factor of 5 to 7, while oxygen is underabundant by a factor of 3 to 6. Nitrogen is overabundant in HD 124448 and HD 168476 while it is about solar in the two other stars. The relevant abundance ratios are given in Table II. Two groups exist with respect to C/N (and also H/He): BD+10°2179 and BD-9°4395 have large C/N ratios while HD 168476 and HD 124448 have lower C/N. Nitrogen and oxygen are equally abundant (N/O$\sim$1 within error limits) in all stars.

3.2.4 **Neon through calcium**. Considering the possible nuclear reactions during helium burning, the abundances of neon and magnesium would be most interesting among the elements of intermediate Z (Ne through Ca). Ne I lines are detectable only in the red spectrum and, therefore, HD 168476 was the only extreme helium star analyzed for Ne since no red spectra were available for the others. Its abundance, however, is regarded as uncertain (Walker and Schönberner, 1981) since NLTE effects are probably large (Auer and Mihalas, 1973). Deviations from LTE might also be important for Mg (Mihalas, 1972, Snijders and Lamers, 1975) and it is thus premature to discuss the LTE Ne/Mg ratio for HD 168476. Additional red spectra and NLTE calculations are required to derive reliable Ne/Mg ratios. The other elements of intermediate Z are more or less normal with the outstanding exception of phosphorus which is strongly overabundant in BD-9°4395 (and to a somewhat smaller extent also in HD 168476). This is reminiscent of the abundance pattern of some helium-weak-line-stars om the main sequence (the so called "phosphorus stars" ι Ori B and 3 Cen A, Baschek, 1975).

Table I: Atmospheric parameters and abundances of four extreme helium stars (Heber, 1983). The atmospheric parameters are based on the set of partially line blanketed model atmospheres as described in section 2.1. L/M is given in solar units. Uncertain abundance values are marked with colons. The abundances of the sun are from Holweger (1979).

	HD 168476	HD 124448	BD+10°2179	BD−9°4395	
T_{eff}/K	13700	15500	16800	23500	
$\log g$	1.35	2.10	2.55	2.50	
$\log (L/M)$	4.6	4.1	3.7	4.4	

Abundances normalized to $\log \sum \mu_i n_i = 12.15$ sun

	HD 168476	HD 124448	BD+10°2179	BD−9°4395	sun
H	<7.8	<7.5	8.5	8.7	12.0
He	11.54	11.53	11.53	11.54	11.0
B			<1.3		2.3
C	9.4	9.46	9.54	9.36	8.67
N	8.9	8.83	8.11	8.0	7.99
O	8.4	8.5	8.1	8.24	8.92
Ne	9.3:				7.73
Mg	7.7	8.2	8.02	7.4:	7.53
Al	7.2:	6.2	6.25	6.25	6.43
Si	7.7	7.51	7.32	7.8	7.50
P	6.3	5.6	5.5	6.8	5.35
S	7.0	7.2	7.12	7.6	7.20
Ar		6.6	6.4		6.83
Ca	7.0	<6.9	<5.9:		6.36
Sc	4.3		<2.0:		2.99
Ti	5.6	6.0	4.06		4.88
V	4.4:				3.91
Cr	6.2		5.0		5.61
Mn	6.2		4.4		5.47
Fe	7.5	7.4	6.49	6.8:	7.46
Co			4.4		4.85
Ni	6.5		5.1		6.18
Cu			4.0		4.24

Table II: Abundance ratios (an uncertain value is marked with a colon)

	HD 168476	HD 124448	BD+10°2179	BD−9°4395
$H/10^4 He$	<2	<1	10	15
C/N	3	4	27	23
N/O	3	2	1	0.6
$Fe/10^5 He$	10	8	0.9	2:

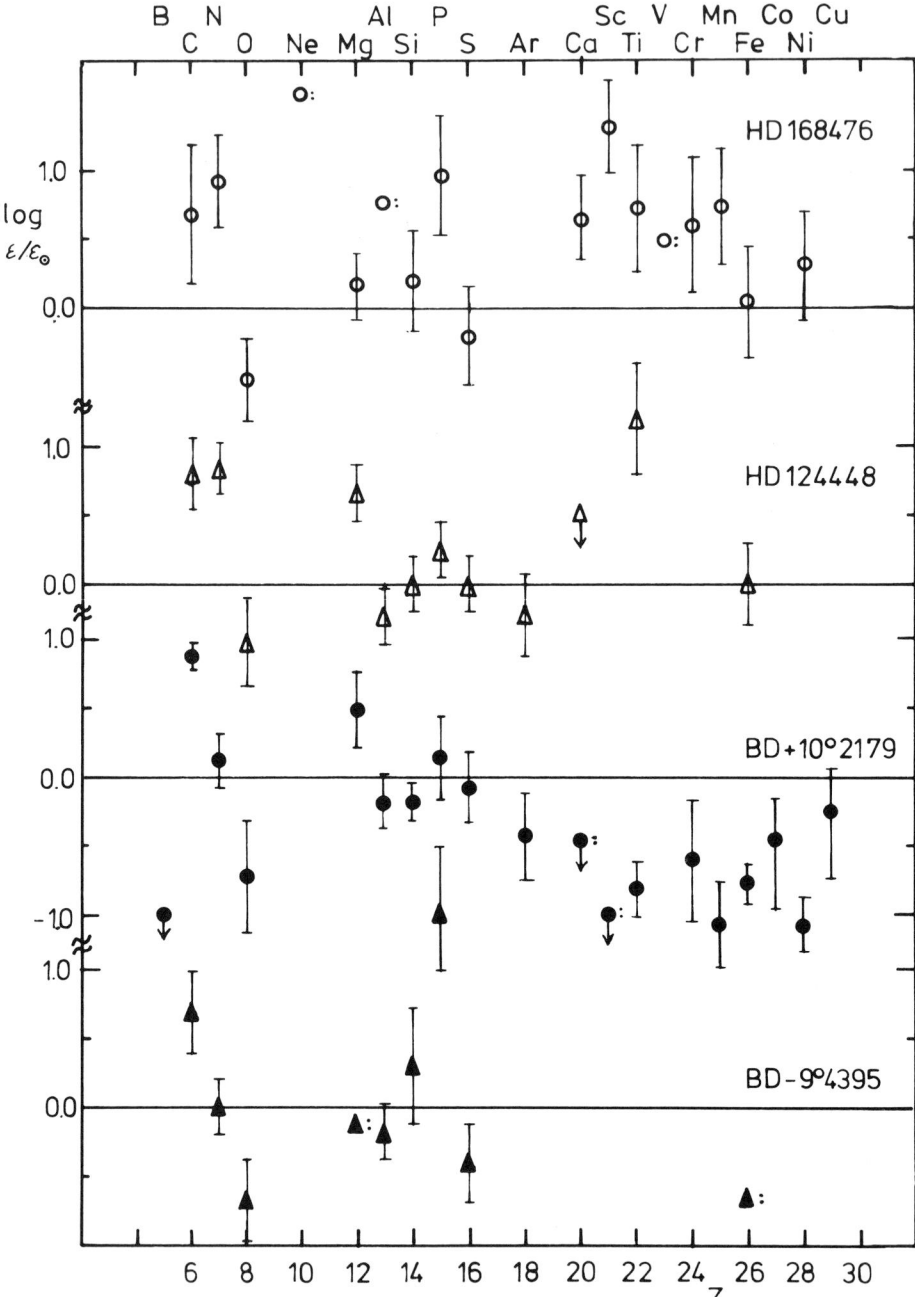

Figure 4: Abundances of extreme helium stars relative to the sun. Open circles: HD 168476; open triangles: HD 124448; filled circles: BD+10°2179; filled triangles: BD-9°4395. Uncertain values are plotted without error bars and are marked with colons. Upper limits are marked with downward arrows.

3.2.5 **The iron group.** The elements scandium through copper are well studied only in BD+10°2179 and HD 168476. For HD 124448, titanium and iron were accessible only. Iron is approximately solar. For BD-9°4395 only a rough estimate of the iron abundance (based solely on one spectral line) was obtained. The abundance pattern for the iron group of BD+10°2179 is strikingly different from those of HD 168476 as can be seen from Figure 4. The whole group of elements is underabundant by 0.75 dex (in the mean) for the former. The odd elements Co and Cu seem to have larger abundances which might not be real since hyperfine structure splitting was neglected in the analyses due to the lack of atomic data. In contrast, HD 168476 has a solar iron abundance and the other iron group elements are even enriched with respect to the sun (mean overabundance of the iron group 0.6 dex.)

3.3 Discussion of abundance patterns

The observed abundances can give important clues to the stars' nuclear history and to the stellar population they belong to. The abundances of the lighter elements can be affected by nucleogenetic processes whereas the heavier elements presumedly have maintained their primordial abundances.

3.3.1 **Population caracteristics.** The abundances of the iron group elements are indicators of the stars' metallicities. HD 168476 and HD 124448 have solar iron abundances whereas BD+10°2179 is iron deficient. Another metallicity indicator is the abundance ratio of intermediate Z elements (Mg through S) to iron. This quantity is independent of any normalization of the abundance scale and was found to be larger than in the sun for old disc and population II stars (Mäckle et al., 1975, Tomkin et al., 1985). In BD+10°2179 the intermediate Z elements are about solar while the iron group is deficient. This abundance pattern is strikingly similar to that of the old disc star Arcturus (α Boo, Mäckle et al., 1975) as can be seen in Figure 5. The abundance pattern of Arcturus is essentially that of the interstellar medium out of which the star formed. (Note that in Arcturus C,N and O are as deficient as Fe). Besides the iron underabundance of BD+10°2179 relative to the sun (which depends on the abundance normalization), this similarity is further evidence for the low metallicity of BD+10°2179. Note that no such gradient in the abundance patterns of HD 168476 and HD 124448 is present, see Figure 4. Hence we can conclude that HD 168476 and HD 124448 belong to the disc population while BD+10°2179 belongs to an old population. Considering also its location in the galaxy, far away from the galactic plane (z = 2.7 kpc, Heber, 1983), it is safe to state that BD+10°2179 belongs to the halo population. The situation is not so clear for BD-9°4395. Although no reliable iron abundance is available for BD-9°4395, the estimated underabundance might be regarded as a slight hint that it belongs to an old poulation too. There are two additional arguments to support this conjecture: (i) The observed C/N and N/O ratios are very similar to those of the low metallicity star BD+10°2179 but distinctly different from those of the metal rich extreme helium stars (HD 168476 and HD 124448).

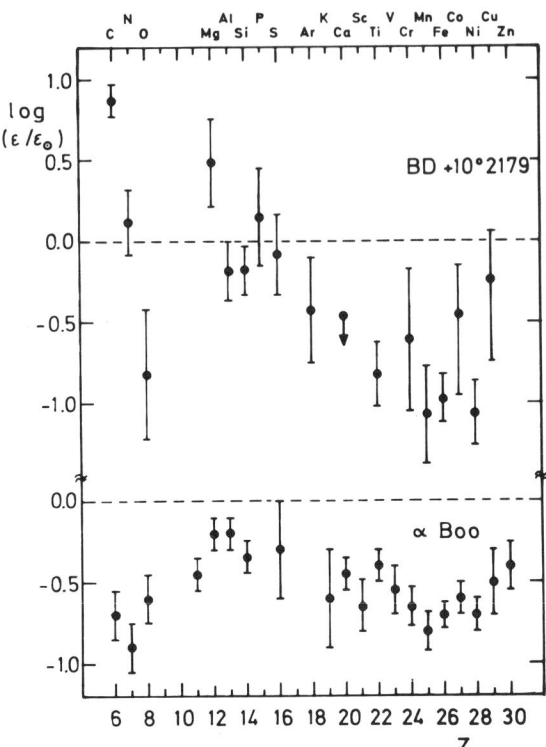

Figure 5:
Chemical composition of BD+10°2179 (top) relative to the sun and of the comparison star α Boo (bottom). the abundances of α Boo are taken from Mäckle et al. (1975).

(ii) Its radial velocity does not follow the galactic rotation (see Drilling and Heber, these proceedings, for a discussion of the extreme helium stars' radial velocities). An abundance analysis for the iron group in BD-9°4395 is urgently needed to unambiguously determine its metallicity. Let us assume in the following that BD-9°4395 is a low metallicity star similar to BD+10°2179.

Apparently, the analyzed extreme helium stars do not belong to one single population.

3.3.2 <u>Nuclear history</u>. The surface composition of the extreme helium stars is obviously dominated by the products of hydrogen and helium burning. The CNO elements (besides H and He) are expected to be the elements primarily affected by these nuclear processes. The composition of the extreme helium stars' atmospheres is that of an helium and nitrogen rich zone in which essentially all hydrogen has been processed by the CNO-cycle. The primordial N abundance of BD+10°2179 (and perhaps also of BD-9°4395) is presumably subsolar and therefore the observed normal abundance is coincidental. The observed N/O ratio of about unity can be explained if ^{16}O, which is the slowest species to come into equilibrium in the CNO bi-cycle, has not reached its (low) equilibrium abundance, or if it has been enriched by helium burning (see below).

Note in passing that the low boron abundance of BD+10°2179 is consistent with this picture, since Boron is rapidly destroyed by protons under conditions that prevail in hydrogen burning shells.

The large carbon abundance indicates that there must be some admixture of helium burnt material: About 1% of the helium has been converted to ^{12}C by helium burning. C/O > 8 (see Table II) indicates that very little ^{16}O has subsequently been produced by α capture. Available data on α capture products Ne and Mg are insufficient for any conclusions to be drawn. However, the high C/O ratio implies that the abundances of these elements will not be enhanced via α captures.

During helium burning subsequent α captures could give rise to the liberation of neutrons via the chains of reactions

$$^{14}N(\alpha,\gamma)^{18}F(\beta^+,\nu)^{18}O(\alpha,\gamma)^{22}Ne(\alpha,n)^{25}Mg$$

$$\text{or } ^{12}C(p,\gamma)^{13}N(\beta^+,\nu)^{13}C(\alpha,n)^{16}O.$$

Large temperatures ($T \gtrsim 3 \cdot 10^8$ K) are required for the former chain of reactions while protons have to be present in the helium burning shell for the latter chain. The abundances of s-process elements (e.g. Ba, Sr) would give important clues to whether the atmospheric material has been exposed by neutrons. These elements, however, are not observable in the extreme helium stars. It is worthwhile to note that elements of intermediate Z are also characteristically enriched if a neutron exposure had taken place. Truran and Iben (1977) calculated the nucleosythesis in thermally pulsing stars due to the operation of the ^{22}Ne neutron source and found that three quarters of the neutrons are absorbed by the ^{22}Ne progeny which, in consequence, leads to an enrichment of abundant isotopes of intermediate Z elements (e.g. ^{26}Mg, ^{31}P, ^{40}Ar, ^{45}Sc, ^{59}Co). Since however the extreme helium stars have rather low masses (0.6 $M_\odot$ to 1.0 $M_\odot$, Schönberner, 1977), the ^{13}C neutron source is more likely than the ^{22}Ne source provided that protons can be mixed into the helium burning shell. Whether the enrichment of some intermediate Z elements (e.g. P in BD-9°4395) could be due to neutron exposures remains to be investigated from detailed evolutionary calculations.

Last but not least it should be mentioned that a small admixture ($\sim 0.1\%$) of the original hydrogen rich material is still present in the atmospheres of BD+10°2179 and BD-9°4395.

4. THE EFFECTIVE TEMPERATURE SCALE FOR EXTREME HELIUM STARS

In view of the inhomogeneity of the extreme helium stars' abundance patterns, it is premature to regard the four stars analyzed so far as being representative of the whole group. It was deemed necessary to analyze all class members. A first step towards this goal was the determination of their effective temperatures. Synthetic colours were calculated from model fluxes by Heber and Schönberner (1981) and the effective temperatures determined from Johnson and Strömgren colours. The IUE satellite allowed the accuracy of the effective temperatures to be improved considerably since the flux maxima can be observed. Drilling et al. (1984) observed 12 extreme helium stars and derived T_{eff}'s from the total fluxes. Results are given in Table III and are plotted in

Figure 6 as a histogram. Most of the extreme helium stars are rather cool (9000 K to 18000 K), while only two stars are hotter than 18000 K. However, this temperature distribution is biased, since hot extreme helium stars – at the same luminosity and distance – are apparently fainter in the visual than cooler ones due to the bolometric corrections and are, therefore, difficult to discover. Indeed, two out of five newly discovered faint helium stars have been found to be hot, too (Heber, Jonas and Drilling, these proceedings). The cool hydrogen deficient stars have T_{eff} between ~ 5000 K (HdC stars) and ~ 7000 K (RCrB stars). Hence, there is a gap in the temperature distribution of hydrogen deficient stars between 7000 K and 9000 K. LS IV–14°109 (T_{eff} = 8400 K) might be a transition object. Since non-variable helium stars in this temperature range are difficult to discover, this gap might be due to a selection effect (Heber and Schönberner, 1981).

Table III. Effective temperatures and interstellar reddening of extreme helium stars as derived from UV fluxes (Drilling et al., 1984)

star	T_{eff}/K	E(B-V)
LS IV-14°109	8400	0.20
LSS 3378	9400	0.35
BD+1°4381	9500	0.10
BD-1°3438	10900	0.40
LS IV-1°002	11900	0.45
HD 168476	12400	0.12
LSE 78	13600	0.10
LS II+33°005	15000	0.22
HD 124448	15500	0.08
BD+10°2179	17700	0.00
BD-9°4395	23000	0.30
HD 160641	31900	0.40

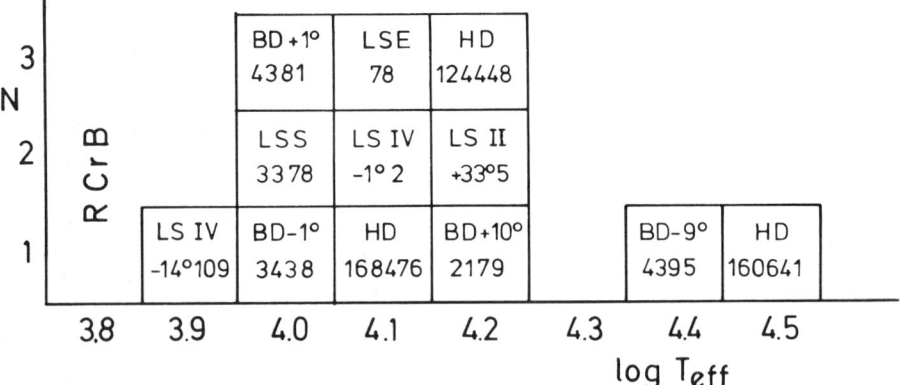

Figure 6: Temperature distribution of the extreme helium stars as derived from UV measurements (Drilling et al., 1984).

5. THE MASS LOSS RATES OF EXTREME HELIUM STARS

The high resolution IUE spectra revealed the existence of expanding envelopes around three extreme helium stars. Some of the resonance lines are obviously formed or affected by a stellar wind. Since high stages of ionization are encountered (N V in HD 160641, C IV in BD-9°4395 and BD+10°2179), the winds cannot be in radiative equilibrium, but are "superionized". This phenomenon is commonly observed in normal early type stars, too, but its physical cause is still unknown. Hamann et al. (1982) constructed empirical wind models for three extreme helium stars by means of the comoving frame formalism. Three different models for the heating of the wind where assumed in the calculations of the ionization balance. The resulting empirical limits on the mass loss rates are given in Table IV. The "corona model", which assumes that the heating is due to a X-ray source, or the "warm wind model", which assumes an electron temperature raised to 10^5 K, can consistently explain the "super-ionization". The unknown heating mechanism is responsible for the large uncertainty of the mass loss rates.

The three extreme helium stars were found to lose mass on rates of the same order of magnitude as normal stars of similar luminosity.

The mass loss rates of extreme helium stars apparently increase with increasing effective temperature, at variance with the results for normal stars (Lamers, 1981), which, at constant luminosity, decrease with increasing T_{eff}.

Table IV: Mass loss rates of three extreme helium stars (Hamann et al., 1982)

star	T_{eff}/K	log $\dot{M}/(M_\odot/yr)$
HD 160641	31900	-8.2.....-7.2
BD-9°4395	23000	-8.5.....-7.7
BD+10°2179	16800	-11.0.....-8.9

6. RELATED OBJECTS

After having discussed the properties of the true class members, we will now draw attention to some helium stars which might be related to the extreme ones. These are the unique object BD+13°3224 and four helium- and carbon rich sdO stars.

6.1. BD+13°3224

BD+13°3224 is carbon weak lined and, therefore, cannot be termed an extreme helium star. Furthermore, it has a L/M ratio smaller than the latter (Hill et al., 1981) and its hydrogen abundance (1% by numbers) is intermediate between the extreme (≤ 0.1%) and the intermediate (50%)

helium stars. BD+13°3224 appears to be more comparable to HD 144941 (Hunger and Kaufmann, 1973) and both stars may define a new sub-group of the hydrogen deficient stars.

BD+13°3224 is of special interest because it is the only helium star for which distance and mass have been determined. BD+13°3224 is a radial pulsator with a well established period (Landolt, 1975). A modified Baade-method based on UV fluxes has been used to determine its mean effective temperature, radius and luminosity (T_{eff} = 23450 K, R = 1.98 $R_\odot$, log $L/L_\odot$ = 3.03±0.12, Lynas-Gray et al., 1984). The distance as derived from the angular diameter is 1.5±0.1 kpc. From the spectroscopically determined gravity (log g = 3.7) the mass $M = 0.7^{+0.4}_{-0.3} M_\odot$ is derived.

6.2 Helium- and carbon rich sdO stars

Some extremely helium rich objects are known among the subluminous O stars (see Hunger, 1975; Berger and Fringant, 1980, Heber et al., these proceedings). It seems to be tempting to search for an evolutionary link between these hot subdwarfs and the B-type extreme helium stars. Most of the subdwarfs are of high gravity (i.e. have small L/M ratios) and cannot be related to the extreme helium stars, since the evolution of an extreme helium star proceeds at constant luminosity (Schönberner, 1977). Recent NLTE analyses revealed that four helium rich subdwarfs have L/M ratios as large as the extreme helium stars: Giddings (1980) analyzed BD+37°442 and derived T_{eff} = 55000 K, log g = 4.0 (i.e. log L/M = 4.35, solar units). He noted that BD+37°1977 displays a spectrum very similar to BD+37°442 and, therefore, must have similar atmospheric parameters. Husfeld et al. (these proceedings) analyzed LSE 153 and LSE 259 and found T_{eff} = 70000 K, log g = 4.75 (i.e. log L/M = 4.0, solar units) and T_{eff} = 75000 K, log g = 4.4 (i.e. log L/M = 4.5, solar units), respectively. Since all four stars are carbon strong lined it has been conjectured, that they are the immediate descendants of the B-type extreme helium stars and should be considered as very hot extreme helium stars.

7. FUTURE DEVELOPMENTS

Ultraviolet spectra obtained with the IUE spectrograph have considerably improved our knowledge of the hot extreme helium stars, since they allowed effective temperatures, abundances of heavy elements and mass loss rates to be determined. However, most of the extreme helium stars are to faint in the UV to be observable with IUE at the required high resolution. Moreover the analyses of the IUE spectra are hampered by strong line blocking. (It might be a fortunate coincidence that BD+10°2179, the only star analyzed from the UV, turned out to be metal deficient and, thus, line blocking was a less severe problem in the analysis than it appears to be for the others). Nevertheless, careful analyses of high resolution IUE spectra can give important information, e.g. the iron abundance of BD-9°4395 (see above). A complete picture of the chemical composition of all extreme helium stars, however, can not be drawn.

Recently, the advent of efficient linear detectors combined with Echelle spectrographs provides a powerful tool for visual spectroscopy. Using the ESO Cassegrain Echelle spectrograph (CASPEC), all but one of the known extreme helium stars have been observed at high resolution (0.25Å). A S/N of 100 was reached for the brightest ones in single short exposures (15 min.). Hence the quality of these spectra is superior to the photographic Coude spectra and will allow weak lines to be measured. These are important for accurate abundance determinations since they depend only weakly on model parameters such as microturbulence. On the CASPEC spectra, lines of equivalent width as low as 10 mÅ can be measured for stars brighter than B=12.0. For the faintest stars (B=13.5) a limiting equivalent width of 50mÅ is reached. Fine analyses of these spectra will certainly give a much clearer picture of the chemical composition of the extreme helium stars and, consequently, will give clues to the origin of their hydrogen deficiency.

ACKNOWLEDGEMENT. I thank Dr. C.S. Jeffery for careful reading of the manuscript. A travel grant from the Deutsche Forschungsgemeinschaft is gratefully acknowledged.

REFERENCES

Aller, L.H.: 1953, in "Les Prossesus Nucléaires dans les Astres", 5me
    Colloque International d'Astrophysique, Liège, p. 337
Auer, L., Mihalas, D.: 1974, Astrophys. J. **184**, 151
Baschek, B.: 1975, in "Problems in Stellar Atmospheres and Envelopes",
    eds. B. Baschek, W. H. Kegel, G. Traving, Springer, Berlin,
    Heidelberg, New York, p. 101
Berger, J., Fringant, A.-M.: 1981, Astron.Astrophys. **85**, 367
Dinger, A. S.: 1970, Astrophys. Space Sci. **6**, 118
Drilling, J.S., Schönberner, D., Heber, U., Lynas-Gray, A.E.: 1984,
    Astrophys. J. **278**, 224
Giddings, J.: 1980, Phd thesis, UCL, London
Hack, M.: 1967, "Hydrogen poor Stars", in Modern Astrophysics, Gordon
    and Breach, New York
Hamann, W.-R., Schönberner, D., Heber, U.: 1982, Astron. Astrophys. **116**, 273
Heber, U.: 1983, Astron. Astrophys. **118**, 39
Heber, U., Schönberner, D.: 1981, Astron. Astrophys. **102**, 73
Hill, P.W.: 1965, Monthly Notices Roy. Astron. Soc. **129**, 137
Hill, P.W., Kilkenny, D., Schönberner, D., Walker, H.J.: 1981, Monthly
    Notices Roy. Astron. Soc. **197**, 81
Holweger, H.: 1979, 22. Liège International Astrophys. Symp. p. 117
Hunger, K.: 1975, in 57
Hunger, K., Klinglesmith, D.: 1969, Astrophys. J. **157**, 721
Hunger, K., Kaufmann, J.P.: 1973, Astron. Astrophys. **25**, 261

Iben, I., Jr., Tutukov, A.V.: 1985, Astrophys. J. Suppl. **58**, 661
Kaufmann, J.P., Schönberner, D.: 1977, Astron. Astrophys. **57**, 169
Klemola, A.R.: 1961, Astrophys. J. **134**, 130
Kurucz, R.L.: 1979, Astrophys. J. Suppl. **40**, 1
Lamers, H.J.G.L.M.: 1981, Astrophys. J. **245**, 593
Landolt, A.U.: 1975, Astrophys. J. **196**, 789
Lynas-Gray, A.E., Schönberner, D., Hill, P.W., Heber, U.: 1984, Monthly Notices Roy. Astron. Soc. **209**, 387
Mäckle, R., Holweger, H., Griffin, R., Griffin, R.: 1975, Astron. Astrophys. **38**, 239
Mihalas, D.: 1972, Astrophys. J. 177, 115
Peach, G.: 1970, Mem. Roy. Astron. Soc. **73**, 1
Scholz, M.: 1972, in Vistas in Astronomy, ed. A. Behr, Pergamon press, Oxford, Vol. **14**, p. 53
Schönberner, D.: 1977, Astron. Astrophys. **57**, 437
Schönberner, D., Wolf, R.E.A.: 1974, Astron. Astrophys. **37**, 87
Snijders, M., Lamers, H.J.G.L.M.: 1975, Astron. Astrophys. **41**, 245
Tomkin, J., Balachandran, S., Lambert, D.L.: 1985, Astrophys. J. **290**, 289
Truran, J.W., Iben, I., Jr.: 1977, Astrophys. J. **216**, 797
Underhill, A.B.: 1966, The early type stars. Reidel, Dordrecht.
Walker, H.J., Schönberner, D.: 1981, Astron. Astrophys. **97**, 291

DISCUSSION

BHATT: Is there any correlation between the heavy element abundances and the excess IR emission from the two stars? Are there IR observations?

HEBER: IR photometry is available for most of the extreme helium stars. Some have also been observed with IRAS. No infrared excess has been found so far.

LAMBERT: Either using the Caspec or Space Telescope which is going up next year, would it be possible to detect the heavy elements and the S-process elements abundances?

HEBER: In the photospheres of the extreme helium stars, the s-process elements like Ba and Sr are two times ionized and, therefore, in a noble gas configuration. The strongest lines lie in the far UV and cannot be obsered from the ground or with the Space Telescope. Other heavy elements can be detected in the UV spectra.

MOHAN RAO: In the comoving frame calculations, what type of velocity law have you chosen?

HEBER: The usually adopted square-root law has been used mostly. Attempts to model the velocity law have been also made.

GARRISON: I see that there is a spread in T_{eff} and a variation of the abundances. Is there any systematic effect? I realize that there are only 4 stars, but, is there any trend, such as the cooler stars having more nitrogen?

HEBER: I don't think so. With only four stars analysed, it is hard to tell whether there is a correlation.

MICHAUD: Can you exclude the importance of NLTE effects?

HEBER: No. In fact, there might be some evidence for deviation from LTE in the outer layers of the atmospheres. In general, LTE-line profile calculations for the strongest helium lines (e.g. 4471 A, He I) cannot reproduce the observed line profiles in the line cores which are formed rather far out in the atmosphere. The observed profiles are too deep. Before NLTE calculations actually have been carried out, we cannot estimate their importance. Since NLTE effect might also influence the ionization equilibria, it is always important to have a second T_{eff} indicator, i.e. IUE flux measurements.

RANGARAJAN: In your analysis why did you use the pure scattering assumption? Why not thermal sources?

HEBER: Thermal sources are unimportant at the low densities prevailing in the expanding envelopes.

RANGARAJAN: Is not the Voigt broadening more appropriate than Doppler broadening?

HEBER: Taking into account Doppler broadening is sufficient since the shells absorbing at a given frequency (so called Sobolev shells) are thin. In fact, in order to match the observed widths of the lines, a microturbulent velocity of the order of 100 km/s has to be invoked.

RANGARAJAN: What is their terminal velocity?

HEBER: The terminal velocities range from 400 km s^{-1} to 600 km s^{-1}.

RANGARAJAN: Why cannot we use the Sobolev method for the expanding envelope?

HEBER: Whether the Sobolev approximation is valid or not depends on the velocity gradients in the expanding envelope. In any case the comoving frame method should give accurate results.

DRILLING: Is it true to say that the metals are like the Pop II abundances?

HEBER: The abundances of the iron group elements are well-established only for HD 168476 and BD +10°2179. These elements are overabundant in HD 168476 whereas they are underabundant (by ≈ 1 dex) in BD +10°2179. Therefore, only in BD +10°2179 the abundance pattern indicates that the star belongs to an old population (but not to an extreme Pop II).

HUNGER: This is rather a question to Dr. Michaud. Do you think it is possible that the abundances for the metals for these different objects might be enhanced by diffusion for these low gravity objects?

MICHAUD: Not if the mass loss rates are 10^{-10} solar mass/yr.

VENUGOPAL: What is the physical significance of the mass loss in these stars being of the same order as in normal stars of the same luminosity?

HEBER: Since radiation pressure is generally believed to drive the stellar winds, it has frequently been argued that the mass loss rate should depend on the chemical composition too. For the extreme helium stars whose chemical compositions differ considerably from normal, this seems not to be the case.

PHOTOMETRIC PROPERTIES OF THE EXTREME HELIUM STARS

Arlo U. Landolt
Louisiana State University
Box BK, LSU Observatory
Baton Rouge, LA 70803-4001, USA

The purpose of this paper is to review the photometric histories of the extreme helium stars.

I. A PHOTOMETRIC OVERVIEW

There exists a small number of hot blue stars whose spectra contain strong helium spectral lines but no, or only very weak hydrogen lines. These stars have had a series of descriptive names, first helium stars, and more recently extreme helium or hydrogen-deficient stars. These names tend to be used interchangeably. Popper (1942) identified the first member of this group, HD 124448. Bidelman (1952) found the second such star, HD 160641, and Thackeray and Wesselink (1952) nearly simultaneously discovered the third, HD 168476 (see Hill, 1969a).

There have been three general reviews of the properties of the helium or hydrogen-deficient stars; in chronological order, the reviews have been by Hack (1967), Dinger (1970), and Hunger (1975).

A tabulation of the original eight extreme helium stars, plus two close binary systems which have quite similar spectra, is given by Hunger (1975). An additional seven extreme helium stars are described by Drilling (1980). Finally, Drilling (1985) recently has discovered three more such objects, thereby more than doubling the number known, and bringing the total number of extreme helium stars to twenty-one. He has found the new members of the class during spectroscopic studies of stars classified as OB+ in the catalogue of Stephenson and Sanduleak (1971).

An object closely related to the extreme helium stars is BD+13°3224 = V652 Her. The words 'closely related' are used because some investigators describe the star as extreme helium and others do not. Even the same individual(s) is not always consistent. Landolt

(1975) found BD+13°3224 to be variable in light with a period of 0.^{d}107995. Since then, a series of papers discussing its characteristics have appeared in the literature, mostly in the Monthly Notices of the Royal Astronomical Society.

Photometric studies of the extreme helium stars go back some twenty years. Both Herbig (1964, 1967) and Hill (1965, 1967a) pointed out the importance of investigating whether or not the helium stars were variable in light. Variability potentially tells one something about the star's structure and evolution. These comments resulted in great part from the light variability of the apparently related star, MV Sgr (Hoffleit 1959, Herbig 1964). Hill (1967b) noted that, other than MV Sgr, no other helium star was known to be variable in light.

Landolt (1968) initiated an observational program which was occasionally to check the then known helium stars for brightness changes. Intensive observations over a period of a few nights on two or three occasions per year would then reveal both short-term and long-term brightness changes, if they occurred. Possible small changes were noted. Hill (1969b) reported on photoelectric observations of HD 124448, HD 160641, and HD 168476. He noted that while HD 124448 appeared to be constant, both HD 160641 and HD 168476 showed 'micro-variation' in magnitude but not in color. In a review lecture, Hill (1969a) said that no helium stars other than MV Sgr were known to vary in brightness by a 'significant' amount; this statement has withstood the test of time. He further noted that all of the original groups of helium stars had high radial velocities. The smallness of their proper motions and the strength of the interstellar lines in their spectra suggested that the helium stars were distant and hence rather luminous objects.

Landolt (1973a) reported that a series of UBV photoelectric observations obtained during the years 1968-72 showed that although some brightness variations had occurred, none were large. The data indicated a decade-long secular brightening of HD 168476 when compared to Hill's data. And, a seven hour monitoring of HD 160641 indicated a definite trend in the sense that the star brightened by $\sim 0.^m1$ over a seven hour period. Already the diversity in the photometric behavior of the extreme helium stars was becoming evident: those nearly and/or constant, those with variations under a day, some of unknown duration, and two, if one includes MV Sgr, with light variations which change on a time scale of years.

One year later, Landolt (1974, 1975) confirmed the variability of HD 160641 and discovered the periodicity of BD+13°3224 (V652 Her). A period near 0.^{d}6 seemed most probable for the former star. BD+13°3224 was found to have a period of 0.107995 days. Not since has an extreme helium star been found to have as simple a light variation as that possessed by BD+13°3224. Investigations of this star continue (Kilkenny 1985).

Stromgren four-color (uvby) photometry was made of hydrogen-deficient stars by Walker and Kilkenny (1980). They found nothing peculiar in these stars' locations in the color-color index diagrams. They did find the hydrogen-deficient stars to have higher c_1 indices than expected for their temperatures. This is so because there is no Balmer discontinuity since there is no hydrogen present. Walker and Kilkenny found that Stromgren photometry indicated the extreme helium stars to be cooler than B stars of the same [u-b] color index. These four-color data further confirm the light variability of HD 160641 and HD 168476. Although Hill (1969b) thought that HD 124448 was constant, Landolt (1973a) and Walker and Kilkenny's data indicated a probable low-amplitude variation.

Both the broad-band photometry (Landolt 1979a) and the intermediate band photometry (Walker and Kilkenny 1980) pointed toward the surmise that to some extent all of the extreme helium stars are variable in light. Most of the variations are complex in nature and will entail strenuous observational and theoretical efforts before our understanding is complete.

Drilling, Landolt, and Schönberner (1984) published broad band UBVRIJHKL photometry of most known extreme helium stars and hydrogen-deficient binaries. Only the latter show an infrared excess. The remaining stars apparently are single objects; none of them shows an infrared excess. The conclusion is that the hydrogen-deficient binaries are encased in dust shells, whereas the extreme helium stars are not.

The observed colors of the extreme helium stars have been used to derive their reddenings and effective temperatures. An initial work by Heber and Schönberner (1981) has been followed by Schönberner et al. (1982) wherein the λ 2200 Å feature in the ultraviolet IUE spectra was used as a criterion in establishing the color excess.

II. THE INDIVIDUAL STARS

The following material is a summary of the photometric histories of the stars now accepted as extreme helium stars. Currently available broadband UBVRI photoelectric data are summarized in Table I. Except for HD 30353 (KS Per), BD+37°1977, LSS 3184, and υ Sgr, all the photometric data are from the author's observations over the past seventeen years. All these data have been tied into the Johnson UBV and the Kron-Cousins UBVRI photometric systems as defined by Landolt (1973b, 1983). A few of the author's earliest extreme helium star UBV photometric measures were tied into Johnson's (1963) standard stars. The only extensive four-color (uvby) photometric data of which the author is aware is that of Walker and Kilkenny (1980), mentioned above. Johnson system JHKL photometry, tied into Elias et al. (1982) standard stars, for more than one-half of these objects has been reported in Drilling, Landolt, and Schönberner (1984). The only two stars which show a pronounced infrared excess are υ Sgr and LSS 4300, each a hydrogen-deficient binary. The conclusion was drawn that some hydrogen-deficient binaries have circumstellar dust shells; however, the single extreme helium stars do not.

Heber and Schönberner (1981) made the first attempt to systematically derive the color excesses of the extreme helium stars. They used a grid of unblanketed model atmospheres together with observations in the literature to derive the stars' color excess and effective temperature. The following year, Schönberner et al. (1982) improved the color excess and effective temperature determinations through use of line-blanketed model atmospheres and identification of the usefulness of the 2200 Å feature in the ultraviolet spectral region as the criterion for zero reddening. The color excesses so determined are listed in Table I. One notes that many of the extreme helium stars are heavily reddened as might have been expected from the strength of the interstellar lines in their spectra and their low galactic latitude.

II.1. BD+37°442

Rebeirot (1966) found that the star BD+37°442 belonged to the helium star group. She found the star to have a radial velocity of -156 km/sec, and estimated the star's luminosity to be perhaps as great as $M_v = -2$.

Darius et al. (1979) found an effective temperature of about 55000 °K. Rossi et al. (1980) discovered P Cyg lines of N V and C IV in IUE high resolution spectra. These observations provided evidence for the presence of mass outflow. Their absorption components extend up to -2200 km/sec. No λ 2200 Å band was detected; hence the color excess E_{B-V} was estimated to be zero.

Landolt (1968, 1973a) published UBV photometry of BD+37°442 extending over 1967-70; in addition, he has unpublished data for the years 1974-77. Although there are maximum differences of $0.^m05$ or so in these data, no obvious periodicity exists. He finds average magnitudes and colors of V = 10.01, B-V = -0.28, and (U-B) = -1.16.

Bartolini et al. (1982) believe that the star that Rebeirot (1966) discovered was BD+37°443, and not +37°442. A recent check by this author at the telescope, though, showed the original identification to be correct. Bartolini et al. (1982) also believed that they found both short period light variations (on the order of 10-15 min.) and longer time scale variations (on the order of months). Their data are not extensive.

Landolt's several hundred unpublished UBV differential photometric observations show no obvious light variations.

II.2. HD 30353 = BD+43°1069 = KS Per

KS Per is a hot hydrogen-deficient star. It is a spectroscopic binary with a period of 360 days. Studies on the UBV photometric system have been reported by Osawa et al. (1963). They find a roughly sinusoidal light curve of amplitude $0.^m14$ and period 30-40 days. The UBV magnitude and colors in Table I are straight averages of data taken from Hiltner (1956), Nariai (1963), and Landolt (1968) in order that the reader have a feel for the star's photometric characteristics. Since it is a binary and will be discussed elsewhere in this colloquium, no further comments will be made.

II.3. LSS 99

The star LSS 99 was recognized as an extreme helium star by Drilling (1985). Landolt obtained UBVRI photometry for LSS 99; these new data are given in Table I. Its colors predict that it is heavily reddened. Too few data are available to say anything about potential variability.

II.4. BD+37°1977

The BD star +37°1977 was recognized as a hydrogen-deficient star by Berger et al. (1974). Wolff et al. (1974) found the star to be a very hot subdwarf, with a temperature of 50,000 °K. The hydrogen lines were weak or absent; the radial velocity was -59 km/sec. Darius et al. (1979) found T_e = 55,000 °K from high resolution IUE spectra. Rossi et al. (1980) reported that P Cygni lines of N V and C IV gave evidence for mass outflow from the star. Absorption components extend up to -2200 km/sec. Since no λ 2200 Å band was detected, they concluded that $E_{B-V} \sim 0.0$. The only photometry known to this author was quoted by Rossi et al. (1980): Johnson V = 10.21, and Stromgren (b-y) = -0.123.

II.5. BD+10°2179

Klemola (1961) found that BD+10°2179 belonged to the small group of objects then known as helium stars. His photometric observations indicated that V = 9.95, (B-V) = -0.18, and (U-B) = -0.90. He determined the star's radial velocity to be +155 km/sec. The proper motion components of μ_α = -0".0285 and μ_δ = +0".001 per year quoted by Klemola were determined by Kopff, Nowacki, and Gondolatsch (1932). Landolt (1968, 1973a, and unpublished) suspected variability at optical wavelengths. Bartolini et al. (1982) confirmed the variability of BD+10°2179. They found an amplitude of 0.08 in V, and were able to phase the data with a period of 0.162645 days. The data of Bartolini et al. (1982) consisted of 38 measures obtained on 7 nights during a 529 day time interval. Subsequent observational programs by Hill, Lynas-Gray, and Kilkenny (1984) and by Grauer, Drilling, and Schönberner (1984) failed to find any light variation. The former had an unstated number of observations taken on 16 nights in the time frame 1979-1982; these data showed no detectable variations above the 0.02 magnitude level. The latter data consisting of nearly 3000 individual five second integrations covered a time interval of 0.21 days on 10 April 1983 U.T. and showed no light variations greater than 0.002 magnitude. Grauer et al. used a two star photometric technique (Grauer and Bond, 1981) for the latter investigation. Landolt's several hundred incompletely analyzed single channel data show no obvious periodicities. There is no doubt that the star was constant during the short interval that it was observed by Grauer et al. Hill et al. also were able to place reasonable limits on its maximum variability, if such exists. BD+10°2179 may be similar to BD-9°4395 which does show on-again, off-again changes in brightness. Additional observations are in order. They should be all-night, several days in a row, in three or four different seasons kind of observations. That way one will have access to as homogeneous a data string as possible. And for the best accuracy, the data should be collected with a two star photometer.

II.6. **LSS 1922 = CPD-58°2721**

Drilling (1980) found LSS 1922 to be an extreme helium star. Its spectrum proved similar to that of υ Sgr. Heber and Schönberner (1981) derived an effective temperature of 14,500°K. LSS 1922 has a close companion ($\Delta\alpha$ = 0^{s}6 west, and $\Delta\delta$ = 11" south; $\rho \sim 13"$, $\theta \sim 195^0$) which may affect photometric observations. Landolt has found the companion's magnitude and color indices to be V = 12.648 ± 0.006, (B-V) = +0.661 ± 0.004, (U-B) = +0.092 ± 0.000, (V-R) = +0.400 ± 0.022, (R-I) = +0.371 ± 0.013, and (V-I) = +0.769 ± 0.018 from three measures. The broad band photoelectric data in Table I for LSS 1922 are the averages of thirteen measures obtained over a thirty eight month (1978-81) interval by Landolt. They indicate that LSS 1922 varies by 0^m.15 in V, and confirm the recent announcement by Hill, Jeffery, and Morrison (1985) that LSS 1922 is a semi-regular light and color variable star. This group found a 0.07 mag variation in V, and a 0.07 mag variation in Stromgren's (u-b), all with a cycle time of about 17 days.

II.7. **LSS 3184**

This star was discovered to be an extreme helium star by Drilling (1985). At the time of writing, modern photometry was not available. The magnitude quoted by Stephenson and Sanduleak (1971) was m_{pg} = 11.9.

II.8. **HD 124448 = CoD-45°9033 = CPD-45°6748**

Popper (1942) discovered and discussed this star's observational characteristics (Popper 1946, 1947), doing so at the McDonald Observatory's 82 inch telescope by sighting southward through the mesquite bushes and cactus. His description of a stellar spectrum dominated by helium spectral lines and lacking any trace of hydrogen marks this star as the first known member of the extreme helium star group. Hill (1964) quoted early UBV photometry by Wesselink: V = 9.98, (B-V) = -0.07 and (U-B) = -0.80. Hill (1964, 1965) also derived relative abundances in HD 124448's atmosphere via a curve of growth analysis. Later photometry (Hill 1969b) found V = 9.99, (B-V) = -0.09, and (U-B) = -0.80, in good agreement with Wesselink's values.

Landolt (1973a) published UBV photometry covering the time interval 1969-72. His limited data indicated possible light variability up to several hundredths of a magnitude. Some hundreds of subsequent differential unpublished observations by Landolt indicate constancy to better than the two percent level. Walker and Kilkenny (1980), based on their observation that the standard deviation in V for HD 124448 is nearly double that for the comparison star that they used, also lean toward believing that the star is a small amplitude variable. On the other hand, Hill et al. (1984) say that their data show no light variation.

II.9. LSS 3378 = CoD-48°10153 = CPD-48°7730

Drilling (1973) found star 3378 in the catalogue of Stephenson and Sanduleak (1971) to be a helium-rich B star. He showed that LSS 3378's spectrum fitted the description of stars in the helium star class (Dinger 1970). The star's magnitude and colors were found to be V = 11.48, (B-V) = +0.43, and (U-B) = -0.31 from two measures obtained at CTIO. Drilling was able to show that the absolute magnitude might fall in the interval $-4 < M_V < -1.5$.

The results of Landolt's long term (1975-81) monitoring of LSS 3378 are shown in Table I, agreeing fortuitously with Drilling's initial values. Forty differential UBV observations obtained on 8 June 1977 U.T. indicate that LSS 3378 varies at optical wavelengths. The data may be interpreted to imply a quasi-periodicity on the order of several minutes superposed on a longer-term variation of perhaps $0^d.243$. The total amplitude is 0.06 magnitude for the shorter term variation and about 0.1 magnitude for the longer term variation. Two other nights with fewer data points also show evidence for light variation, but in a more subdued manner. The star deserves detailed photometric attention.

Drilling et al. (1984) found LSS 3378 to be one of the coolest of the extreme helium stars, with an $T_e \sim 9200$ °K. Heber and Schönberner (1981) and Drilling, Landolt, and Schönberner (1984) found the star to be heavily reddened by appreciably more than one magnitude of absorption.

II.10. BD-9°4395

MacConnell et al. (1972) found BD-9°4395 to be a hydrogen-deficient star during an objective prism survey. An abundance analysis by Kaufmann and Schönberner (1975, 1977) showed that BD-

9°4395 was an extreme helium star. They found the star's mass to fall in the range 0.6 - 0.9 solar mass and its luminosity to be on the order of log $L/L_\odot$ = 4.25. Schönberner et al. (1982) determined the color excess to be E_{B-V} = 0.30 and found an effective temperature of 23,500 °K.

A summary of the star's photometry is in Table I. Landolt found the star to be variable in light in data taken on 13 June 1978 UT; subsequent observations by Grauer and Landolt using a two-star photometer failed to find any hint of photometric changes. Landolt and Grauer (1985) have determined that the light variations showing a maximum amplitude of 0.06 magnitude are quite complex. Quotes in the literature of other unpublished data confirm this (see Jeffery and Malaney, 1985).

II.11. LSS 4300 = CoD-35°11760 = HDE 320156 = CPD-35°7069

Drilling (1980) found LSS 4300 to be an extreme helium star during a spectroscopic survey of OB+ stars. Schonberner and Drilling (1984) showed LSS 4300 to be a high-temperature analogue of the hydrogen-deficient binaries υ Sgr and KS Per. Broad-band JHKL photometry revealed an infrared excess nearly identical to that of υ Sgr. Schönberner and Drilling suggested that LSS 4300 also is a close binary system containing a helium supergiant component along with a less luminous secondary. They believe that the latter is accreting matter from the primary.

Landolt has accumulated fifteen multicolor UBVRI observations of LSS 4300 over a 1,229 day interval. The data are summarized in Table I. The suspected binary ranges in brightness from V = 9.71 to 9.86. The color, too, changes by about 0.1 magnitude. The data indicate a maximum change in brightness in a time interval as short as one day. While the data hint at a period of 65 days, two data points do fall well off the phased light curve. Hence, like both υ Sgr and KS Per, LSS 4300 may be a low amplitude eclipsing system.

LSS 4300 is star number 8849 in the Cape Photographic Catalogue (Cape Annals 18, p. 177). One finds therein the star's proper motion values to be $\tilde{\tilde{\mu}}_\alpha$ = -0.4 and μ_δ = -2.7 per 100 years.

II.12. HD 160641

The second helium star to be discovered was found by Bidelman (1952). Hill (1969b) found the star's magnitude and colors to be V = 9.86, (B-V) = +0.15 and (U-B) = -0.85. Extensive monitoring of HD 160641 in the interval 1968-72 showed the star to be variable in light (Landolt 1973a, 1975). An amplitude up to a tenth of a magnitude was evident. A period of perhaps 0.6 days was suggested. Additional unpublished data obtained in the intervening years corroborate the discovery results. Clean-cut periodicities, however, do not exist. Other observers' unpublished data, as quoted by Jeffery and Malaney

(1985), confirm these conclusions. Walker and Kilkenny (1980) have uvby data which suggests a 0.71 day "period", but they, too, note that the concept of a simple period most likely does not apply to HD 160641. A preprint (Lynas-Gray et al. 1985) indicates that "observed pulsation frequencies have been found which are consistent with $\ell = 4$ fundamental mode pulsation of a one solar mass extreme helium star".

II.13. LSS 4357

This extreme helium star was discovered by Drilling (1985). The available photometry is presented in Table I, and was obtained by Landolt in September, 1985 at the CTIO telescopes. The data are too few to address the question of light variability. They do indicate, however, by analogy to other stars in the table, a reddening of perhaps 0.3 magnitude.

II.14. LSIV-1°2

The star LSIV-1°2 was discovered by Drilling (1980) to be an extreme helium star. He found its spectrum to be nearly identical to that of HD 168476, for which Schonberner and Wolf (1974) give an effective temperature of 13,500°K. Drilling (1975) acquired broad band photometry which provided V = 10.99, (B-V) = +0.36, and (U-B) = -0.47. These values agree well with Landolt's data in Table I in this paper. The star should be observed for possible optical variations.

II.15. BD-1°3438 = LSIV-1°3

The extreme helium star BD-1°3438 was discovered by MacConnell et al. (1972) on blue objective prism plates taken with the Curtis Schmidt telescope at CTIO. In addition to pointing out BD-1°3438's similarity to other then known hydrogen deficient stars, MacConnell and colleagues provided the first UBV photometry: V = 10.42, (B-V) = +0.43, and (U-B) = -0.29. They noted that this star suffers appreciable absorption, when compared to other extreme helium stars. This was verified by Heber and Schönberner (1981) and refined by Schönberner et al. (1982); the latter group found E_{B-V} = 0.40.

Landolt has 66 unpublished UBV observations on three nights which by themselves are not sufficient to establish variability. His average magnitude and colors, though, of V = 10.328, (B-V) = +0.460 and (U-B) = -0.246 differ substantially from the discovery values determined by MacConnell et al. (1972). Hill (1985) writes that there is evidence for a mixture of long and short periods. In the sense of long term photometric behavior, BD-1°3438 may be similar to HD 168476. These two stars also are similar in temperature.

II.16. HD 168476 = CoD-56°7300 = CPD-56°8755

Thackeray and Wesselink (1952) found HD 168476 to be a helium star. The latter's photometry, quoted by Hill (1964), showed the star's brightness and colors to be V = 9.37, (B-V) = -0.01, and (U-B) = -0.67. Later photometry (Hill, 1969b) showed the star to be somewhat brighter at V = 9.30 but unchanged in colors: (B-V) = -0.01 and (U-B) = -0.69. Spectroscopic investigations (Hill, 1964, 1965) were carried out in the same time interval. HD 168476 is catalogued as CPC 5911 in the Cape Annals, volume 20, wherein one finds proper motion values of μ_α = -0.8 and μ_δ = +1.1 per 100 years. Landolt's (1973a) UBV data seemed to indicate a small secular brightening for HD 168476 when compared to Hill's results. Walker and Kilkenny (1980) published Stromgren uvby data which, together with previous published photometry, indicated variability on a long time scale. The author's unpublished data tend to reinforce the idea of a long-term trend. These photometric variations appear to be complex. Walker and Hill (1985) also have found the radial velocity variations of HD 168476 to be of a complex nature.

II.17. LSS 5121

Star number 5121 of the Stephenson and Sanduleak (1971) catalogue was discovered to be an extreme helium star by Drilling (1985). Photometry done by Landolt at the CTIO telescopes in September, 1985 resulted in the magnitude and colors in Table I. One can estimate from these data that LSS 5121 has a color excess E_{B-V} of 0.2 - 0.3 magnitude.

II.18. LSIV-14°109

The star LSIV-14°109 was discovered to be an extreme helium star by Drilling (1979). He found its effective temperature to be a bit less than 13000°K. Drilling's (1975) broadband magnitude and color indices for LSIV-14°109 are V = 11.19, (B-V) = +0.31, and (U-B) = -0.31. These compare to Landolt's measures in Table I herein of V = 11.152, (B-V) = +0.33, and (U-B) = -0.277. There really are not sufficient data to indicate variability, although similar differences for other objects among the extreme helium star group are deemed indicative of variability.

II.19. υ Sgr = 46 Sgr = HR 7342 = BD-16°5283 = HD 181615

The star υ Sgr is one of the classical members of the hydrogen-deficient binary star group; the other is KS Per. The UBV photometry for this star in Table I was taken from Johnson et al. (1966). Other Johnson photometric system colors are (V-R) = +0.27 and (R-I) = +0.13.

υ Sgr was discovered to be an eclipsing binary by Gaposchkin (1944). Eggen, Kron, and Greenstein (1950) obtained the first photoelectric light curve. The binary has a period of 137.939 days.

Table I

Broad-band Photometry for the Extreme Helium Stars

Star	V	B-V	U-B	V-R	R-I	E_{B-V}
BD+37°442	9.991	-0.294	-1.149			
HD 30353	7.85	+0.49	-0.16			
LSS 99	12.289	+0.700	-0.295	+0.445	+0.474	
BD+37°1977	10.21					
BD+10°2179	9.948	-0.191	-0.859			0.00
LSS 1922	10.495	+0.721	-0.169	+0.562	+0.596	0.70
LSS 3184	11.9					
HD 124448	9.980	-0.097	-0.775			0.08
LSS 3378	11.483	+0.440	-0.318	+0.329	+0.306	0.35
BD-9°4395	10.535	+0.055	-0.833			0.30
LSS 4300	9.779	+0.839	-0.133	+0.614	+0.654	
HD 160641	9.825	+0.144	-0.802			0.40
LSS 4357	12.620	+0.412	-0.521	+0.288	+0.284	
LSIV-1°2	11.009	+0.375	-0.485	+0.264	+0.284	0.45
BD-1°3438	10.328	+0.460	-0.246			0.40
HD 168476	9.268	-0.012	-0.666			0.13
LSS 5121	13.253	+0.316	-0.699	+0.212	+0.227	
LSIV-14°109	11.152	+0.331	-0.277	+0.298	+0.250	0.20
υ Sgr	4.61	+0.10	-0.53			
LSII+33°5	10.307	+0.160	-0.754	+0.086	+0.092	0.25
LSIV+2°13	9.557	+0.188	-0.536	+0.134	+0.107	0.15

The amplitude of the light variation is about 0.1 magnitude. Irregularities occur in the light variation. Photometry of the system also indicates the presence of a large infrared excess.

A modern accurate multicolor light curve would be desirable. The task will be difficult since the period is long, and the system is bright.

II.20. LSII+33°5

The star LSII+33°5 was found to be a hydrogen-deficient star by Drilling (1978). He found this object to differ from other extreme helium stars in that numerous, strong O II lines were observed in its spectrum. Drilling (1975) obtained UBV photometry of LSII+33°5, and found V = 10.43, (B-V) = +0.13, and (U-B) = -0.75. As one can see in Table I, the V magnitude obtained by Landolt is quite discrepant, although two of his four measures were obtained at an extreme air mass. Even neglecting the high air mass observations, however, the magnitude difference still remains about 0.1 magnitude. There is a reasonable chance, then, that LSII+33°5 is variable in light. Additional evidence for light variability may have been recorded in small differences noted between two sets of IUE observations (Drilling et al. 1984).

II.21. LSIV+2°13 = BD+1°4381

BD+1°4381 was discovered to be a hydrogen-deficient star by Drilling (1979). He concluded that it had an effective temperature similar to that of υ Sgr ($\sim$ 13,000°K, Heber and Schönberner 1981), but a smaller hydrogen abundance. Photometry by Drilling (1975) found the broadband magnitude and colors of BD+1°4381 to be V = 9.56, (B-V) = +0.18, and (U-B) = -0.56. These values agree with similar photometry by Landolt in Table I in this paper. An effective temperature of 9500 °K (Drilling et al. 1984) makes BD+1°4381 one of the coolest known extreme helium stars. Jeffery and Malaney (1985) have published Stromgren uvby photometry for BD+1°4381 which appears to show light variations with an amplitude of $0\overset{m}{.}04$ on a time-scale of perhaps 20 days. Their mean V magnitude, 9.525 ± 0.018 resulting from 47 observations, is $0\overset{m}{.}032$ brighter than Landolt's mean V = 9.557 ± 0.025 in Table I herein, a value resulting from six measures over two years. If one was not expecting the star to be variable in light, one might discount the observed differences, especially since the zero point of the uvby y magnitude historically has not been well established. On the other hand, the mean errors of a single observation are about twice what one would expect for a star of this brightness when measured at 0.4 - 0.9-m telescopes. As Jeffery and Malaney suggest, additional data covering a time span much longer than the suggested 20 day period are needed. Further, given the tiny apparent amplitude of light variation, a two star photometer ought to be used for the data acquisition.

III. SUMMARY

Three stars sometimes included in the extreme helium star group and certainly related, have not been discussed in this paper: BD+13°3224, MV Sgr, and V 348 Sgr. The latter two stars are hot R CrB stars. A review of V 348 Sgr is given in a recent paper by Heck, et al. (1985). It is noted therein that V 348 Sgr ranges from V = 12th to 18th magnitude, values confirmed by the author's unpublished photometry. MV Sgr has been undergoing small scale light variations during recent years. The author's UBVRI photometry indicates that V = 13.09, (B-V) = +0.26, (U-B) = -0.57, (V-R) = +0.23, (R-I) = +0.41, and (V-I) = +0.67. The total range in the V magnitude has been 0.26 magnitude. It should be noted that the fainter than normal observation of MV Sgr reported by the author (Landolt 1979b) almost certainly was a misidentification at the telescope.

ACKNOWLEDGEMENTS

I wish to thank Professor G. H. Herbig for introducing me to this subject in 1967. I also acknowledge many useful conversations with Professor J. S. Drilling and Dr. Alan Uomoto. The author's initial studies of helium stars were carried out under grants from the National Science Foundation. This review was written with the support of the Air Force Office of Scientific Research (grant no. 82-0192).

REFERENCES

Bartolini, C., Bonifazi, A., D'Antona, F., Fusi Pecci, F., Oculi, L., Piccioni, A., and Serra, R. 1982, Ap. and Space Sc. 83, 287.
Berger, J., Fingant, A. M., Rebeirot, E. 1974, Compt. Rend. Serie 278, 227.
Bidelman, W. P. 1952, Ap. J. 116, 227.
Darius, J., Gidding, J. R., and Wilson, R. 1979, The First Year of IUE, p. 36.
Dinger, A. S. 1970, Ap. and Space Sc. 6, 118.
Drilling, J. S. 1973, Ap. J. Letters 179, L31.
Drilling, J. S. 1975, A. J. 80, 128.
Drilling, J. S. 1978, Ap. J. Letters 223, L29.
Drilling, J. S. 1979, Ap. J. 228, 491.
Drilling, J. S. 1980, Ap. J. Letters 242, L43.
Drilling, J. S. 1985, private communication.
Drilling, J. S., Landolt, A. U., and Schönberner, D. 1984, Ap. J. 279, 748.
Drilling, J. S., Schönberner, D., Heber, U., and Lynas-Gray, A. E. 1984, Ap. J. 278, 224.
Eggen, O. J., Kron, G. E., and Greenstein, J. S. 1950, Pub. A. S. P 62, 171.

Elias, J. H., Frogel, J. A., Matthews, K., and Neugebauer, G. 1982, A. J. 87, 1029.
Gaposchkin, S. 1944, A. J. 51, 109.
Grauer, A. D., and Bond, H. E. 1981, Pub. A. S. P. 93, 388.
Grauer, A. D., Drilling, J. S., and Schönberner, D. 1984, Astron. and Astrophys. 133, 285.
Hack, M. 1967, Modern Astrophysics: A Memorial to Otto Struve, ed. M. Hack (Paris: Gauthier-Villars), p. 163.
Heck, A., Houziaux, L., Manfroid, J., Jones, D. H. P. and Andrews, P. J. 1985, Astron. Astr. Suppl. 61,
Heber, U., and Schönberner, D. 1981, Astron. Astrophys. 102, 73.
Herbig, G. H. 1964, Ap. J. 140, 1317.
Herbig, G. H. 1967, Trans. I.A.U. XIIIA, 546.
Hill, P. W. 1964, M.N.R.A.S. 127, 113.
Hill, P. W. 1965, M.N.R.A.S. 129, 137.
Hill, P. W. 1967a, Trans. I.A.U. XIIIB, 152.
Hill, P. W. 1967b, Obs. 87, 210.
Hill, P. W. 1969a, M.N.A.S.S.A. 28, 56.
Hill, P. W. 1969b, I.A.U. Infor. Bull. Var. Stars, No. 357.
Hill, P. W. 1985, private communication.
Hill, P. W., Jeffery, C. S., and Morrison, K. 1985, I.A.U. Circular 4097.
Hill, P. W., Lynas-Gray, A. E., and Kilkenny, D. 1984, M.N.R.A.S. 207, 823.
Hiltner, W. A. 1956, Ap. J. Suppl. 2, 389.
Hoffleit, D. 1959, A. J. 64, 241.
Hunger, K. 1975, Problems in Stellar Atmospheres and Envelopes, ed. B. Baschek. W. H. Kegal, and G. Traving (New York: Springer-Verlag), p. 57.
Jeffery, C. S., and Malaney, R. A. 1985, M.N.R.A.S. 213, 61P.
Johnson, H. L. 1963, in Basic Astronomical Data, edited by K. Aa. Strand (Chicago: Univ. of Chicago Press), p. 204.
Johnson, H. L., Mitchell, R. I., Iriarte, B., and Wisniewski, W. Z. 1966, Comm. Lunar Planetary Labs., 4, 99.
Kaufmann, J. P., and Schönberner, D. 1975, Mitt. Astron. Ges. 38, 198.
Kaufmann, J. P., and Schönberner, D. 1977, Astron. and Astrophys. 57, 169.
Kilkenny, D. 1985, private communication.
Klemola, A. R. 1961, Ap. J. 134, 1961.
Kopff, A., Nowacki, H., and Gondolatsch, F. 1932, A. N. 244, 385.
Landolt, A. U. 1968, Pub. A. S. P. 80, 318.
Landolt, A. U. 1973a, Publ. A. S. P. 85, 661.
Landolt, A. U. 1973b, A. J. 78, 959.
Landolt, A. U. 1974, Bull. A. A. S. 6, 324.
Landolt, A. U. 1975, Ap. J. 196, 789.
Landolt, A. U. 1979a, Bull. A. A. S. 11, 208.
Landolt, A. U. 1979b, I.A.U. Circular No. 3419.
Landolt, A. U. 1983, A. J. 88, 439.
Landolt, A. U., and Grauer, A. D. 1985, Pub. A. S. P., 97, submitted.

Lynas-Gray, A. E., Kilkenny, D., Skillen, I., and Jeffery, C. S. 1985, preprint.
MacConnell, D. J., Frye, R. L., and Bidelman, W. P. 1972, Pub. A. S. P, 84, 388.
Nariai, K. 1963, Pub. A. S. Japan 15, 7.
Osawa, K., Nishimura, S., and Nariai, K. 1963, Pub. A. S. Japan 15, 313.
Popper, D. M. 1942, Pub. A. S. P. 54, 160.
Popper, D. M. 1946, Pub. A. S. P. 58, 370.
Popper, D. M. 1947, Pub. A. S. P. 59, 320.
Rebeirot, E. 1966, Pub. Obs. Hte.-Provence 8, No. 19.
Rossi, L., Viotti, R., Darius, J., and D'Antone, F. 1980, Proceedings of Second European IUE Conference (ESA SP-157), p. 323.
Schönberner, D., and Wolf, R. E. A. 1974, Astr. Ap. 37, 87.
Schönberner, D., and Drilling, J. S. 1984, Ap. J. 276, 229.
Schönberner, D., Drilling, J. S., Lynas-Gray, A. E., and Heber, U. 1982, in Advances in Ultraviolet Astronomy: Four Years of IUE Research (NASA CP-2238), p. 593.
Stephenson, C. B., and Sanduleak, N. 1971, Pub. Warner and Swasey Obs. 1, 1.
Thackeray, A. D., and Wesselink, A. J. 1952, Obs. 72, 248.
Walker, H. J., and Kilkenny, D. 1980, M.N.R.A.S. 190, 299.
Walker, H. J., and Hill, P. W. 1985, Astron. and Astrophys. Suppl., 61, 303.
Wolff, S. C., Pilachowski, C. A., and Wolstencroft, R. D. 1974, Ap. J. Letters, 194, L83.

HIGH RESOLUTION SPECTROSCOPY OF SIX NEW EXTREME HELIUM STARS

U. Heber, G. Jonas and J.S. Drilling*
Institut für Theoretische Physik und Sternwarte
der Universität Kiel
Olshausenstr. 40, 2300 Kiel, F.R.G.

*Department of Physics and Astronomy
Louisiana State University
Baton Rouge, LA 70803-4001, U.S.A.

ABSTRACT. High resolution spectra of six newly discovered extreme helium stars are presented. LSS 5121 is shown to be a spectroscopical twin of the hot extreme helium star HD 160641. A preliminary LTE analysis of LSS 3184 yielded an effective temperature of 22000 K and a surface gravity of $\log g = 3.2$. Four stars form a new subgroup, classified by sharp-lined He I spectra and pronounced O II spectra, and it is conjectured that these lie close to the Eddington limit. The whole group of extreme helium stars apparently is inhomogenous with respect to luminosity to mass ratio and chemical composition.

1. INTRODUCTION

The group of extreme helium stars at present comprises 17 stars. Only four of them were analyzed, model atmosphere techniques being used. The results indicate that the group is inhomogenous with respect to chemical composition (see Heber, these proceedings). It was therefore deemed necessary to analyze all members of the group and an observational campaign was started to secure UV and visual spectra. In the meantime, 12 stars have been observed with IUE (low resolution mode) in the UV and high resolution visual spectra of 16 stars were obtained with the CTIO 4 m and ESO 3.6 m telescopes (see also Drilling and Heber, these proceedings). The UV energy distributions had already been analyzed for effective temperatures (Drilling et al., 1984) which revealed that hot extreme helium stars are rare: only two stars (HD 160641 and BD-9°4395) are hotter than 18000 K. For two other stars (LSE 78 and LS II+33°005), derived effective temperatures are considerably lower than indicated by their intrinsic UBV colours. Four newly discovered stars (LSS 99, LSS 3184, LSS 4357 and LSS 5121) are also intrinsically blue (see Table I). Since UV measurements are not available, we estimated their effective temperature by comparing the dereddened colour index Q to model predictions. Intrinsic colours were calculated from model fluxes as described by Heber and Schönberner (1980). The calibration of Matthews and Sandage (1963) was used. The interstellar extinction was derived by comparing the measured B-V to a model prediction. The results (given in Table I) indicate that these stars are hot ($T_{eff} > 18000$ K) and that some are highly reddened. Owing to the rarity of hot extreme helium stars and the above-mentioned discrepancy for LSE 78 and LS II+33°005, a high resolution study of these six stars appears to be of high priority. Presented here are high resolution spectra taken with the ESO Cassegrain Echelle spectrograph.

TABLE I: Effective temperatures and reddening estimated by comparison of observed and synthetic colours

star	B	B-V	Q	T_{eff} (Q)	E(B-V)
LSS 99	12.99	0.70	− 0.80	18700	0.89
LSS 3184	12.63	0.03	− 0.86	21700	0.26
LSS 4357	13.02	0.41	− 0.80	18700	0.60
LSS 5121	13.57	0.32	− 0.92	28300	0.61
LSE 78	11.28	0.06	− 0.88	23300	0.25
LS II+33°005	10.68	0.14	− 0.84	20200	0.33

2. OBSERVATIONS

The program stars were observed with the Cassegrain Echelle spectrograph (CASPEC) attached to the ESO 3.6 m telescope at La Silla, Chile. An Echelle grating with 52 lines/mm was used and a spectral resolution of 0.25 Å achieved. The spectral range from 3900 Å to 4800 Å was recorded with a CCD detector. The spectrum of LSS 5121 was binned and, thus, the spectral resolution degraded to 0.5 Å. Data reduction proceeded in two steps: first, the raw data were wavelength calibrated and the signal and background of each order extracted with a numerical slit using the Midas Software at ESO, Garching. Since the Echelle orders are well-separated, even in the blue, the background correction was straightforward. The main problem with the data reduction process was the correction for the Echelle blaze function ("ripple correction"). The best results were obtained when standard stars were used to define empirical "ripple functions". Subluminous O stars are well-suited for this purpose since very few absorption lines are present in their spectra (apart from Balmer lines). Empirical "ripple functions" were derived by least square fits of the (almost) line-free Echelle orders. Those orders containing Balmer lines were excluded and, instead, their "ripple functions" were derived from fits of the continuum of the extreme helium star BD-9°4395 which does not show strong Balmer lines. Vice versa, an extreme helium star can be efficiently used as a standard star to reduce Echelle orders containing strong Balmer lines.

3. INDIVIDUAL OBJECTS

3.1. LSS 5121

The high resolution spectrum is almost identical to that of the hot extreme helium star HD 160641 (T_{eff} = 31900 K; Drilling et al., 1984). The spectral range from 4600 Å to 4700 Å for both stars is plotted in Figure 1 demonstrating the strengths of C III and He II lines. The high effective temperature, indicated by the colours, is thus confirmed by the CASPEC spectrum.

3.2. LSS 3184

The spectrum of LSS 3184 is characterized by broad He I absorption lines, similar to those of BD+10°2179. However, their spectra differ considerably as far as the weak lines are concerned. As can be seen from Figure 2, He II, λ 4686 Å, the C III and O II lines are much stronger in LSS 3184 than in BD+10°2179, this pointing to a higher effective temperature of the former. An analysis was started using line-blanketed LTE model atmospheres. In the first step, the gravity, as a function of effective temperature, was determined by matching the line wings of He I λ 4471 Å. The effective temperature was derived from the ionization equilibria of N II/N III and Si II/Si III/Si IV. T_{eff} = 22000 K and log g = 3.2 result. Nitrogen is over-abundant by a factor of about 50. These results are still in the preliminary stage and additional calculations will have to be made to accurately determine the atmospheric parameters. Nevertheless, it is safe to state that LSS 3184 is indubitably as hot as is indicated by its intrinsic colours.

3.3. LSS 99, LSS 4357, LSE 78 and LS II+33°005

The high resolution spectra of these four objects are unique amongst the extreme helium stars: no He II lines are present; the He I lines appear to be as sharp as in the cool extreme helium stars, e.g. LSS 3378 (T_{eff} = 9400 K; Drilling et al., 1984). However, the metal line spectra, as well as the intrinsically blue colours, are not consistent with such a low temperature. Besides the ions (e.g. C II, N II, Al III, Si II/III/IV) usually observed in the hotter extreme helium stars, e.g. HD 124448, the O II line spectrum is well-developed and much stronger than in other extreme helium stars. These absorption line spectra cannot be reproduced from our LTE-model grid (Drilling et al., 1984). We, therefore, conjecture that these stars have very low gravities or, more precisely, very large luminosity to mass ratios not covered by our model grid. If this were true, their L/M ratio would be larger than $10^{4.6}$ $L_\odot/M_\odot$ (upper limit of the model grid). It is then doubtful whether the assumption of plan-parallel geometry (as in our models) would still apply.

The region in the (log T_{eff}, log g)-plane, where the LSS 99, LSS 4357, LS II+33°005 and LSE 78 stars can probably be found, is shown in Figure 3 (hatched area). The dashed lines indicate the approximate

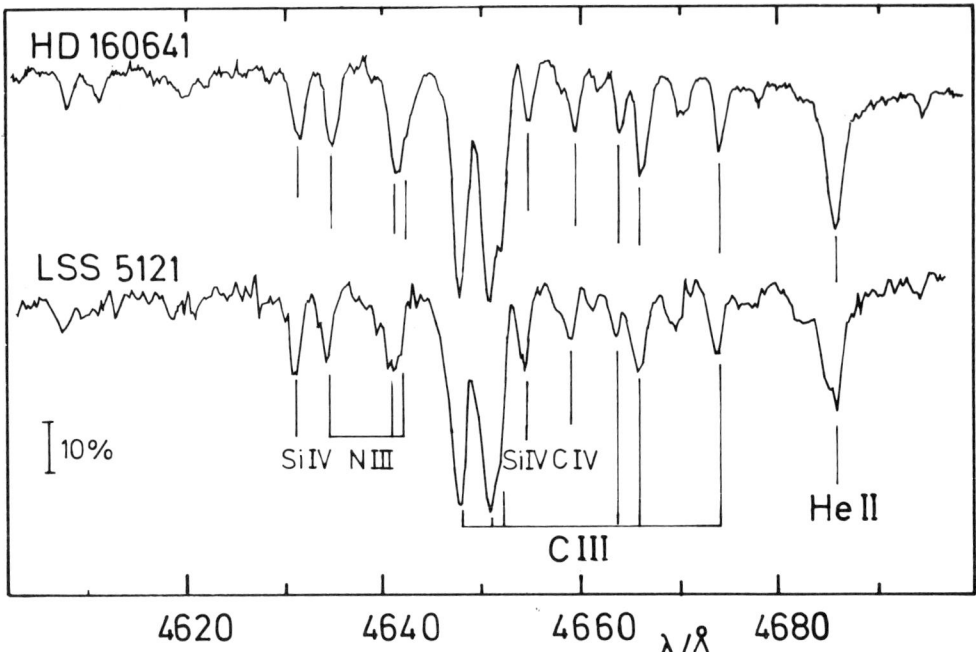

Fig. 1 CASPEC spectra of HD 160641 (top) and LSS 5121 (bottom).

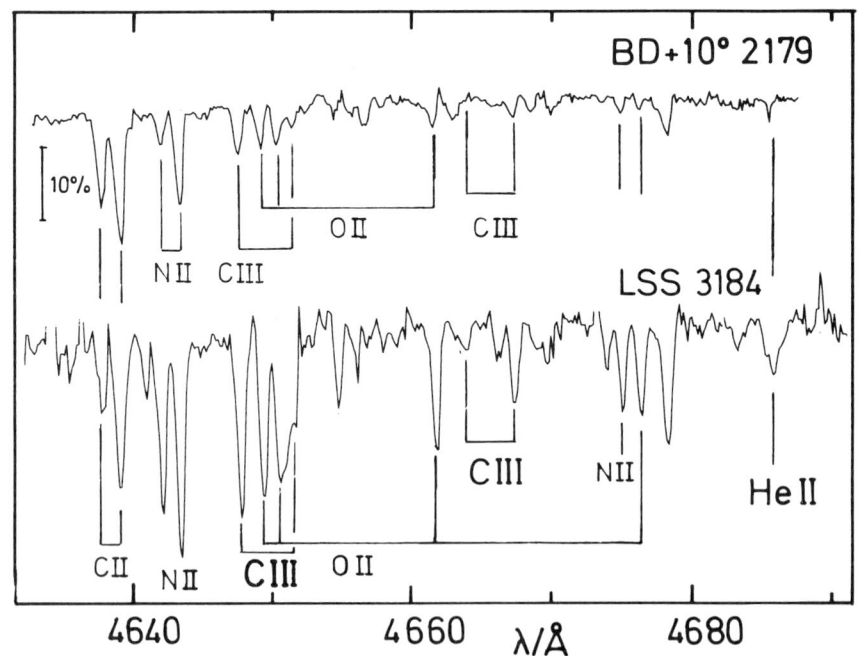

Fig. 2 CASPEC spectra of BD+10°2179 (top) and LSS 3184 (bottom)

limits of our model grid ($10^{3.6} L_\odot/M_\odot \leq L/M \leq 10^{4.6} L_\odot/M_\odot$). The hatched area is bound by the Eddington limit for pure electron scattering (see Mihalas, 1978, p.554) and the upper model L/M, the temperature being restricted by colour- and UV-temperature estimates.

4. DISCUSSION

In this presentation, high resolution spectra of six intrinsically blue extreme helium stars have been discussed. Two stars, LSS 5121 and LSS 3184, have been shown to have effective temperatures in excess of 20000 K. Only two other extreme helium stars (HD 160641 and BD-9°4395) are known to have such high T_{eff}. Four stars (LSS 99, LSS 4357, LSE 78 and LS II+33°005) apparently form a subgroup classified by sharp-lined He I spectra and pronounced O II spectra and it is conjectured that these star lie close to the Eddington limit (see Fig. 3).

In Figure 3, the positions of LSS 3184 and four previously analyzed extreme helium stars are shown. It can be seen that LSS 3184 has L/M smaller than the previously analyzed extreme helium stars while the stars close to the Eddington limit have a larger L/M ratio.

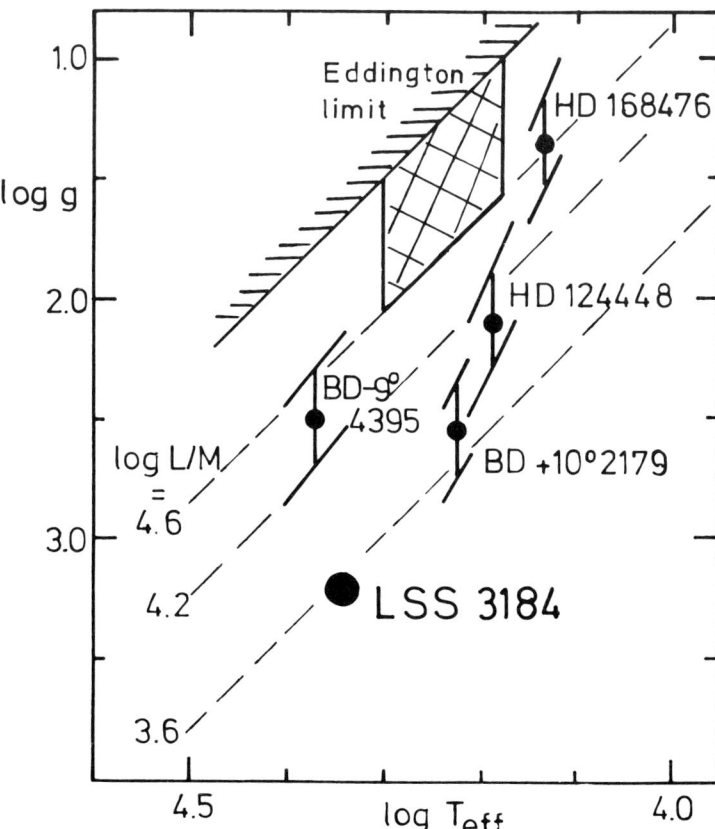

Fig. 3

Position of extreme helium stars in the ($\log T_{eff}$-$\log g$)-plane. The hatched line is the Eddington limit. The hatched area indicates the most likely position of LSS 99, LSS 4357, LSE 78 and LSS II+33°005 (see text). Lines of constant L/M are dashed.

We conclude, therefore, that there is a larger spread of the luminosity to mass ratio amongst the extreme helium stars than previously assumed.

ACKNOWLEDGEMENTS. We thank D. Ponz (ESO, Garching) for his kind assistance in the data reduction. This research was supported in part by NSF Grants AST 8018766, AST 8514574 and INT 8219240 and by NASA Grant NAG 5-71.

REFERENCES

Drilling, J.S., Schönberner, D., Heber, U., Lynas-Gray, A.E.: 1984, Astrophys. J. **278**, 224
Heber, U., Schönberner, D.: 1981, Astron. Astrophys. **102**, 73
Matthews, T.A., Sandage, A.R.: 1963, Astrophys. J. **138**, 80
Mihalas, D.: 1978, Stellar Atmospheres, 2nd ed., Freeman, San Francisco

EMISSION LINES IN HIGH RESOLUTION SPECTRA OF EXTREME HELIUM STARS

U. Heber
Institut für Theoretische Physik und Sternwarte
der Universität Kiel
Olshausenstr. 40, 2300 Kiel, F.R.G.

ABSTRACT. The occurrence of emission lines in high resolution blue spectra of four extreme helium stars is reported. This phenomenon is most apparent in the case of BD-9°4395 which displays emissions in He I, λ 3889 Å, C II, λ 4267 Å and in the C II multiplet No. 1 (λ 4740 Å). The latter is also present in LSE 78 and found to be variable. The C II emission lines are regarded as evidence for extended moving envelopes around BD-9°4395 and LSE 78. A possible relation to the hot R CrB star V348 Sgr is discussed.

HD 160641 and LSS 5121 show unidentified emission lines ($\lambda\lambda$4485.6 Å, 4504.0 Å) which are also observed in some Of stars. No emission lines were found in the blue spectra of 12 extreme helium stars.

1. INTRODUCTION

Pecularities in the spectrum of an extreme helium star were first reported for BD-9°4395 (Kaufmann and Schönberner, 1977). The lines He I, λ 3889 Å and C II, λ 4267 Å appeared to be filled in by emissions. In the case of He I, λ 3889 Å, the emission occurred at the central wavelength and seemed to be non-variable, while C II, λ 4267 Å was suspected to be variable. The latter has been confirmed by recent observations (Jeffery et al., 1985) which revealed that the line occasionally occurs as a P Cygni profile. Jeffery et al. (1985) suggested that the general variability of C II, λ 4267 Å indicates non-uniform mass loss which may derive from the underlying stellar pulsation (see also Jeffery, these proceedings). Since pulsational instabilities are observed in many extreme helium stars, it would be worthwhile to search for emission line features in other extreme helium stars.

Recently, high resolution spectra of all known (except one) extreme helium stars were obtained at La Silla using the ESO Gassegrain Echelle spectrograph (CASPEC) attached to a 3.6 m telescope (see also Heber, Jonas and Drilling; Drilling and Heber, both in these proceedings). These spectra covered the wavelength range from ~3850 Å

to ~4800 Å in most cases and careful examination revealed that five
stars out of 17 showed emission lines. LSS 1922, which showed variable
H β, emission is a hydrogen-deficient binary and is discussed elsewhere
(Morrison et al., these proceedings). The following extreme helium
stars were found to have emission lines in their CASPEC spectra:
BD-9°4395, LSE 78, HD 160641 and LSS 5121. The emission line spectra
are described in detail in the next sections.

2. HELIUM AND CARBON EMISSION LINES IN BD-9°4395 AND LSE 78

2.1. BD-9°4395

One CASPEC spectrum of BD-9°4395 was obtained on April 9th, 1985. As
already known from previous investigations (Kaufmann and Schönberner,
1977; Jeffery et al., 1985) emission line features can be observed in
He I, λ 3889 Å and C II, λ 4267 Å. The former line has an unshifted
central emission, while the emission component of C II, λ 4267 Å occurs
in the blue wing (shifted by -58 km/s) of the (unshifted) absorption
line (see Fig. 1). In previously published studies of spectra, the
emission occurred either in the core or in the red wing of the
underlying absorption line.

The high quality of the CASPEC spectrum allowed another peculiarity in BD-9°4395 to be detected: four spectral lines ($\lambda\lambda$ 4735.46 Å,
4737.97 Å, 4744.77 Å and 4747.28 Å), which arise from the two electron
transition $2p^2\ ^2P - 3p\ ^2P°$ C II (Multiplet No. 1) occur in emission, as
demonstrated in Figure 2 where the spectrum of BD-9°4395 is compared to
that of BD+10°2179. The latter shows C II, multiplet No. 1, in
absorption. Note that 4735.46 Å is severely blended by an Ar II
absorption line in both stars. The C II emission lines in BD-9°4395 are
marked in Figure 2. They are shifted by -17 km/s with respect to the
absorption line spectrum (see Table I).

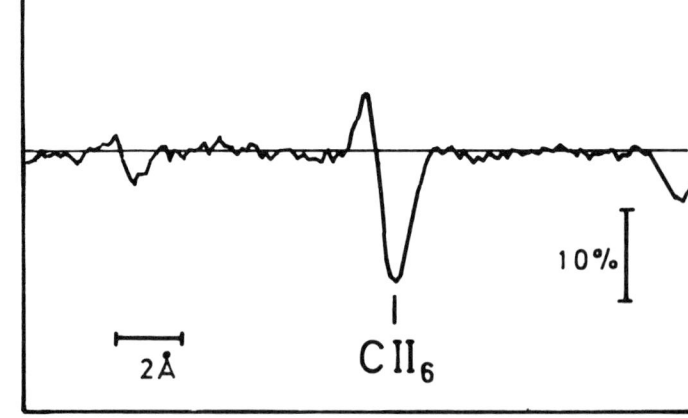

Fig. 1:

C II, λ 4267 Å
in the CASPEC
spectrum of
BD-9°4395. 10%
continuum height
is indicated by
a vertical bar.

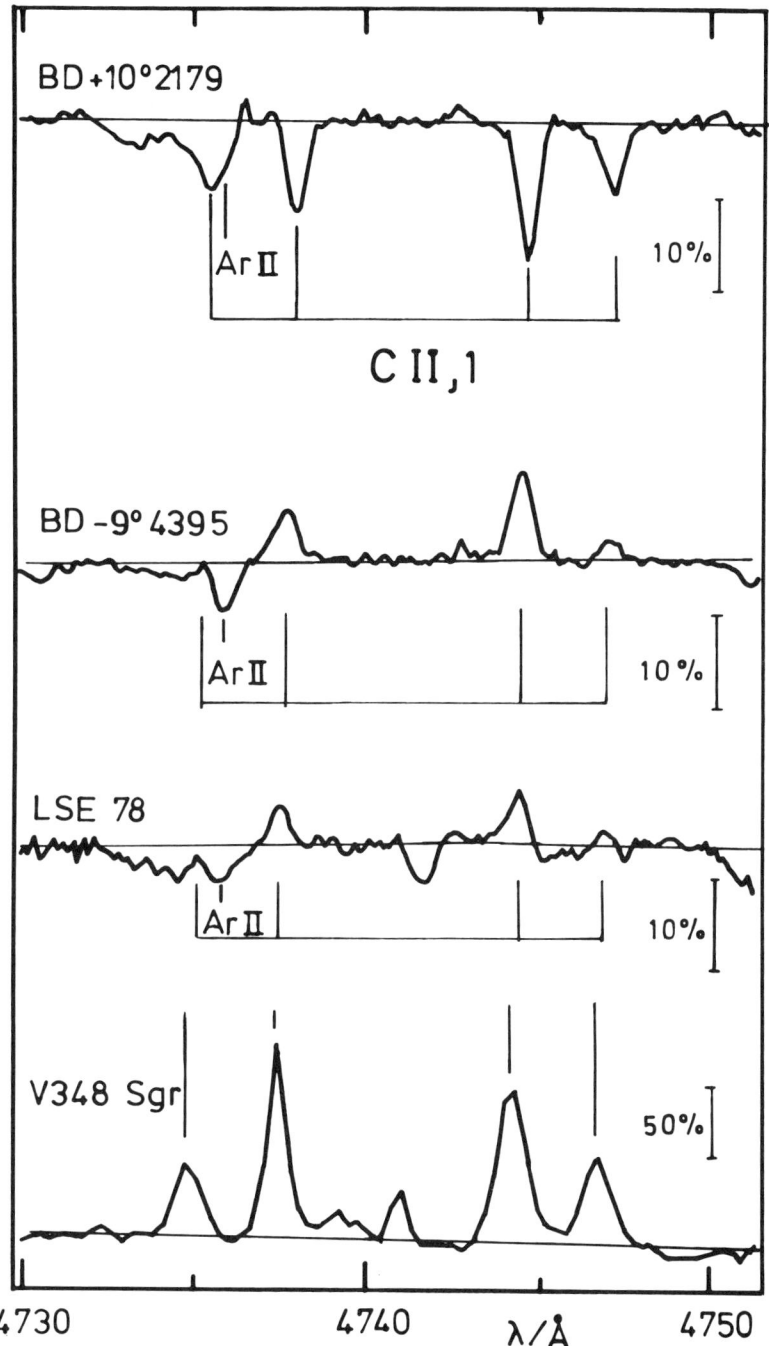

Fig. 2 C II, multiplet No. 1 in the CASPEC spectra of three extreme helium stars (BD+10°2179, BD-9°4395, LSE 78) and the hot R CrB-variable V348 Sgr (bottom).

2.2. LSE 78

Three CASPEC spectra of LSE 78 were obtained which revealed pecularities in He I, λ 3889 Å, C II, λ 4267 Å and C II, multiplet No. 1 similar to BD-9°4395: He I, λ 3889 Å is filled in by a central emission. C II, 4267 Å is unusually weak but, unlike BD-9°4395, did not show an emission component. C II, multiplet No. 1, is occasionally found in emission (see Fig. 2) and appears to be variable in strength: It is present in a spectrum obtained on April, 8, 1985, but was absent in a spectrum taken three days earlier. Radial velocities are given in Table I.

2.3. Discussion and comparison to V348 Sgr

The variable C II emission lines in the spectra of BD-9°4395 and LSE 78 can be regarded as evidence for an extended moving envelope around these stars and are probably formed by recombination at low envelope densities.

Since the extreme helium stars appear to be closely related to the R CrB stars, it would be worthwhile comparing the envelope spectra of BD-9°4395 and LSE 78 to the spectrum of a hot R CrB star. V348 Sgr is known to be such a star ($T_{eff} \approx 20000$ K; Schönberner and Heber, these proceedings) with a carbon-rich envelope spectrum (Dahari and Osterbrock, 1984). Recently, Hunger (private communication) observed V348 Sgr with CASPEC when the star was faint ($V \approx 15$). The spectrum covers approximately the same wavelength range as for BD-9°4395 and reveals emission lines superimposed on a weak stellar continuum. The carbon emission line spectrum consists of exactly the same lines as in BD-9°4395, namely C II, λ 4267 Å and Multiplet No. 1. The latter is displayed in Figure 2. (Note that the plot scale for V348 Sgr differs from the others.) Radial velocities (see Table I) can be measured only for the emission lines since no photospheric absorption lines are visible. (The results are in good agreement with previous measurements of Dahari and Osterbrock, 1984, see Table I.) Houziaux (1968) observed V348 Sgr at maximum light (V=12) with high spectral resolution and derived a radial velocity of 174 km/s from the absorption line

Table I. Heliocentric radial velocities (km/s)

star	date	C II emission lines		absorption lines
		4267Å	4740Å	
BD-9°4395	8 Apr. 85	-111	-73	-56
LSE 78	4 Oct. 84	-	-107	-90
LSE 78	8 Apr. 85	-	-117	-92
V348 Sgr	Oct. 85	158	135	--
V348 Sgr			140[a]	174[b]

[a] Dahari and Osterbrock (1984)
[b] Houziaux (1968)

spectrum. As in the case of BD-9°4395 and LSE 78, the C II emission lines are blueshifted with respect to the absorption lines. Hence, the C II emission line spectra of BD-9°4395 of LSE 78 apparently are formed in an envelope under physical conditions similar to that of the envelope of V348 Sgr.

This strikingly similar envelope spectrum can be regarded as additional evidence for a link between R CrB stars and extreme helium stars.

3. UNIDENTIFIED EMISSION LINES IN HD 160641 AND LSS 5121

Emission features at $\lambda = 4485.6$ Å and $\lambda = 4504.0$ Å were found in two spectra of HD 160641 (taken on April 3rd, 1984 and April 9th, 1985, respectively) and were found to have the same strengths in both spectra. LSS 5121 showed an emission only at $\lambda 4485.6$ Å. A line at $\lambda 4504.0$ Å occurred in absorption. These lines could not be identified. Wolf (1963) described spectra of Of stars and reported unidentified emission lines at $\lambda = 4485$ Å and 4503 Å, along with numerous emission lines of hydrogen, helium, nitrogen and silicon. Unlike in the Of stars, no other emission lines are present in HD 160641 and LSS 5121.

REFERENCES

Dahari, O., Osterbrock, D.E.: 1977, Astrophys. J. **277**, 648
Houziaux, L.: 1968, Bull. Astron. Inst. Czechoslwakia **19**, 265
Jeffery, C.S., Skillen, I., Hill, P.W., Kilkenny, D., Malaney, R.A.,
    Morrison, K.: 1985, Monthly Notices Roy. Astron. Soc. **217**, 701
Kaufmann, J.P., Schönberner, D.: 1977, Astron. Astrophys. **57**, 169
Wolf, R.J.: 1963, Publ. Astron. Soc. Pacific **75**, 485

DISCUSSION

N.K. RAO: Regarding the inverse P-Cygni profiles that is observed in the spectrum of BD −9°4395 is that part of the expanding atmosphere or emission superposed on absorption line?

HEBER: The observed C II profile in BD −9°4395 appears to consist of an emission superimposed on photospheric absorption line. The emission probably arises from recombination at low envelope densities.

THE PECULIAR SPECTRUM OF THE EXTREME HELIUM STAR BD-9°4395

C.S.Jeffery
University Observatory
Buchanan Gardens
St Andrews
Fife KY16 9LZ
Scotland

ABSTRACT. The spectrum of the extreme helium star BD-9°4395 has already been noted for emission lines of HeI 3889Å and CII 4267Å. Further anomalies in the HeI and CII spectra have been observed, including the identification of additional emission lines, asymmetry in the He I line profiles and variability in all features. It is suggested that these features may be associated with other evidence for non-radial pulsations in BD-9°4395.

1. INTRODUCTION

The extreme helium star BD-9°4395 has an effective temperature of 23 000 K (Drilling et al. 1984) and a surface gravity log g=2.6 (Kaufmann & Schönberner 1977). With helium and carbon abundances $n_{He}=0.994$, $n_C=0.006$, it resembles other hot carbon-rich extreme helium stars (e.g. HD124448, BD+10°2179: Heber 1983) and appears to be the remnant of a post-asymptotic giant-branch star evolving to become a white dwarf (Schönberner 1977). Some spectroscopic anomalies, notably emission in CII 4267Å and HeI 3889Å, were reported in early high resolution spectra.

Jeffery et al. (1985) have interpreted multiperiodic photometric and spectroscopic variability in BD-9°4395 as evidence for non-radial pulsations, identifying periods of 3.5 and 11.2 day. They confirmed the presence of variable emisiion lines, interpreting the behaviour of CII 4267Å as evidence for variable mass-loss. Indeed this line was observed on one occasion with a P Cygni profile. A comparison with theoretical subordinate-line profiles in stellar winds (Olson 1981) provided an upper limit of 10^{-7} $M_\odot$ yr^{-1} to the instantaneous mass-loss rate. This is appreciably larger than that obtained from the UV resonance-line profiles by Hamann et al (1981) (10^{-8} $M_\odot$ yr^{-1}). In order to investigate the time-scale of the spectroscopic variability, as well as to examine the behaviour of other HeI and CII lines, new high-resolution spectra have been obtained with the Anglo-Australian Telescope. These spectra show several interesting anomalies and suggest a more detailed interpretation is required.

2. OBSERVATIONS

Spectra of BD-9°4395 were obtained in 1985 March at the AAT with the IPCS and the 82cm camera of the RGO spectrograph at reciprocal dispersions of 10, 5 and 2.5 Å mm^{-1}. Repeated observations in the vicinity of CII 4267Å, HeI 3889Å and HeI 4144Å were made, as well as around HeI 5876Å and other strong CII lines. The data were reduced in the manner described by Jeffery et al. (1985).

Comparisons of the line profiles of CII 4267Å and HeI 4121Å/4144Å on several nights are shown in Figs. 1 and 2, which include AAT spectrograms already reported (Jeffery et al. 1985). Whilst CII 4267Å showed a P Cygni profile in 1979 February and was apparently absent in 1979 July and 1980 May, its apparent development from 1985 March 11 to March 13 gives a better idea of the time-scale of the variation. On 1985 March 11 the blue wing of the CII 4267Å line is steeper than the red wing, with a hint of emission, resembling a reverse P Cygni profile. Indeed a pronounced reverse P Cygni profile was observed on an ESO echelle spectrogram (Heber, private communication). On March 12 the blueward emission is replaced by an additional blue-shifted absorption component, and on March 13 CII 4267Å appears to have a nearly normal profile. An unidentified absorption feature with a blue-shift of 350 km s^{-1} may be related to CII 4267Å since it has no counterpart in the 'quiet' spectrum.

The sequence of spectrograms of HeI 4121Å/4144Å (Fig. 2) presents a new dimension to the variations already documented in BD-9°4395. Although the data are of variable quality and resolution, and the three day sequence available for CII 4267Å is absent, we note significant changes in the HeI line profile. In 1979 February, HeI 4121Å and OII 4119Å were fully resolved, but in later years and at higher spectral resolutions these lines have been increasingly blended. The increasing width of HeI 4144Å is documented in table I, but an extended blue wing observed in 1985 March consistent with the blending of HeI 4121Å and OII 4119Å suggests that the line symmetry is also variable.

In the cases of other lines in the spectrum of BD-9°4395, the number of repeated observations is insufficient for detailed comparisons. However two members of CII multiplet 1 appear in emission, while another is in absorption: all may be variable. Since HeI 3888Å is seen in emission HeI 5876Å was also examined. Both lines are often found to be peculiar in Be stars (Underhill 1966). In the one spectrogram available (1985 March 13), HeI 5876Å shows an emission component, but it is not clear whether the line profile is a P Cygni type or a normal atmospheric absorption line with a superimposed emission core.

THE PECULIAR SPECTRUM OF THE EXTREME HELIUM STAR BD-9°4395

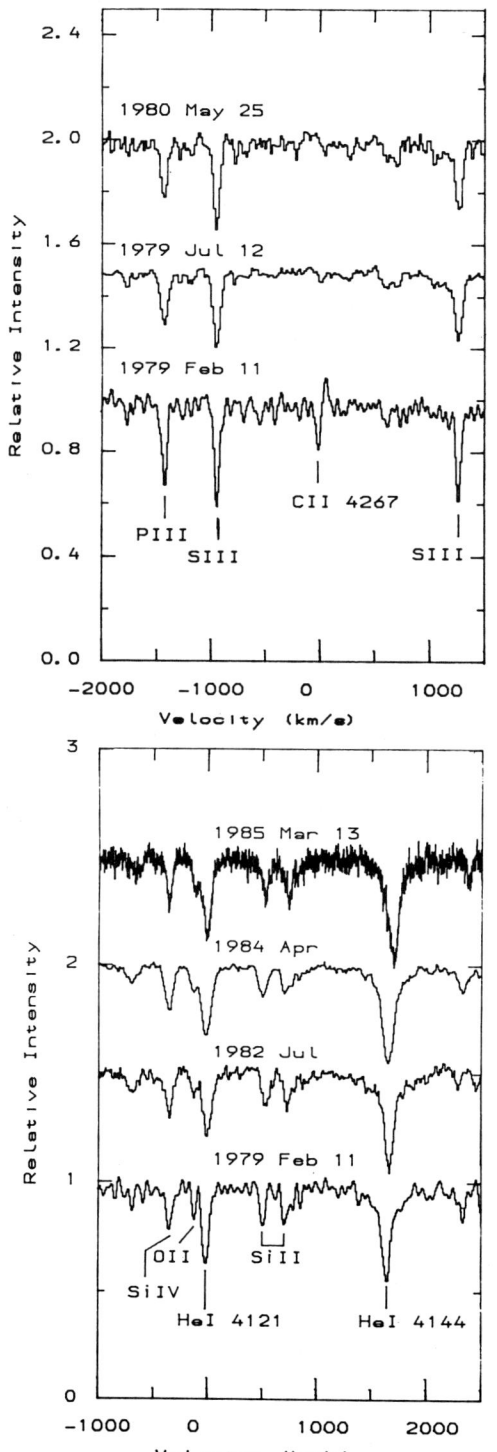

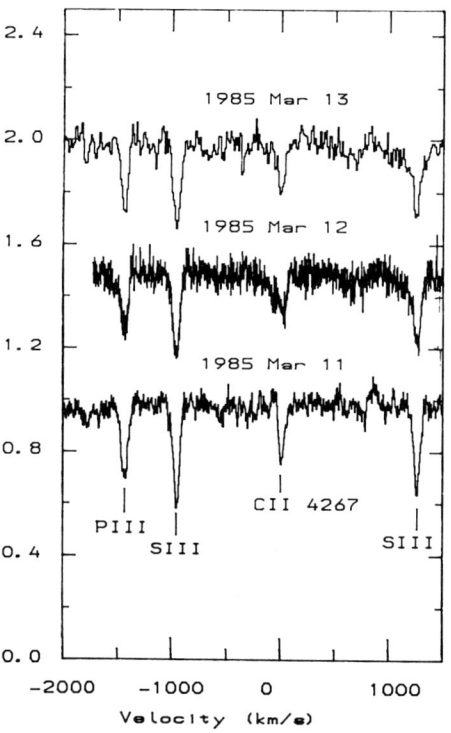

Figure 1. Variations in the line profile of CII 4267Å in BD-9°4395. Spectral dispersions (in Å mm^{-1}) were 8 (1979 Feb), 16 (1979 Jul & 1980 May), 5 (1985 Mar 11), 2.5 (1985 Mar 12) and 10 (1985 Mar 13).

Figure 2. Variations in the line profiles of HeI 4121Å and HeI 4144Å in BD-9°4395. Spectral dispersions (in Å mm^{-1}) were 8 (1979 Feb), 10 (1982 Jul & 1984 Apr) and 2.5 (1985 Mar).

Table I. Width of HeI 4143Å

Date	FWHM (Å)	W (Å)
1979 February	1.36	0.873
1982 July	1.39	0.853
1984 April	1.76	0.970
1985 March	1.61	0.963

3. DISCUSSION

It has been established that BD-9°4395 has an emission spectrum which is variable on a time-scale of days. The emission spectrum provides ample evidence for the presence of circumstellar material. The time-scale of the variations is comparable with that of the light variations which are believed to arise in non-radial pulsations (Jeffery et al. 1985).

It is necessary to determine whether the circumstellar material takes the form of an envelope as in the classical Be stars, or clouds of C-rich material as in the R CrB stars, or a stellar wind. The emission spectrum shows many properties normally found in Be stars, but the low rotational velocity (v sin i<30 km s^{-1}) and short period variations of BD-9°4395 appear to preclude this interpretation. A spherically symmetric uniform stellar wind is precluded by the observation of both reversed and normal P Cygni profiles at different times. Feast's (1979) description of the R CrB phenomenon being caused by the ejection of C-rich clouds into the line of sight gives some satisfaction since emission lines could be formed in clumps of circumstellar material with different temperatures and velocities. However the hot R CrB star MV Sgr shows emission lines of H, FeII and CaII (Herbig 1975) which are absent in BD-9°4395. Moreover BD-9°4395 shows no infra-red excess (Walker 1985) characteristic of the R CrB stars.

It is proposed that the emission spectrum in BD-9°4395 is related to g-mode non-radial pulsations. Amongst the extreme helium stars and like the emission spectrum, these have so far only been observed in BD-9°4395. It has been known for some time that moving shells are seen in the radial β Cephei pulsators (Smith 1983). Typically an optically thick shell appears with a radial velocity of some 30 to 50 km s^{-1} which decelerates, becoming transparent as it moves out to some 25% of the stellar radius. It eventually returns, but some 1% of the shell may be lost to the star (Burger et al. 1982). Penrod & Smith (1984) suggest that similar phenomena occur in nearly all B stars as a result of non-radial pulsations. Some evidence to support this theory exists in the cases of ρ Leo (Smith & Ebbetts 1981) and λ Eri (Penrod & Smith 1984).

This conjecture must be tested with a more homogeneous set of high resolution spectra whilst the behaviour of CII 4267A must be related to UV observations of the stellar wind.

ACKNOWLEDGMENTS

This paper is based on observations made at the Anglo-Australian Observatory. The author is undebted to Drs. U.Heber, P.W.Hill and A.E.Lynas-Gray for the use of some of their spectroscopic material. Financial support has been provided by the UK Science and Engineering Research Council.

REFERENCES

Burger,M., de Jager,C., van den Oord,G. & Sato,N., 1982. *Astr.Astrophys.*, 107,320.
Drilling,J.S., Schönberner,D., Heber,U. & Lynas-Gray,A.E., 1984. *Astrophys.J.*, 278,224
Feast,M.W, 1979. *IAU Colloq. No.46,p.246*, eds Bateson,F.M., Smak,J. & Urch,I.H., University of Waikoto, Hamilton, New Zealand.
Heber,U., 1982. *Astr.Astrophys.*, 118,39
Herbig,G.H., 1975. *Astrophys.J.*, 199,702.
Jeffery,C.S., Skillen,I., Hill,P.W., Kilkenny,D., Malaney,R.A., & Morrison,K., 1985. *Mon.Not.R.astr.Soc.*, 217,710
Kaufmann,J.P. & Schönberner,D., 1977. *Astr.Astrophys.*, 57,169.
Olson,G.L., 1981. *Astrophys.J.*, 245,1054.
Odgers,G.J., 1955. *Publs.Dominion astr.Obs.*, 10,215.
Penrod,V. & Smith,M.A., 1985. *NASA Conf.Publ.No.2358*, 'The Origin of Nonradiative Heating/Momentum in Hot Stars', p.53, eds. Underhill,A.B. & Michalitsianos,A.G.
Schönberner,D., 1977. *Astr.Astrophys.*, 57,437.
Smith,M.A. & Ebbetts,D., 1981. *Astrophys.J.*, 247,158.
Smith,M.A., 1983. *Astrophys.J.*, 265,338.
Underhill,A.B., 1966. 'The Early Type Stars', D.Reidel, Dordrecht, Holland.
Walker,H., 1985. *Astr.Astrophys.*, 152,58.

NON-RADIAL PULSATIONS IN THE EXTREME HELIUM STAR HD 160641

A.E. Lynas-Gray[1] D. Kilkenny[2] I. Skillen[3] & C.S. Jeffery[3]

(1) Department of Physics and Astronomy, University College London, Gower Street, London WC1E 6BT, England.

(2) South African Astronomical Observatory, P.O. Box 9, Observatory, 7935 Cape Town, South Africa.

(3) University Observatory, Buchanan Gardens, St. Andrews, Fife, Scotland.

ABSTRACT. Simultaneous radial velocity and photometric observations are reported for the variable extreme helium star HD 160641. Pulsation in the l = 4 mode could explain the observed variations; the corresponding Wesselink radius would be 8 ± 2 $R_\odot$. The consequent luminosity of log $(L/L_\odot) = 4.8 \pm 0.2$ would be consistent with Schönberner's evolutionary model for a 1 $M_\odot$ extreme helium star.

1. INTRODUCTION

Landolt (1975) and Walker & Kilkenny (1980) suggest a period in the range $0\overset{d}{.}6 - 0\overset{d}{.}7$ for the extreme helium star HD 160641. HD 160641 is therefore a candidate for radial velocity monitoring to establish whether or not it pulsates. Hill et al. (1981) and Lynas-Gray et al. (1984) have determined a pulsation mass for the radially pulsating extreme helium star BD +13°3224, but it has a surface gravity (log g = 3.7 $\pm$ 0.2) and composition (n(H)/(n(He)+n(H)) = 0.01 by numbers) inconsistent with other extreme helium stars found on Schönberner's (1977) evolutionary track. Aller (1954) determined carbon and helium abundances in HD 160641 roughly consistent with those found for HD 124448 (Schönberner & Wolf 1974), BD -9°4395 (Kaufmann & Schönberner 1977), HD 168476 (Walker & Schönberner 1981) and BD +10°2179 (Heber 1983), which are found to lie on Schönberner's evolutionary track. HD 160641 might also lie on Schönberner's evolutionary track, a fine analysis has not yet proved feasible because of the difficulty in computing non-LTE model atmospheres composed primarily of helium and carbon (Husfeld et al. 1984).

This paper presents simultaneous radial velocity and photometric observations of HD 160641. A mass of 1 $M_\odot$ seems consistent with l = 4 mode pulsation and Schönberner's (1977) evolution scheme. A more detailed account of this work will be given by Kilkenny & Lynas-Gray (1986) and Lynas-Gray et al. (1986).

2. OBSERVATIONS

Photoelectric observations were made in 1979 June and 1982 June with a "People's Photometer" attached to the Cassegrain focus of the 0.5-m telescope at the South African Astronomical Observatory (SAAO). Strömgren filters were used, the standard stars being selected from Crawford & Barnes (1970) and Grønbech et al. (1976). HD 160641 and Landolt's (1975) comparison star (BD -17°4880) were observed with six 20-s integrations (star) and one 20-s integration (sky); 10^5 counts per observation were thereby obtained, corresponding to a theoretical photometric accuracy of $0.^m003$. Figure 1 shows the photometric variation observed on a few nights in 1982; no single period is identifiable over this 9-day interval.

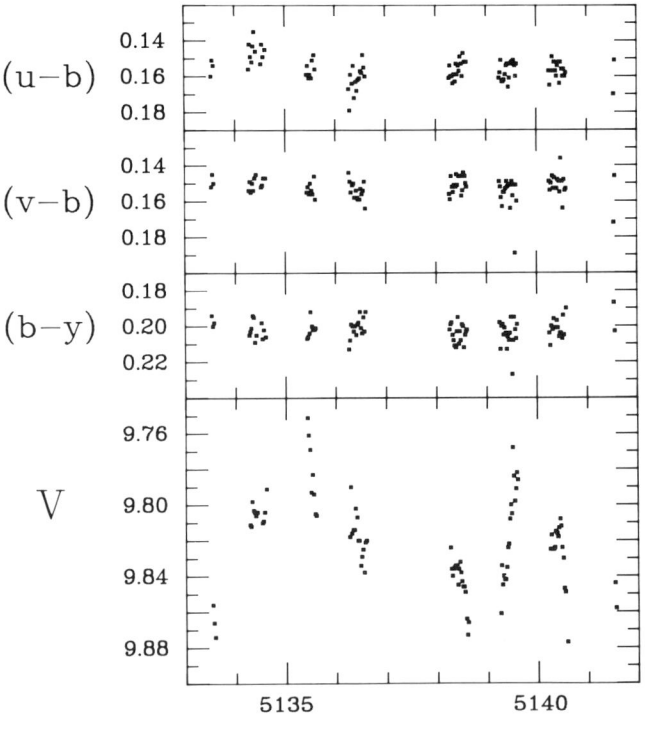

Figure 1. 1982 Strömgren four-colour photometry of HD 160641.

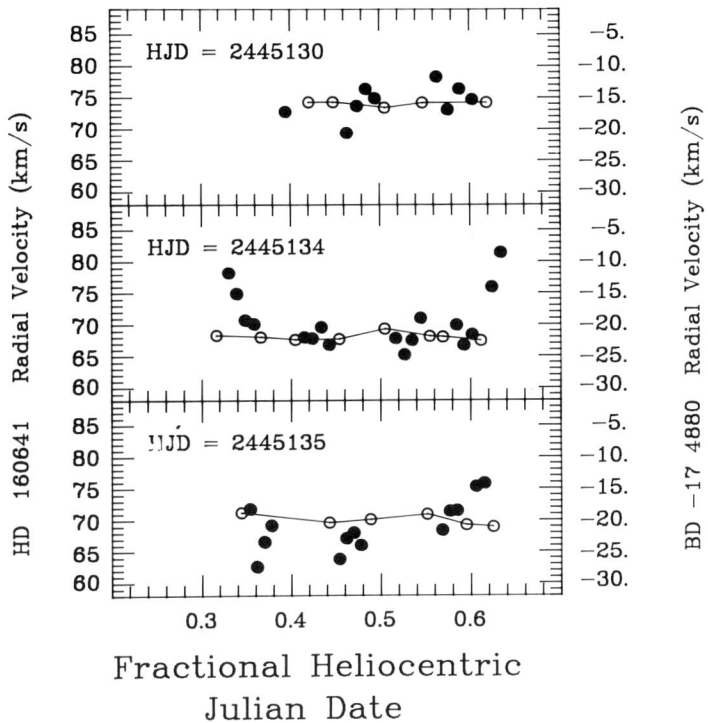

Figure 2. Radial velocities of HD 160641 (filled circles) compared with BD -17°4880 (connected open circles).

Spectra were obtained with the Image Tube Spectrograph, and Reticon Photon Counting System, attached to the Cassegrain focus of the SAAO 1.9-m telescope. A dispersion of 30Å/mm was used, with an integration time of 1000-s. Simultaneity with 0.5-m photometry was achieved on the nights of 1982 June 9th/10th, 13th/14th and 14th/15th. As a check on instrumental stability, BD -17°4880 was also observed at frequent intervals. Radial velocities are determined with the cross-correlation method of Tonry & Davis (1979). BD -17°4880 has a constant radial velocity (-19 ± 7 km/sec) within errors of the template velocity (Figure 2); the comparatively large standard deviation is due to the paucity of lines available for velocity measurement. HD 160641 velocity variations seen in Figure 2 are therefore considered to be real.

Low resolution ultraviolet spectra of HD 160641 were obtained with the International Ultraviolet Explorer (IUE) Satellite on 1979 July 14th/15th and 1983 May 14th. All other available IUE low resolution images for HD 160641 were retrieved from the data bank. Geometric and photometric corrections are applied, as necessary, with standard procedures described by Boggess et al. (1978) and absolute fluxes

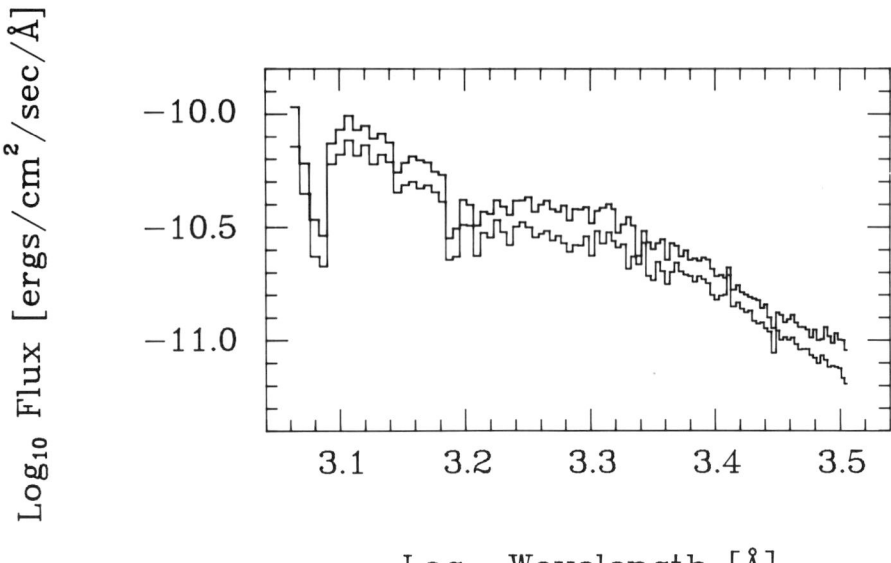

Figure 3. Maximum (SWP 5801 + LWR 5051: thick line) and minimum (SWP 19982 + LWR 15942: thin line) ultraviolet fluxes observed for HD 160641, de-reddened with E(B-V) = 0.40 and presented in 20Å bins.

extracted using IUEDR (Giddings 1983). Adopting E(B-V) = 0.40 (Drilling et al. 1984) and using Seaton's (1979) formula, all ultraviolet spectra are de-reddened and compared with each other; the maximum difference is illustrated in Figure 3.

3. FREQUENCY ANALYSIS

Photometric data from the two seasons (1979 and 1982) are interpreted using Skillen's (1985) power spectrum analysis code. The optimum set of frequencies, with corresponding amplitudes and phases, deemed to be present in the data, are listed in Table I. Figure 4 shows the comparison between observations and the fitted light curve for the 1982 season.

4. DISCUSSION

From optical photometry (Figure 1) and ultraviolet spectrophotometry (Figure 3) there appears to be no observable colour change associated with the magnitude variations. Jeffery et al. (1985) have found a similar phenomenon in BD −9°4395. The absence of colour variations, particularly in the ultraviolet, indicates that a constant effective temperature prevails during the pulsation; this precludes the possibility of radial pulsation being responsible for the light

Table I

Optimum Frequencies Derived for HD 160641

Data	Frequency (cycles/day)	Amplitude (magnitudes)	Phase (cycles)
1979	0.89278	0.046	0.3595
	0.49741	0.049	0.3799
	1.41691	0.040	0.8475
	2.64278	0.018	0.4590
1982	1.41540	0.029	0.6632
	0.89446	0.021	0.3591
	0.58110	0.018	0.5849
	2.95494	0.009	0.1241

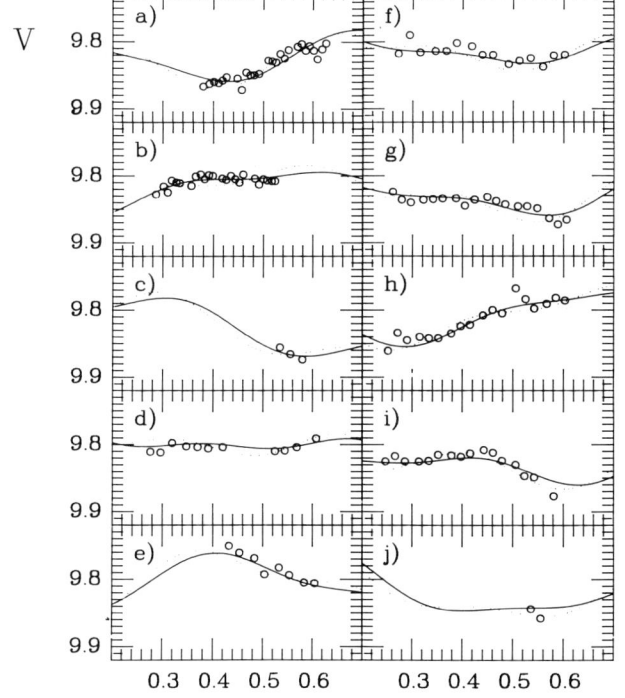

Figure 4. 1982 June Johnson V-magnitude observations (derived from Strömgren y) compared with the Fourier series representation for the following heliocentric Julian dates a) 2445129, b) 2445130, c) 2445133, d) 2445134, e) 2445135, f) 2445136, g) 2445138, h) 2445139, i) 2445140 and j) 2445141.

variations. It also seems implausible that the irregular light curve could have originated from a binary system. Light variations in HD 160641 are considered to arise entirely from geometrical distortions of the projected stellar disk. HD 160641 would therefore seem to be a non-radial pulsator with the mode l being even and non-zero; odd values of l are precluded because they occur when the light variation is entirely due to surface brightness changes (Balona & Stobie 1979).

The Wesselink radius of HD 160641 is derived following Balona & Stobie (1979), adopting Buta & Smith's (1979) expression for the dimensionless frequency (ω_o) defined as $\omega_o^2 = \omega^2 R^3/GM$, from simultaneous photometry and radial velocities obtained on 1982 June 13th/14th. In addition to being even, l cannot be 2 since a negative radius would result. For l = 4 or 6 the respective radii are 8 ± 2 $R_\odot$ and 69 ± 14 $R_\odot$; the latter is improbable because of the implied luminosity.

HD 160641 appears, therefore, to be a non-radial pulsator oscillating with mode l = 4 and having R = 8 ± 2 $R_\odot$. With the effective temperature and angular radius obtained by Drilling et al. (1984), the corresponding luminosity and distance are log $L/L_\odot$ = 4.8 ± 0.2 and 3.3 ± 0.6 kpc. Errors larger than those quoted may be present because of the assumption of Eddington limb-darkening. Evolutionary tracks computed by Schönberner (1977) show a 1 $M_\odot$ extreme helium star to have log $L/L_\odot$ = 4.6. The luminosity derived in this paper is therefore consistent with Schönberner's model.

Assuming a mass of 1 $M_\odot$ for HD 160641, the fundamental pulsation frequency for the l = 4 mode is 0.65 cycles/day. Frequency resolutions of 0.2 cycles/day (1979) and 0.08 cycles/day (1982) are present in reported observations. Consequently, the lowest observed frequencies of 0.497 cycles/day (1979) and 0.581 cycles/day (1982) are entirely consistent with fundamental l = 4 mode pulsation. Lower frequencies could be present in the data, but remain undetected because of the limited duration of observing runs. Accordingly, it is not possible to ascertain as to whether or not the higher frequencies are overtones or the result of rotation splitting.

Acknowledgements

We are indebted to Dr. P.W. Hill for suggesting that radial velocities be obtained simultaneously with the photometric data in 1982. Telescope time and travel grants were awarded by the United Kingdom Science and Engineering Research Council (SERC) and the South African Council for Scientific and Industrial Research. Presentation of this paper at the 87th Colloquium of the International Astronomical Union (IAU) was made possible by travel grants received from the IAU and University College London. The work was carried out during the tenure of SERC research assistantships held by two of us (IS and CSJ). AELG is supported by the SERC Collaborative Computational Project No. 7.

References

Aller, L.H. 1954. In Colloquium on "Les Processus Nucléaires dans les Astres", Liège Mem. 15, 337.
Balona, L.A. & Stobie, R.S. 1979. Mon. Not. R. astr. Soc. 187, 217.
Boggess, A. et al. 1978. Nature 275, 377.
Buta, R.J. & Smith, M.A. 1979. Astrophys. J. 232, 213.
Crawford, D.L. & Barnes, J.V. 1970. Astron. J. 75, 978.
Drilling, J.S., Schonberner, D., Heber, U. & Lynas-Gray, A.E. 1978. Astrophys. J. 278, 224.
Giddings, J.R., 1983. IUE ESA Newsletter 17, 53.
Grønbech, B., Olsen, E.H. & Strömgren, B. 1976. Astron. Astrophys. 26, 155.
Heber, U., 1983. Astron. Astrophys. 118, 39.
Hill, P.W., Kilkenny, D., Schönberner, D. & Walker, H.J. 1981. Mon. Not. R. astr. Soc. 197, 81.
Husfeld, D., Kudritzki, R.P., Simon, K.P. & Clegg, R.E.S. 1984. Astron. Astrophys. 134, 139.
Jeffery, C.S., Skillen, I., Hill, P.W., Kilkenny, D., Malaney, R.A., & Morrison, K. 1985. Mon. Not. R. astr. Soc 217, 701.
Kaufmann, J.P., Schönberner, D. 1977. Astron. Astrophys. 57, 169.
Kilkenny, D. & Lynas-Gray, A.E. 1986. SAAO Circulars (In preparation).
Landolt, A.U. 1975. Astrophys. J. 196, 789.
Lynas-Gray, A.E., Kilkenny, D., Skillen, I. & Jeffery, C.S. 1986. Mon. Not. R. astr. Soc. (to be submitted).
Lynas-Gray, A.E., Schönberner, D., Hill, P.W. & Heber, U. 1984. Mon. Not. R. astr. Soc. 209, 387.
Schönberner, D. 1977. Astron. Astrophys. 57, 437.
Schönberner, D. & Wolf, R.E.A. 1974. Astron. Astrophys. 37, 87.
Seaton, M.J. 1979. Mon. Not. R. astr. Soc 187, 73p.
Skillen, I. 1985. Ph.D. Thesis, University of St. Andrews.
Tonry, J. & Davis, M. 1979. Astron. J. 84, 1511.
Walker, H.J. & Kilkenny, D. 1980. Mon. Not. R. astr. Soc 190, 299.
Walker, H.J. & Schönberner, D. 1981. Astron. Astrophys. 97, 291.

THE PERIOD OF THE EXTREME HELIUM STAR BD+1°4381

C.S.Jeffery, P.W.Hill and K.Morrison
University Observatory
Buchanan Gardens
St Andrews
Fife KY16 9LZ
Scotland

ABSTRACT. Photometry of the variable extreme helium star BD+1°4381 is used to improve the 21.2 day period found by Jeffery & Malaney (1985). The amplitude of the light-curve appears to be variable, resembling that of RY Sgr. Variability in the extreme helium star BD-1°3438 is also reported.

1. INTRODUCTION

The extreme helium star BD+1°4381 was found to be variable with an amplitude of 0^{m}04 and a provisional period of 21.2 days by Jeffery and Malaney (1985). It was suggested that the variations were probably due to radial pulsations such as those found in the R Coronae Borealis stars (Alexander et al. 1972). This discovery is important because masses can sometimes be obtained for pulsating stars (e.g. V652 Her: Hill et al. 1981, Lynas-Gray et al. 1984). The effective temperature of the star is 9 500 K (Drilling et al. 1984) making it one of the coolest known extreme helium stars and placing it midway between the variable hot extreme helium stars (e.g. BD-9°4395: Jeffery et al. 1985) and the R CrB stars (e.g. R CrB: Fernie 1982). The discovery of a singly periodic variation in one of the extreme helium stars would enable, in time, possible period changes to be investigated and hence provide a test for theories regarding the evolutionary status of the star. Experience with determining period changes in V652 Her (Kilkenny & Lynas-Gray 1982, 1984) shows that it is important to monitor the period of the star regularly in order to avoid ambiguities. Further observations have been made to ensure these conditions are satisfied for BD+1°4381 and to search for similar variables amongst the cooler extreme helium stars.

2. OBSERVATIONS

Strömgren photometry has been carried out in 1985 with the SAAO 0.5m telescope. The observing programme contained the known variable BD+1°4381 and two new candidates, BD-1°3438 (McConnell et al. 1972) and CPD-58°2721 (Drilling 1980). The last of these is discussed in another paper (Morrison et al. 1986). BD-1°3438 was first reported as variable by Landolt (Walker & Kilkenny 1980). Data were obtained on 25 nights during a 6 week observing run, data acquisition and reduction techniques follow precisely those described by Jeffery et al. (1985).

3. RESULTS

3.1. BD+1°4381.

The light and colour curves of BD+1°4381 for the 1985 observing season are shown in Fig. 1. We note that the shape of the light curve appears to be variable, a feature also present in the light curves of the R CrB stars (Alexander et al. 1972, Fernie 1982). This introduces a difficulty when attempting to determine the period of the star. Assuming the slow variations in the light curve to be independent of the fundamental period, they were removed by applying a high-pass filter to the data. The data from both 1984 and 1985 observing seasons were combined. A single sine function was then swept over a range of periods between 15 and 30 days in order to obtain the best fit to the data. The goodness of fit was determined from a number of statistical parameters, including the multiple correlation coefficient. The data are still insufficient to determine the period unambiguously, an uncertainty of 1 cycle per year remaining in the frequency. The two best values for the period are 21.529 day and 23.026 day, for which ephemerides are given in Table I. It is now possible to determine the mean amplitudes of the light and colour variations more precisely. The amplitude of the light curve is $0^m057 \pm 0^m003$ and in u-b the variation is $0^m015 \pm 0^m002$. The variations in b-y and v-b are somewhat smaller and at $0^m005 \pm 0^m002$ are comparable with the scatter in the data. The maximum change in either surface gravity or effective temperature that can be produced by the observed colour changes may be estimated from model atmospheres for extreme helium stars (Heber & Schönberner 1981). Taking the reddening free colour index [u-b] (ibid), the maximum expected change in T_{eff} would be 200 K, while the maximum change in log g would be 0.22.

While further photometry is necessary to resolve the period ambiguity, radial velocity measurements covering the pulsation cycle are a matter of urgency.

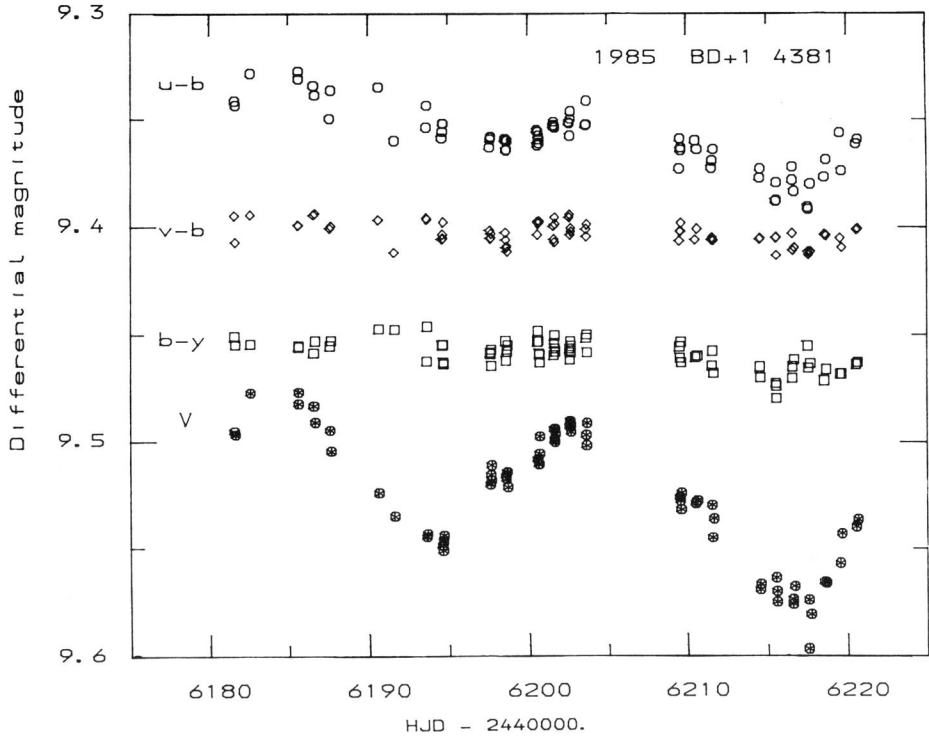

Figure 1. Differential Strömgren photometry of the extreme helium star BD+1°4381 obtained at SAAO during 1985.

TABLE I. Ephemerides for BD+1°4381.

$V = \langle V \rangle + a \cdot \sin(2\pi(HJD - t_0) / P)$

$\langle V \rangle = 9^m530 \pm 0^m001$
$a = 0^m026 \pm 0^m001$

	P (day)	t_0 (HJD)
1.	21.529 ± 0.05*	2445860.122 ± 0.161
2.	23.026 ± 0.05*	2445859.223 ± 0.168

* estimated from experiments with modified data samples

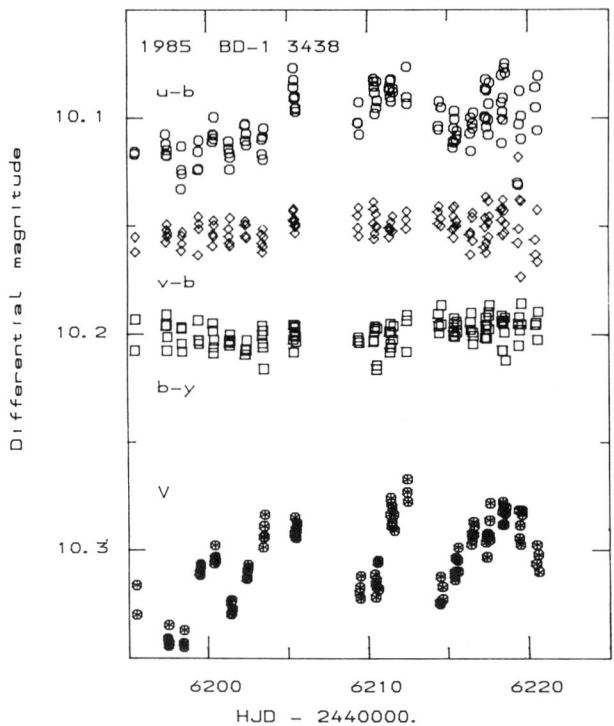

Figure 2. Differential Strömgren photometry of the extreme helium star BD-1°3438 obtained at SAAO during 1985.

3.2. BD-1°3438.

The light and colour curves of BD-1°3438 are shown in Fig. 2. Being nearly a magnitude fainter than BD+1°4381 the photometry was somewhat noisier, particularly when the moon was bright. Its mean magnitude and colours and those of two comparison stars were:

	$\langle V \rangle$	$\langle b-y \rangle$	$\langle m_1 \rangle$	$\langle c_1 \rangle$	n
BD-1°3438	10.304	0.381	0.028	0.248	138
BD-1°3435	7.966	0.152	0.064	0.940	138
BD-0°3406	10.518	0.358	0.072	0.124	131

The star is clearly variable by 0^m07 in V. Colour variations are much smaller, but appear at least to be present in (u-b). The timescale of the light variations is of the order of 5 to 8 days, with some evidence for systematic trends within a night. A frequency analysis may be premature for the number of data points in our sample, likewise it would be unwise to attempt to attribute the light variations to any particular type of stellar oscillation at this stage.

4. DISCUSSION

Drilling et al. (1984) derived effective temperatures for 12 extreme helium stars. These temperatures are listed in table II for those stars for which variability studies have been published. It would appear that a period - temperature sequence is emerging, which is consistent with theoretical models for pulsation in cool luminous helium stars (Wood 1976, Saio & Wheeler 1985), but theoretical studies of pulsation in hot luminous extreme helium stars have not been published. BD+1°4381 appears to represent a straightforward extension of radial pulsation in RCrB stars to higher temperatures, but the absence of confirmed variablility at temperatures between 15 000 K and 20 000 K and the mode of oscillation in the hottest extreme helium stars require further attention.

TABLE II.

Star		T_{eff} (K)	Light curve	Period (days)
HD160641	V2076 Oph	31 900[1]	Multiperiodic	0.7,1.1,...[4,5]
BD-9°4395		23 000[1]	Multiperiodic	3.5,11.2,...[6]
BD+10°2179		17 700[1]	Not variable[7,8]	
HD124448		15 500[1]	Suspected variable[4,9]	
HD168476	PV Tel	12 400[1]	Irregular	?[4,10]
BD-1°3438		10 900[1]	Irregular	5 - 8 [11]
BD+1°4381		9 500[1]	Periodic	21.5[11,12]
	RY Sgr*	7 100[2]	Roughly periodic	38.6[13]
	R CrB*	7 000[3]	Roughly periodic	46[14]

Notes:
1 Drilling et al. (1984)
2 Schönberner (1975)
3 Cottrell & Lambert (1982)
4 Walker & Kilkenny (1980)
5 Lynas-Gray et al. (1986)
6 Jeffery et al. (1985)
7 Hill et al. (1984)
8 Grauer et al. (1984)
9 Hill (1969)
10 Walker & Hill (1985)
11 This paper
12 Jeffery & Malaney (1985)
13 Kilkenny (1982)
14 Fernie (1982)

* Excluding deep RCrB-type minima

ACKNOWLEDGMENTS

The authors are grateful to the UK Panel for the Allocation of Telescope Time and to the SAAO for generous awards of observing time, to SAAO staff for assistance with the data reduction, to Mr.I.Skillen for the use of his frequency analysis program and to Dr.A.J.Adamson for assistance with the presentation.

REFERENCES

Alexander,J.B., Andrews,P.J., Catchpole,R.M., Feast,M.W., Lloyd Evans,T., Menzies,J.W., Wisse,P.N.J. & Wisse,M., 1972. *Mon.Not.R.astr.Soc.*, 158,305.
Cottrell,P.L. & Lambert,D.L., 1982. *Astrophys.J.*, 261,595.
Drilling,J.S., 1980. *Astrophys.J.*, 242,L43.
Drilling,J.S., Schönberner,D., Heber,U. & Lynas-Gray,A.E., 1984. *Astrophys.J.*, 278,224
Fernie,J.D., 1982. *Publs.astr.Soc.Pacif.*, 94,172.
Grauer,A.D., Drilling,J.S. & Schönberner,D., 1975. *Astr.Astrophys.*, 133,285.
Heber,U., & Schönberner,D., 1981. *Astr.Astrophys.*, 102,73.
Hill,P.W., 1969. *Inf.Bull.Variable Stars No. 357.*
Hill,P.W., Kilkenny,D., Schönberner,D. & Walker,H.J., 1981. *Mon.Not.R.astr.Soc.*, 197,81.
Hill,P.W., Lynas-Gray,A.E. & Kilkenny, D., 1984. *Mon.Not.R.astr.Soc.*, 207,823.
Jeffery,C.S. & Malaney,R.A., 1985. *Mon.Not.R.astr.Soc.*, 213,61p.
Jeffery,C.S., Skillen,I., Hill,P.W., Kilkenny,D., Malaney,R.A., & Morrison,K., 1985. *Mon.Not.R.astr.Soc.*, 217,710
Kilkenny,D., 1982. *Mon.Not.R.astr.Soc.*, 200,1019.
Kilkenny,D. & Lynas-Gray,A.E., 1982. *Mon.Not.R.astr.Soc.*, 198,873.
Kilkenny,D. & Lynas-Gray,A.E., 1984. *Mon.Not.R.astr.Soc.*, 208,673.
Lynas-Gray,A.E., Schönberner,D., Hill,P.W. & Heber,U., 1984. *Mon.Not.R.astr.Soc.*, 209,387.
Lynas-Gray,A.E., Kilkenny,D., Skillen,I. & Jeffery,C.S., 1986. in 'Hydrogen-deficient stars and related objects', eds. K.Hunger, N.K.Rao & D.Schönberner.
MacConnell,D.J., Frye,R.L. & Bidelman,W.P., 1972. *Publs.astr.Soc.Pacif.*, 84,388.
Morrison,K., Drilling,J.S, Heber,U., Hill,P.W., & Jeffery, C.S., 1986. in 'Hydrogen-deficient stars and related objects', eds. K.Hunger, N.K.Rao & D.Schönberner.
Saio,H. & Wheeler,J.C., 1985. *Astrophys.J.*, 295,38.
Schönberner,D., 1975. *Astr.Astrophys.*, 44,383.
Walker,H.J. & Hill,P.W., 1985. *Astr.Astrophys.Suppl.*, 61,303.
Walker,H.J. & Kilkenny,D., 1980. *Mon.Not.R.astr.Soc.*, 190,299.
Wood,P.R., 1976. *Mon.Not.R.astr.Soc.*, 174,531.

A PRELIMINARY ANALYSIS OF THE PULSATING EXTREME HELIUM STAR V652 HER
(BD+13°3224)

[1]C.S.Jeffery, [2]U.Heber, and [1]P.W.Hill.

[1]University Observatory, Buchanan Gardens, St Andrews,
Fife KY16 9LZ, Scotland.
[2]Institut für Theoretische Physik und Sternwarte der
Universität, Olshausenstrasse 40, D-2300 Kiel, Federal
Republic of Germany.

ABSTRACT. A preliminary fine analysis of the atmosphere of the
pulsating extreme star V652 Her (=BD+13°3224) is reported. The mean
effective temperature and temperature variation have been determined
from the ionisation equilibria of Si and N. Both of these elements are
strongly overabundant, whilst C appears to be underabundant.

1. INTRODUCTION

The extreme helium star V652 Her (=BD+13°3224) discovered by Berger
and Greenstein (1963) was found to be variable by Landolt (1975).
Radial velocity studies confirmed that the variations were due to
radial pulsations (Hill et al 1980; Paper I) and enabled a pulsation
mass to be derived. Absolute flux measurements and new velocity data
enabled Lynas-Gray et al. (1984; Paper II) to improve the mass
determination to 0.7^{+8}_{-3} $M_\odot$, the major uncertainty being the
surface gravity. The high surface gravity (log g=3.7±0.2) and the
period decrease found by Kilkenny & Lynas-Gray (1982, 1984) led to the
proposal (Paper II & Jeffery 1984) that V652 Her is unlike the extreme
helium stars discussed by Hunger (1975), but that it is contracting
onto the helium main-sequence. As such, V652 Her occupies a unique
position in the Hertzsprung-Russell diagram. An abundance analysis is
required to understand the origin of the star.

2. OBSERVATIONS

High-resolution spectroscopic data obtained with the IPCS and the
Anglo-Australian Telescope in 1979, 1980, 1982 and 1984 in the
wavelength range 3400-4600Å has been described in Papers I and II and
by Jeffery & Hill (1986, Paper III). In addition to the integrated
spectrum obtained from the sum of all spectra used in radial velocity
measurements, we chose two phase bins of width 0.2 cycles centred at

phases 0.68 and 0.02, corresponding to the minimum and maximum of the T_{eff} curve determined in Paper II. All unnormalised spectra obtained within these phase bins (as determined by Kilkenny & Lynas-Gray's 1984 cubic ephemeris) were added. Rectification and normalisation were carried out as the final stage of data reduction.

3. ANALYSIS

We have begun an analysis of these spectra based on LTE model atmospheres. Here we report preliminary results for calculations of the ionisation equilibria of silicon (SiII/III/IV) and nitrogen (NII/III) and hydrogen line profiles.

3.1. Model Atmospheres

LTE model atmospheres have been constructed with a code specially designed for helium-star atmospheres. Details can be found in Heber and Schönberner (1981) and references therein. It is important to note here that line blanketing is not taken into account. Models have been calculated with gravities of log g=3.5 and log g=4.0, which bracket the paper II result (log g=3.7), and a few effective temperatures. Hydrogen, nitrogen and silicon line profiles have been calculated with the code of Schönberner (1973). Atomic data are the same as those used by Heber (1983). We have assumed a fixed microturbulent velocity of 5 km s^{-1} throughout.

3.2. Ionisation Equilibria.

3.2.1. <u>Effective temperature and abundances</u>. Equivalent widths of as many lines of NII, NIII, SiII, SiIII, SiIV as possible were measured for the integrated spectrum and for each of the two phase-binned spectra; these are given in Table I. Only the lines SiII 4128Å and SiIV 4116Å are sufficiently free of blends to be representative of these species in an equilibrium analysis, but several lines of SiIII are available. Equivalent width variations in the NII and SiIII lines were less than 10%. The run of the ionisation equilibria of NII/III, SiII/III and SiII/IV in the (T_{eff}, log g)-plane is shown in Fig. 1. Since the gravity is known we can read off the effective temperature of the model, obtaining T_{eff}=28100$\pm$1100 K. Simultaneously the nitrogen and silicon abundances are determined from the ionisation eqilibria. Both elements are strongly overabundant with respect to the Sun. Silicon is enhanced by 0.9$\pm$0.1 dex and nitrogen by an even larger factor (1.6$\pm$0.1 dex).

TABLE I

Ion/Multiplet		λ(A)	⟨T⟩	W_λ (mA) Tmax	Tmin
NII	5	4601.5	235		
	5	4607.2	230		
	12	3995.0	341		
	15	4447.0	174		
	33	4227.7	159		
	38	4082.3	124		
	39	4056.9	100		
	48	4241.8	357		
	48	4236.9	196		
	55	4442.0	110		
NIII	1	4097.3	173	212	149
	1	4103.4	133	157	103
SiII	3	4128.1	58	48	83
SiIII	2	4567.9	323		
	2	4574.8	232		
	2	4552.7	372		
SiIV	1	4116.0	142	186	128

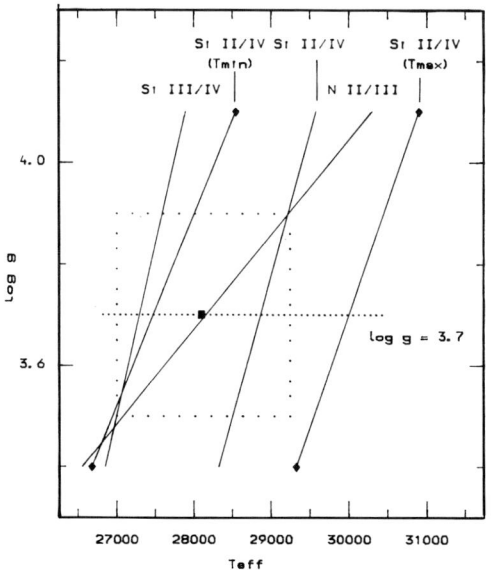

Figure 1. The ionisation equilibria of SiII/III, SiII/IV and NII/NIII calculated from the integrated spectrum of V652 Her. The variation of the SiII/IV equilibrium between minimum and maximum effective temperature is also shown.

3.2.2. Temperature variations. The SiII/SiIV line ratios are very sensitive to the effective temperature. We can use these line ratios to investigate temperature change during a pulsation cycle. The SiII/SiIV equlibria were calculated for the spectra near T_{min} and T_{max} and are also shown in Fig. 1. The amplitude of the T_{eff} variation is 2400±500 K for log g=3.7.

3.3. The Hydrogen Line Profiles

Paper I reported that the the Hγ/HeI4471 equivalent-width ratio gives an abundance ratio n_H/n_{He}=0.02, whereas the adopted model has n_H/n_{He}=0.01. In the current data two hydrogen lines (Hγ, Hδ) are available, Hε can also be identified. It has proved impossible to fit both line profiles simultaneously with the current models. in both cases the best result was obtained with T_{eff}=25000K and log g=3.5. However to fit Hγ, n_H/n_{He}=0.02 is required, whilst Hδ requires n_H/n_{He}=0.01.

4. DISCUSSION

Paper II determined the mean effective temperature of V652 Her and its variation from observations of the total flux emitted by the star, based mainly on the ultraviolet flux distributions. We have to compare our spectroscopic results with those of Paper II. The latter derived T_{eff}=23450±1320 K and the amplitude of the T_{eff} variation was 2850±110 K when the mean observations obtained within ±0.1 cycles of T_{min} and T_{max}, respectively, were considered.

While our results for the temperature variation are in good agreement (the difference is only about 350 K) the effective temperature of our model is considerably larger than the empirical T_{eff} of paper II (ΔT_{eff}=4600 K). This discrepancy can be explained (at least partially) by the fact that our model does not account for line blanketing. The observed ultraviolet spectrum, however, displays strong line blocking in the wavelength region 1200-2000Å, as can be seen in Fig. 3 of Paper II. Including the UV line blocking in the models would result in backwarming of the deeper layers. Consequently, a lower model T_{eff} has to be assigned to fit the ionisation equilibria.

5. CONCLUSIONS

We have presented preliminary results of an analysis of visual spectra of V652 Her. We derived T_{eff} and T_{eff} variations in reasonable agreement with previous investigations of the ultraviolet flux distribution.

The abundances of hydrogen, nitrogen and silicon have also been determined. The hydrogen deficiency is probably due to CN burning, since we find nitrogen strongly enriched while the carbon lines are weak. Only CII 4267Å can be identified (W_λ=158mÅ) in our spectra. The silicon enrichment (by ~1 dex) with respect to the Sun cannot be explained by nuclear processing in the stellar interior.

ACKNOWLEDGMENTS.

We are grateful to Dr.J.P.Kaufmann for preliminary calculations suggesting that both Silicon ionisation equilibria should be observable. This paper is based on observations made with the Anglo-Australian Telescope. The project is supported by the UK Science and Engineering Research Council.

REFERENCES

Berger,J. & Greenstein,J.L, 1963. *Publs.astr.Soc.Pacif.*, 75,336.
Heber,U. & Schönberner,D., 1981. *Astr.Astrophys.*, 102,73.
Heber,U., 1983. *Astr.Astrophys.*, 118,39.
Hill,P.W., Kilkenny,D., Schönberner,D. & Walker,H.J., 1981.
    Mon.Not.R.astr.Soc., 197,81, (Paper I).
Hunger.K., 1975. *Problems in Stellar Atmospheres and Envelopes*, p. 57,
    eds Baschek,B., Kegel,W.H. & Traving,G., Springer-Verlag, Berlin.
Jeffery,C.S., 1984. *Mon.Not.R.astr.Soc.*, 210,731.
Jeffery,C.S. & Hill,P.W., 1986. *Mon.Not.R.astr.Soc.*, (Paper III)
    (submitted).
Kilkenny,D. & Lynas-Gray,A.E., 1982. *Mon.Not.R.astr.Soc.*, 198,873.
Kilkenny,D. & Lynas-Gray,A.E., 1984. *Mon.Not.R.astr.Soc.*, 208,673.
Lynas-Gray,A.E., Schönberner,D., Hill,P.W. & Heber,U., 1984.
    Mon.Not.R.astr.Soc., 209,387, (Paper II).
Schönberner,D., 1973. *Astr.Astrophys.*, 28,433.

DISCUSSION

LYNAS-GRAY: What is the reason for the discrepancy between IUE and ionization equilibrium mean T_{eff} variations?

JEFFERY: It is the model atmosphere which we used. Yours is from a line blanketed model atmosphere. These ones are from continuum model atmosphere. The mean effective temperature derived from the IUE fluxes is an empirical one, i.e. essentially independent of the model. The T_{eff} derived from the ionization equilibrium is a model T_{eff}. The model does not take into account line blanketing. Since the UV line blocking is extra-ordinarily large, it can explain the difference between the two effective temperatures derived. The large UV line blocking might be related to the over abundances we find.

TUTUKOV: What is the time scale for the change of the pulsation period?

JEFFERY: It is roughly 10^{-10} days/cycle.

SAIO: Did you get the oxygen abundances?

JEFFEREY: We have not measured the oxygen abundances yet.

THE RADIAL VELOCITY CURVE OF V652 HER (BD+13°3224)

P.W.Hill and C.S.Jeffery
University Observatory
Buchanan Gardens
St Andrews
Fife KY16 9LZ
Scotland

ABSTRACT. New radial velocity data for the pulsating extreme helium star V652 Her (BD+13°3224) have been obtained with a time resolution of 100 s. High frequency structure in the radial velocity curve is detected, and a comparison with previous data suggests that the detailed shape of the velocity curve is variable. The data imply that the effective surface gravity must increase by a factor of 4 at minimum radius.

1. INTRODUCTION

The hot extreme helium star V652 Her (=BD+13°3224) has both higher surface gravity for its temperature and higher hydrogen abundance, $n_H=10^{-2}$ (Hill et al. 1981; Paper I) than the true extreme helium stars. Likewise it is too hydrogen-deficient to be regarded as an intermediate helium star (Hunger 1975). The discovery of radial pulsations with a period of 0.108 day by Landolt (1975), confirmed spectroscopically in Paper I, enabled a mass of $0.7^{+0.3}_{-0.2}M_\odot$ to be derived (Lynas-Gray et al. 1984; Paper II). The location of V652 Her in the HR diagram and the discovery that its period is decreasing on a secular time-scale (Kilkenny & Lynas-Gray 1982, 1984) are consistent with the hypothesis that the star is contracting rapidly onto the helium main-sequence (Jeffery 1984). It is not known to be a binary, and yet appears to have lost its hydrogen-rich envelope either before or simultaneous with the ignition of core-helium burning.

The radial velocity curve of V652 Her is characterised by a slow rise to maximum radial velocity followed by a rapid transition to minimum velocity (Paper I). Previous studies have found some evidence for additional structure in the velocity curve (Paper II), but the data suffer from poor time resolution at minimum radius and from comparatively large standard errors (± 5 km/s). The short period makes V652 Her an ideal target from which to obtain radial velocities with good S/N and complete phase coverage for detailed comparison with theoretical pulsation models when these become available. A fine

analysis to determine the surface gravity and elemental abundances of
V652 Her is in progress (Jeffery et al. 1986). This is essential to
improve the mass determination and understand its evolutionary history
and pulsational properties.

2. OBSERVATIONS

New spectroscopic data have been obtained with the RGO Cassegrain
spectrograph on the Anglo-Australian Telescope (AAT) in 1982 and 1984.
Full details will be published elsewhere (Jeffery & Hill 1986). Over
300 individual heliocentric radial velocities determined by the
cross-correlation technique of Kilkenny et al. (1981) are shown in
Figs. 1 and 2 with phases derived from the cubic ephemeris of Kilkenny
& Lynas-Gray (1984). Increasingly high precision and time resolution
have enabled us to examine closely the detailed behaviour of the radial
velocity curve.

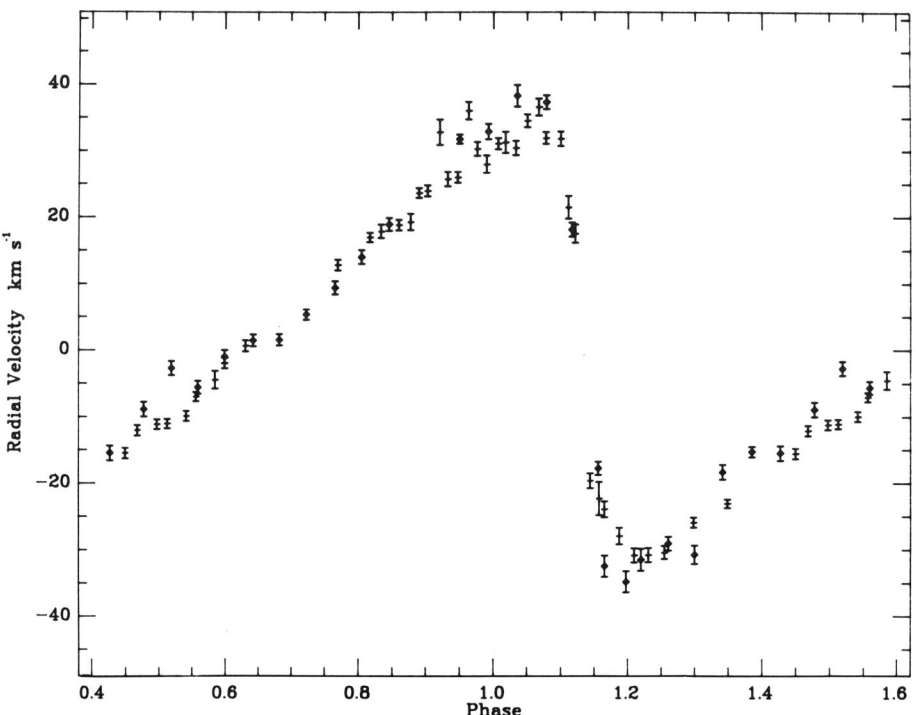

Figure 1. The radial velocity curve of V652 Her obtained with the AAT
in 1982. Exposures were for 300 s at 10 A/mm with the IPCS on
July 2 (horizontal bars) and for 200 s at 26 A/mm with a CCD
on July 3 (diamonds).

Two features of the 1982 radial velocity curves are noteworthy. In the IPCS data, the scatter of the data is greater at phase 0.95 (just before radial velocity maximum) than during the rest of the cycle. At phase 0.2 the approach to minimum velocity is much shallower in the IPCS data than in the CCD data. These data raise particular questions which we attempt to answer in the following analysis; namely: 1) Are small amplitude high frequency oscillations superimposed on the radial velocity curve, particularly around radial velocity maximum? 2) Does the shape of the radial velocity curve vary, e.g. from cycle to cycle? 3) What is the acceleration at minimum radius?

The general shape of the radial velocity curve of V652 Her comprises a rapid acceleration phase covering 0.1 cycles at minimum radius followed by a relaxation phase covering 0.8 cycles. Previous analyses suffered from a paucity of observations in the rapid acceleration phase. Moreover, exposure times were long compared to this phase of the pulsation cycle. The 1982 data go some way toward ameliorating this situation, but only with the 1984 data was continuous coverage of the radial velocity curve achieved with good time resolution.

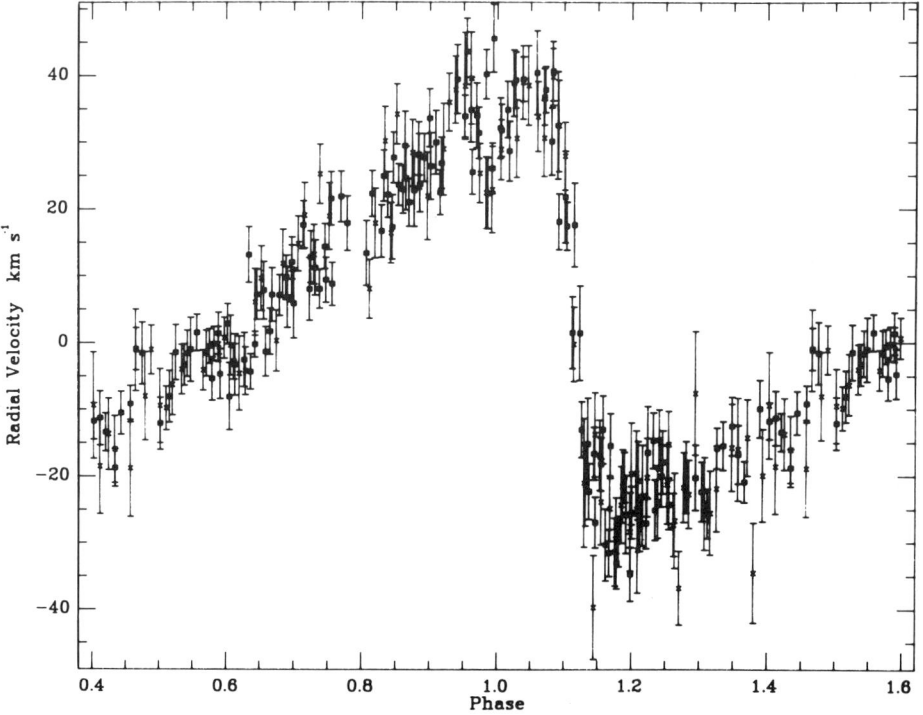

Figure 2. The radial velocity curve of V652 Her obtained with the AAT in 1984 on April 15 (solid squares) and April 17 (crosses) from 100 s exposures at 10 Å/mm using the IPCS.

In this paper the data are represented by models defined by passing a Gaussian filter through the observed values. Although noise is reduced at the expense of time resolution, the latter can be optimised to the noise level in the data by selecting an appropriate value for the FWHM of the filter. The combined radial velocity data from both 1984 runs were used to obtain a model which satisfied a χ^2 test at the 95% confidence level. The standard deviation of the adopted Gaussian smoothing function was 0.02 cycles. All the data obtained prior to 1984 (including that from Papers I and II) were combined to produce a similar model. There is a small difference between the mean velocities of the two models, 1.3 km/s (pre-1984) and 3.5 km/s (1984), but this is within the absolute error obtained from measurements of radial velocity standards made with the same instrumental configurations. It is best removed by transforming the heliocentric velocities to the stellar rest frame (Parsons 1972). The resulting surface velocity curves (Fig. 3) will be those referred to in the rest of this paper.

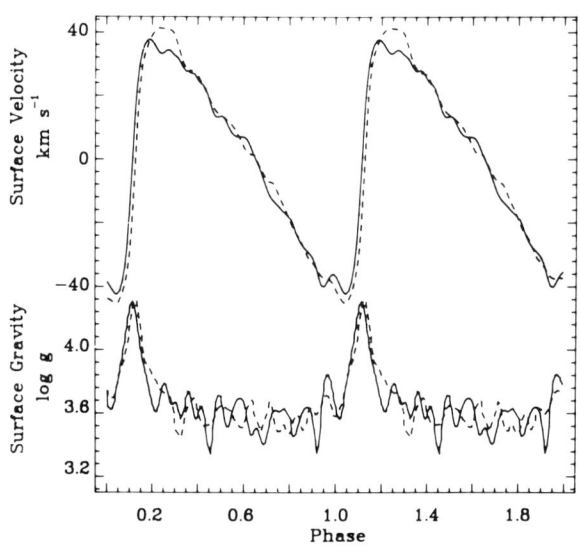

Figure 3. The transformed stellar surface velocity curves and 'effective' surface gravity derived from the two smoothed radial velocity models (1979-82: broken line, 1984: solid line).

3. DISCUSSION

The two surface velocity curves obtained in section 2 provide some useful information about the pulsation of V652 Her. The smoothed 1984 velocity curve (Fig. 3, solid line) shows a dip just 0.1 cycles before minimum velocity, with amplitude 6 km/s. The χ^2 test also implies that the small amplitude oscillations on the falling part of the velocity curve are physically significant, but these features do not seem to persist over many cycles. In the 1979-82 data (Fig. 3, broken line) these features are not so pronounced.

Minimum velocity occurs at the same phase in both models. The small displacement between the curves is accounted for by a new ephemeris (Lynas-Gray & Kilkenny 1986) including the period of these observations. The phase discrepancy at maximum velocity ($\sim$0.1 cycles) cannot be attributed to the adopted ephemeris and the amplitudes of the velocity curves differ by 9 km/s, implying that there are real long-term variations in the shape of the velocity curve.

In Paper II a mean radius, $1.98R_\odot$, was obtained and a mass of $0.7M_\odot$ derived from the best available spectroscopic value for the surface gravity, $\log g=3.7$. The absolute radius variation around the cycle is derived from the integrated velocity curve and this mean radius, but with an amplitude of <0.1 mean stellar radii this alone has little effect on the surface gravity of the star as given by $g=GM/R^2$. However the effective gravity experienced by a particle in the stellar atmosphere also includes the acceleration due to pulsation. Differentiating the velocity curve and combining this with the small gravity changes due to radius variations yields an 'effective' surface gravity as a function of phase (Fig. 3). A startling feature of this 'effective' gravity curve is the amplitude of the peak at minimum radius which represents an increase in the surface gravity by a factor of at least 4 during a short part of each pulsation cycle.

The magnitude of the accelerative forces in V652 Her raises the question of whether a shock wave develops in the atmosphere at minimum radius. This question can be resolved by comparing the time for a pressure wave to propagate through the photosphere, t_p, given by the ratio of the pressure scale height to the photospheric sound speed, with the actual time taken for the photosphere to respond to the accelerative forces of the pulsation, given simply by the time between minimum and maximum velocity, t_a=1400 s. Making the simplifying assumptions of small radiation pressure, ionisation roughly uniform throughout the photosphere, and mean molecular weight 2, with T=23500 K (Paper II), then t_p=150 s. Since $t_a>>t_p$, a shock wave is unlikely to be produced by the pulsation. This conclusion is supported by the apparent absence of other spectroscopic phenomena normally associated with the presence of atmospheric shock waves such as emission-lines and absorption-line splitting (e.g., Willson 1975).

Models of pulsation in luminous helium stars have only been calculated for the cooler R Coronae Borealis (R CrB) variables (Wood 1976, Saio & Wheeler 1985). With lower temperatures and surface gravities (Schönberner 1975, Cottrell & Lambert 1982) than those of V652 Her, pulsation in the comparitively tenuous envelopes of the R CrB variables may only be compared with that in V652 Her via the period mean density relation (Jeffery 1984). One non-linear pulsation model (Saio & Wheeler 1985, model 8) for a $0.7M_\odot$ R CrB star with T_{eff}=7000 K and L=1.5x$10^4 L_\odot$ gives a period of 0.15 day for a star with similar surface properties to V652 Her (T_{eff}=23500 K, L=1.1x$10^3 L_\odot$, Paper II) and an internal structure homologous with that of the R CrB stars.

Although the agreement with the observed period (0.11 day) is reasonable, the comparison is of limited value because the homology condition is almost certainly violated by stars with such different temperatures. Indeed, the shock waves present in the R CrB models appear to be absent in V652 Her. Linear and non-linear studies of pulsation in *hot* helium stars are urgently needed in order to understand the pulsation in V652 Her. The detailed features of the radial velocity curve presented in this paper provide sensitive tests which models should endeavour to satisfy.

ACKNOWLEDGMENTS

We are indebted to Mr S.Lee for applying first aid to the data acquisition software at a crucial moment. The data were reduced with VAX computers of the AAO, STARLINK and the University of St. Andrews. We acknowledge support from the SERC in the form of observing time awards and research grants.

REFERENCES

Cottrell,P.L. & Lambert,D.L., 1982. *Astrophys.J.*, 261,595.
Hill,P.W., Kilkenny,D., Schönberner,D. & Walker,H.J., 1981. *Mon.Not.R.astr.Soc.*, 197,81, (Paper I).
Hunger.K., 1975. *Problems in Stellar Atmospheres and Envelopes*, p. 57, eds Baschek,B., Kegel,W.H. & Traving,G., Springer-Verlag, Berlin.
Jeffery,C.S., 1984. *Mon.Not.R.astr.Soc.*, 210,731.
Jeffery,C.S., Heber,U. & Hill,P.W., 1986. *IAU Colloquium 87*.
Jeffery,C.S. & Hill,P.W., 1986. *Mon.Not.R.astr.Soc.*, submitted.
Kilkenny,D., Hill,P.W. & Penfold,J.E., 1981. *Mon.Not.R.astr.Soc.*, 194,429.
Kilkenny,D. & Lynas-Gray,A.E., 1982. *Mon.Not.R.astr.Soc.*, 198,873.
Kilkenny,D. & Lynas-Gray,A.E., 1984. *Mon.Not.R.astr.Soc.*, 208,673.
Landolt,A.U., 1975. *Astrophys.J.*, 196,789.
Lynas-Gray,A.E. & Kilkenny,D., 1986. *IAU Colloquium 87*.
Lynas-Gray,A.E., Schönberner,D., Hill,P.W. & Heber,U., 1984. *Mon.Not.R.astr.Soc.*, 209,387, (Paper II).
Parsons,S.B., 1972. *Astrophys.J.*, 174,57.
Saio,H. & Wheeler,J.C., 1985. *Astrophys.J.*, 295,38.
Schönberner,D., 1975. *Astron.Astrophys.*, 44,383.
Willson,L.A., 1976. *Astrophys.J.*, 205,172.
Wood,P.R., 1976. *Mon.Not.R.astr.Soc.*, 174,531.

DISCUSSION

SCHÖNBERNER: What is the range of the variations in effective gravity?
HILL: 3.6 to about 4.2; 3.7 would be roughly what we had before. This of course is only the variation, we do not have the absolute limit at that.
LYNAS-GRAY: Is the old radial velocity curve the one published by Lynas-Gray et al. (1984)?
HILL: No. The "old" curve includes the 1979 and 1980 velocities from Lynas-Gray et al. (1984) along with additional velocities obtained in 1982.
LIEBERT: Can you put useful limits on a binary companion at a separation close enough to affect the period change determination?
HILL: Yes, we looked at the mean velocities to see if there is anything obvious that could be due to a close binary. There seems to be no evidence for large changes in the systemic velocity which could be due to a binary.
JEFFERY: I think we can put an upper limit of $\pm$ 1 kms^{-1} to the velocity variations.

THE LIGHT CURVE OF THE PULSATING EXTREME HELIUM STAR BD +13°3224: FURTHER EVIDENCE OF A DECLINE IN THE PERIOD DECREASE RATE

A.E. Lynas-Gray[1] & D. Kilkenny[2]

(1) Department of Physics and Astronomy, University College London, Gower Street, London WC1E 6BT, England.

(2) South African Astronomical Observatory, P.O. Box 9, Observatory, 7935 Cape Town, South Africa.

ABSTRACT. Further optical and infrared (J-filter) observations are presented for the extreme helium star BD +13°3224. No infrared excess, which could have been attributed to a cool companion, is found. Very recent times of maximum light confirm the ephemeris cubic term already proposed. Current models for the evolution of BD +13°3224, while approximately accounting for the period decrease rate ($\dot{P}$), do not explain the decreasing $\dot{P}$ indicated by the cubic term.

1. INTRODUCTION

The hydrogen-deficient hot star BD +13°3224 (V652 Her, FB 168) was discovered to be a short-period variable by Landolt (1975), who derived a period of 0.107995 ± 0.000004 days from the observation of 10 maxima distributed over more than 200 cycles. Radial velocity measurements and spectral analyses by Hill et al. (1981) and Lynas-Gray et al. (1984), show BD +13°3224 to be a radial pulsator with mean values of $\log L/L_\odot$ = 3.03 ± 0.12, $R/R_\odot$ = 1.98 ± 0.21, T_{eff} = 23450 ± 1320°K, $\log g$ = 3.7 ± 0.2, $M/M_\odot$ = $0.7^{+0.4}_{-0.3}$ and a hydrogen abundance (by numbers) of $n(H)/(n(H)+n(He)) \simeq 0.01$.

2. EPHEMERIS

Kilkenny & Lynas-Gray (1982, hereafter KLG1) discovered the pulsation period of BD +13°3224 to be decreasing at a rate of about 46 x 10^{-10} days/cycle. The $\dot{P}$ is consistent with evolution towards higher effective temperatures and smaller radii, as suggested by Schönberner's (1977) models for a post-giant helium star evolving towards a DB white dwarf. Schönberner's models are not applicable to BD +13°3224, which has non-zero hydrogen abundance, but Kilkenny

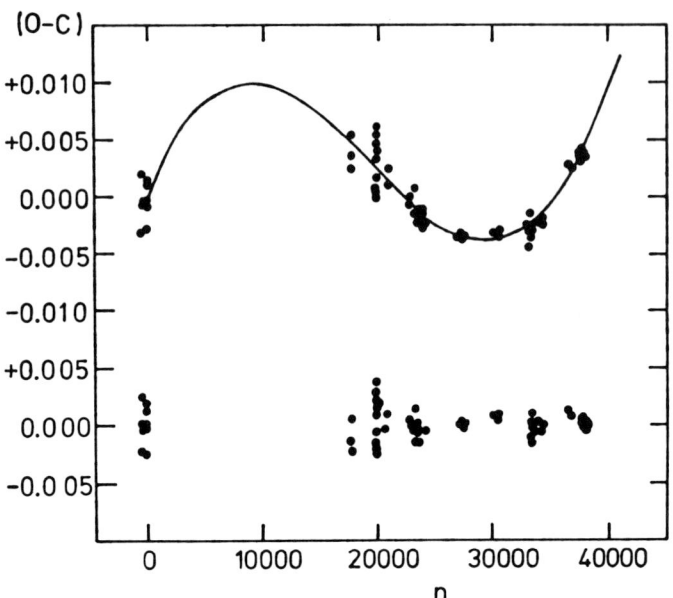

Figure 1. Observed minus calculated residuals for 62 maxima of BD +13°3224. The lower diagram is for the cubic ephemeris given in Section 2; the upper diagram is for the best-fit quadratic with the same 'n' values as the cubic. The solid line is the difference (cubic minus quadratic) between the two solutions.

(1982) discovered the R Coronae Borealis variable RY Sgr to have a $\dot{P}$ entirely consistent with Schonberner's models.

Subsequently, Kilkenny & Lynas-Gray (1984, hereafter KLG2) noted that the KLG1 ephemeris needed revision to fit observations of maxima during 1982-83. In particular, it was suggested that a quadratic ephemeris (ie. constant $\dot{P}$) was insufficient and that a cubic ($\dot{P}$ decreasing) equation was a better representation of the observations. A binary hypothesis was also proposed by KLG2. As Jeffery (1984) is able to explain the quadratic (but not the cubic) term in the KLG2 ephemeris with evolution sequences for horizontal branch stars having very low envelope hydrogen abundance (X < 0.01 by numbers), we obtained a further 18 timings of maxima of BD +13°3224 in 1984 and 1985. Using the KLG2 ephemeris to determine cycle numbers (n) for new maxima, and adopting KLG2 nomenclature where

$$T_{max} = T_0 + nP_0 + n^2 k_1 + n^3 k_2$$

we find from 62 maxima, including 10 from Landolt (1975):

THE LIGHT CURVE OF THE PULSATING EXTREME HELIUM STAR BD +13°3224

$$T_o = 2442216.80405 \pm 0.00032 \text{ day}$$
$$P_o = 0.10799295 \pm 0.00000014 \text{ day}$$
$$k_1 = (-44.711 \pm 0.095) \times 10^{-10} \text{ day}$$
$$k_2 = (+3.23 \pm 0.16) \times 10^{-15} \text{ day}$$

The standard deviations are considerably improved over KLG2, although the ephemeris parameters are essentially unchanged. A simple quadratic fit is much less satisfactory (see Fig. 1).

It is interesting to note that if the cubic term is correct and if it is a result of evolutionary effects, then BD +13°3224 must be in a phase of very rapid evolution because the k_2 term will dominate the k_1 term in about 7×10^5 cycles ($\simeq 200$ years). At this stage, the period will start increasing again, presumably as a result of the star reversing its direction of evolution. That BD +13°3224 is in a rapid phase of evolution is supported by the fact that it appears to be unique insofar as its hydrogen abundance ($\simeq 1\%$) is between the extreme and intermediate helium stars (with $n(H)/(n(H)+n(He)) < 10^{-3}$ and > 0.06 respectively).

Again, if the cubic term is the correct interpretation, the cubic and quadratic solutions, which presently differ in predicted T_{max} by less than $0^d.01$ (0.1 cycles) will within about 4 years differ by one complete cycle and within 12 years the difference will be about a day ($\simeq 10$ cycles). Hence the risk of aliasing will increase rapidly with time and so it is important to monitor BD +13°3224 more frequently in future if we wish redetermine ephemeris parameters and thus provide constraints for the evolutionary models.

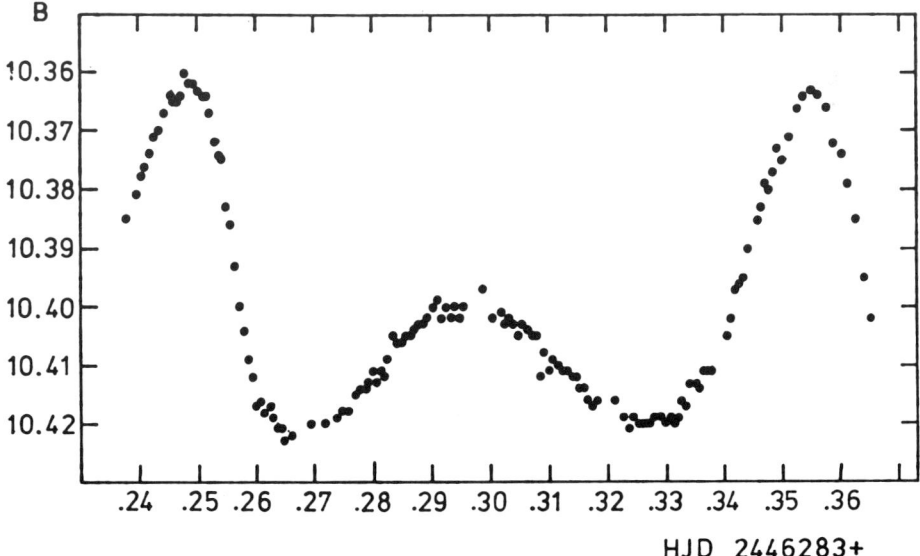

Figure 2. B-Light Curve obtained on 1985 August 5th/6th.

3. THE B-LIGHT CURVE

Previous observations (Kilkenny 1983) have suggested that small amplitude variations (< $0\overset{m}{.}01$) might be superimposed on the mean light curve of BD +13°3224. In our latest observations (1985 August) obtained with facilities used by KLG2, we made an almost continuous series of 60-s integrations over 3 hours using a Johnson B-filter. Only occasionally was the sky or comparison star (HD 151862, HR 6246) measured. The resultant light curve, the best we have, is shown in Fig. 2 and it can be seen that there is no evidence for any superimposed fluctuations larger than about $0\overset{m}{.}002$.

4. THE J-LIGHT CURVE

Johnson J-filter photometry of BD +13°3224 was obtained in 1985 July/August using the Mk II Infrared Photometer, see Glass (1984) for a brief description, attached to the 0.75-m telescope of the South African Astronomical Observatory (SAAO). The observing procedure was similar to that adopted for optical photometry, HD 151862 being used as a comparison star. Integration times were fixed at 10-s with repeat integrations being used as necessary to reach a precision of $\Delta J \simeq \pm 0\overset{m}{.}05$ for each observation of BD +13°3224. Comparison star observations are presented in Table I, where 'N' is the number of observations in a night.

After differential correction, J-filter observations of BD +13°3224 are presented in Fig. 3. Phases are computed using the ephemeris presented in this paper. It can be seen that there is little evidence of any variation. The solid line in Fig. 3 is the result of convolving J-filter observations with a Gaussian having FWHM = 0.07 cycles, the mean time taken to obtain an observation. From 166 observations, we obtain a mean value of J = 10.963 ± 0.039 (sd).

Lynas-Gray et al. (1984) adopt a static plane parallel model atmosphere having T_{eff} = 25000°K, log g = 3.5 and n(H)/(n(H)+n(He)) = 0.01 as representing the atmosphere of BD +13°3224 at a reference phase of ϕ = 0.856; the corresponding angular radius and interstellar reddening were found to be (2.97 ± 0.10) x 10^{-11} radians and E(B-V) = 0.07 respectively. The predicted model atmosphere physical flux at the stellar surface is 1.682 x 10^7 ergs/cm^2/sec/Å at 1.25µ; applying Howarth's (1983) galactic extinction law (for optical and infrared spectral regions) and Wamstecker's (1981) calibration for J-filter photometry, this leads to a prediction of J = 10.89 ± 0.07 for BD +13°3224. It is therefore clear that BD +13°3224 has no infrared excess larger than $\Delta J \simeq 0\overset{m}{.}05$.

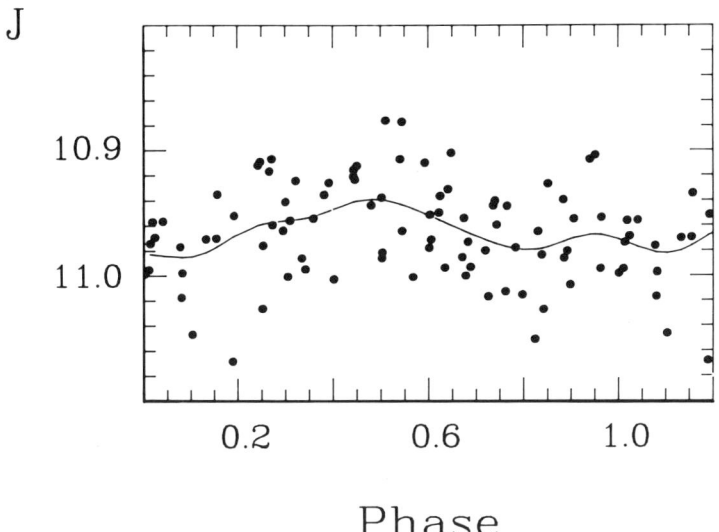

Figure 3. Phased J-Filter observations of BD +13°3224 (see text).

Table I

J-Filter Photometry of Comparison Star HD 151862

HJD	Mean J	N
2446278	5.825 ± 0.009	6
2446279	5.864 ± 0.009	5
2446281	5.863 ± 0.018	6
2446282	5.837 ± 0.006	7
2446283	5.799 ± 0.023	7
2446288	5.869 ± 0.004	7
Mean	5.853 ± 0.023	

5. CONCLUDING REMARKS

From temperature and radius curves derived by Lynas-Gray et al. (1984), and by analogy with the Johnson-V and Strömgren-u light curves given by KLG1, the J-light curve would be expected to have a more pronounced secondary maximum than the V curve. Higher precision infrared photometry is clearly needed to establish the JHKL light curves for comparison with predictions made using a series of plane-parallel static model atmospheres by Lynas-Gray et al. (1984), and as a further search for an infrared excess which might indicate the presence of a cool companion needed to explain departures from the quadratic ephemeris. Times of maximum optical light will continue to be obtained on a regular basis, with the view to continued monitoring of the evolution.

Acknowledgements

Telescope time and travel grants were awarded by the United Kingdom Science and Engineering Research Council (SERC) and the South African Council for Scientific and Industrial Research. Presentation of this paper at the 87th Colloquium of the International Astronomical Union (IAU) was made possible by travel grants received from the IAU and University College London. AELG is supported by the SERC Collaborative Computational Project No. 7.

References

Glass, I.S. 1984. Mon. Not. R. astr. Soc. 211, 461.
Hill, P.W., Kilkenny, D., Schönberner, D. & Walker, H.J. 1981. Mon. Not. R. astr. Soc. 197, 81.
Howarth, I.D. 1983. Mon. Not. R. astr. Soc. 203, 301.
Kilkenny, D. 1982. Mon. Not. R. astr. Soc. 200, 1019.
Kilkenny, D. 1983. SAAO Circulars 7, 70.
Kilkenny, D. & Lynas-Gray, A.E. 1982. Mon. Not. R. astr. Soc. 198, 873 (KLG1).
Kilkenny, D. & Lynas-Gray, A.E. 1984. Mon. Not. R. astr. Soc. 208, 673 (KLG2).
Jeffery, C.S. 1984. Mon. Not. R. astr. Soc. 210, 731.
Landolt, A.U. 1975. Astrophys. J. 196, 789.
Lynas-Gray, A.E., Schönberner, D., Hill, P.W. & Heber, U. 1984. Mon. Not. R. astr. Soc. 209, 387.
Schönberner, D. 1977. Astron. Astrophys. 57, 437.
Wamstecker, W. 1981. Astron. Astrophys. 97, 329.

DISCUSSION

HILL: Calculations based on black body energy distributions show that at wavelengths longer than about 10 microns the flux from BD+13°3224 will be in the Rayleigh-Jeans region and simply proportional to temperature. Magnitude changes arising from temperature and radius changes are about the same. As temperature and radius vary in antiphase these magnitude changes will tend to cancel each other out so that nowhere in the spectrum will the radius variation dominate the light curve. Towards shorter wavelengths the temperature effect gradually begins to dominate, but any infrared photomerty would need to be very precise to measure details of the light curve.
LYNAS-GRAY: Yes, at 12 microns, I think the situation is considerably improved over B.
HILL: Around J it might be fairly flat.
LYNAS-GRAY: This is right. But further and better observations are needed.

IV COOL HYDROGEN DEFICIENT STARS

THE CHEMICAL COMPOSITION OF COOL STARS: II - THE HYDROGEN DEFICIENT STARS

David L. Lambert
Department of Astronomy
The University of Texas
Austin, TX 78712 USA

ABSTRACT

The chemical composition of the R Coronae Borealis and cool hydrogen deficient carbon stars is reviewed. Similarities and differences between these stars and the hot He stars are noted. Proposed origins for the hydrogen deficient stars are sketched. Recent claims that normal (spectral type N) cool carbon stars are hydrogen deficient are shown to be unfounded. Attention is drawn to the curious case of pop. II variables (RV Tauri, W Virginis, and RR Lyrae stars) whose atmospheres show striking deficiencies of heavy elements and may be hydrogen deficient.

1. Introduction

In a Universe in which hydrogen is so supremely abundant, the few stars with atmospheres deficient in hydrogen deserve careful study. Their compositions should offer clues to the events leading to the birth of a hydrogen deficient star. In this brief review, I discuss the cool hydrogen deficient giant and supergiant (Hd) stars. The R Coronae Borealis (RCrB) stars define the high temperature limit of my sample.

Investigators of Hd stars hope to obtain clues to the origin of the H deficiency and to the identity of the stars' progenitors and descendants. Processes capable of transforming a star with an atmosphere of normal composition to a Hd star include severe mass loss via a stellar wind or an explosion, Roche-lobe transfer of mass across a binary system, and deep prolonged mixing leading to hydrogen burning at the base of the convective envelope. Throughout this review, I ignore the possibility that diffusion may create a hydrogen deficient atmosphere around a star with a normal (integrated) composition; my definition of the adjective 'cool' does not embrace the chemically peculiar stars of the upper main sequence.

The review begins with a full discussion of the two well-known classes of Hd stars - the RCrBs and the hydrogen-deficient carbon (HdC) stars. Then, I comment on claims that the atmosphere of the typical (spectral type N) cool carbon star is also significantly deficient in hydrogen; I show that there is no compelling evidence for these claims. I conclude the review by highlighting several varieties of peculiar red giants which may be hydrogen deficient.

TABLE I Abundance Analyses of RCrB Stars

Star	Ref[a]	Method[b]	Major Results
R Cr B	S61	CG rel. to δ CMa (F0Ia)	log C/H = 1.1, [C/Fe] = 1.4[c], Mg to Sm (13 el) rel. to Fe ~ solar.
	S75	FA: LTE models, W_λ's from S61	See Table 2 log C/H = 2.2, log C/He = -1.5 ± 0.5, and [C/Fe] = 1.7. Near-solar metallicity (Mg, Ti, Fe).
	HSS	FA: LTE models, new spectra	Li up 0.5 ± 0.4 dex on cosmic. Ba (i.e., s-process) up 1.2 ± 0.3 dex
	CL	FA: LTE (Schön.) models, Reticon spectra	See Table II. log C/H = 2.5, -2.5 < log C/He < -1.5 (= -2.4?), [C/Fe] = 1.5. $^{12}C/^{13}C > 40$. Al to Nd (12 el.) rel. to Fe ~ solar. Na overabundant? Li up 0.7 dex on cosmic, $^7Li/^6Li > 7$.
XX Cam	OR74	CG	[C/Fe] = 2.0 rel. to Sun. Ca to Zr (7 el.) rel. to Fe ~ solar.
	S75	As for RCrB with W_λ's from OR74	See Table II. log C/H > 3.5, log C/He = -1.5 ± 0.5, and [C/Fe] = 1.3. Near-solar metallicity (Ti, Fe).
	HSS	As for RCrB	Li not present: -1.1 dex or more sub-cosmic. Ba (i.e., s-process) up +0.7 ± 0.2 dex.
	CL	As for RCrB	See Table II. log C/H > 5.1, -2.7 < log C/He > -2.0 (= -2.5?), [C/Fe] = 1.2. Al to Nd (12 el.) rel. to Fe ~ solar. Na overabundant? Li absent: -0.6 dex or more subcosmic.
RY Sgr	D65	CG rel. to β Aqr (G0Ib)	log C/H = 1.4, [C/Fe] = 1.5[d] $^{12}C/^{13}C > 50$. Li up 0.8 dex on cosmic. Mg to Sm. (19 el.) rel. to Fe ~ solar. Na overabundant?
	S75	As for RCrB: W_λ's from D65	See Table II. log C/H = 2.6. log C/He = -1.5 ± 0.5, and [C/Fe] = 1.2. Near-solar metallicity (Mg, Ti, Fe)
U Aqr	BLN	CG rel. to HdC HD182040	Remarkable s-process pattern [X/Fe] = 1.7 (Sr), > 2.4 (Y) but 0.1 (Ba), also -0.1 (Ca).
	M85	As above	[X/Fe] = 1.4 (Sr), 1.2 (Y), but 0.4 (Zr) and -0.1 (Ba), also 0.0 (Ca).
UV Cas	OR81a	CG rel. to RCrB	[C/Fe] = +1.2 Sc, Ti, and Ba rel. Fe ~ solar.
SU Tau	OR81b	As for UV Cas	[C/Fe] = +0.3 Sc, Ti and Cr rel. Fe ~ solar.

Notes to Table I

    [a] BLN = Bond, Luck, and Newman (1979).
    CL = Cottrell and Lambert (1982a).
    D65 = Danziger (1965).
    HSS = Hunger, Schönberner, and Steenbock (1982).
    M85 = Malaney (1985)
    OR74, OR81a, OR81b = Orlov and Rodriguez (1974) (1981a) (1981b)
    S75 = Schönberner (1975)
    S61 = Searle (1961)
   [b] CG = Curve of Growth
    FA = Fine Analysis
   [c] Rel. to δ CMa. [C/Fe] = 0.9 rel. to Sun since [C/Fe] = -0.5 for δ CMa (Luck and Lambert 1981).
   [d] Rel. to β Aqr. [C/Fe] = 0.8 rel. to Sun since [C/Fe] = -0.7 for β Aqr (Luck and Lambert 1981).

Notation: [X] = log X_{star} - log $X_{standard}$

2. The R Coronae Borealis Stars

2.1 Introduction

In 1985, we celebrate the fiftieth anniversary of the publication of Berman's (1935) pioneering analysis of RCrB. Modern quantitative spectroscopy of RCrBs began with Searle's (1961) differential curve of growth analysis of RCrB. The principal characteristics of the chemical composition of RCrBs are summarized in Table I. This table and the ensuing discussion draws upon no more than a dozen papers - the total production over fifty years by a handful of spectroscopists.

Here, I comment on the following abundance ratios and the associated questions of nucleosynthesis:

- C/He	How is this to be determined when He may be undetectable in the spectrum?
- C/H	Is a trace of H always present?
- Li	Is Li present in RCrBs?
- CNO	How are these key ratios determined?
- Na to Ni	Is the ratio of odd and even (α) light elements to the Fe-group representative of young or old disk stars?
- Heavy (>Fe) elements	Are s-process elements (e.g. Ba) overabundant (rel. to Fe)?

2.2 The C/He Ratio

The atmosphere of a RCrB star is severely H deficient (Bidelman 1953). A plausible assumption is that the atmosphere is now dominated by nuclear processed material in which the initial H was converted to He. The presence of an extraordinary number of C I lines in the spectra would seem to imply that some of the synthesized He may have been burnt to C and, perhaps, O and other light α-rich elements. Since there is no *a priori* argument by which to determine the C/He ratio, a spectroscopic derivation is necessary.

Beginning students of stellar spectroscopy quickly appreciate that the abundance ratio X/H is obtainable from a line of element X in the spectrum of a normal star without recourse to a measurement of a hydrogen line. Of course, the continuum is that "line" because hydrogen controls the continuous opacity. What is the corresponding opacity in

RCrBs? Searle's (1961) tentative identification of the photoionization of neutral carbon was confirmed by Schönberner (1975) who constructed model atmospheres. Helium is a minor source of opacity unless the C/He is very small. Furthermore, photospheric He may not contribute absorption lines to the spectra of these rather cool stars. In short, He may be undetectable, but yet the most abundant element. With C as the major source of continuous opacity (κ_c), it is the abundance ratio X/C that is provided by analysis of lines of element X. Unless photospheric He I lines are identifiable, one is restricted to setting upper and lower limits on the C/He ratio (S75).

The plausible assumption that the mass fraction of intermediate elements ($Z \geq 16$) has not been changed by nuclear burning is employed to set an upper limit on the C/He ratio. Line analysis gives the ratio $r_X(CrB) = f_X/f_C$, where f denotes a mass fraction and X is an element heavier than oxygen. Since the RCrB stars do not belong to the extreme Pop. I, we may suppose that their initial metal content did not exceed the solar value. Their Galactic distribution - a marked concentration towards the Galactic center - suggests an old-disk population. Then

$$f_X(RCrB) = r_X(RCrB)f_C(RCrB) \leq f_X(\odot),$$

or

$$f_C(RCrB) \leq f_X(\odot)/r_X(RCrB).$$

In this way, Schönberner (1975) set a limit C/He $\lesssim$ 0.10. Later, Cottrell and Lambert (1982a, here CL), who had access to high S/N Reticon spectra, gave lower limits: C/He $\leq$ 0.03(RCrB) and $\leq$0.01(XX Cam) where $f_{He} + f_C + f_N + f_O \approx 1$. The difference in part reflects the lower continuous opacity employed by CL. However, the major factor is their use of much weaker lines, which are less dependent upon the microturbulence.

A lower limit to C/He is provided by the observed equivalent widths of C I lines of known oscillator strength. If C/He exceeds some minimum value, then C is the dominant contributor to κ_c and the equivalent width of a C I or C II line is then effectively independent of C/He. The equivalent width is, of course, affected by the effective temperature, the surface gravity, and microturbulence. If C/He is smaller than a certain value, the contribution of C to κ_c is negligible and the strength of a C line is dependent on the C/He ratio. Observed equivalent widths ($W_\lambda s$) of weak C lines then demand a minimum value of C/He. Schönberner set C/He $\geq$ 0.01 in RCrB, XX Cam, and RY Sgr from observations of strong C I lines. With access to weak C I lines, CL obtained lower limits: C/He $\geq$ 0.003 (RCrB) and 0.002 (XX Cam). These limits are almost independent of the adopted microturbulent velocity and the atmospheric parameters.

The C/He ratio also affects the atmospheric structure. Hunger, Schönberner, and Steenbock (1982, here HSS) note that the residual line intensity in the cores of strong lines is sensitive to the C/He ratio through its influence on the atmospheric structure. A fit to the Li I 6707 Å doublet in RCrB gave log C/He = -2.0 (HSS). Earlier, Schönberner had adopted log C/He = -1.5 ± 0.5. CL's limits were -2.5 ≤ log C/He < -1.5 for RCrB.

Keenan and Greenstein (1963) reported the presence of the He I line 5876 Å in the spectrum RCrB. CL confirmed this identification and showed that the line is considerably stronger in XX Cam and RY Sgr. If this is a photospheric line and formed in LTE, it gives a direct estimate of the C/He ratio. This abundance would seem to be very dependent on the adopted effective temperature, but this dependence is reduced considerably by analyzing the He I lines relative to the C II lines. The resulting C/He ratios of 0.004(RCrB) and 0.003(XX Cam) given by CL are close to their above lower limits. Unfortunately, the chromosphere may contribute to the 5876 Å triplet. Querci and

Querci (1978) reported the He I 10830 Å line to have a P Cygni profile in a spectrum of RCrB taken 1 mag below maximum light. They attributed the line to the circumstellar shell. In contrast, Zirin (1982) reports the He I 10830 Å line to be absent ($W_\lambda \lesssim 70$ mÅ) in his higher dispersion spectra taken in 1978 July when RCrB was at maximum light. The fact that Zirin's upper limit for the W_λ (10830 Å) is inconsistent with the Querci and Querci (1978) measurement and CL's W_λ (5876 Å) may indicate a variable chromospheric structure. The chromospheric (i.e., NLTE) contribution should be greatly reduced for the singlet He I lines. Unfortunately, the 6678 Å line, which is the most promising candidate for detection, is severely blended. CL's rough estimates of its equivalent width correspond to C/He ratios only slightly larger than the above lower limits. Further observations of the He I lines in RCrB stars may uncover a variable chromospheric component of the triplet (D_3 and 10830 Å) lines. A NLTE study of photospheric helium should be attempted. These lines may yet provide a direct measure of the C/He ratio.

2.3 The C/H Ratio

This ratio is provided by analysis of the Balmer lines. Figure 1 from CL shows the Hα region in XX Cam, RCrB, RY Sgr, and the normal F8 Ia star δ CMa. Hα is undetectable in XX Cam; the line at 6562.1 Å, which may be taken for Hα on lower quality spectra, was identified by HSS82 as due to Ti I. The Ti I line is probably responsible for the apparent blueshift of Hα in RCrB and RY Sgr. Results for the C/H ratio (see Table 1) are available only for the three brightest RCrBs.

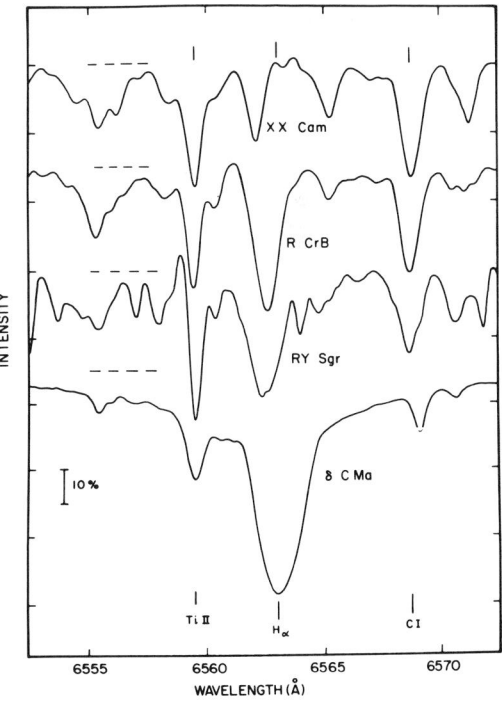

Fig. 1 - Spectra of three RCrBs near Hα. The star δ CMa is a normal F8 supergiant. Continuum levels are denoted by the short dashed lines. Note the great strength of the CI line and the weakness of Hα in the RCrBs.

Published C/H ratios are entered in Table II where abundances are normalized such that $\sum \mu_i n_i = 12.15$ where μ_i is the mean atomic weight of element i having a number density n_i. The numerical value is that for a solar mixture with H/He = 10 and log n_H = 12.0. Table II shows that H in RCrBs is between 4 to more than 8 orders of magnitude less abundant than in normal stars. It is a trace element with a variable abundance from star to star.

2.4   Lithium

Keenan and Greenstein (1963) tentatively identified the Li I 6707 Å resonance doublet in RCrB - an identification confirmed by HSS and CL (Figure 2). Lithium is not present in XX Cam or RY Sgr. The 6707 Å doublet ion RCrB is sufficiently strong to permit an estimate of the ^{7}Li/^{6}Li ratio; the wavelength shift between ^{7}Li and ^{6}Li is 0.12 Å. CL provided a limit ^{7}Li/^{6}Li > 7 which suffices to exclude the possibility that the lithium in RCrB was produced by spallation reactions on the stellar surface, as suggested by HSS following Canal, Isern, and Sanahuja (1977): ^{6}Li/^{7}Li ~ 4 is expected from spallation. The isotopic ratio is consistent with ^{7}Li-production by the ^{7}Be-transport mechanism: ^{3}He(^{4}He,γ)^{7}Be(e$^-$,ν)^{7}Li (Cameron and Fowler 1971). Since there is no comparable mechanism for ^{6}Li production, ^{7}Li/^{6}Li ~ ∞ is predicted.

The observed ratio Li/H ~ 10^{-4} for RCrB excludes the possibility that the H and Li are tracers of material which escaped nuclear processing in the star or which was accreted at a late stage. Since the cosmic ratio, Li/H ~ 10^{-9}, is less than the observed value and normal burning of H necessarily destroys Li, production of Li is required to account for a

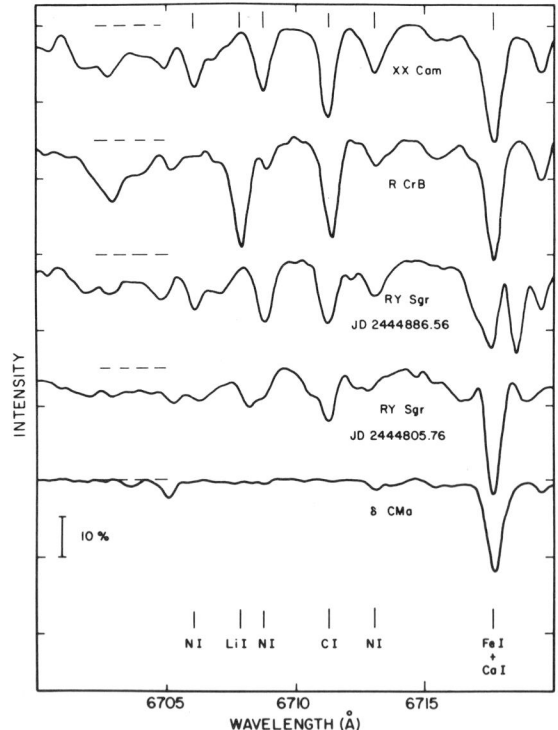

Fig. 2 - Spectra of three RCrBs and δ CMa near 6710 Å. The Li I 6707 Å is present only in RCrB. The two spectra of RY Sgr show that line doubling at certain phases (the lines are split on JD2444886) may compromise an abundance analysis.

ratio Li/H ~ 10^{-4}. Since the ^{3}He seed nuclei are produced in fairly copious amounts in the outer envelopes of main sequence stars, the ^{7}Be-transport mechanism is a likely source of ^{7}Li.

2.5 Carbon, Nitrogen, and Oxygen

Since CNO along with H and He are expected to be the elements primarily, perhaps exclusively, affected by H and He burning, whose products surely dominate the RCrB envelope, the relative abundances of CNO are clues to the nuclear history of the stars. Both molecular and atomic lines are available for a CNO analysis. To date, a thorough study of the molecular (C_2, CN) lines is unavailable; Searle (1961) and Danziger (1965) gave preliminary accounts now needing modernization. Searle obtained neither N nor O from atomic lines. Danziger in a differential analysis of RY Sgr relative to β Aqr (G0 Ib) gave [C/Fe], [N/Fe], and [O/Fe] from C I, N I, and O I lines. Five O I lines gave [O/Fe] = -0.6 with the faint O I lines in β Aqr providing a major uncertainty. For nitrogen, [N/Fe] = 0.0, his analysis is based apparently on one N I line referred to the solar spectrum. However, the chosen N I line (λ6484.88) is <u>not</u> present in the solar spectrum. Schönberner's (1975) model atmosphere analysis of RY Sgr includes a discussion of Danziger's C I and O I lines, but not the N I line. These results are given in Table II.

TABLE II Light Elements in RCrB Stars

Star	Ref.	C/He	H	He	C	N	O	Fe
					log ε[a]			
RCrB	S75	0.03	7.8	11.5	10.0	---	---	7.1
	CL	0.004	6.6	11.5	9.1	8.1	8.8	6.5
		0.03	7.5	11.5	10.0	8.9	9.7	7.4
XX Cam	S75	0.03	<6.5	11.5	10.0	---	---	7.5
	CL	0.003	<3.9	11.5	9.0	8.5	8.9	6.6
		0.01	<4.4	11.5	9.5	9.0	9.4	7.1
RY Sgr	S75	0.03	7.4	11.5	10.0	8.8[b]	8.5[c]	8.3
Hot He Stars								
BD+10° 2179	H83	0.01	8.5	11.5	9.5	8.1	8.1	6.5
HD124448	H83	0.01	<7.5	11.5	9.5	8.8	8.5	7.4
Sun	G84	0.005	12.0	11.0	8.7	8.0	8.9	7.7

Notes to Table II
  G84 = Grevesse (1984), H83 = Heber (1983). Other references listed at end of Table I

  [a] Normalized to $\Sigma\mu_i n_i = 12.15$

  [b] from Danziger's (1965) estimate that [C/Fe] = +1.5 and [N/Fe] = 0.0 for RY Sgr relative to β Aqr, Luck and Lambert's (1981) CNO abundances for β Aqr, and a normalization to $\log \varepsilon_C = 10.0$. This same procedure gives $\log \varepsilon_O = 8.3$.

  [c] See text for comment on the O abundance.

For RCrB and XX Cam, CL measured many C I (and a few C II), N I, and O I lines including the low excitation forbidden lines of C I and O I. Abundances obtained using Schönberner's model atmospheres are summarized in Table II. The relative abundances are markedly non-solar:

	C	N	O
RCrB	10	1	5
XX Cam	10	3	5
Sun	10	2	18

For RY Sgr, Schönberner obtained a lower O/C ratio (O/C = 0.03) from Danziger's equivalent widths. CL's measurements of a selection of C I, N I, and O I lines in RY Sgr indicate that its O/C ratio is comparable to that in the other two stars. Additional studies of CNO would be of interest.

The presence of nitrogen suggests that the atmosphere of a RCrB star contains N-rich material from layers exposed to the H-burning CNO cycles. The low N/O ratio serves to exclude the possibility that the nitrogen is a tracer of primordial (unprocessed) material. Of course, the dominant fraction of the atmospheric carbon must be a product of He-burning, a conclusion supported by the absence of ^{13}C: ^{12}C/^{13}C > 40 for RCrB (CL), > 50 for RY Sgr (D65). An apparently higher limit may be guessed for U Aqr, the peculiar RCrB star (see below), whose spectrum contains strong ^{12}C$_2$ bands and no detectable ^{12}C/^{13}C bandheads (see BLN). A rough limit, ^{12}C/^{13}C > 100, may be similarly given for SV Sge (Bidelman 1953).

During the progressive heating of a stellar core prior to He-burning the ^{14}N synthesized from ^{12}C and ^{16}O by the CNO cycles in H-burning is converted to ^{18}O whose abundance is then equal to the sum of the initial CNO abundances. Since some ^{18}O nuclei may survive He-burning and the ensuing mixing, a detection of this isotope could set constraints on models of the birth of RCrBs and related stars. Warner (1967) drew attention to the possible presence of ^{18}O ("nearly all of the O in the Hd C stars is predicted to be in the form of ^{18}O") and suggested that the infrared vibration-rotation bands of CO would provide the ^{16}O/^{18}O ratio. To date, these bands have not been detected in RCrB stars: high resolution spectra at 2.3 μm of RCrB near maximum and near minimum show a continuous spectrum with no evidence of the CO bands (K. H. Hinkle and D. L. Lambert, unpublished observations from the KPNO 4m FTS). The cooler Hd C star HD182040 shows CO in absorption with ^{12}C^{16}O not ^{12}C^{18}O as the most abundant species; the ^{16}O/^{18}O ratio is probably rather high.

2.6 Sodium to Nickel

With the persistent exception of sodium, the elements in this group have been shown repeatedly to have solar-like abundance ratios - see Table I. This is not surprising because unevolved (and evolved) disk and halo stars show only minor differences in abundance ratios for this group. For example, abundances of the light α-nuclei (e.g. Mg, Si, Ca) increase to [α/Fe] ~ +0.5 for the halo ([Fe/H] < -1) with a smooth transition to this value occurring between -1.0 < [Fe/H] < -0.3 (see Tomkin, Lambert, and Balachandran 1985, Nissen, Edvardsson, and Gustafsson 1985; François 1985). CL's analysis of RCrB and XX Cam offers a hint ([α/Fe] ~ 0.3) that the α-nuclei have the higher abundance characteristic of the old disk. A thorough reexamination using lines showing minimal or similar sensitivities to the atmospheric structure is recommended because a precise estimate of the [α/Fe] offers a new probe of the RCrB stars' initial metallicity.

The odd-mass elements Na and Al are predicted to be less abundant in metal-poor stars. For RCrB and XX Cam, the Al abundance (CL) matches that seen in normal dwarfs with [Fe/H] > -1.0. The Na overabundance for both stars, [Na/Si] ~ 0.3, is somewhat at odds with results for the normal dwarfs. Since Na overabundances have been reported for a variety of normal supergiants, we presume that the overabundance reflects a systematic error in the abundance analysis (e.g. the appearance of unaccounted for non-LTE effects).

2.7 Beyond the Iron Group

The remarkable star U Aqr apart, the RCrB stars analysed to date show normal abundances (relative to Fe) of the heavy elements; i.e. [s/Fe] ~ 0.0 because the selection of elements is dominated by those whose normal origin is ascribed to the s-process. A similar claim for the r-process must be regarded as tentative because it is based apparently on a single line of Eu II measured in RY Sgr D65. A claim by HSS of a Ba overabundance in RCrB and XX Cam comparable to that seen in the classical barium stars was disputed by CL who argued that lines chosen by HSS were too strong for a reliable analysis. Weaker lines of Y II and Nd II allowed CL to show that the s-process elements are not overabundant.

Although one can envisage forms of s-processing that produce abundance patterns not yet tested by the abundance analyses, it is clear that the C-rich material in the RCrBs has not come from a cool carbon star in which [s/Fe] ~ 1 is found for elements from Sr to the rare-earths. Perhaps one can not yet exclude non-standard s-processing such as the 'mild' s-process suggested by Holweger and Kovács (1984) to account for Sr and Ba enrichment of certain G and K supergiants.

The cool high velocity star U Aqr is a RCrB star with a most unusual pattern of s-process enhancements. BLN discovered and discussed the pattern; the Sr overabundance is so large that the Sr II 4077 Å line rivals the Ca II H and K lines in strength. Further observations and discussion are given by Malaney (1985). In U Aqr (see Table I), the progression of overabundances in the consecutive elements Sr, Y and Zr is remarkable; Sr and Y are greatly overabundant, but Zr is only slightly enhanced. Furthermore, Ba is not enhanced.

Such a pattern for the overabundances suggests that exposure of the Fe-group seed nuclei to star neutrons has been slight. If the integrated exposure is assumed to follow an exponential distribution $\exp(-\tau/\tau_o)$ where

$$\tau = \int N(n) \, v \, dt$$

and τ_o is an adjustable scale factor and $N(n)$ is the neutron density, Malaney (1985) finds that $\tau_o \sim 0.1$ mb^{-1} fits the observed pattern of abundances. This mild exposure converts few of the Fe-group 'seed' nuclei to heavier elements so that high Sr, Y and normal Fe-group abundances are possible even though the entire atmosphere may have been exposed to neutrons. BLN consider a single exposure rather than an exponential distribution and find that $\tau \sim 0.6$ mb^{-1} fits their observed overabundances. At this exposure, the Fe-group are depleted significantly at the s-process site. Since U Aqr is presumed to have a normal Fe abundance, BLN's choice of a single exposure requires that the present atmosphere be a mixture of neutron-processed and unprocessed material.

Identification of the s-processing event with a location on a stellar evolutionary track is tantamount to speculation. BLN suggest that an exceptionally violet helium core flash in a low-mass star stimulated mixing leading to ignition of a neutron source (presumably ^{13}C $(\alpha,n)^{16}$O), and ejection of the H-rich envelope. Malaney prefers to locate the birth of U Aqr on the asymptotic giant branch (AGB) where He-shell flashes are supposed to be responsible for the s-processing. Continued evolution with mass loss via a stellar wind on the AGB or envelope ejection to form a planetary nebula result in a white dwarf having a thin hydrogen-rich envelope. A final He shell flash is predicted to consume the envelope and to restore the star for a short time to a position close to the AGB with, perhaps, a conposition characteristic of RCrBs in general and U Aqr in particular (Renzini 1979, 1981; Iben 1984). The s-processing charachteristic of U Aqr may occur in this final He shell flash.

To constrain these and future speculations, it will be necessary, as noted by BLN and Malaney, to complete a thorough abundance analysis of U Aqr. Both published studies combine just a few very strong lines and a crude differential analysis relative to the Hd C star HD182040. A thorough search for additional heavy elements could yield new information. The Rb/Sr ratio could indicate the neutron density at the s-process site (Tomkin and Lambert 1983). While technetium may be present if the s-processing occurred in a He-shell flash, it should have decayed if the core flash was responsible for the birth of a He main sequence star which subsequently evolved to the RCrB stage. The Fe-group and lighter elements should be investigated for abundance anomalies arising from mild s-processing. Such an extended analysis should be made of the other high velocity and the fainter low velocity stars.

2.8   Nuclear History and Stellar Evolution

The history of nuclear processing, mixing, and mass loss is recorded in the chemical composition of the atmosphere. I discuss this composition in terms of processing wrought by the CNO cycles and the 3α process. Next, I comment on the similarities and differences in composition between the RCrB's and the hot He stars. Finally, I sketch possible origins for RCrB's.

Qualitatively, the composition of the RCrB stars (Table II) is that of a He and N-rich zone (in which essentially all the H has been processed through the CNO cycle), into which 3α processed material has been mixed. A production scenario for the RCrB variables must account for the paucity of these objects in the Galaxy; i.e., the evolution through this phase must be extremely rapid and/or the RCrB structure is produced in only a small percentage of stars.

The remarkable H deficiency shows that the fraction of pristine material in the atmosphere is very slight. I assume that nuclear-processed material dominates the atmosphere, and that the iron-group nuclei have not been affected by processing. I assume that the abundances relative to iron were initially solar. This initial composition is summarized in Table III with the observed abundances for RCrB and XX Cam.

A stellar core is first affected by H-burning through the p-p chain and the CNO cycles. Prior to the onset of He burning, the CNO cycles have run to equilibrium even though the p-p chain may have been the dominant energy source. A low mass star will reach $T \sim 60 \times 10^6$ K in the core where the equilibrium abundances are $^{12}C/^{14}N \approx 0.03$, $^{16}O/^{14}N \approx 0.005$, and $^{12}C/^{13}C = 3.4$ (Fowler, Caughlan, and Zimmerman 1975). Li is completely destroyed. The composition of the He-rich core prior to He ignition is given in Table III under the heading "CNO"; the key point is that it is N rich and C and O poor.

TABLE III Observed and Predicted Compositions (after CL)

Element	Sun	CNO	Stellar Core[a] He'	He"	RCrB	XX Cam
H	4.5	XXX[b]	XXX	XXX	0.2	<-2.6
He[c]	3.3	4.0	3.9	5.0	5.1(4.2-5.2)	4.8(4.0-5.0)
^{12}C	1.2	0.1	2.6	2.6	2.7	2.4
^{13}C	-0.7	-0.4	-0.5	-1.5	<1.1	...
^{14}N	0.5	1.6	1.5	1.9	1.6	1.9
^{16}O	1.4	-0.7	2.2	2.2	2.3	2.2

Column header spans "Abundance Ratio log (X/Fe)" across CNO, He', He", RCrB, XX Cam.

[a] The predicted composition is shown for the following exposures to nuclear burning: CNO = post H-exhausted core collapse with CNO cycles run to equilibrium for $T \sim 60 \times 10^6$ K. He' = post He-core flash with the 3α process converting 20% of the He to C and O in the proportions 2 to 1. He" = as for He' but the H-exhausted core was originally metal poor (i.e., an old disk star) by a factor of 10 with a modest O overabundance (Clegg, Lambert, and Tomkin 1981). 2% of the He is burnt to C and O.
[b] XXX denotes a very small abundance ratio X/Fe.
[c] For RCrB and XX Cam, upper and lower limits (see § 2.2) are shown in parentheses.

The onset of He burning converts He to ^{12}C, destroys ^{14}N by the $^{14}N(\alpha,\gamma)^{18}F(e^+,\nu)^{18}O$ reactions, and may produce ^{16}O. When ^{16}O production occurs, the ^{18}O will be destroyed by $^{18}O(\alpha,\gamma)^{22}Ne$ and $^{18}O(\alpha,n)^{21}Ne$ reactions. Since the burning may be inhomogeneous through the core, some ^{18}O may survive and be present in H-poor, C-rich atmospheres. The ^{16}O production depends on the core temperature and the duration of He burning. The rate for the crucial $^{12}C(\alpha,\gamma)^{16}O$ reaction is uncertain. Since ^{14}N is readily destroyed in He-burning, the observed high N abundance demands that the atmosphere now contain a mixture of CNO-cycle exposed and 3α-processed material. Table III shows predictions for two cases of partial He burning with the arbitrary stipulation that the C and O be produced in the proportions 2 to 1. The (uncertain) C/He ratio derived from the 5876 Å line would seem to imply that the intial star was moderately metal deficient. The quantitative correspondence between the He' and He" predictions in Table III and the observed abundance is not as significant as the general constraints set by the abundances; the atmospheres contain a large amount of He previously exposed to the CNO cycles with a small mixture of the products of He burning under conditions adequate to produce ^{16}O. Li production is attributed to the ^{7}Be-transport mechanism (see § 2.4).

A nuclear history must also explain why s-processing has not occurred in RCrB and XX Cam. The s-processing in a He core or shell is most probably controlled by the neutron sources ^{22}Ne$(\alpha,n)^{25}$Mg and ^{13}C$(\alpha,n)^{14}$O. The initial ^{22}Ne abundance is low, and an insufficient neutron flux (< 1 per Fe seed nucleus) is supplied. Additional ^{22}Ne nuclei may be created from ^{14}N via ^{18}O and ^{18}O$(\alpha,\gamma)^{22}$Ne. Then, the maximum supply is 30 to 60 neutrons per Fe seed nucleus; the higher figure pertains to metal-poor cores. It is interesting to note that BLN report that an exposure of 30 neutrons per Fe seed nucleus is required to explain the s-process enhancements in U Aqr. Consumption of ^{22}Ne is preventable by limiting the core temperature; if $T \leq 2 \times 10^8$ K, ^{16}O production seems possible without significant neutron release from ^{22}Ne [note that at low temperatures ^{22}Ne$(\alpha,\gamma)^{26}$Mg is favored over ^{22}Ne$(\alpha,n)^{25}$Mg].

The residual abundance of ^{13}C in the He burning core or shell is too low to cause significant s-processing by neutrons released through the ^{13}C$(\alpha,n)^{16}$O reaction at He burning. The neutron supply may be enhanced by mixing in a little H-rich material so that ^{13}C is regenerated and neutrons created from fresh ^{12}C by the chain ^{12}C$(p,e^+)^{13}$C$(\alpha,n)^{16}$O. As long as the number of protons added is small, such mixing will not produce an inadmissible reduction of the ^{12}C/^{13}C ratio. This latter proviso is physically plausible because the large energy release, which must accompany the addition of protons, may be presumed to quench the mixing. For RCrB and XX Cam, I suggest that protons were not mixed into the He-burning region and, hence, s-processing did not occur.

This sketch of the nuclear history of the RCrB atmosphere must be integrated into a full evolutionary history for these unusual stars, which probably consist of a low mass ($M \sim 1 M_\odot$) star with a thin He-rich envelope around a carbon-oxygen core. One may reasonably suppose that the RCrB are related to the hot extreme He stars. Schönberner (1975) notes that evolutionary tracks for pure He stars evolving to the red giant region follow similar paths in the H-R diagram on their ascending and descending branches. With present uncertainties in the stars effective temperatures and surface gravities, one cannot rely on location in the H-R diagram to determine whether the hot He stars are the progenitors or descendants of RCrBs. Heber and Schönberner (1981) claimed that the observed galactic density of RCrB and He B stars is consistent with evolution from supergiant (RCrB) to dwarf (He B star) as modeled by Schönberner (1977). The changing pulsational period of RY Sgr confirms this suggestion (Kilkenny 1982).

Whatever the precise relation between the RCrB and He B stars, one might expect them to have similar surface compositions. However, one should recognize that very few stars (3 RCrBs, 4 He Bs) have been analysed in detail; we may not yet know the composition of the typical RCrB or He B star. In Table II, I include data from Heber (1983) for two He Bs: HD124448 and the metal-poor BD + 10° 2179. For the He B stars, the C/He ratio is directly obtainable from absorption lines. Heber's (1983) four stars have log C/He = -2.0 to -2.2, a value slightly higher than CL's D$_3$-based results for RCrB (-2.4) and XX Cam (-2.5) and less than the compromise (-1.5) adopted in S75. The presence of a metal-poor star (initial [Fe/H] ~ -1) in the sample of four He B stars certainly encourages the view that some RCrBs may also be metal-poor. On the assumption that their C/He ratio is close to CL's derived lower limit, we see that the compositions of RCrB, XX Cam, and BD +10° 2179 are similar except for H and O.

Hydrogen is evidently variable from star-to-star in both classes. Oxygen is apparently rather more abundant in RCrBs than in BD +10° 2179; the earlier and lower estimate for RY Sgr (S75) is in good agreement with that for BD + 10° 2179. If the RCrBs are assumed to have solar metal abundance, their composition matches that of He B stars (HD168476, 124448, and BD -9° 4395) except for oxygen which is about an order of magnitude higher in RCrBs according to CL (but not S75). Clearly, oxygen deserves further study.

What is the origin of RCrB stars? The AGB is an attractive location for the first appearance of He-rich stars. Envelope instabilities (He shell flashes) affecting AGB stars are possibly responsible for mass ejection leading to the formation of planetary nebulae. Härm and Schwarzschild (1975) shows that such instabilities could leave the star with a very thin H envelope. Continued mass loss either as a steady wind or in bursts would produce a He-rich star. In addition, H may be consumed between shell flashes (Paczyński 1971). Thus one could envisage a connection between the RCrB stars and the He and C-rich ejecta of two planetary nebulae, Abell 30 and 78 (Jacoby and Ford 1983).

Ejection of the H-rich envelope through explosion may also occur at the He core flash, as a low-mass (< 2.5 $M_\odot$) star reaches the tip of the first red giant branch. Deupree and Cole (1981) in a hydrodynamical analysis of the He-core flash show that, although a prediction of the nucleosynthesis is beset with considerable uncertainties, a substantial amount of ^4He is processed to ^{12}C, and a fraction is further processed to ^{16}O. They suggest that significant mass loss could occur at the He core flash. The stellar remnant is probably a He star which evolves to the RCrB stage.

The above scenarios presume that RCrBs and He B stars evolve from single stars. Recently, Iben and Tutukov, (1985, here IT) proposed that these He-rich stars are created by the merger of a pair of white dwarfs with the less massive star having a He core with a thin surface layer of H, and the more massive white dwarf having a C-O core. IT argue that the merger results in a star with a surface composition like that of a RCrB: the He white dwarf forms a coating around the C-O dwarf. The N comes from the He white dwarf where the N was synthesized during H burning from the initial CNO. The surface C and O of the RCrB star is presumed to be dredged up during the merger from the underlying C-O dwarf. Abundance variations from star to star are attributable to variations in stellar masses and in the dredge-up of C and O which, in turn, is presumed to be a function of the stellar masses. The immediate product of the merger is expected to be a RCrB star which later evolves to a He B star, as suggested by Heber and Schönberner's (1981) study. Although a massive and short-lived star is needed to create the C-O white dwarf, the much longer time scale for production of a RCrB is set by the orbital decay induced by emission of gravitational waves. IT note that their scenario can account for the observed number of giant He stars. This attractive feature of the IT model sets it apart from the speculations assigning the birth of Hd's to either the AGB or the tip of the He-core flash at the tip of the red giant branch. In such alternative speculations, special and as yet unidentified selection effects must be invoked to ensure that Hd stars are created in the observed numbers as the vast majority of stars evolve through the He-core flash and along the AGB without major changes in surface H content.

2.9 Concluding Comments

My present emphasis on chemical composition should not be misinterpreted as an indication that the photospheric structure of a RCrB star is now devoid of interest for spectroscopists. Their abrupt and unpredictable declines to minimum light remain a puzzle. The key to the puzzle may lie in spectroscopic studies of the photospheric pulsations (Cottrell and Lawson, this conference) which appear as line shifts, asymmetries and even line doubling over narrow phase intervals (Cottrell and Lambert 1982b).

With the advent of large telescopes equipped with efficient spectrometers and a resurgence of theoretical work on the creation and evolution of peculiar stars, abundance analyses are certain to be extended from the brighter RCrBs to the majority of the 40 or so galactic members of the class and to the brighter examples now identified in the Magellanic Clouds. These analyses should be comprehensive in order that the nuclear history can be unravelled; in particular, He, Li, C, N,) and the s-process elements should be included. Similarly thorough studies should be made of the Hd C and He B stars.

3. The Hydrogen-Deficient Carbon (Hd C) Stars

In 1953, Bidelman (1953) coupled the RCrB stars with four non-variable cooler stars to form a class of "apparently hydrogen-deficient stars which appear also to be rich in carbon". I refer to the non-variables as Hd C stars. The prototype may be taken to be HD182040 whose spectral peculiarities were suspected first by Rufus (1915). Today, five stars comprise the class (HD137613, 148839, 173409, 175893, and 182040); HD148839 was identified as a Hd C star by Warner (1967). XX Cam ("The Inactive RCrB Star" - Rao, Ashok, and Kulkarni 1980), which has shown just one minimum in 90

TABLE IV Composition of Hd Cs (after Warner, 1967)

Quantity	Result	Comments
$\dfrac{C}{He}$	$\sim 10^{-2}$	Uncertain - see comments on this ratio for RCrBs.
$\dfrac{C}{H}$	$\gtrsim 10^3$	For HD148839, which Warner proposed as intermediate between normal cool carbon stars and the Hd Cs C/H $\sim$ 0.05.
$\left[\dfrac{C}{Fe}\right]$	0.8 dex	The four extreme Hd Cs have [C/Fe] $\approx$ 0.7 to 1.0 dex. HD148839 has [C/Fe] = 0.5 dex. RCrBs (CL) have [C/Fe] $\sim$ 1.2 to 1.5.
Z	(2 to 0.3)$\odot$	"Most indications are that the Hd C stars are slightly metal deficient, in agreement with their large space velocities."
X/Fe	solar	With the possible exception of Na (see also § 2.6), 16 elements (X) from Ca to Sm have [X/Fe] = 0 to within errors of measurements; i.e., no s-process enhancement. The lack of an enhancement for HD148839 suggests that it is not a close relative of normal carbon stars for which [s/Fe] $\sim$ 1.
$^{12}C/^{13}C$	$\gtrsim$ 100	See Climenhaga (1960), Lambert (unpublished).

years, and that just 1.7 mag., deep may be a hybrid, part RCrB and part Hd C. If the photospheric pulsations trigger the deep minima characteristic of RCrBs, their relatives lying just outside the instability strip will show very similar spectral characteristics, but lack the variability associated with RCrBs. Although the data on chemical composition are incomplete, the evidence supports Bidelman's bold coupling of RCrBs and Hd Cs.

The standard (and only) abundance analysis of Hd Cs is by Warner (1967) who gave a curve of growth analysis of the five members of the class. Equivalent widths of about 500 lines per star spanning the intervals 4225-4570 Å and 4737-4960 Å (the gap is dominated by C_2 lines) were measured in each star off 15.6 Å/mm IIa O plates. Table IV summarizes the principal results which show clearly the similarities between RCrB and Hd C stars. The relative abundances of CNO are unknown. Warner's prediction that ^{18}O is the dominant oxygen isotope is not confirmed - see § 2.5.

I look forward to new analyses of the Hd Cs using broad spectral coverage and model atmospheres.

4. Are All Carbon Stars Helium Rich?

The question posed in the section's title was asked first by Vardya (1966) in a letter to *The Observatory*. Struck by the fact that hot He stars and RCrB are C-rich (i.e. C.O > 1) and He-rich, Vardya wondered about other carbon stars. An assertion - "It is very likely that they are also hydrogen deficient and helium enriched" - was followed by an inconclusive discussion of CH-stars. (I shall not discuss CH stars and the related Barium stars - see Lambert [1985]). I shall comment on the cool carbon stars and the evidence presented recently for a substantial H deficiency in these AGB stars. To the observer, H deficiency and He-enrichment seems a plausible malady of an AGB star and, especially, a carbon-enriched star which is presumably contaminated with carbon and helium from the He-burning shell. Severe mass loss on the AGB will reduce the thickness of the H-rich envelope whose base may be sufficiently hot to burn hydrogen. The pragmatist would not be surprised to find a range of H deficiencies on the AGB whose members may include the immediate progenitors of the RCrB and Hd C stars. In short, the opening question is worthy of examination.

Recent claims that the cool carbon stars are H deficient are based on model atmosphere based interpretations of observations of H_2 quadrupole vibration-rotation lines and the H$^-$ flux peak near 1.6 μm. Goorvitch, Goebel, and Augason (1980) pointed out that the 1-0 S(1) H_2 line was not detectable in high resolution spectra of the carbon stars UU Aur and S Cep. Johnson et al. (1983) expanded the search for the H_2 line, detected it in C-rich Mira variables (i.e., the coolest stars in their sample), but reported it absent from several warmer carbon stars. In our analysis (Lambert et al. 1986) of infrared spectra of 30 galactic carbon stars, we detected the weaker 1-0 S(0) H_2 line in almost all of the sample - see Figure 3. The S(0) line unlike the S(1) line is essentially unblended. Our measurements of the S(0) line are consistent with the S(1) detections reported by Johnson et al. (1983). Here, comparisons with predicted equivalent widths are for the S(0) line. Across the temperature range spanned by these carbon stars, the dominant form of hydrogen changes from atomic to molecular. The predicted H_2 column density as sampled by a H_2 quadrupole line may be reduced by either introduction of a H deficiency or a warming of the atmosphere such as might occur through additional opacity in the upper atmosphere. (Other possibilities for reducing the H_2 density include departures from LTE and a deep chromosphere.) In the coolest stars, the temperatures favor H_2 so strongly that severe warming of an extensive atmosphere is demanded in order to reduce the molecule's column density. Hence, one should not be surprised that the H_2 lines are seen in the Mira variables with about the predicted strength. All of Johnson et al.'s predictions of H_2 lines are based on model atmospheres computed from standard assumptions and including diatomic molecules (principally C_2, CN, and CO) as contributors to the line blanketing. Polyatomic molecues (HCN, C_2H_2) resident in the upper atmosphere also contribute to the blanketing (Eriksson et al. 1984) and, hence, to the backwarming of the photosphere. Models constructed using the diatomic and polyatomic contributions to the line opacity provide significantly smaller H_2 equivalent widths for the carbon stars. This is illustrated in Figure 3. Our predictions for models lacking HCN and C_2H_2 agree well with those provided by Johnson et al. from similar but independently computed models. With the inclusion of the polyatomics, the predicted and observed H_2 lines are of similar strength; the H_2 is seen to be an indicator of the C/O ratio, a controlling influence on the abundance of the polyatomics, and the inferred C/O ratio is in fair agreement with the ratio determined directly from C_2 and CO lines. Note how the influence of the polyatomics is reduced in the cooler stars where association of H into H_2 is more complete. These calculations, as summarized in Figure 3, show quite clearly that the H_2 line is a sensitive probe of the atmospheric structure of these carbon stars. Our improved model atmospheres show it is not necessary to invoke H deficiency in order to account for the observed strengths of the H_2 lines. (Tsuji [1983] has noted that the predicted H_2 strengths exceed the observed strengths for late M giants. We suspect that a

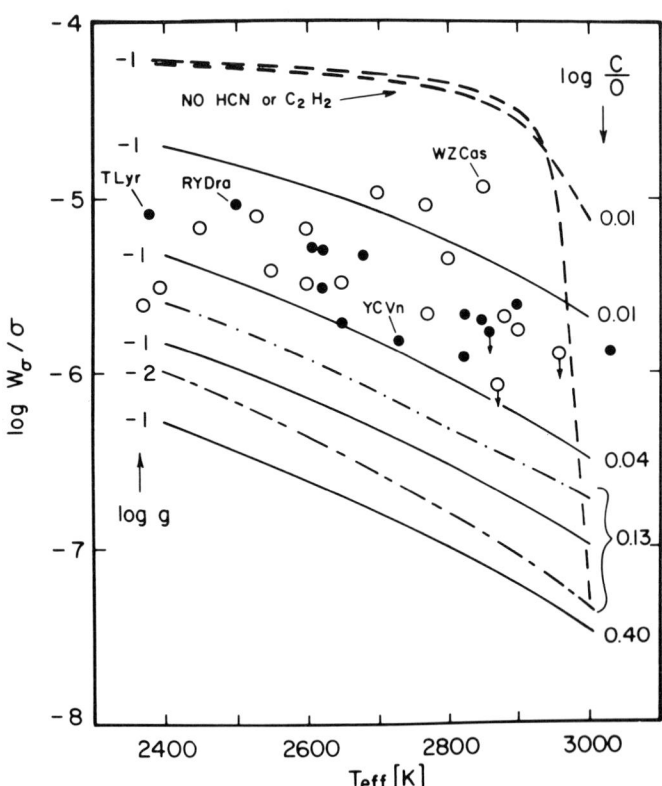

Fig. 3. - Observed and predicted intensity (log W_σ/σ) of the H_2 1-0 S(0) line in cool carbon stars. Observations are represented by filled (T_{eff} from Tsuji, 1981, and the infrared flux method) and open (T_{eff} by interpolation) circles. The four stars identified by name are ^{13}C-rich. Predictions for models constructed without line blanketing by HCN and C_2H_2 for log C/O = 0.01 (C/O = 1.02) and log C/O = 0.13 (C/O = 1.35) are given by the broken lines and identified. All other predictions are from models with HCN and C_2H_2 blanketing and for log g = -2, -1, or 0 and with the log C/O ratios indicated on the right-hand side. Carbon stars have $-1 \le \log g \ge 0$.

more complete accounting for the line blanketing in the construction of the model atmospheres may remove this discrepancy, too.)

The variation with wavelength of the emergent flux from a star betrays the identities of the principal contributors to the continuous opacity. For a cool star of normal composition, the H⁻ ion dominates the opacity and, hence, the continuum brightness temperature peaks near the 1.6 μm opacity minimum. Goebel and Johnson (1984) and Johnson et al. (1985) report that the temperature/flux peak is absent for cool carbon stars and suggest that a H deficiency of a factor of approximately 10 is required to reconcile predicted and observed fluxes. This deduction is based on predictions for model atmospheres to which diatomic, but not polyatomic, molecules provide line blanketing. Lambert et al. (1986) show that the H-flux peak is disguised by overlying molecular bands and an altered atmospheric structure when lines from polyatomic molecules are included in the blanketing: observed flux curves may be fit by predictions for atmospheres having a normal H abundance.

In a cool atmosphere, hydrogen deficiency and helium enrichment lower the continuous opacity such that the various atomic and molecular lines from elements other than H are strengthened even when the elemental mass fractions are unchanged - see Böhm-Vitense (1979) for an elegant discussion. Conversely, if the spectrum of a H-poor star is interpreted on the assumption that the star has a normal H abundance, the abundance - here X/H - will be overestimated. Lambert et al. (1986) derived the metal abundance of their cool carbon stars from several atomic (Ca I, FeI) lines: the individual and mean abundances ([Fe/H] ~ -0.1 dex) derived on the assumption that H is not deficient are consistent with that expected from the kinematics of the stars and with the metal abundance derived from comparable samples of G, K, and M giants among whom must be found the progenitors of the cool carbon stars. Dominy, Wallerstein, and Suntzeff (1986) eliminate with thsi same argument the hypothesis that SC stars are H deficient.

If the AGB stars were H deficient, their descendants should suffer the same malady. I identify the planetary nebulae (PN) as the descendants; the ionized shell is part of the AGB star's outer envelope. Some PNs are carbon-rich and I suppose that these are the descendants of the carbon stars; the claim that C/O > 1 for such PNs rests on analyses of the emission line spectrum and/or the detection of the infrared signature of circumstellar graphite grains. Comparisons between the composition of carbon stars and PNs are restricted by the lack of common elements unaffected by nucleosynthesis and dredge-up on the AGB. Oxygen is one possible example. Lambert et al. (1986) show that the mean O/H ratio and its dispersion is the same for M giants (Smith and Lambert 1985, 1986), carbon stars, and the C-rich PNs (Aller and Czyzak 1983). The stellar analyses provide O/H from observations of CO lines and an assumption about the H abundance. Analyses of PN provide the O and the H abundances directly without an assumption about the He/H ratio. Indeed, this latter ratio is measured and shows directly that the PNs are not markedly H poor: He/H = 0.11as the mean for C (and O) rich PNs (Aller and Czyzak 1983). Few PNs are significantly enriched in He. A ratio He/H = 0.10 ± 0.01 is exhibited by local H II regions and young stars (Dufour, Shields, and Talbot 1982; Nissen 1983). Both the O/H and He/H ratios reinforce our direct arguments that hydrogen is not significantly depleted in a cool carbon star's atmosphere. An enthusiast favoring H-deficient carbon stars is forced to assert that these stars do not evolve into PNs.

In our view, the direct analyses of carbon stars and the indirect appeals to progenitors (G, K, M giants) and descendants (PNs) provide no evidence for a significant (> a factor of two) H deficiency in the atmospheres of the carbon stars. Of course, the fact that they are not H deficient and He enriched translates to a requirement on the C/He ratio in the material and the mass of the material added to the envelope from the He burning shell. It appears that such requirements are met by presently available theoretical calculations of the dredge-up (see Lambert et al. 1986). Our answer to Vardya's (1966) question is 'NO'.

5. An Observer's Ragbag

This concluding section was stimulated by an inference aired by Luck (1981) in reviewing the implications of his abundance analysis of R Sct, a RV Tauri variable. A standard analysis using a model atmosphere gave T_{eff} = 4000 K, log g = 0.0, and [Fe/H] = -0.9 and [X/Fe] ~ 0.0 for the eight elements between Na and Ni with suitable lines on the single echellogram. For the heavy elements - Y, Zr, Ba, La, Ce, Pr and Nd = [s/Fe] ~ -1.2 with significant scatter which is plausibly attributed to the small number of available lines - a total of 12 distributed over the 7 elements. Luck noted that W Virginis stars show a similar anomaly according to three published analyses (Rodgers and Bell 1963; Barker et al. 1971; Anderson and Kraft 1971). RV Tauri stars lie on an extension of the period-luminosity relation followed by the W Virginis stars. These stars may be on blue loops from the AGB following a He-shell flash. I note that RR Lyrae stars apparently show a similar overdeficiency of the heavy elements over a wide range of metallicities (-2.2 < [Fe/H] < 0.2) - see Butler (1975). Unevolved and other evolved stars with [Fe/H] ~ -1 do not show an overdeficiency of the heavy elements: the several published analyses show that [s/Fe] ~ 0 for [Fe/H] ≳ -1.5 with the overdeficiency in more metal-poor stars reaching [s/Fe] ~ -1 at about [Fe/H] = -2.5. The over-deficiency of the heavy elements at [Fe/H] ~ -1 sets RSct, W Vir and RR Lyr stars apart from normal stars. Luck suggested that the anomalous combination of [Fe/H] and [s/Fe] arises because the atmospheres are hydrogen poor thanks to the ravages of stellar evolution: the [s/Fe] for R Sct implies [Fe/H] ~ -2.5 at birth and a current hydrogen deficiency of a factor of 30. The hydrogen converted to helium gives a He mass fraction Y ~ 0.99. Direct determination of the He content is difficult. I shall not catalog additional arguments supporting or challenging Luck's inference. The acid test should come, as Luck noted, from analyses of these variable stars in globular clusters where the metallicities of variable stars and "normal" red giants may be compared.

The perception that hydrogen deficiency is a common characteristic of peculiar giants appears to be contagious! Campbell (1985) in an imaginative paper on the origin of the super-metal rich stars proposes that they result from the coalescence of stars in a binary system. The envelope of the resulting single star is expected to be helium-rich. Fortunately, Campbell describes a couple of spectroscopic signatures that such a star should possess; e.g., an enhanced ^{17}O abundance. A search for these signatures will test the proposal.

Other peculiar red giants have been shown recently not to be hydrogen deficient. Sneden and Pilachowski (1984) exploiting precepts outlined by Böhm-Vitense (1979) combined observations of CO and CH lines in weak G-band K giants to show that the atmospheres are hydrogen-rich (Y < 0.3). Böhm-Vitense (1979) shows that H deficiency and He enrichment influences the pressures and, hence, the partial pressure of neutral atoms of low ionization potential. Dominy (1984) in an analysis of warm (early-R type) carbon giants used Ca I lines to show that their atmospheres are hydrogen-rich, i.e., the derived Ca/Fe ratio was solar, as expected, provided that the He/H ratio was approximately normal.

In 'A Prefatory Observation on Modern Biography', Samuel Taylor Coleridge, the English essayist and poet, wrote "How mean a thing a mere fact is except as seen in the light of some comprehensive truth" (Coleridge 1810). To the biographer of the stars, the chemical composition of the stellar atmosphere is a mere fact to be seen in the light of stellar evolution and nucleosynthesis. Both mere facts and comprehensive truth are the biographer's responsibility. I stumbled across Coleridge's dictum in a critical biography of A. E. Housman, the English classical scholar and poet, where the author remarks "there are times when a comprehensive truth must defer to mere fact" (Page 1983). Today, our knowledge of the mere facts about the cool hydrogen deficient stars is so incomplete that we can only guess at the comprehensive truth. While biographers of poets as reticent as Housman may be permanently inhibited in their search for the comprehensive truth by a lack of mere facts, we can assemble a full set of the mere facts about cool hydrogen deficient stars by observing them with telescopes and spectrometers. I hope that the next reviewer of these rare stars may be able to shed more light on 'the comprehensive truth'.

I thank several colleagues for helpful discussions: Dr. P. L. Cottrell, C. Sneden, V. V. Smith, and J. Tomkin. I am especially grateful to Drs. K. Eriksson, B. Gustafsson, and K. H. Hinkle for their essential and substantial contributions to the analyses of the carbon stars which form the basis for §4. This research has been supported in part by the National Science Foundation (grant AST 83-16635) and the Robert A. Welch Foundation.

REFERENCES

Aller, L. H., and Czyzak, S. J. 1983, *Ap. J. Suppl.*, **51**, 211.
Anderson, K. S., and Kraft, R. P. 1971, *Ap. J.*, **167**, 119.
Barker, T., Baumgart, L. D., Butler, D., Cudworth, K. M., Kemper, E., Kraft, R. P., Lorre, J., Rao, N. K., Reagan, G. H., and Soderblom, D. R. 1971, *Ap. J.*, **165**, 67.
Berman, L. 1935, *Ap. J.*, **81**, 369.
Bidelman, W. P. 1953, *Ap. J.*, **117**, 25.
Böhm-Vitense, E. 1979, *Ap. J.*, **234**, 521.
Bond, H. E., Luck, R. E., and Newman, M. J. 1979, *Ap. J.*, **233**, 205 (BLN).
Butler, D. 1975, *Ap. J.*, **200**, 68.
Cameron, A. G. W., and Fowler, W. A. 1971, *Ap. J.*, **164**, 111.
Campbell, B. 1985, preprint.
Canal, R., Isern, J., and Sanahuja, B. 1977, *Ap. J.*, **214**, 189.
Clegg, R. E. S., Lambert, D. L., and Tomkin, J. 1981, *Ap. J.*, **250**, 262.
Climenhaga, J. L. 1960, *Pub. D. A. O. Vic.*, **11**, 307.
Coleridge, S. T. 1810, *The Friend*, No. 21.
Cottrell, P. L., and Lambert, D. L. 1982a, *Ap. J.*, **261**, 595 (CL).
_____. 1982b, *Observatory*, **102**, 149.
Danziger, I. J. 1965, *M.N.R.A.S.*, **130**, 199 (D65).
Deupree, R. G., and Cole, P. W. 1981, *Ap. J. (Letters)*, **249**, L35.
Dominy, J. F. 1984, *Ap. J. Suppl.*, **55**, 27.
Dominy, J. F., Wallerstein, G., and Suntzeff, N. B. 1986, *Ap. J.*, in press.
Dufour, R. J., Shields, G. A., and Talbot, R. J., Jr., 1982, *Ap. J.*, **252**, 461.
Eriksson, K., Gustafsson, B., Jørgensen, U. G., and Nordlund, Å 1984, *Astr. Ap.*, **132**, 37.
Fowler, W. A., Caughlan, G. R., and Zimmerman, B. A. 1975, *Ann. Rev. Astr. Ap.*, **13**, 69.
François, P. 1985, *Astr. Ap.*, submitted.

Goebel, J. H., and Johnson, H. R. 1984, *Ap. J. (Letters)*, **284**, L39.
Goorvitch, D., Goebel, J. H., and Augason, G. C., *Ap. J.*, **240**, 588.
Grevesse, N. 1984, *Phys., Scr.*, **T8**, 49 (G84).
Härm, R., and Schwarzschild, M. 1975, *Ap. J.*, **200**, 234.
Heber, U. 1983, *Astr. Ap.*, **118**, 39 (H83).
Heber, U., and Schönberner, D. 1981, *Ap. J.*, **102**, 73.
Holweger, H., and Kovács, N. 1984, *Astr. Ap.*, **132**, L5.
Hunger, K., Schönberner, D., and Steenbock, W. 1982, *Astr. Ap.*, **107**, 93 (HSS).
Iben, I., Jr. 1984, *Ap. J.*, **277**, 333.
Iben, I., Jr., and Tutukov, A. V. 1985, *Ap. J. Suppl.*, **58**, 661.
Jacoby, G. H., and Ford, H. C. 1983, *Ap. J.*, **266**, 298
Johnson, H. R., Goebel, J. H., Goorvitch, D., and Ridgway, S. T. 1983, *Ap. J. (Letters)*, **270**, L63.
Johnson, H. R., Alexander, D. R., Bower, C. D., Lemke, D. A., Luttermoser, D. G., Petrakis, J. P., Reinhart, M. D., Welch, K. A., and Goebel, J. H. 1985, *Ap. J.*, **292**, 228.
Keenan, P. C., and Greenstein, J. L. 1963, *Contr. Perkins Obs.* **2**, 13.
Kilkenny, D. 1982, *M.N.R.A.S.*, **200**, 1019.
Lambert, D. L. 1985, in *Cool Stars with Excesses of Heavy Elements*, ed. M. Jaschek, and P. C. Keenan (Dordrecht: Reidel), p. 191.
Lambert, D. L., Gustafsson, B., Eriksson, K., and Hinkle, K. H. 1986, *Ap. J. Suppl.*, in press.
Luck, R. E. 1981, *P.A.S.P.*, **93**, 211.
Luck, R. E., and Lambert, D. L. 1981, *Ap. J.*, **245**, 1018.
Malaney, R. A. 1985, *M.N.R.A.S.*, **216**, 743 (M85).
Nissen, P. E. 1983, in Proc. ESO Workshop on *'Primordial Helium'*, ed. P. A. Shaver, D. Kunth, and K. Kjär (Garching: ESO), p. 163.
Nissen, P. E., Edvardsson, B., and Gustafsson, B. 1985, in Proc. ESO Workshop on *"Production and Distribution of C, N, O Elements"*, in press.
Orlov, M. Ya., and Rodriguez, M. H. 1974, *Astr. Ap.*, **31**, 203 (OR74).
—————————————. 1981a, *Soviet Astr. Lett.*, **7**, 126 (OR81a).
—————————————. 1981b, *Soviet Astr. Lett.*, **7**, 382 (OR81b).
Paczynski, B. 1971, *Act. Astr.*, **21**, 1.
Page, N. 1983, in 'A. E. Housman, a critical Biography', (New York: Schocken Books), p. 5.
Querci, M., and Querci, F. 1978, *Astr. Ap.* **70**, L45.
Rao, N. K., Ashok, N. M., and Kulkarni, P. V. 1980, *J. Astr. Ap.*, **1**, 71.
Renzini, A. 1979, in *Stars and Star Systems*, ed. B. E. Westerlund (Dordrecht: Reidel), p. 155.
—————————. 1981, in *Effects of Mass Loss on Stellar Evolution*, ed. C. Chiosi, and R. Stalio (Dordrecht: Reidel), p. 319.
Rodgers, A. W., and Bell, R. A. 1963, *M.N.R.A.S.*, **125**, 487.
Rufus, W. C. 1915, *Pub. Michigan Obs.*, **2**, 103.
Schönberner, D. 1975, *Astr. Ap.*, **44**, 381 (S75).
—————————. 1977, *Astr. Ap.*, **57**, 437.
Searle, L. 1961, *Ap. J.*, **133**, 531 (S61).
Smith, V. V., and Lambert, D. L. 1985, *Ap. J.*, **294**, 326.
—————————————. 1986, *Ap. J.*, submitted.
Sneden, C., and Pilachowski, C. A. 1984, *P.A.S.P.*, **96**, 38.
Tomkin, J., Balachandran, S., and Lambert, D. L. 1985, *Ap. J.*, **290**, 289.
Tomkin, J., and Lambert, D. L. 1983, *Ap. J.*, **273**, 722.

Tsuji, T.1981, *J. Astr. Ap.*, **2**, 95.
————. 1983, *Astr. Ap.*, **122**, 314.
Vardya, M. S. 1966, *Observatory*, **86**, 162.
Warner, B. 1967, *M.N.R.A.S.*, **137**, 119.
Zirin, H. 1982, *Ap. J.*, **260**, 255.

DISCUSSION

GARRISON: When you presented the table with the C/He values of the R CrB stars you said that helium lines are measurable, but you did not quote the values. What were the values?

LAMBERT: If you believe that the helium D_3 line is a simple photospheric line, then the numbers are very close to the lower limits.

WING: Would either you or Dr. Liebert care to add a comment about the low-luminosity carbon star G77-61?

LIEBERT: G77-61 is the only star I am aware of with carbon greater than oxygen that is firmly tied to the lower main sequence (M_v = +11). It is analyzed in a forthcoming paper (Dearborn, Liebert et al., 1986 Jan. I. Astrophys. J.). In addition to the near uniqueness of the C/O ratio, it is shown in a forthcoming Ph.D. thesis by H. Gass at Heidelberg (working with R. Wehrse) to be one of the most metal-poor stars known, with [Fe/H] = -3. There is no evidence for H deficiency. Dearborn et al. show that G77-61 is a single line spectroscopic binary with a period of 245 days. The most likely evolutionary history invokes mass transfer when the up to now undetected (presumably cool white dwarf) companion transferred a modest amount of C-rich material, probably near maximum radius (AGB phase). The extreme metal deficiency makes it easier to invert the C/O ratio, with a minimal amount of mass transfer. This object is obviously in the halo. The kinematics warrants and should receive little more investigation. Perhaps this phenomenon might turn out to be rather less freakish than one thought and the lower main sequences of Omega Centauri and so on might be some of the things accessible.

VARDYA: Is this the star which Greenstein discovered about 1968?

LIEBERT: No. I believe this refers to a paper by Greenstein and Oinas, referring to one or two G dwarfs with enhanced CH and carbon features and, as I recall, somewhat abnormally high carbon abundances. There are other possible cases - e.g. GH7-21 (Liebert, 1976 AP. J. Letters) - which may relate to subgiant CH stars (Sneden and Bond), which are too low in luminosity to be explained by post-helium evolution of a single star. G77-61 at M_v = +11 may just be the lowest luminosity known "CH Star" whose evolution requires a binary mass transfer scenario. (Greenstein is a co-author of our paper, Dearborn et al., 1986, Ap. J.).

FEAST: I was a little surprised that the carbon stars are going to be planetary nebulae. Do we really know enough about the order of evolution and where the carbon stars stand in relation to planetary nebulae. You obviously think so.

LAMBERT: There are some genuinely carbon rich planetary nebulae; for example, they have IR excesses that demonstrate that they contain graphite. So they must come from somewhere. They must come from carbon rich photospheres or envelopes.

FEAST: This is all very nice, but I would not trust it, I even would doubt it.

LAMBERT: There are two points to that argument. One is you may wish the planetary nebula to be carbon rich because the oxygen has disappeared in the nebula. The other point is that the metal contents of carbon stars are consistent with kinematics when you analyse with normal hydrogen atmospheres. If you make the atmosphere helium rich you loose that consistency. If you are worried about planetary nebulae then I will ask you to take the second point.

THE RCB STARS AND THEIR CIRCUMSTELLAR MATERIAL

M.W. Feast
South African Astronomical Observatory
P O Box 9
Observatory 7935 Cape
South Africa

SUMMARY. RCB stars are surrounded by circumstellar dust and gas moving radially outwards at ~200 km/sec. The circumstellar shell is made up of discrete puffs of matter, a typical puff occupying an area ~0.03 of a complete shell. On the average puffs are ejected about once every 40 days (comparable with the known pulsation periods of RCB stars). The reddening law of the dust indicates that it is composed of small carbon particles (radii ~100A). The flux from the shell at L typically varies by 1 to 3 mags over periods of 1000-2000 days. The average mass loss rate is ~$10^{-6} M_\odot$/yr.

1. INTRODUCTION AND MODEL

The RCB stars are a particularly spectacular subgroup of the hydrogen deficient stars. They are all carbon rich objects and their most striking property - the deep minimum that they undergo at random times - has been attributed for over 50 years to obscuration by soot. Despite this long time there remains very considerable uncertainty as to the place of these stars in stellar evolution and in the physical processes - including particle formation - taking place in their extended atmospheres.

The only viable model that has been proposed for the deep minima of RCB stars involves the ejection by the star of puffs of soot in random directions. Occasionally one puff is ejected in the line of sight causing obscuration. The main pieces of evidence that force one to such a model are as follows (cf. Feast 1975 and (especially) 1979 and references there).

(1) During a "typical" decline the normal absorption spectrum of the star is replaced by a "chromospheric" emission spectrum which changes with time in a manner analogous to the changes in the solar chromosphere with height above the limb (as at an eclipse). The spectroscopic changes (e.g. Alexander et al. 1972) are difficult to explain unless we are witnessing a real eclipse, first of the main body of the star and then of more and more of the chromosphere.

(2) On the rapid decline, one sees absorption components of the D lines and of H and K displaced by about -200 km/sec. Evidently matter is moving rapidly away from the star. It is reasonable to suppose (e.g. Hartmann and Apruzese 1976) that soot, formed above the stellar surface, is being blown away from the star by radiation pressure and is dragging gas with it. It is this expanding puff of soot which, if in our line of sight, causes the eclipse phenomena.

The forms of the light curves of RCB stars are quite diverse from one deep minimum to another. However they frequently consist of a very rapid initial drop (time scale ~<5 days for RY Sgr in 1967), a slower fall (time scale ~20 days) and then a very gradual return to maximum. The total drop in visual light is often ~8 magnitudes and the time to recover from minimum perhaps 500-1000 days. The deep minimum of RY Sgr in 1967-1969 which remains the event with the most extensive spectroscopic and photometric coverage, was of this type and for definiteness I shall take it as "typical". The initial drop seems to coincide with the fading of the continuum and may be identified as the time taken for the cloud to expand to cover the main body of the star. Coupled with the measured ejection velocity of ~200 km/sec we find that we can make a crude model in which matter is ejected radially from a limited area near the stellar surface. In our "typical" case the semi-angle of the cone of matter so formed (measured at the centre of the star) is about $20°$. That is, the cloud occupies about 1/30 the area of a complete shell. The slower decline follows as the extensive chromosphere is covered.

(3) When RCB stars are very faint their spectra are found to contain broad emission lines - the D lines, H and K and 3888 HeI. Structure in these lines has been reported by a number of workers (Alexander et al. 1972, Feast 1979, Herbig unpublished, Spite & Spite 1979). The half width of these lines is roughly constant from star to star and minimum to minimum and is about the same as the expansion velocity of the cloud in the line of sight - though with the structure in the lines, these statements are difficult to quantify. The simplest interpretation of these broad lines is that, with the star obscured, we are seeing the integrated emission from gas being carried away from the star by puffs of dust moving in all directions (radially outwards from the star). A detailed study of the structure of these lines and its variation with time would obviously be very valuable since it would give us information on the structure of the expanding material. Two points are clear from the limited data we already have. Firstly, the broad lines are roughly centred on the stellar velocity. Thus there must be a roughly symmetrical group of puffs around an RCB at any one time - or at least at any one minimum. The puff causing the eclipse is simply one of a group - not of course all necessarily ejected simultaneously. Secondly the presence of structure in the lines and the variation from time to time of the central wavelengths - indicating varying structure - shows that the number of puffs at any one time is relatively small. One might hope in future studies to be able to estimate the number of discrete puffs emitting at a given time. This is of obvious importance for estimating the total mass loss rate. The relative intensities of the various components in a line might also help in estimating the range in the masses of individual puffs.

Studies of 10830 HeI are also of importance in this connection. On the rise of RCrB itself back to normal in January 1978, Querci & Querci (1978) found the line to have a P Cygni profile. Presumably this was the combined effect of emission from the various puffs and absorption from the one in the line of sight (the absorption component had a velocity of -240 km/sec). Similar P Cygni profiles in the D lines and H and K were found earlier at a minimum of RY Sgr (Alexander et al. 1972). Some months after the Querci work, in July 1978 when RCrB had already been back at maximum several months, Zirin (1982) found no normal 10830A (which one would have expected to be strong since HeI 5876 is seen in absorption). However he reported a "huge absorption" at 10822A. It seems reasonable to interpret this as 10830A displaced by -220 km/sec. Presumably the P Cygni profile is now such as to just fill the photospheric 10830A line. The observation is of importance since it shows we can detect and study the still expanding soot and gas puff well after it has become too thin to appreciably dim the star. At low resolution (Alexander et al. 1972) the shell absorption in the D lines disappeared just prior to maximum light but perhaps these lines too could be followed after maximum at high resolution.

(4) On the rise from minimum (at least in the "typical" case of RY Sgr) the normal stellar spectrum is seen (as at maximum) but the star is faint and reddened. We are seeing the gradual dispersal of the soot cloud and the colour can be used to derive the law of reddening of the shell (Alexander et al. 1972, Feast 1979).

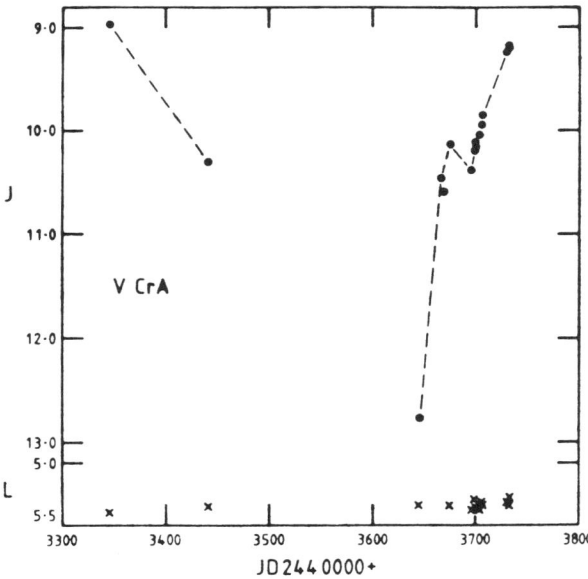

Fig. 1. A minimum of V CrA. The J flux (due to the star) drops by a large amount during minimum whereas the L flux (from the cool circumstellar matter) remains unchanged.

(5) Infrared excesses are a characteristic of RCB stars (e.g. Feast & Glass 1973, Glass 1978 etc). The exception is XX Cam (Rao et al. 1980), a star for which only a single, relatively shallow (1.7 mag) minimum is known in 80 years; so that its membership of the class is tenuous. The observed excesses correspond roughly to blackbody emission at 700-900K (Feast & Glass 1973, Glass 1978, Kilkenny & Whittet 1984, Walker 1985). It is natural to associate this excess with reradiation from the puffs of soot. The earlier result of Forrest et al. 1972, Glass 1978, Feast 1979 that the excess does not vary significantly (either up or down) when the star goes into deep minimum is fully substantiated by more recent work. One can see this by looking at light curves in J and in L. Rough calculations indicate that the flux at J is almost entirely due to the star whilst the flux at L is from the circumstellar material. Figure 1 shows for instance a minimum in J for VCrA with L remaining constant. The minima in J are of course also seen in V. Similar results apply to several other RCB stars (SAAO observations, to be published).

The results show that the size of the cloud ejected at a minimum cannot be too great otherwise the reradiated flux would increase significantly. Evidently there are enough puffs reradiating at any one time (about 10 or more would be sufficient) so that one new one does not significantly add to the flux.

In all this I am implicitly adopting a model in which the soot is directly heated by the star itself. Evidence that this is indeed so was given some years ago for RY Sgr (Feast et al. 1977). RY Sgr pulsates with a period of 38 days and the variations of the stars luminosity in this period should lead to the flux from circumstellar material showing similar periodicity. The earlier work showed that the flux at L was indeed primarily from the circumstellar dust and that this varied in the 38 day period. Further evidence for this is shown in Figure 2 where it is clear that the flux from the shell (at L) varies in the pulsation period of the star (as seen at J).

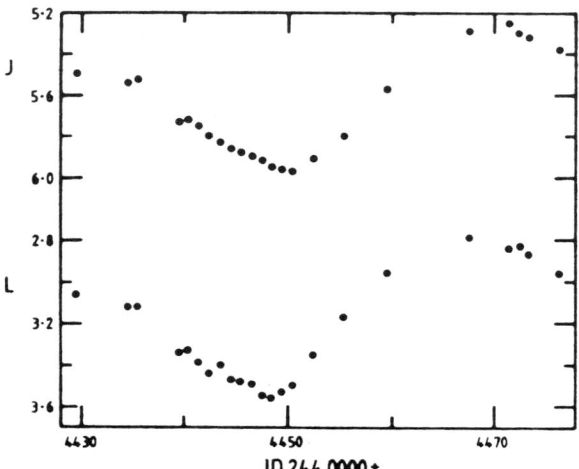

Fig. 2. Observations of RY Sgr showing that both the flux from the star itself (at J) as well as that from the shell (at L) vary in the 38 day pulsation cycle.

2. THE RCB STARS AS A GROUP

Before discussing the dust shell further it is useful to stress the similarity in the physical characteristics and behaviour of the various RCB stars - at least if one omits the three unusual hot objects MV Sgr, DY Cen and V348 Sgr. This is seen for instance in looking at an infrared two colour diagram. Figure 3 is the composite $(J-H)_0/(H-K)_0$ diagram for 12 RCB stars. All the stars (especially if we omit the crosses) tend to lie along a quite narrow path. This locus is that expected for the combination of a star with emission from a dust shell. Individual stars move along this path as they undergo obscuration phases or when the shell changes luminosity. We have observed several of the stars to track practically the whole length of this path.

These results suggest that the RCB stars have quite a limited range of colour temperatures for both the central stars and for the shells. The open symbols show the position of the non-RCB HdC stars (Catchpole & Feast, unpublished). These are apparently typical of what RCB stars would be if the circumstellar shells were removed.

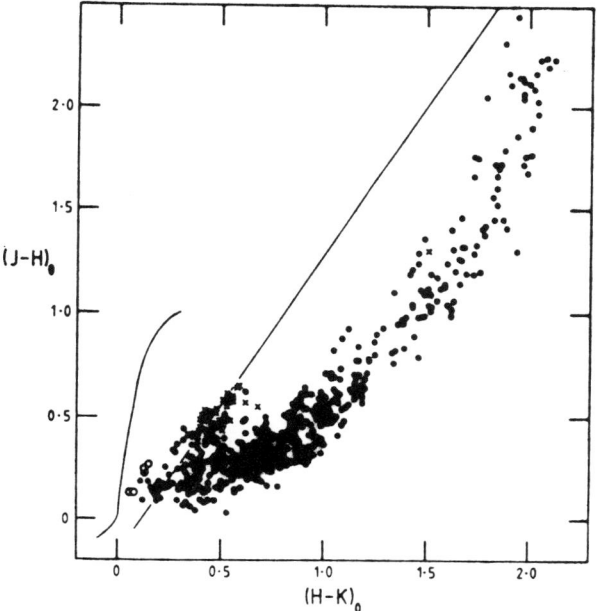

Fig. 3. $(J-H)_0/(H-K)_0$ diagram for 12 RCB stars. Crosses refer to S Aps, open circles to HdC stars. The (straight) blackbody line is shown. The curve is the locus of normal stars.

The position of the HdC stars themselves is of some interest. The deviation of normal stars from the blackbody line in this plot was shown by Catchpole & Glass (1974) to be due to the effects of various sources of continuous opacity - the main one being H^-. Obviously H^- opacity is not expected to be significant in HdC stars so that a position nearer the blackbody line might be anticipated for these stars. It would be useful to have models of HdC stars extending into the infrared with which one could compare the observations. Until we have models we cannot use this sort of diagram to estimate temperatures.

Schonberner (1975) has derived T_{eff} ~7000K for RY Sgr and RCrB. There is a range of C_2 band strengths in RCB stars. S Aps - which is the star shown as crosses in Figure 3 has very strong bands. It is quite likely that this range in bandstrength indicates a range of temperatures amongst these stars - down to say 5000K. However it should be stressed that the existence of such a range of temperatures is by no means certain. Stronger C_2 bands could perhaps indicate higher carbon abundance or the bands might arise in a cool outer shell. The position of S Aps in the $(J-H)_o/(H-K)_o$ diagram could be a temperature effect but could also be due to relatively mild circumstellar extinction. The visual observations of Espin (1890, 1894, 1900) at the end of the last century are perhaps relevant to this problem (cf. Ludendorf 1908, Berman 1935). He found that for a period of a week or so RCrB itself developed strong C_2 bands when the star was at maximum and well away from obscuration minimum. RS Tel has sometimes been mentioned as a very cool (R8) RCB star (Payne-Gaposchkin 1936, 1963). However Bidelman (1953) found C_2 weak and the UBVRI data (Kilkenny & Whittet 1984) do not make the star unusually cool, furthermore the type was given as Ro by Payne (1928). The early spectroscopic observations should be checked to see whether there has really been a large change in C_2 strength.

The range of temperatures covered by the RCB stars is thus uncertain. It is of importance if we want to compare with evolutionary predictions. Furthermore the temperatures are important for the problem of the pulsation of the stars. Saio & Wheeler (1985) were unable to keep the amplitudes finite for temperatures lower than about 7000K.

Are all RCB stars pulsating variables? RY Sgr is, with a mean period of 38.6 days which is decreasing at a rate of about one second per day (Kilkenny 1982) consistent with Schonberner's evolutionary calculations (1977) (cf. also Lloyd Evans 1985). All RCB stars seem to be at least slightly variable near maximum so they could all be pulsating variables but this is not yet certain. A 44-day period for RCrB itself was suggested by Fernie et al. (1972) and extensive radial velocities by Griffin (1985 analysed by Dr L Balona) show a periodicity of ~49 days. A period of ~43 days has been suggested for UW Cen (Bateson 1972, Kilkenny & Flanagan 1983). S Aps has apparently now got a period near 40 days (Kilkenny 1983) though a 120-day periodicity seemed to exist previously. Much more work is needed on this matter but we cannot at present rule out the hypothesis that all RCB stars pulsate with periods near 40 days.

I have stressed the rather narrow range in temperatures both of stars and shells in these stars. This similarity extends to the relative flux of star and shell.

Values of $(J-L)_0$ (= 2.48 ± 0.12 averaged over 10 RCB stars) are remarkably similar from one star to another. Here L is a mean value for each star and J is the value appropriate to maximum light. This suggests that if the stars have similar luminosities the amount of circumstellar matter is, in the mean, the same for each star within a factor of 3 or less.

3. THE CIRCUMSTELLAR PARTICLES

The size of the soot particles can be estimated from the reddening law. The only really safe way of deriving this law is to follow Alexander et al. (1972) and use only epochs at which the spectrum is observed to be normal. Unfortunately the amount of simultaneous photometry and spectroscopy is limited. However if all RCB stars follow an analogous pattern it is possible to choose phases at which the reddening law can be estimated even without spectroscopic information. Preliminary values have been obtained for 8 RCB stars at UBVRIJ (though not at all wavelengths for all stars) (the data are from the SAAO group (to be published), Kilkenny et al. (1975), Eggen (1985), Böhme (1984). Agreement between different stars is generally good.

Mean values are plotted in Figure 4 together with published ultraviolet results for RY Sgr (Holm et al. 1982). The dotted lines are from recent laboratory work on extinction by amorphous carbon smoke (Borghesi et al. 1985). The mean particle radii of the two samples shown are 40A (upper curve) and 150A (lower curve). One should not be too concerned about the failure to fit in the region of the 2200A graphite bump. Borghesi et al. stressed that the absorption in the ultraviolet is sensitive to the shapes and size distribution of the particles and this appears to be illustrated by theoretical computations for 100A graphite spheres (Draine 1985, Wickramasinghe 1973) which give a huge bump going up to $E \sim 15$ on the scale of Figure 4.

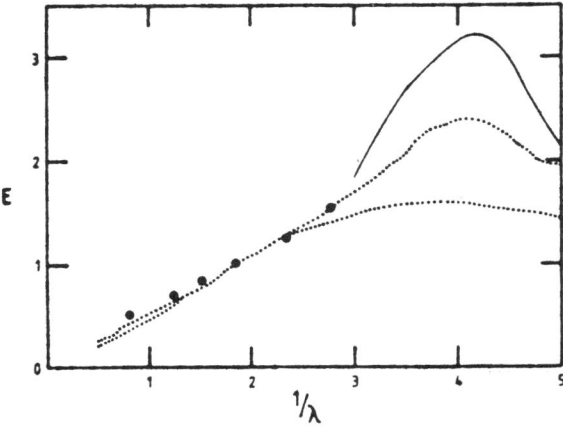

Fig. 4. Extinction for RCB stars as a function of reciprocal wavelength (μm^{-1}). Points are UBVRIJ results. The curve is from IUE data. The dotted curves are laboratory results for amorphous carbon smokes. See text for discussion.

Thus although the agreement shown is certainly not exact it would appear that the observations can be satisfied by the presence of small carbon particles and in the following I adopt 100A (0.01μm) as a typical size. We cannot fit the data with the much larger particles (0.1 - 0.2 μm) which observations suggest are around the carbon Mira R For (Feast et al. 1984). Such large particles give essentially neutral extinction shortward of ~1 μm.

On the model I sketched earlier the initial drop of an RCB star is caused by an optically thick (essentially opaque) cloud which is at first smaller than the star. The cloud expands as it moves out, eclipsing more and more of the star. Thus apart from centre to limb effects we would expect the initial drop to be at nearly constant colour. As Alexander et al. showed the chromospheric emission at a later stage, complicates the interpretation of observed colours.

Neutral extinction during the initial phase of a decline of RCrB itself was found by Fernie et al. (1972). UBVRIJHKL observations by R M Catchpole and I M Coulson of RS Tel soon after the start of a recent decline also suggest a very small variation of extinction with wavelength. In view of the complications likely in the real situation these results are probably as good a verification of the model as one could hope for. The alternative of postulating much larger particles on the decline than on the following rise does not seem at all attractive.

Variations in the flux from the dust shell was first reported for RCrB itself (cf. Strecker 1975 and references there). These suggested a periodicity of 1100 days. The few later observations (Ashok et al. 1984 and SAAO unpublished) suggest a more irregular variation but with at least a timescale for change of the order of 1000 days. In the case of RY Sgr the 38 day pulsations introduce considerable scatter. However there is again evidence of a variation in the flux from the shell with a timescale of ~1000 days (cf. Menzies 1985). Extensive observations of 10 RCB stars shows that variations at L of 1 to 3 magnitudes are typical. The times of rise or decline last typically 1000-2000 days (SAAO observations to be published). Menzies (1985) has pointed out that for RY Sgr the last two (or three) obscuration minima have followed maxima in the flux at L. This does not however appear to be a general phenomenon amongst RCB stars.

4. FURTHER CONSIDERATIONS OF MODEL

We can sketch a very crude model which seems to fit the existing data and gives us something to test with future observations.

(1) On the eclipse model for the RY Sgr minimum of 1967, soot is being ejected from near the star at ~200 km/sec in a cone of semi angle ~20°. The material begins to become optically thin (i.e. we begin the rise from mimimum) ~200 days after the ejection. These numbers together with a particle size of 0.01μ allow us to estimate that the mass of solid carbon per ejection event is ~$10^{-8} M_o$. If essentially all the ejected carbon is in solid form and if it drags along all the gas out of which it has condensed, then assuming the material has photospheric composition the total mass ejected per event is ~$10^{-7} M_o$.

(2) RCB obscuration minima seen to occur at random. The three stars which have been studied statistically (Sterne 1935, Howarth 1976, 1977) are RCrB, SU Tau and S Aps. In these three cases the average time bvetween fades (initial drops from maxima) are remarkably similar, 1026 ± 156 days, 1143 ± 220 days, 1249 ± 184 days (Mean = 1139 days). In the case of the first two stars the mean time between minima is roughly half these values, presumably due to new minima occurring before total recovery. An important point that can be inferred from those numbers is that because there is nothing preferred about our line of sight, ejection events must be occurring very frequently. If we take our figures for the 1967 ejection event in RY Sgr as typical then one puff of soot covers ~1/30 of the stellar surface. This suggests that averaged over the star (and taking the frequency of fades as above) the average time between ejections is 1139/30 = 38 days. If we took the time between minima we would get double the frequency. Evidently the model predicts that one or two puffs are, on the average, ejected per pulsation cycle of the star. These figures make plausible the idea that ejection is somehow connected with pulsation - a possibility also suggested by the fact that drops of RY Sgr tend to occur within a narrow range of pulsation phases (Pugach 1977).

One ejection event every 30 days leads to a total mass loss of ~$10^{-6} M_\odot$ per year. A relatively modest amount.

It is obviously possible to estimate the expected infrared emission from a circumstellar model of this kind. Dr P Whitelock has calculated a preliminary model which suggests that the observed mean flux at L can be maintained by an ejection rate of one puff every 30-40 days. Thus the preliminary indications are that the model will account both for the deep minima and the infrared excess.

(3) If the rise from minimum starts ~200 days after the first drop then our idealized model predicts that the rate of thinning of the cloud will be such that we shall reach ~1.5 mag below maximum ~400 days after the first drop and that we shall be near maximum again 800-1000 days after the first drop. These numbers were roughly correct for the 1967 minimum of RY Sgr.

(4) Since the flux at L shows slow increases or decreases lasting 1000-2000 days, there must be a variation of the mean mass ejection rate on this time scale. It would be important to check whether the pulsation amplitude changes with L flux (Is greater pulsational amplitude associated with greater mass loss?) On our model we would expect to see obscuration events more frequently at times of higher mass loss.

(5) The crude model we have considered predicts too high a colour temperature at K-L for the integrated flux from the dust, ~1000K as against ~800K observed. This is almost certainly due to the oversimplifications adopted. The model predicts that the colour temperature should fall with increasing wavelength - the larger wavelength radiation coming on the average, from further from the star.

A model of the type we have discussed can probably be made to work. However amongst the questions that still need answering are the following:

1. At what height above the photosphere does soot form?
2. Why is soot formation limited to patches (cf. Wdowiak 1975)?
3. Why does the amount of dust formed vary over periods of ~1000 days?
4. Is it true that all RCB stars pulsate? What are their periods if so.

ACKNOWLEDGEMENTS

Much of this work depends on SAAO infrared data which is being prepared for publication. I am grateful to my collaborators on this project for use of the data. I am indebted to Drs Whitelock, Kilkenny, Menzies and Balona for helpful discussions on various aspects of this work. Drs Griffin and Herbig kindly allowed me to see their unpublished radial velocity results on RCrB and extensive visual observations were supplied by the AAVSO (Dr Mattei) and the Variable Star Section of the Royal Astronomical Society of New Zealand (Dr Bateson).

REFERENCES

Alexander, J.B., Andrews, P.J., Catchpole, R.M., Feast, M.W., Lloyd Evans, T., Menzies, J.W., Wisse, P.N.J. & Wisse, M., 1972. Mon. Not. R. astr. Soc., 158, 305.
Ashok, N.M., Chandrasekhar, T. & Bhatty, H.C., 1984. Inf. Bull. Var. Stars, 2510.
Bateson, F.M., 1972. Inf. Bull. Var. Stars, 661.
Berman, L., 1935. Astrophys. J., 81, 369.
Bidelman, W.P., 1953. Astrophys. J. 117, 25.
Böhme, D., 1984. Inf. Bull. Var. Stars, 2646.
Borghesi, A., Bussoletti, E. & Colangeli, L., 1985. Astr. Astrophys., 142, 225.
Catchpole, R.M. & Glass, I.S., 1974. Mon. Not. R. astr. Soc., 169, 69p.
Draine, B.T., 1985. Astrophys. J. Suppl., 57, 587.
Eggen, O.J., 1985. Preprint.
Espin, T.E., 1890. Mon. Not. R. astr. Soc., 51, 11.
Espin, T.E., 1900. Astr. Nachr., 152, 139.
Espin, T.E., 1894. Astr. Nachr., 134, 127.
Feast, M.W. & Glass, I.S., 1973. Mon. Not. R. astr. Soc., 161, 293.
Feast, M.W., 1975. Variable Stars and Stellar Evolution, IAU Symp. 67, p. 129, eds. Sherwood, V.E. and Plaut, L., Reidel, Dordrecht.
Feast, M.W., 1979. Changing Trends in Variable Star Research, IAU Coll. 46, p. 246, eds. Bateson, F.M., University of Waikato, New Zealand.
Feast, M.W., Catchpole, R.M., Lloyd Evans, T., Robertson, B.S.C., Dean, J.F. & Bywater, R.A., 1977. Mon. Not. R. astr. Soc., 178, 415.
Feast, M.W., Whitelock, P.A., Catchpole, R.M., Roberts, G. & Overbeek, M.D., 1984. Mon. Not. R. astr. Soc., 211, 331.
Fernie, J.D., Sherwood, V. & Du Puy, D.L., 1972. Astrophys. J., 172, 383.

Forrest, W.J., Gillett, F.C. & Stein, W.A., 1972. Astrophys. J., **178**, L129.
Glass, I.S., 1978. Mon. Not. R. astr. Soc., **185**, 23.
Griffin, R., 1985. Private communication.
Hartmann, L. & Apruzese, J.P., 1976. Astrophys. J., **203**, 610.
Holm, A.V., Wu, C.C. & Doherty, L.R., 1982. Publs. astr. Soc. Pacif., **94**, 548.
Howarth, I.D., 1976. Publ. Var. Star Sec. R. astr. Soc. N.Z., No. **4**, 4.
Howarth, I.D., 1977. Acta. Astron., **27**, 65.
Kilkenny, D. & Flanagan, C., 1983. Mon. Not. R. astr. Soc., **203**, 19.
Kilkenny, D., 1982. Mon. Not. R. astr. Soc., **200**, 1019.
Kilkenny, D., 1983. Mon. Not. R. astr. Soc., **205**, 907.
Kilkenny, D. & Whittet, D.C.B., 1984. Mon. Not. R. astr. Soc., **208**, 25.
Kilkenny, D., Coulson, I.M., Laing, J.D., Spencer Jones, J. & Engelbrecht, C., 1985. Sth. Afr. astr. Obs. Circ., No. **9**, 87.
Lloyd Evans, T., 1985. Mon. Not. R. astr. Soc. In press.
Ludendorff, H., 1908. Publ. Astrophys. Obs. Potsdam, **19**, No. 57.
Menzies, J.W., 1986. This meeting.
Payne, C.H., 1928. Harvard Bull., **861**, 11.
Payne-Gaposchkin, C., 1936. Harvard Bull., **903**, 35.
Payne-Gaposchkin, C., 1963. Astrophys. J., **138**, 320.
Pugach, A.F., 1977. Inf. Bull. Var. Stars, **1277**.
Querci, M. & Querci, F., 1978. Astr. Astrophys., **70**, L45.
Rao, N.K., Ashok, N.M. & Kulkarni, P.V., 1980. J. Astrophys. Astr., **1**, 71.
Saio, H. & Wheeler, J.C., 1985. Astrophys. J., **295**, 38.
Schönberner, D., 1975. Astron. Astrophys., **44**, 383.
Schönberner, D., 1977. Astr. Astrophys., **57**, 437.
Spite, F. & Spite, M., 1979. Astr. Astrophys., **80**, 61.
Sterne, T.E., 1935. Harvard Bull. **896**, 17.
Strecker, D.W., 1975. Astron. J., **80**, 451.
Walker, H., 1985. In press.
Wdowiak, T.J., 1975. Astrophys. J., **198**, L139.
Wickramasinghe, N.C., 1973. Light Scattering Functions for Small Particles with Application in Astronomy, Adam Hilger, London.
Zirin, H., 1982. Astrophys. J., **260**, 655.

DISCUSSION

LAWSON: One point I would like to make. There is some evidence that XX Cam has a period of about 42 days. The amplitude of the periodic variations of the most of the R CrB stars is 0.1 or 0.2 magnitude. So detecting the pulsations in these stars is hard. The pulsations of RY Sgr are easier to analyse because the amplitude is about 0.5 magnitude.

FEAST: The pulsation problem is very difficult. I work in an institute where one member of the staff probably favours periodicities for a number of these stars and some body else claims that the evidence for periodicities is very unconvincing. So I think in the case of something like UW Cen you do see these cycles. They are isolated. The problem is that for the majority of these stars the visual observations that have been made are just not good enough. What one really has to have is a very large body of high accuracy photoelectric data.

N.K.RAO: Regarding the broad line spectrum in R CrB, the Na D lines also show P-Cygni profiles with the absorption going below the continuum and with a shift of around 300 km s^{-1}.

FEAST: I mentioned the 10830 feature probably seen in absorption from the matter coming towards you, when the star goes back to maximum. This is important, for you can study the matter even at maximum light. I think you have the same thing in the D lines too.

N.K.RAO: Yes. I saw an absorption component in λ 3889 He I also very close to the rise. During the minimum, the He I line λ 3889 is strong in emission; λ 7065 and λ 6678 also are strong in emission, but surprisingly, λ 5876 is not seen in emission. These lines are collisionally excited. They reflect the optical depth effects. Such anaomalies in He I lines are seen in V 348 Sgr.

FEAST: I don't think they are collisionally excited. I think they are radiatively excited.

N.K.RAO: If you are driving the gas along with dust wouldn't you expect to see strong absorption bands characteristic of this cool gas associated with the dust?

FEAST: I think one of the problems is, if the material were optically thick at an early stage, you hope to see strong C_2 bands. But I think you can get around this by saying that in the initial stages of the expansion the material is in fact not optically thick enough to be seen at all and you have to say by the time it does become optically thin, the carbon has already gone into particles.

N.K.RAO: At no stage do we see the presence of cool gas condensing into particles. We do not see an intermediate phase for the presence of cool gas.

FEAST: The main condensation phase you don't really see because, it is essentially optically thick at that stage. There may be phases, in rare cases, in which you do see some kind of condensation. There is a remarkable observation which has neve been properly explained. Visual observations made by a very experienced observer, Espin, at the end of the last century showed that R CrB itself showed extremely strong carbon bands when it was near maximum light. That has never been seen by anybody alse.

RANGARAJAN: Do you think there are differences between the dust and gas temperatures in the shell?

FEAST: That is not entirely clear to me.

BHATT: Is there any CO emission from the dust shell? What is the dust to gas ratio?

FEAST: CO emission has not been detected. The dust to gas ratio is perhaps a factor 1/10.

LAMBERT: We have two IR spectra for R CrB, one at maximum and one at minimum. The spectra are continuous in that range, CO is not seen in absorption.

WING: I wonder what picture you have in mind as to the size of the region producing the chromospheric emission. We are used to thinking of chromospheres as thin shells just outside the photosphere as in the case of the sun, but IUE studies of chromospheric densities indicate that giant and supergiant stars have very extensive chromospheres occupying perhaps ten times the volume of the star. In this case we should be imagining the soot particles forming under the region we are calling the chromosphere.

FEAST: Well, I think that it introduces all kinds of difficulties. The problem with the early model was that you have a spherical shell of carbon forming near the star and so the chromosphere is seen. I do not think detailed spectroscopic data will fit that, and I don't think the IR will fit that either. I think you have to have a proper sort of limited sized cloud causing a real eclipse. Now where that starts, near the star or far away, I think is a very uncertain question. The rough estimates of size of the chromosphere from the eclipse curve show it to be vey extensive. In this case it was only called chromosphere for want of a better word. The conditions are roughly chromospheric so to speak. It does not have to be the same as in ordinary stars. For example if the material is ejected from outside the chromosphere, then you have the problem of how the ejected material is replaced. Conceivably that is replaced by rather slowly ejected material through the chromospheric region. In fact chromospheric lines, I seem to remember, are displaced slightly by one km s^{-1} or so; is that not so, Dr. Rao?

N.K.RAO: Yes about 3 km s^{-1} in R CrB.

FEAST: Which would agree to a gentle outflow through the region that we call chromospheric.

DISCUSSION

N.K.RAO: May I add something to that. The chromosphere of R CrB seems to be something more like that of a normal F type supergiant because the IUE spectra show for R CrB the MgII core emission width etc. similarly as for γ Cyg (Rao, Bappu MNRAS, 1981). In addition it shows the CII line of λ 1335 very strongly in emission, even though the continuum does not appear or is extremely weak below λ 1800. I understand it indicates a temperature of about 2×10^4 K; hence there is need for energy input to increase the temperature. There are other emissions poking through the strong resonance lines of Fe II, Mn II etc. even at maximum light which again show similarities to normal giant or supergiant chromospheres.

VARDYA: I was wondering whether you observe any SiC feature at 11.3 microns in R CrB stars as we see in cool carbon stars in the IRAS data?

FEAST: I believe not.

N.K.RAO: Yes, agreed. Do you think this identification with amorphous carbon is established, particularly with the difficulties of chromospheric emission line interference during a minimum.

FEAST: If you want to be absolutely certain you must have simultaneous spectroscopy to see the spectrum of the star. In most cases I have shown that this is not the case.

GURM: In a pulsating star, depending upon certain conditions, a particular mode of radial oscillations becomes unstable and leads to supersonic velocities, thus puffing off some material. The pulsatons will slow down and build up again. This will introduce sort of cycles which seem to be present in the observational data. Are there any quantitative studies?

FEAST: It does seem that the mass loss rate is variable in a time scale of some thousands of days but the basic mechanism responsible for this time scale is not understood.

GURM: Do these stars have magnetic fields? If so, the presence of magnetic fields could lead to a sun like phenomenon of coronal holes. Signatures of the coronal holes in case of the sun are represented by the He I 10830 line. It is this line which has an anomalous behaviour in R CrB stars. Has anything been attempted in this direction?

FEAST: I don't think anything is known about magnetic fields in these stars.

WALKER: R. Wolstencroft and I have looked at RY Sgr with a spectropolarimeter and found intrinsic polarisation in the starlight. There is a feature in the curve which may correlate with a CN band head at about 3880 Å. The polarisation in the blue is very different from interstellar, but the data is of low signal to noise ratio. We have not yet worked out what this is trying to tell us about magnetic fields.

FEAST: These results could be related to a magnetic field; on the other hand they may be telling one something about the structure of the atmosphere which is not necessarily spherically symmetrical (at least as far as the distribution of molecules is concerned).

WALKER: I am worried about some sort of small particles, which may be of spherical graphite. But there are no spherical graphite grains as graphite forms thin long flakes.

FEAST: I agree entirely that the λ 2200 hump is very sensitive to the shapes of the particles.

WALKER: What is the most significant fact for the particle size?

FEAST: I don't know. I think it is important to know the sizes of the particles because you have a problem when you are trying to form the particles very quickly and if you are told that the particles are of 1000 Å in size. You have a much more serious problem than for the size of 100 Å.

GARRISON: That leads to a question I have. It was not clear to me whether one wants to exclude the 1000 Å particles or whether one has in addition the 100 Å particles.

FEAST: One cannot say anything significant at present about the range of particle size. However, the dominant particles cannot be in the range of 1000-2000 Å since these particles produce neutral extinction shortward of the visible. A range of sizes around 100 Å seems to be demanded by the observations.

POLARIMETRIC OBSERVATIONS OF HYDROGEN DEFICIENT STARS

A.V.Raveendran, N.Kameswara Rao
Indian Institute of Astrophysics
Bangalore 560034,India

M.R.Deshpande,U.C.Joshi and A.K.Kulshrestha
Physical Research Laboratory
Ahmedabad - 380009,India

ABSTRACT. Polarimetric observations of HD 30353,SU Tau, XX Cam, R Cr B,UV Cas, BD+13 3224,BD+10 2179 and HD 124448 are presented. The linear polarization of HD 30353 is found to vary appreciably at Hα over a time scale as short as one day. It is also found that SU Tau shows significant variation in polarisation even at light maximum.

1.INTRODUCTION

Hydrogen deficient stars are, generally,thought to be the remnants of red giants which have lost their hydrogen envelopes during their evolution along the asymptotic giant branch. Some of these objects are surrounded by circumstellar dust and polarimetric observations are very important for understanding the nature of the materials which constitute these dust shells. In this paper we present the results of linear polarisation measurements of a few hydrogen deficient stars belonging to the different subgroups. The observations were made with a view to detect and monitor the changes in the polarisation exhibited by them.

2.OBSERVATIONS

The programme stars were observed with a polarimeter attached to the 102-cm Telescope of Kavalur Observatory during the period from October 1984 to February 1985. The polarimeter,which works on rapid modulation principle,consists of a half-wave retarder rotated at 10.41 Hz acting as the polariser and a Wollaston prism as the analyser. For data acquisition and on-line processing a micro-computer system built around a Z-80 microprocessor was employed. A set of UBVRI broad band filters were used for the study of the wavelength dependence of linear polarisation. On some occasions an Hα interference filter was also made use of. Depending on the faintness of the objects in each wavelength band, the integration times were increased and the observations repeated 3 to 5 times to bring down the errors in the measurement of polarisation. A detailed description

TABLE 1.

STAR	JULIAN DAY OF OBSERVATION	POLARISATION (P% and θ°)										
		U		B		V		R		I		Hα
	JD 2446000.+											
HD 30353	004.42	1.6±0.3	155	1.7±0.3	154	1.8±0.3	154	1.7±0.3	156	1.4±0.2	154	2.5±0.8 140
	005.34	2.0±0.3	149	2.2±0.4	152	2.1±0.3	153	2.0±0.4	153	1.7±0.4	156	2.9±0.7 150
	007.32	1.6±0.1	149	1.7±0.1	147	1.8±0.1	147	1.7±0.1	146	1.5±0.1	148	1.5±0.2 139
	055.24	1.5±0.2	149	1.9±0.1	143	1.9±0.1	147	1.9±0.1	153	1.5±0.1	149	1.5±0.8 153
	115.23	1.8±0.1	142	1.9±0.1	149	2.0±0.1	147	1.7±0.1	141	1.5±0.1	146	-
SU Tau	054.24	1.7±0.5	160	2.0±0.2	148	2.2±0.1	163	2.1±0.1	163	1.8±0.1	148	
	116.26	2.6±0.9	142	1.5±0.2	154	2.0±0.1	160	1.9±0.1	155	1.6±0.1	159	
XX Cam	004.38	2.7±0.6	139	2.6±0.4	127	2.7±0.3	130	2.6±0.2	130	2.1±0.3	130	
	117.10	2.8±0.5	126	2.8±0.1	114	2.7±0.1	121	2.4±0.1	126	2.1±0.1	117	
R Cr B	115.38	0.2±0.1	136	0.2±0.0	99	0.2±0.0	103	0.1±0.0	102	0.1±0.0	94	
UV Cas	055.18	-		5.0±0.6	52	4.2±0.2	48	3.9±0.2	48	3.9±0.2	53	
BD+13° 3224	117.46	0.2±0.3	-	0.2±0.1	-	0.1±0.1	-	0.3±0.2	-	0.5±0.1	-	
BD+10° 2179	115.32	0.4±0.2	-	0.1±0.1	-	0.1±0.1	-	0.2±0.2	-	0.4±0.2	-	
HD 124448	116.38	0.4±0.2	57	0.5±0.1	45	0.6±0.1	69	0.6±0.2	53	0.5±0.2	52	

of the polarimeter and the method of calibration are given in Deshpande et al.(1985).

3. RESULTS

In Table 1,the results of our polarimetric observations are summarised. It gives the name of the object,Julian day of observation, percentage of linear polarisation (P%) and position angle ($\theta°$). The measured quantities are given for each of the filter used.

3.1. Hydrogen deficient binary HD 30353

HD 30353 (KS Per) is a single-lined spectroscopic binary with an orbital period of 360 days and exhibits light variations apparently unrelated to the orbital motion (Heard 1962,Osawa et al.1963). Bidelman (1950) has reported the presence of $H\alpha$ in emission in its spectrum. HD 30353 was observed on five different nights spanning an interval of about 110 days. In addition to the UBVRI filters,this object was observed through an $H\alpha$ interference filter also. Our observations indicate that the linear polarisation of HD 30353 is fairly constant at all wavelengths but there is large variation at $H\alpha$. In Fig.1,we have plotted the linear polarisation and the corresponding position anagles obtained on two nights,namely,JD 246005.34 and 2446007.32 and it can be seen clearly that significant changes in polarisation occurred even on a time-scale as short as one day.

3.2 R Coronae Borealis stars

These objects are characterised by irregular and large drop in brightness, sometimes,by several magnitudes. Four well-known members of this group SU Tau,XX Cam,R Cr B and UV Cas - were observed polarimetrically at their light maxima.

The measurements of SU Tau, obtained on two nights separated by about 60 days show appreciable changes in the polarisation at all wavelengths. In Fig.2,we have plotted these two sets of observations along with those obtained by Coyne (1974) and we note that there is a slight indication of a change in the wavelength dependence. Unlike all the other members of its class, XX Cam shows no infrared excess and is a very inactive member of the R Cr B group (Kameswara Rao et al.1980). We observed XX Cam on two occasions separated by about 110 days and both sets of data agree well with each other and those obtained at earlier epochs (Zhilyaev et al.1978).

The other two objects, namely R Cr B and UV Cas, could be observed only on one occasion each. It is known already that R Cr B, polarimetrically a comparatively well studied object, shows less polarisation at the light maximum. The present measurements,which are also obtained near maximum, showed linear polarisation ranging from 0.2% in U to about 0.1% in I.

As far as we know there is no information in the literature on the wavelength dependence of polarisation shown by UV Cas. The few measurements obtained by Orlov and Rodriguez (1977) at an effective wavelength λ_{eff} = 4500 Å are less than the value which we obtained in B by about

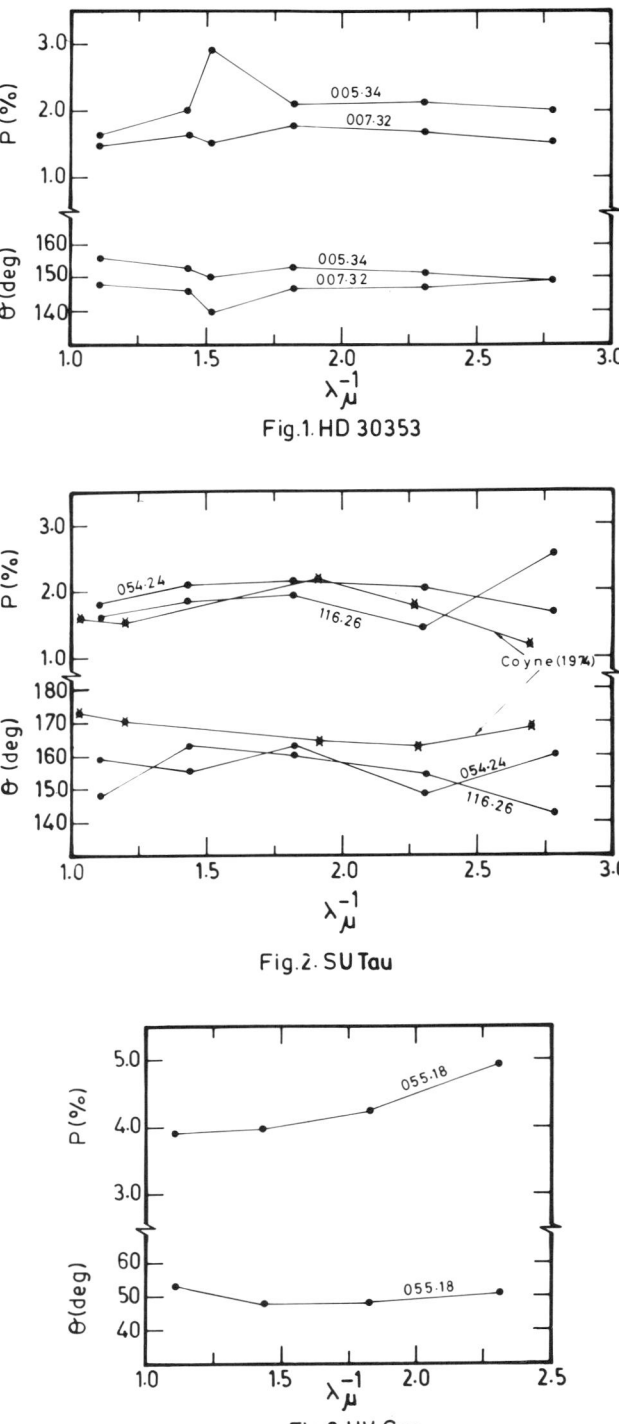

Fig.1. HD 30353

Fig.2. SU Tau

Fig.3. UV Cas

1%, indicating variability in linear polarisation. Our results are plotted in Fig.3.

3.3 Extreme helium stars BD + $10°$ 2179, HD 124448 and BD + $13°$ 3224

Schonberner and Wolf (1974) find that both BD + $10°$ 2179 and HD 124448 have similar physical characteristics; $T_{eff} \sim 16000$ K, log g ~ 2.2, and m~ 1 $M_\odot$ and Landolt (1973) has reported that both stars are light variables with a range of about 0.1 mag in V. These stars were observed only on one occasion each. BD+10 2179 did not show any significant polarisation, whereas, HD 124448 showed slightly significant linear polarisation. Further polarimetry and analysis are needed to know the contribution by the interstellar medium in the observed quantities. No significant polarization is seen in the pulsating extreme helium star BD+13 3224.

REFERENCES

Bidelman,W.P.1950,Astrophys.J.111,333.
Coyne,G.V.1974,Inf.Bull.Var.Stars.No.914.
Deshpande,M.R.,Joshi,U.C.,Kulshrestha,A.K.,Banshidhar,Vadher,N.M.,
    Mazumdar,H.S.,Pradhan,S.N. & Shah,C.R. 1985,Bull.astr.Soc.India,13, 157
Heard,J.F.1962,Publ.David Dunlap Obs.2,267.
Kameswara Rao,N.,Ashok,N.M. & Kulkarni,P.V.1980,J.Astrophys.Astr.1,71.
Landolt,A.U.,1973,Publ.astr.Soc.Pacific,85,661.
Orlov,M.Y. & Rodriguez,M.H.1977,(in Russian) Astron.Circ.No.969,PP 3- 4.
Osawa,K.,Nishimura,S.& Nariai,K.1963,Publ.astr.Soc.Japan,15,313.
Schonberner,D.& Wolf,R.E.A.1974,Astron.Astrophys.37,87.
Zhilyaev,B.E.,Orlov,M.Y.,Pugach,A.F.,Rodriguez,M.H.,Totochava,A.G.,1978,
    (in Russian) in Nankava dumka,Kiev,ed.M.Y.Orlov.

DISTRIBUTION OF LIGHT MINIMA OF R CORONAE BOREALIS TYPE STARS

A.E.Rosenbush
Main Astronomical Observatory Academy of Sciences
of the Ukrainian SSR, Kiev, U.S.S.R.

Deep light minima of R CrB type stars are considered to occur due to the carbon abundance excess in the atmospheres of these stars, and condensation of carbon into graphite dust leads to light decrease (Zhilyaev et al., 1978). The cause and conditions of dust formation are not clear. Therefore, it is necessary to find out regularities in the occuring of light minima and to search for correlations between different parameters of this type stars. This paper deals with the time intervals between the series of consecutive light minima.

A long series of observations of the R Coronae Borealis variables brightness allowed to study repeatedly their light curves in order to obtain their periodicity or other parameters in occuring of deep light minima. The work by Sterne (1935) was the first, in which the conclusion was made that R CrB is an irregular variable. Next investigations contained the analyses of light curves and calculations of the number of minima planned in the definite intervals, f.ex., 0-300, 300-600 days etc. (Howarth, 1977; 1978). Other papers (Tempesti et al., 1975) considered the duration of minima and maxima states. In all these investigations the time was counted off from the moment when a star reached a definite light decrease magnitude, f.ex., $\Delta m = 1^m$. The conclusions obtained were similar: the considered parameters were in agreement with Poisson statistics.

The procedure which led to this conclusion is wrong. This can be seen from the following fact established by Pugach (Zhilyaev et al., 1978): the start of deep light minima of RY Sgr falls always on one and the same phase of light pulsation with the mean period $P = 38.6^d$. Hence, it is necessary to consider the moment of the beginning of light decrease as the starting moment in a given minimum, and δ analyse the actually observed time interval between the consecutive minima.

In this investigation we accepted as minima the light decreases of 0.5^m and more as well as light decreases occurring during the phase when the star returned from the previous minimum, e.g. at JD 2439380 for R CrB. We used the combined light curves of R CrB (Isles, 1973; Mayall, 1960; Schweitzer, 1982) and of RY

Sgr (Mayall, 1972; Circ IAU, 1977, 1982a). The accuracy of determination of the minima starting date using these light curves is not worse than 10-15^d. Histograms of the time intervals T (for R CrB) and the number of pulsations K = ΔT/P (for RY Sgr) between the consecutive light minima are given in Figs 1 and 2.

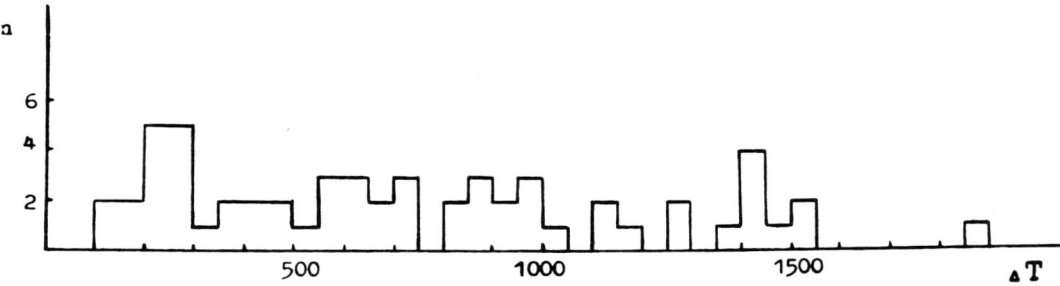

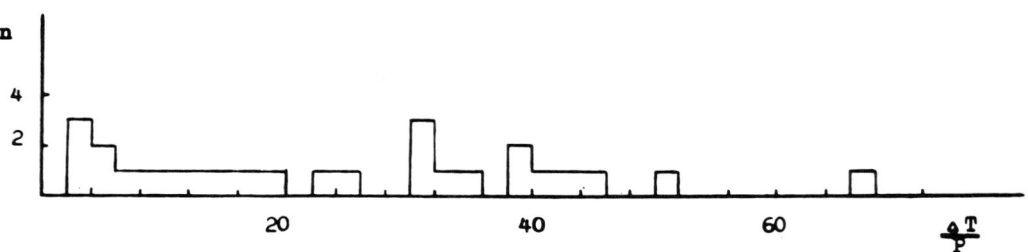

Calculation of the number of light minima for R CrB is carried out within the intervals of 101-150, 152-200 days etc., for RY Sgr within the intervals of 2.1-4.0, 4.0, 4.1-6.0 number of pulsations etc. "Instantenous" value of pulsation period (according to Marraco et al., 1982) was used in the case of RY Sgr. The number of pulsations was not an integor value, because the observed period of pulsations varies in the of 8^d. Figs 1 and 2 show that the time intervals ΔT upto 1540^d, or the number of pulsations K upto 46 are distributed regularly between the consecutive minima. The use of statistical criteria testifies this conclusion. A conclusion may be drawn that there are maximum time intervals between the consecutive light minima: about 1800^d for RY Sgr and about 1540^d for R CrB. Figs 1, 2 show that there are 1 or 2 time intervals for each of two stars that are larger than those values, and that may be to omitted small, not observed light minima. This may indicate that there exists a mechanism that impels the star to diminish its brightness not later than after this maximum interval of time.

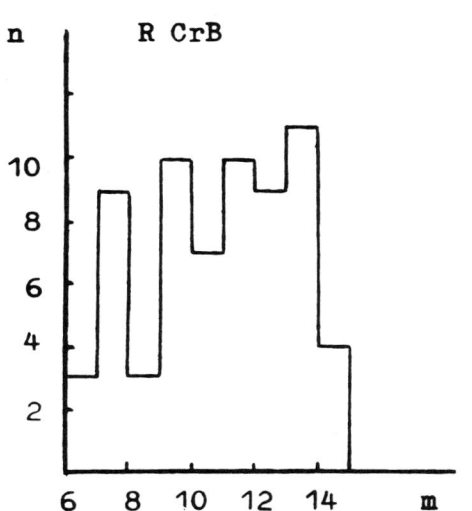
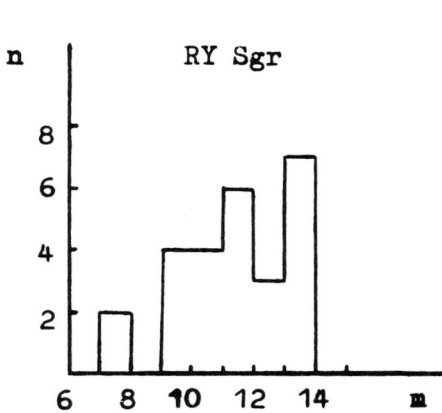

Another interesting fact can be noted if we consider the number of light minima of a definite depth (Fig.3): deeper minima occur more often. This dependence is more clear for RY Sgr. This dependence is not connected with the selection of small light minima observed out of deep minima, because the minima, at least, with $\Delta m > 1^m$ cannot remain not observed. Some shortage of minima can be supplied with the small ones occured at some phase of deeper light minimum, f.i. unaccounted $\Delta m \simeq 0.8^m$ weakening of brightness of R CrB at JD 2442800 at the stage when its brightness was increasing after minimum with $\Delta m \simeq 5^m$. The value of the deepest brightness decreases is quite definite: it is approximately 8^m for R CrB and 7^m for RY Sgr.

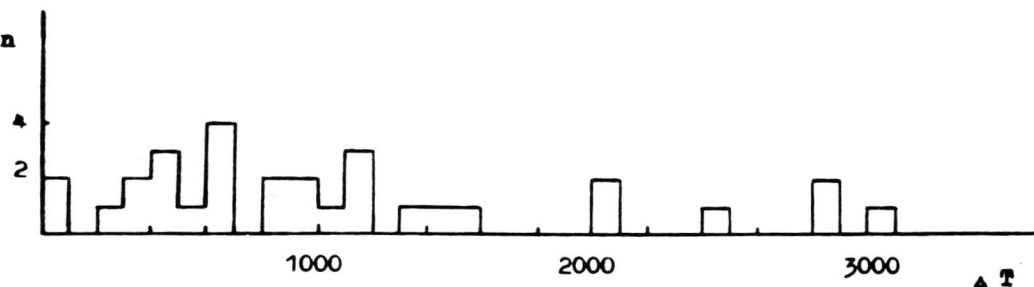

The similar time interval distribution between the consecutive minima has place also for SU Tau (Fig.4). The initial data were taken from Doroshenko et al. (1978); Campbell (1940); Howarth (1977); Jacchia (1933); IAU Circ. (1979; 1982b). The maximum time interval between the consecutive light minima appears to be about 1500^d, because for some minima with $\Delta T > 2000^d$ it is not known whether there occurred other minima during these intervals.

REFERENCES

Campbell L. 1940. Harv.Observ.Reprint.No. 250,1.
Doroshenko V.T. et al. 1978. Astrophysik 14,5.
Howarth J.D. 1977. Asta Astron, 27,65.
Howarth J.D. 1978. J.Brit.Astron.Assos. 88,145.
Isles J.E. 1973. J.Brit.Astron.Assos. 83, 368.
Jacchia L.1933. Publs.Observ.Astron.Univ.Bologna. 2, 242.
Marraco H.G., Milesi G.E. 1982. Astron.J. 87, 1775.
Mayall M. 1960. J.Roy.Astron.Soc.Can. 54, 194.
Mayall M. 1972. J.Roy.Astron.Soc.Can. 66, 233.
Circ.Cent.Astron.Telegrams.Astron.Union. 1977, No. 3098; 1979, 3407; 1982a, No 3663; 1982b, No 3740.
Schweitzer E. 1982. L'Astronomie 96, 356.
Sterne T.E. 1935. Bull.Harv.Observ. No 896, 17.
Tempesti P. De Santis R. 1975. Mem.Soc.Astron.Ital. 46, 451.
Zhilyaev et al. 1978. R Corona Borealis stars. Naukova dumka. Kiev.

ABUNDANCE ANALYSIS OF R CrB VARIABLE UW CEN

Sunetra Giridhar and N. Kameswara Rao
Indian Institute of Astrophysics
Bangalore 560034, India

ABSTRACT: Burrell (1976) performed an abundance analysis of the southern R CrB star UW Cen and reported a metal deficiency by a factor of 660 compared to the sun. However, his analysis was made employing very uncertain gf values, which resulted in lower estimates for T_{exc} and T_{ion} etc. We have reanalysed his data using new values and model atmospheres calculated by Schönberner (1975). Our analysis shows that the star has nearly solar abundance.

1. INTRODUCTION

Among the group of R CrB stars which are hydrogen deficient stars with excess of helium and carbon, only very few have been studied in detail. The prototype of this group, R CrB was studied using high dispersion spectra by Searle (1961), Schönberner (1975), and Cottrell & Lambert (1982). Abundance analysis of R CrB, RY Sgr and UW Cen was done by Burrell (1976) who reported all the three stars to be metal deficient in general, but UW Cen to be deficient by a factor of 660. The metal deficiency of this order has not been observed in any other R CrB stars studied so far. Moreover, the observed radial velocity of UW Cen (34 km s^{-1} :Herbig 1985) and its galactic position do not support such extreme metal deficiency.

An examination of the line data of Burrell (1976) revealed that the gf values used in his investigation (Corliss and Bozmann 1962, Corliss and Tech 1968, Tech 1972 etc) are highly uncertain.

As pointed out by Foy (1972), these gf estimates suffered large errors, particularly for ionised lines. It is likely that these systematic errors in gf values might have led to an erroneous estimation of excitation temperature and consequently to very uncertain abundance estimates. We, therefore, undertook to reanalyse the line data of UW Cen.

2. EQUIVALENT WIDTHS AND EXCITATION

We have used equivalent width data of Burrell (1976) measured on spectrograms of 10 A mm^{-1} dispersion in 5400 A to 6400 A spectral region. We have applied a systematic correction as pointed out by Burrell (1976) of -0.16 in log (Wλ/λ) to the equivalent width of Burrell to bring these equivalent widths to the scale of Searle (1961) so that we could compare the derived abundance of UW Cen with R CrB.

Excitation Temperature: A preliminary curve of growth analysis of Fe I lines indicated an excitation temperature in the range 6500 K - 6800 K. These equivalent widths when plotted on theoretical curve of growth of Van der Held as computed by Powell (1969), gave a microturbulent velocity (V_t) of 3 kms^{-1} ± 1 kms^{-1} for Fe I lines. A higher value of V_t = 7 kms^{-1} ±1 kms^{-1} was obtained for Fe II lines. Burrell (1976) however had derived an excitation temperature. T_{exc} = 5040 K and V_t = 3.0 kms^{-1} for Fe I and 7.0 kms^{-1} for Fe II lines.

Atmospheric Parameters: We have summarized in Table 1 broad band colours for UW Cen at maximum light observed by different workers. After applying the correction for reddening adopted by Glass (1978) and Kilkenny and Whittete (1984) the observed colours of UW Cen are similar to those of R CrB indicating a similar temperature, thus the initial guess of T_{eff} is adopted as 6800 K ± 300 K which is same as the adopted value of R CrB.

Table 1.

	UW Cen			R CrB
	MW	Glass $A_v = 0.99$	KW $A_v = 0.44$	
V	9.2			
B-V	0.72	0.43	0.59	0.59
U-B	0.19	-0.03	0.10	0.10
R-I	0.26	0.01	0.15	

MW   Moreno & Walker (1971)
KW   Kilkenny & Whittete (1984)

We have used model atmospheres of Schönberner (1975) in the effective temperature range 6000 - 7200 K and log g = 0.0 to 1.0. We modified the spectrum synthesis code of Sneden (1974) to meet the requirements of R CrB type atmosphere by scaling the abundances to total number of H + He + C, instead of hydrogen alone. Also the opacities due to C I, C$^-$, Mg I, Al I and Si I are included. Details of our theoretical computations and also the adopted gf values for the lines, compiled from various sources have been described elsewhere.

The T_{eff} for UW Cen was derived by requiring the [Fe/H] from all the lines of the sample to be independent of their excitation potentials; the gravities were derived using ionization equilibrium for Fe II/Fe I, Si II/Si I and V II/V I. We arrived at following atmospheric parameters for UW Cen: T_{eff} = 6800 K, log g = 0.0 He=97%. Figures 1 and 2 demonstrate the fact that there is no dependence of derived [Fe/H] on excitation potentials and equivalent widths of the lines. We adopted V_t = 7.0 kms^{-1} although Fe I lines suggest V_t = 4.0 kms^{-1} whereas Fe II lines give V_t = 9.0 kms^{-1}.

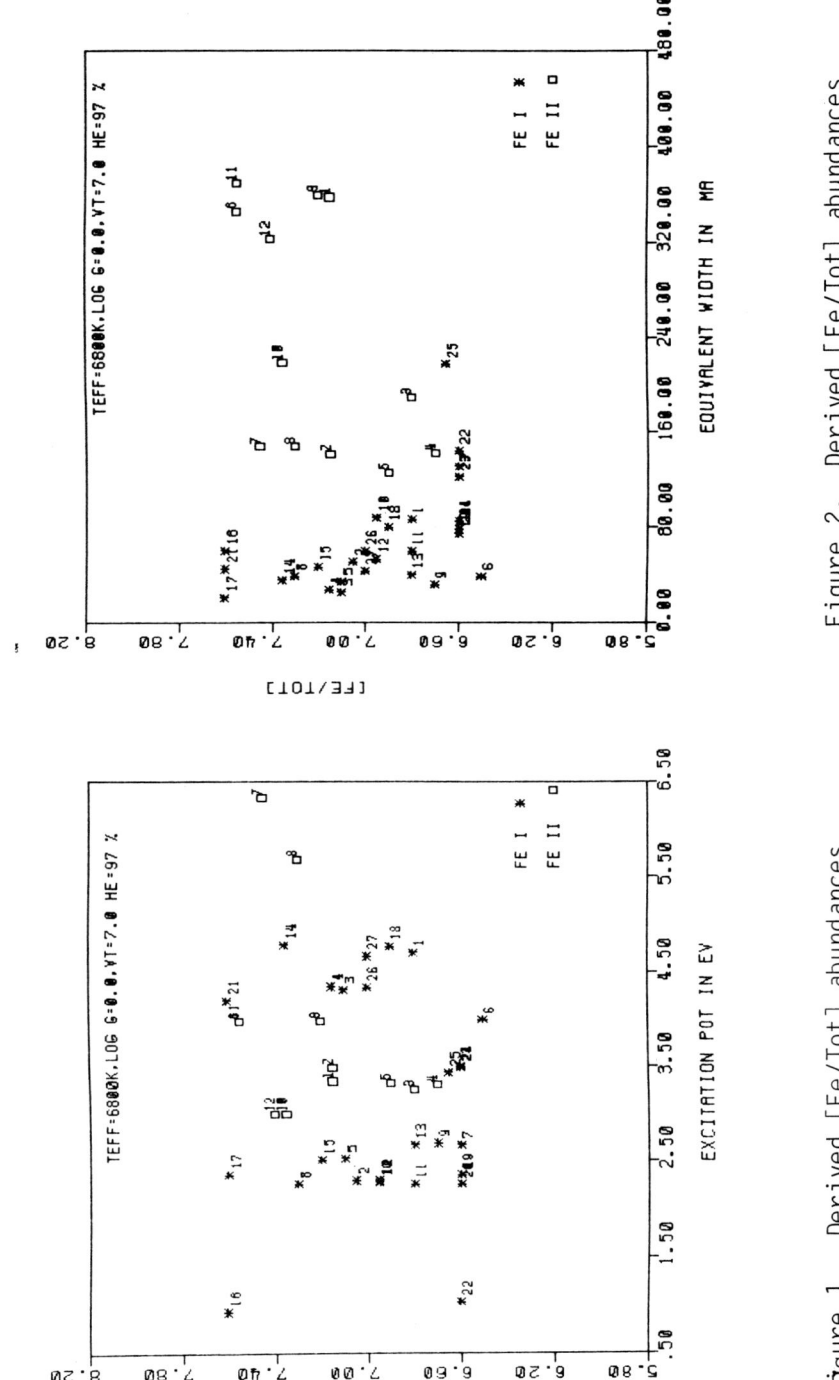

Figure 1. Derived [Fe/Tot] abundances as a function of excitation potentials of the lines.

Figure 2. Derived [Fe/Tot] abundances as a function of equivalent widths of the lines.

Similar analysis was carried out for comparison star δ CMa which was analysed in the investigation of Burrell (1976) and also of Searle (1961). The abundance analysis for δ CMa was done using the lines that were common in the two investigations and a model atmosphere of $T_{eff} \simeq 6250$ K, log g=1.0 of Kurucz (1979) with V_t of 7.0 kms^{-1}. δ CMa has been recently analysed by Castely and Desikachary (1984).

3. RESULTS

The derived atmospheric parameters and the abundances of different elements for UW Cen and δ CMa are presented in the Table 2. The table also contains the abundances for R CrB derived by Schönberner (1975). For δ CMa there is a good agreement between abundances derived by Castely and Desikachary and by us. For UW Cen we arrive at almost solar abundances for most of the elements.

Table 2.

Element	UW Cen		R CrB	δ CMa		
	No. lines	Present work	Schon-berner	No. lines	Present work	Castley & Desikachary
He I	5	11.51	11.51			
Si I	2	7.11±.58		2	7.30±.2	
Si II	2	7.60		2	7.75±.15	7.70±.16
Ca I	5	5.44±.4		5	6.01±.25	6.24±.11
V I	2	4.1				
Cr II	4	5.42		4	5.69±.17	6.00±.23
Fe I	26	6.95±.51	7.1	10	7.55±.33	7.40±0.20
Fe II	12	7.26±.18		2	7.50±.15	7.40±.18
Ni I				3	6.03±.17	5.82±.19
Y II	2	3.65		3	2.28±.20	1.70±.57

It should also be noted that for UW Cen T_{eff} estimated in the present investigation is about 6800 K indicating an earlier spectral type against the spectral type K assigned to it by Gaposhkin (1936).

Acknowledgement:

We express our gratitude to Detlef Schönberner for providing a grid of model atmospheres suitable for R CrB type stars.

REFERENCES

Burrell, J.F., 1976, Ph.D. Thesis, Australian National University, Canberra.
Castley, J.C., and Desikachary, K., 1984, Preprint.
Corliss, C.N., and Bozman, W.R., 1962, NBS Monograph, No.53.
Corliss, C.N., and Tech, J.L., 1968, NBS Monograph No.108.
Cottrell, P.L., and Lambert, D.L., 1982, Astrophys. J. **262**, 595.
Foy, R., 1972, Astr. Astrophys. **18**, 26.
Gaposhkin, Payne C., 1936, Bull. Harvard Obs. No.903, 35.
Glass, I.S., 1978, Mon. Not. R. astr. Soc. **185**, 23.
Herbig, G.H., 1985, Private Communication.
Kilkenny, D., and Whittete, D.C.B., 1984, Mon. Not. R. astr. Soc. **208**, 25 (KW).
Kurucz, R.L., 1979, Astrophys. J. Suppl. Ser. **40**, 1.
Moreno, B.F., and Walker, W.S.G., 1971, Circ. R. astr. Society, NewZealand, **184**, 1 (MW).
Powell, A.L.T., 1969, Roy. Obs. Bull. No.152
Schönberner, D., 1975, Astr. Astrophys. 44, 383.
Searle, L., 1961, Astrophys. J. 133, 531.
Sneden, C.A., 1974, Ph.D. Thesis, The University of Texas, Austin.
Tech, J.L., 1972, NBS Monograph No.119.

DISCUSSION

FEAST: The spectral type of K given in the literature is based on very old data and need not be taken seriously.

SPECTROPHOTOMETRIC OBSERVATIONS OF R CrB DURING 1972, 74 MINIMA

N. Kameswara Rao., R. Vasundhara and B.N. Ashoka
Indian Institute of Astrophysics
Bangalore 560034, INDIA

ABSTRACT: The spectrophotometric observations obtained in the wavelength range 3400A to 8000A during 1974, 1972 minima of R CrB have been fitted with Mie scattering calculations assuming sperical graphite particles and various size distributions. Power law type distributions fit the observations fairly well, and indicate that there is decrease in mean particle size as the star comes out of minimum.

1. INTRODUCTION

It was suggested by Loreta (1934) and O'Keefe (1934) that the minima of R CrB are caused due to extinction by carbon particles ejected by the star (see Feast 1985 for a review). Since the particles are supposed to have been formed near the star, a growth in particle size may be expected during the light minimum. Because the extinction properties as a function of wavelength depend upon the size and chemical composition of the particles, the change in the particle size can be inferred from the change in the energy distribution of the star during the light minimum assuming the chemical composition. With this in view we obtained scanner energy distributions during 1972 and 1974 light minima of R CrB using 36 inch Crossley reflector in the wavelength range of 3400A to 8000A with band passes of 16 and 20A (shortward and longward of 5200A respectively). The 1972 minimum observations are displayed in Fig.1.

One of the main problems in the study of energy distribution for estimating the differential extinction is the contamination due to chromospheric emission lines. We have selected regions with least contamination with emission lines except for few wavelengths 4290A, 4395A, etc. in which the emission line behaviour was monitored. The chromospheric emission seems to decay with a time constant ~ 20 days similar to that seen in RY Sgr (Alexander et al. 1972). To study the differential extinction curve we selected the observations obtained in the recovery branch of the light curve (3, 12 June and 10 July, 1972) corresponding to 3, 2 and 1.5 mag below maximum) in which the emission line contribution is assumed not to be there. The

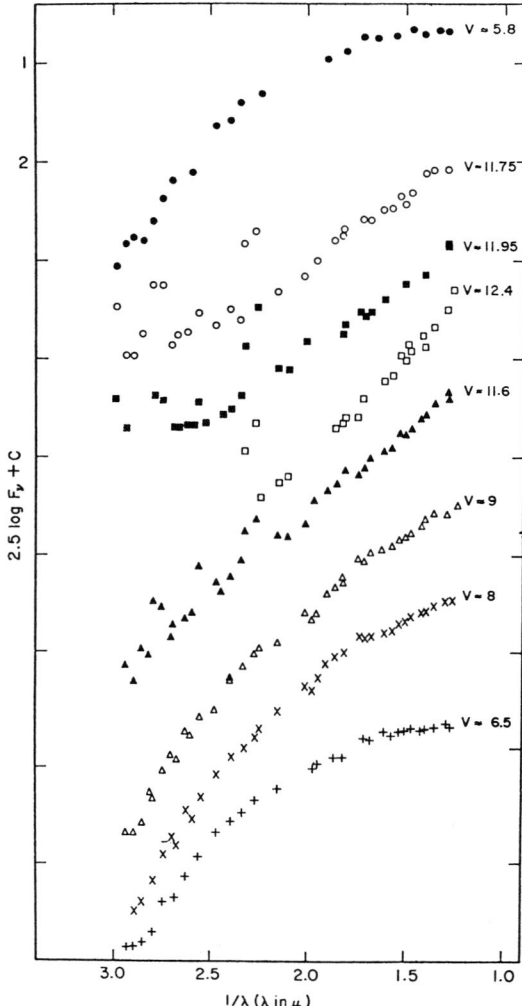

Fig. 1. 1972 minimum observations of R CrB using 36 inch crossley reflector. Pass band: 16A shortward of 5200A and 20A longward of 5200A.

1 February 1974 scan was obtained at an earlier phase (V ~ 10.8) and corrections for the presence of emission lines have been made using coude spectrograms obtained during 1962 minimum by Herbig, assuming that the emission line intensities are roughly the same at a given V magnitude. These observations are differenced with the normal light maximum observation (obtained June 10, 1973) and normalised at 7760 A These are displayed in Fig.2. Since the pulsation period and amplitude of R CrB are quite uncertain no pulsation phase matching has been done.

The extinction curves are matched with theoretical curves computed using Mie theory (Shah 1977) assuming various distributions of spherical graphite particles. Even though Hecht et al. (1985) suggested amorphous or glassy carbon from the ultraviolet data, the composition of the particles is assumed to be of graphite for the following reasons. It was shown by Hecht et al. that the fit for the extinction curves to R CrB in the wavelength range 1750 to 3250A is better for graphite of 40 nm size than for glassy carbon. Secondly the extinction bump at 2400-2500A attributed to amophrous carbon was not present in the 1983 decline in light of R CrB (Holm et al. 1985). Further, amorphous carbon (similar to glassy carbon) is expected to show an emission peak at about 6-8 μm (Koike et al. 1980, Borghesi et al. 1985) which is not seen in R CrB infrared spectra. Since no single sized particle could match the observed extinction curves we have assumed power law type size distributions similar to the distribution proposed by Mathis, Rumple and Nordsieck (1977) for interstellar medium. Refractive indices were calculated using the dielectric functions tabulated by Draine (1985).
In fitting the theoretical curve with observations, three parameters were allowed to vary: the lower and upper limits of grain size distribution and the power law index q. These fits are shown in Fig.2. There seems to be an increase in the power law index as the star is coming out of the deep minimum indicating steady decrease in the mean size of the particles (such a trend is also seen in the behaviour of R the ratio of total to selective absorption estimated from UBV photometry of the recovery branch of minima). This could result if the larger grains are selectively ejected or the larger grains break up into smaller grains. The efficiency for radiation pressure QPR for the bigger grains is high with a maximum around 800-900A size, as such they could be ejected selectively. Thus there is production and to an extent dispersal or distruction of grains during a light minimum.

REFERENCES

Allexander, J.B., Andrews, P.J., Catchpole, R.M., Feast, M.W., Lloyd Evans, T., Menzies, J.W., Wisse, P.N.J., and Wisse, M., 1972, M.N.R.A.S. **158**, 305.
Borghesi, A., Bussoletti, E., and Colangeli, L., 1985, Astron. Astrophys. **142**, 225.
Draine, B.T., 1985, Astrophys, J. Supplement Series **57**, 587.
Feast, M.W., 1985: IAU Colloquium 87.
Hecht, J.H., Holm, A.V., Donn, B., and Wu, C.C., 1985, Ap. J. **280**, 228.
Holm, A.V., Hecht, J.H., Wu, C.C., Donn, B., 1985, Future of ultraviolet astronomy based on six years of IUE Research NASA Conf. Publ. 2349, ed. Mead, J.M., Chapman, R.D., and Kondo, Y. p.338.
Koike, C., Hasegawa. H., and Manabe, A., 1980, Astrophys. Space Sci. 67, 495.

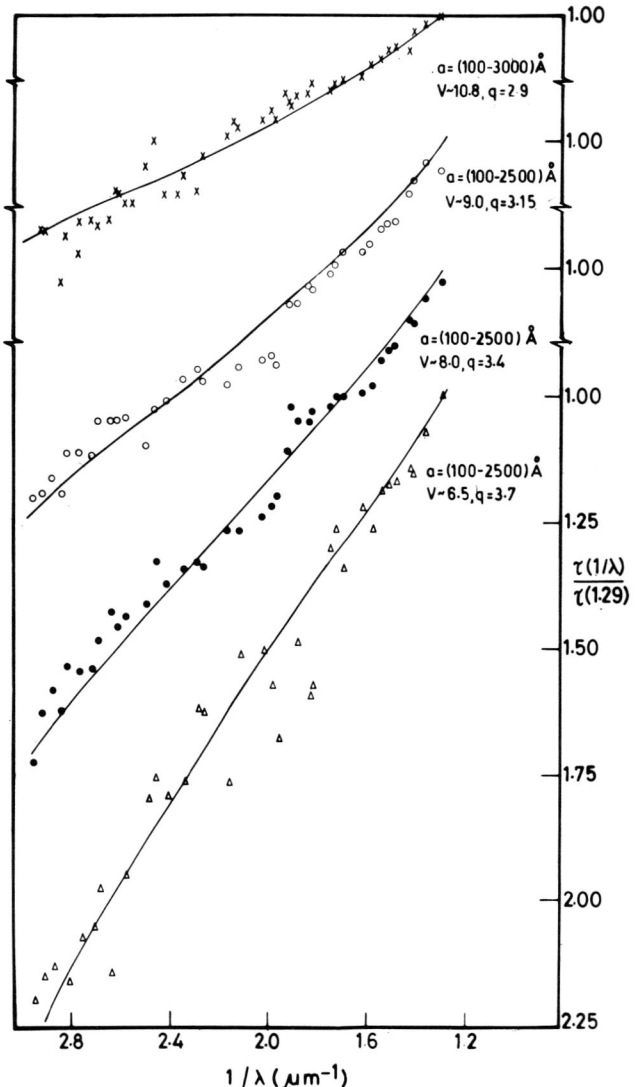

Fig.2. Extinction curves normalised at λ7760 Å compared with theoretical ones computed for MRN size distribution. Crosses: 1 Feb. 1974, open circles; 3 June 1972, filled circles 12 June 1974, and open triangles: 10 July 1972.

Loreta, E., 1934, A. N. **254**, 151.
O'Keefe, J.A., 1939, Ap. J. **90**, 294.
Mathis, J.S., Rumple, W.L., and Nordsieck, K.H., 1977, Ap. J. **217**, 425.
Shah, G.A., 1977, Kodaikanal Obs. Bull. Ser.A, 2, 42.

DISCUSSION

FEAST: You could say that the particle size decreases as the star comes out of minimum. I prefer to say that the spectrophotometry does not tell you the size. But whether you can get around this problem of starting off with big particles and ending up with small ones, I don't know. I mean, this model worried me; it seems to be the wrong way round.

N.K. RAO: Maybe what we can really do is to try to look for features at $\lambda 2200$ or at $\lambda 2470$, which can say something about the size. But there one really runs into trouble because of the emission lines. Probably the best thing is to observe in the IR, but apparently it doesn't show any features.

VARDYA: There was a contention in the IRAS meeting as to whether the particle density distribution has gradients of -1 or -2. And here you are getting steeper gradients. I don't know whether it indicates anything or not.

N.K. RAO: Well, maybe we are dealing with smaller particles.

PHOTOMETRIC AND RADIAL VELOCITY VARIATIONS OF RCrB NEAR MAXIMUM LIGHT

A.V. Raveendran, B.N. Ashoka, N. Kameswara Rao
Indian Institute of Astrophysics
Bangalore 560034, India

ABSTRACT: Fourier analysis of the light curves of RCrB in V band near maximum shows that in addition to several significant short periods there is a modulation of the visual light with a period around 1170 day, similar to that of L band flux, noticed by Strecker. This indicates that there is some contribution to the visual light variations of the star from the pulsating circumstellar dust. Radial velocities of R CrB obtained at Kavalur during February-May 1985 show variations with a period around 47 days.

1. INTRODUCTION

In the case of R CrB, the light near maximum fluctuates with an amplitude in the range of about 0.1-0.3 mag and appears very irregular in shape, amplitude and period. In order to check the possibility that the apparent irregular behaviour is as a result of the superposition of several independent waves, we decided to Fourier analyse the available light curves of R CrB obtained near maximum. The technique of Discrete Fourier Transform as developed by Deeming (1975) was employed, since the information on aliasing due to the different data spacings can be obtained directly from the spectral window derived from the times of observations. All the available measurements in V band from 1971 to 1983, except those belonging to either a primary or a secondary fading, were included in the analysis.

2. POWER SPECTRUM

Several peaks of significant and comparable amplitudes could be identified in the power spectrum. We made a least square solution for the determination of the optimum periods and their corresponding amplitudes and phases of the dominant peaks. We found that with these quantities the observed light curves could not be reproduced satisfactorily. This prompted us to suspect that the various periodicities had their origins from different data segments obtained at different epochs.

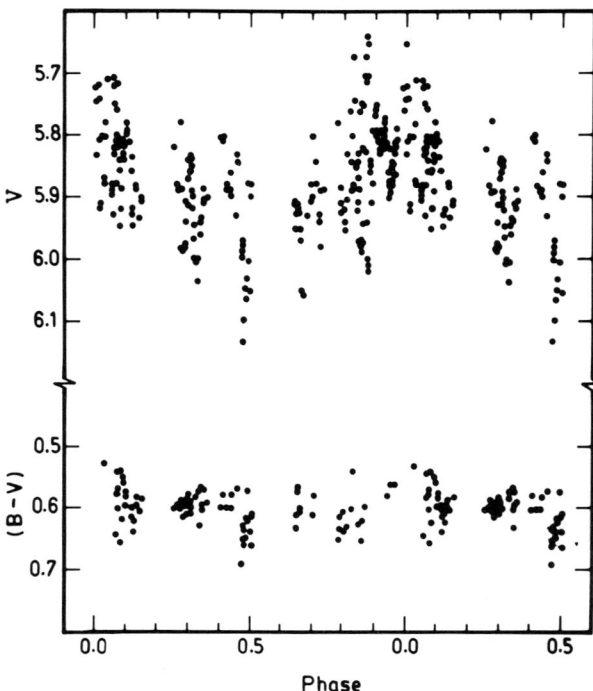

Fig. 1a. V and B-V values of R C rB near maximum during 1971-83 folded over a 1170 day period.

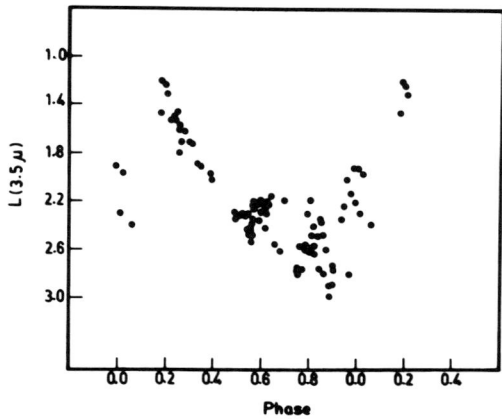

Fig. 1b. L magnitudes available during 1968-82 folded over a 1170 day period.

Hence, we made a piece-wise analysis of the data which confirmed our suspicion. Further, we could find no single period which is present in all the different data segments. These results convincingly rule out the possibility that the observed irregular light fluctuations are caused by the superposition of several independent oscillations.

3. LONG-TERM BEHAVIOUR

3.1 Visual light

Of the several periods present in the power spectrum, the most dominant is a long-term modulation with a period around 1200 day. In Fig. 1a, we have plotted the V and B-V magnitudes of R CrB obtained near maximum against the photometric phases calculated using the following ephemeris:

$$JD = 2444294.0 + 1170^d.0 \text{ E}$$

The initial epoch corresponds to the time of maximum brightness in V for the 1170 day modulation. Both the period and the initial epoch are derived using least square method assuming a sinusoidal form for the variation. The total amplitude in V is found to be around 0.15 mag. Apparently, (B-V) does not show significant variation over this period. It is clear from Fig.1a, that the short-term fluctuations are superposed on this long-term behaviour.

3.2 L Band (3.5µ) flux

The evidence of long-term modulation of the light draws one's attention towards the variability of infrared excess at L band in R CrB, noticed first by Humphreys and Ney (1974), and confirmed later by Strecker (1975) who derived a characteristic time of about 1100 days for the variation. The available L band observations from 1968 to 1982 are plotted in Fig.1b with the above ephemeris. It is evident from the figure that the period fits the infrared data also fairly well. The variability in infrared excess is attributed to the pulsating circumstellar dust around R CrB (Feast et al. 1977). If true, this implies that there is some contribution to the optical variation from the circumstellar dust.

4. RECENT RADIAL VELOCITY DATA

In order to look for the suspected pulsation, R CrB was observed spectroscopically on 27 nights during February-May 1985 with the 75-cm Telescope at Kavalur at a dispersion of 26 A° mm^{-1}. In Fig.2a, we have plotted the radial velocities derived by us against the corresponding Julian Days of observation. It is clear from the figure that the radial velocity of the star is variable, but no definite pattern for the variation could be seen. The total amplitude of variation is around 8 kms^{-1} in about 50 days. We have obtained

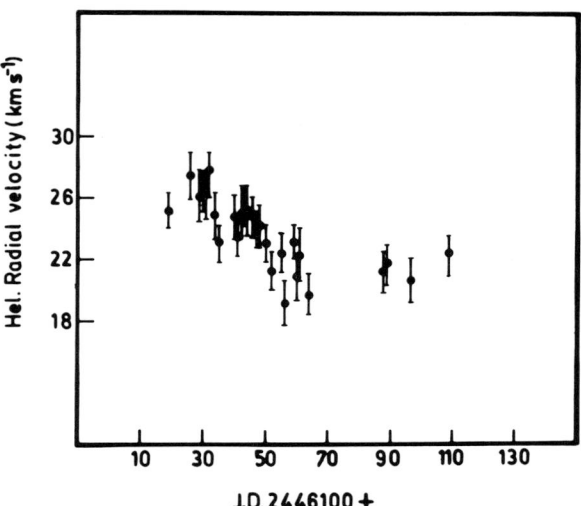

Fig. 2a. Radial velocities of R CrB obtained at Kavalur during February-March 1985.

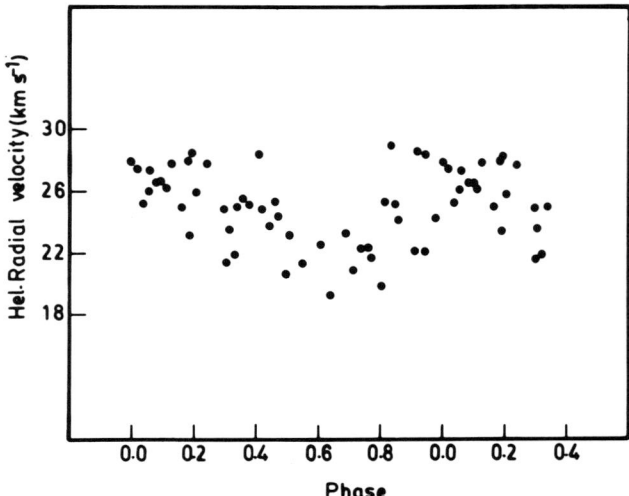

Fig. 2b. Radial velocities of R CrB near maximum available since 1972 folded over a 47.18 day period.

a few radial velocity measurements during 1983 also. A period of 47.18 day is found to fit fairly well all our observations along with those given by Fernie et al. (1972), as shown in Fig.2b. However, the period is a poor presentation if observations earlier to 1972 are also included. An extended version of this paper will be published elsewhere.

REFERENCES

Deeming, T.J. 1975, Astrophys. Sp. Sci. 36, 137.
Feast, M.W., Catchpole, R.M., Lloyd Evans, T., Robertson, B.S.C., Dean, J.F. & Bywater, R.A. 1977, Mon. Not. R. astr. Soc., 178, 415.
Fernie, J.D., Sherwood, V. & DuPuy, D.L. 1972, Astrophys. J. 172, 383.
Humphreys, R.M. & Ney, E.P. 1974, Astrophys. J. 190, 339.
Strecker, D.W. 1975, Astr. J. 80, 451.

DISCUSSION

FEAST: Two points. One: there is some rough time timescale for the L mag. I think it is unlikely that the period given by Strecker will hold. I mean, later observers don't seem to confirm that period. So I think that all one wanted to say is that there is a timescale almost that long. It is interesting if you tie it to that sort of timescale in the visual, presuming that is due to the absorption in the shell. There seems to be an indication in the case of S Aps, there are peiods where the V light is down for a period of some thousands of days when it is at maximum. The other point that we agree on is that the radial velocities show periods of the order of 47 or 48 days, derived from Griffin's data which I mentioned. What I did not mention was that Dr. Balona has analyzed a whole lot of Dr. Herbig's data but here we don't see any periods in these early observations. If there are pulsations, then they seem to be of very low amplitude.

PRELIMINARY ANALYSIS OF THE BROAD He I EMISSION LINES IN R CrB

R.Surendiranath, K.E.Rangarajan and N.Kameswara Rao
Indian Institute of Astrophysics
Bangalore 560034
INDIA

ABSTRACT:
During light minimum phase, R CrB shows a broad emission line spectrum of He I including λ 10830, λ 7065, λ 7281, λ 3889, λ6678 and λ 3188. But λ5876 is very weak. The observed intensity ratios of I(λ 3889)/I(λ 5876) and I(λ 7065)/I(λ5876) were greater than 1. The anomalous intensities of these lines appear to be due to optical depth efects. Peliminary analysis is presented to derive the physical conditions of the emitting gas.

I. INTRODUCTION

During the visual light minimum, R CrB shows three types of spectra: a) an absorption line spectrum similar to that observed at maximum light, b) A sharp emission line spectrum mainly due to singly ionized metals; the spectrum is displaced to the blue by 3 to 10 kms^{-1} with reference to the absorption spectrum observed at maximum, c) A broad emission line spectrum consisting of lines of He I, H and K lines of Ca II and the D lines of Na I. The He I lines seen are: λ 10830, λ 7065, λ 7281, λ3889, λ6678 and λ3188. But λ 5876 is very weak. (see Fig.1). Typical line widths are as follows: He I λ 3889 extends from +270 kms^{-1} to -270 kms^{-1}. The Ca II H and K lines extend from +310 kms^{-1} to -310 kms^{-1} (Gaposchkin, 1963; Rao, 1981).

RY Sgr also shows such emission lines during minimum. Such anomalies in He I lines are also seen in V348 Sgr (Dahari and Osterbrock, 1984). It is likely that other R CrB type stars also exhibit similar phenomenon. These lines change their profiles as the minimum progresses.

We envisage a possible scenario as follows: A highly excited and electron collision-dominated gas is ejected at high velocities during light minimum. The emission lines probably arise from this and we attempt model calculations described in Sec. II to derive physical conditions of the gas cloud.

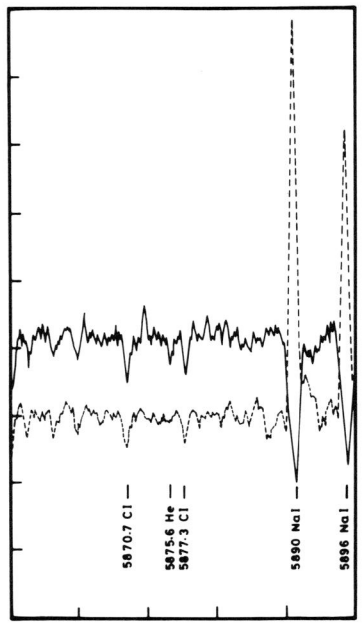

Fig. 1. Spectra of R CrB near λ 5876 region. The lower one was taken at minimum (11 July 1962) by G.H. Herbig when the star's V magnitude was 10.0; the upper one was taken at maximum (11 April 1973).

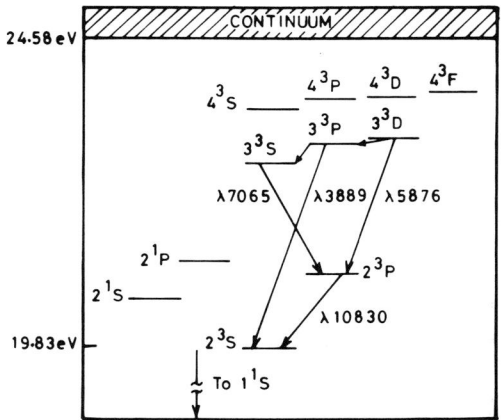

Fig. 2. Energy level diagram for the model He I atom used for the NLTE calculations. Radiative transitions solved in detail are shown by arrows. For the calculations using the photon escape probability formalism, the continuum and the level 1'S were not included.

II. MODEL CALCULATIONS

Following Feldman and MacAlpine (1978), we first solved the equations of statistical equilibrium for an 11 level He I atom, and derived the line intensities. The effect of finite optical depth τ was introduced, by means of escape probability for the photon $\varepsilon(\tau)$, into the equilibrium equations. The coupled equations were solved for the level populations by using the Gauss-Jordan method. The intensity ratios $I(\lambda 3889)/I(\lambda 5876)$ and $I(\lambda 7065)/I(\lambda 5876)$ were found to be >1 for values of election temperature (Te), electron density (Ne) and optical depth τ_0 (10830) in the range.

$$Te \sim 10^4 \,°K, \quad Ne : 10^{11} \text{ to } 10^{12} \text{ cm}^{-3} \text{ and } \tau_0(10830): 300\text{-}500$$

Next, using the complete linearization technique of Auer et al (1972), we did Non-LTE line transfer calculations. The model slab of gas for which the transfer calculations were done was divided into 50 layers characterized by Te, electron pressure Pe, and column mass. The upper and lower boundaries had a difference in Te of $3 \times 10^3 \,°K$ and the change over the 50 layers in between was uniform. Likewise, Pe values changed unformly and the boundary values differed by a factor around 5.

The composition of the gas was 98% He (by number) and the rest C, H and heavy elements. A 13 level He I model atom was used (see Fig.2). Doppler broadening was assumed. The stellar radiation field was assumed to have no effect on the gas cloud. The emission line profiles were computed for $\lambda 10830$, $\lambda 3889$, $\lambda 7065$, $\lambda 5876$, $\lambda 4.3\mu$ and $\lambda 18.6\mu$. The fluxes (integrated over the line profile) for these lines were calculated.

Various model slabs with T_e, P_e, and τ were tried in the range Te : 5×10^3 to 2×10^4, Ne 10^7 to 10^{15} cm^{-3} and τ_0 ($\lambda 5876$) of 40 to 100. The following values of $\tau_0(\lambda 5876) \sim 10^2$, Ne $\sim 10^{11}$ cm^{-3} and Te $\sim 1.6 \times 10^4 \,°K$, seem to give the flux ratios $F(\lambda 3889)/F(\lambda 5876)$ and $F(\lambda 7065)/F(\lambda 5876)$ as 3.2 and 1.1 respectively, close to the observed ratios. The above calculations do not incorporate velocity fields. The main conclusion is that large optical depths are needed to cause the inversion in the line intensity of $\lambda 5876$.

ACKNOWLEDGEMENTS

Two of us wish to thank D.Mohan Rao for useful discussions and kind help in checking some of the Fortran routines.

REFERENCES

Auer, L.H., Heasley, J.N., and Milkey, R.W., 1972, KPNO contribution No.555.

Dahari, O., and Osterbrock, D.E., 1984, Astrophys. J., **277**, 648.
Feldman, F.R., and MacAlpine, G.M., 1978, Astrophys. J. **221**, 486
Gaposchkin, C.P., 1963, Astrophys. J., **138**, 320.
Rao, N.K., 1981, in "Effects of mass loss on stellar evolution", IAU Coll.59, chiosi, C., and Stalio, R., (eds), D. Reidel, p.469.

3.0 TO 3.5 MICRON SPECTRUM OF V348 SGR AND R CrB

K. Nandy[1], N. Kameswara Rao[2], D.H. Morgan[1]

1. Royal Observatory, Edinburgh
2. Indian Institute of Astrophysics, Bangalore 560034

1. INTRODUCTION

The circumstellar dust in R CrB stars is often thought to be due to graphite, because of the high carbon abundance in the stars. Further, the spectra of these stars in the infrared show featureless smooth continuum (Forrest 1974, Roche and Aitken 1984) which was also thought to be characteristic of graphite. However, the recent comparison of the ultraviolet spectra obtained at maximum and minimum light of R CrB showed an extinction peak in the region of 2400 to 2500A (Holm, Wu & Doherty 1982, Hecht et al. 1984) which was identified as due to amorphous or glassy carbon particles. According to Duley and Williams (1981,83) amorphous carbon is supposed to show spectral features in the 3.3-3.4 μm region. Further many dust emission features are also supposed to appear in the spectral region 3 to 3.5 μm (Aikten 1981). The previous studies in this spectral region in R CrB (Forrest 1974) and in the hotter star V348 Sgr (Allen et al. 1982) showed smooth continuum. V348 Sgr shows spectroscopically many similarities with other WC 11 stars and was grouped with CPD-56°8032, He 2-113 and M4-18 (Webster and Glass 1974). All these three stars show strong dust emission features at 3.3, 8.6 11.25 μm. With a view to search for weaker dust spectral features we obtained the spectrum of R CrB and V348 Sgr with higher resolution than employed before.

2. OBSERVATIONS

The observations were obtained on 1985 July 21, with the 3.8 meter U.K infrared telescope at Mauna Kea using the 7-channel cooled grating Spectrometer CGS 11 (Wade 1983). The resolution is $\lambda/\Delta\lambda \simeq 350$. The entrance aperture used corresponds to 5.4 arc sec. The data points were spaced every half-width of each detector (the spectrum is over sampled by a factor of two). The standard stars BS 6863 and BS 6063 were observed for atmospheric corrections before and after the observations of V348 Sgr & R CrB respectively. Figs.1a and 1b show the spectrum of V348 Sgr relative to BS 6863 and Fig.1c shows the ratio of BS 6863-2 over-1. These observations were obtained when the star was at maximum. There is an indication

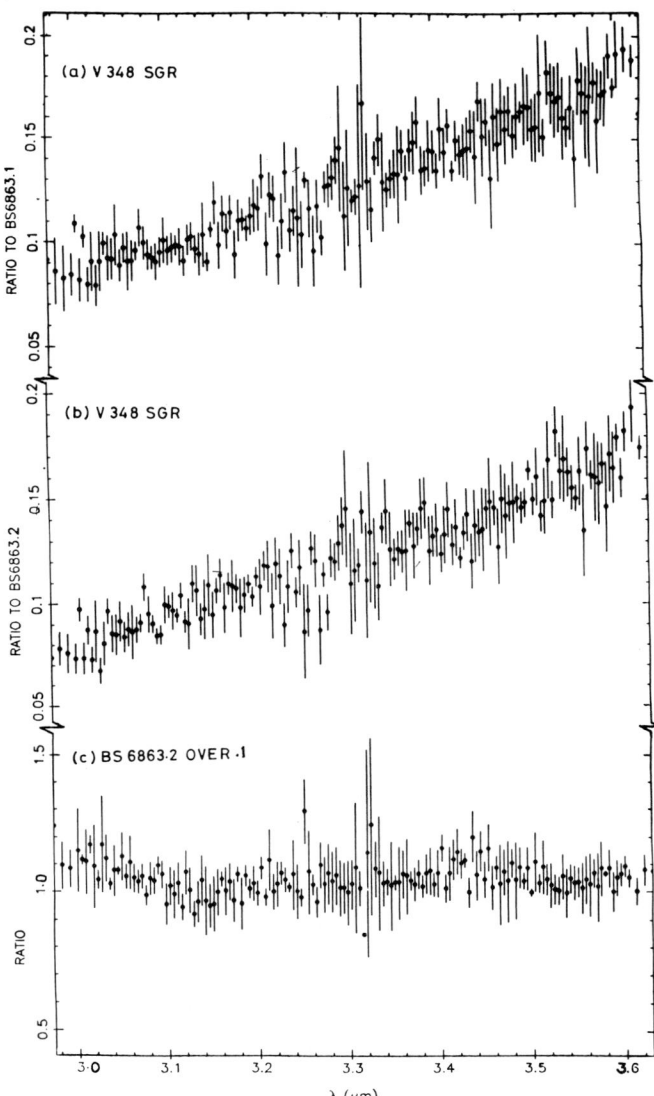

Fig. 1(a): Spectrum of V 348 Sgr relative to BS 6863-1. (b) Spectrum of V 348 Sgr relative to BS 6863-2. (c) Spectrum of BS 6863-1 relative to BS 6863-2.

of the presence of weak emission at 3.3 μm in V348 Sgr whereas the spectrum of R CrB (Fig.2) shows a smooth featureless spectrum. Further observations of V348 Sgr are being planned to confirm the 3.3 μm emission. Observations of these stars at the time of infrared minimum would be useful for detection of the spectral features.

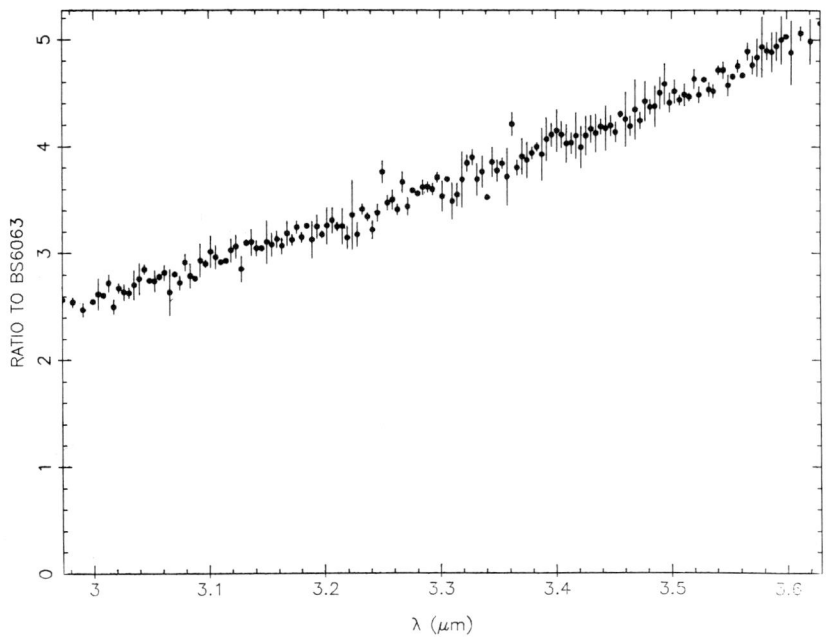

Fig. 2. Spectrum of R CrB relative to the standard star BS 6063.

REFERENCES

Allen, D.A., Baines, D.W.T., Blades, J.C., Whittet, D.C.B., 1982: MNRAS **199**, 1017.
Aitken, D.K., 1981: IAU Symp. **96**, p.207.
Duley, W.W., and Williams, D.A., 1981: MNRAS **196**, 269.
Duley, W.W., and Williams, D.A., 1983: MNRAS, **205**, p.67.
Forrest, W.J., 1974: Thesis Univ. Calif. San Diego.
Hecht, J.H., Holm, A.V., Donn, B., and Wu, C.C., 1984: Ap. J. **280**, 228.
Holm, A.V., Wu, C.C. and doherty, L.R., 1982: PASP, **94**, 548.
Roche, P.F., and Aitken, D.K., 1984: MNRAS, **208**, 481.
Wade, R., 1983: UKIRT Users and Operations Manual.
Webster, B.L., and Glass, I.S., 1974: MNRAS, **166**, 491.

RY SGR: CAN THE TIME OF THE NEXT DEEP MINIMUM BE PREDICTED?

J W Menzies
SAAO
P O Box 9
Observatory 7935
South Africa

ABSTRACT. The L magnitude of the RCB star, RY Sgr, varies with an amplitude of up to 2.0 mag on a time scale of about 1500 days. The last two deep optical minima have occurred approximately one year after a maximum in L. This behaviour may be characteristic of the star and may allow the time of onset of the next deep minimum to be predicted, in which case it will occur in the second half of 1986.

1. INTRODUCTION

RY Sgr has been observed in the infrared (JHKL) since about 1968, although there are only a few sporadic measurements before 1975 when fairly regular observations began at SAAO. It is evident from the data collected thus far that there are long-term trends in the flux from the dust shell which is presumed to surround the star. Virtually all of the flux in the L band comes from the dust shell while at J the flux at maximum is mostly stellar continuum. The J and L light curves differ markedly.

2. J, L LIGHT CURVES

In J the same range of phenomena are seen as in the visual region, viz, (a) periodic variations with P ~ 39 day and amplitude ~ 0.8 mag, presumably due to pulsation of the star, and (b) deep minima in which a decline of about 3 mag in a few tens of days is followed by a recovery to the pre-minimum brightness in about 2 years. During the recovery, the pulsation is still evident, so the event causing the deep minimum appears not to affect the photosphere in any marked way.

Although the pulsation is also evident in L (where it is presumably the result of periodic heating and cooling of the dust shell) there are no deep minima. On the contrary, the L magnitude tends to be near a local maximum when an optical minimum occurs and declines only slowly on a substantially longer time scale than is occupied by the recovery in the optical.

Figures 1(a) and 1(b) show all the available J and L photometry plotted against Julian date. To make the variations in L a little clearer, the observations have been lumped into 38.57 day bins as shown in Figure 1(c) and a smooth curve has been drawn freehand amongst the points. This period was chosen as a compromise and is appropriate to the middle of the time interval considered here, according to the ephemeris derived by Kilkenny (1982).

Remarkably, the last two minima (1977, 1983) occurred about one year after a maximum in L. The earlier minima (1967, 1972) could plausibly have also occurred in the same relationship to L maxima but there are not enough data available to say. (The first infrared observations of RY Sgr were only made towards the end of the recovery from the 1967 minimum). At least, the L magnitude was very bright during this early period.

3. (J-L) COLOUR CURVE

It is somewhat more instructive to consider the behaviour of the star in the (J-L) colour, which shows both the phenomena of interest here. Again, the data have been binned into 38.57 day intervals and are plotted in Figure 1(d). The vertical lines in the figure indicate the times of onset of deep minima. The 1977 minimum occurred about 600 days after bluest (J-L), i.e. L minimum. The shell contribution at L had increased by about 1 mag in this time. On recovery from this optical minimum the (J-L) colour reached an extremum again at about JD 2444400 and some 600 days later the 1982 deep optical minimum occurred. Since then the star has returned to the optical pre-minimum level. Another extremum in (J-L) appears to have occurred around JD244600. If the earlier pattern of behaviour continues, another deep optical minimum can be expected near JD2446600, i.e., 1986 June.

The degree of obscuration of the star seems to be related to the maximum L magnitude achieved before a deep optical minimum. The minimum in 1977 (ΔV, ΔJ ~ 2 mag) was relatively weaker than that in 1982 (ΔV ~ 6 mag, ΔJ ~ 3 mag) while those in 1967 and 1972 both had $\Delta V > 6$ mag. In the last three cases, L_{max} was about 2.0 mag while before the 1977 minimum it was only ~ 2.8 mag.

4. MECHANISM

The mechanism by which the deep minima occur is not yet well understood. Feast (1978) has discussed a model in which carbon-rich clouds are shot off the star in random directions at irregular intervals. In this case, deep minima are thought to be produced by clouds which happen to be ejected along the line of light to the star. The L magnitude would be due to the ensemble of clouds present around the star at any given time and would be little affected by the ejection of a new cloud. It is difficult to see how the timing of L maximum and the following deep minimum could be as closely tied together as has been suggested above.

RY SGR: CAN THE TIME OF THE NEXT DEEP MINIMUM BE PREDICTED?

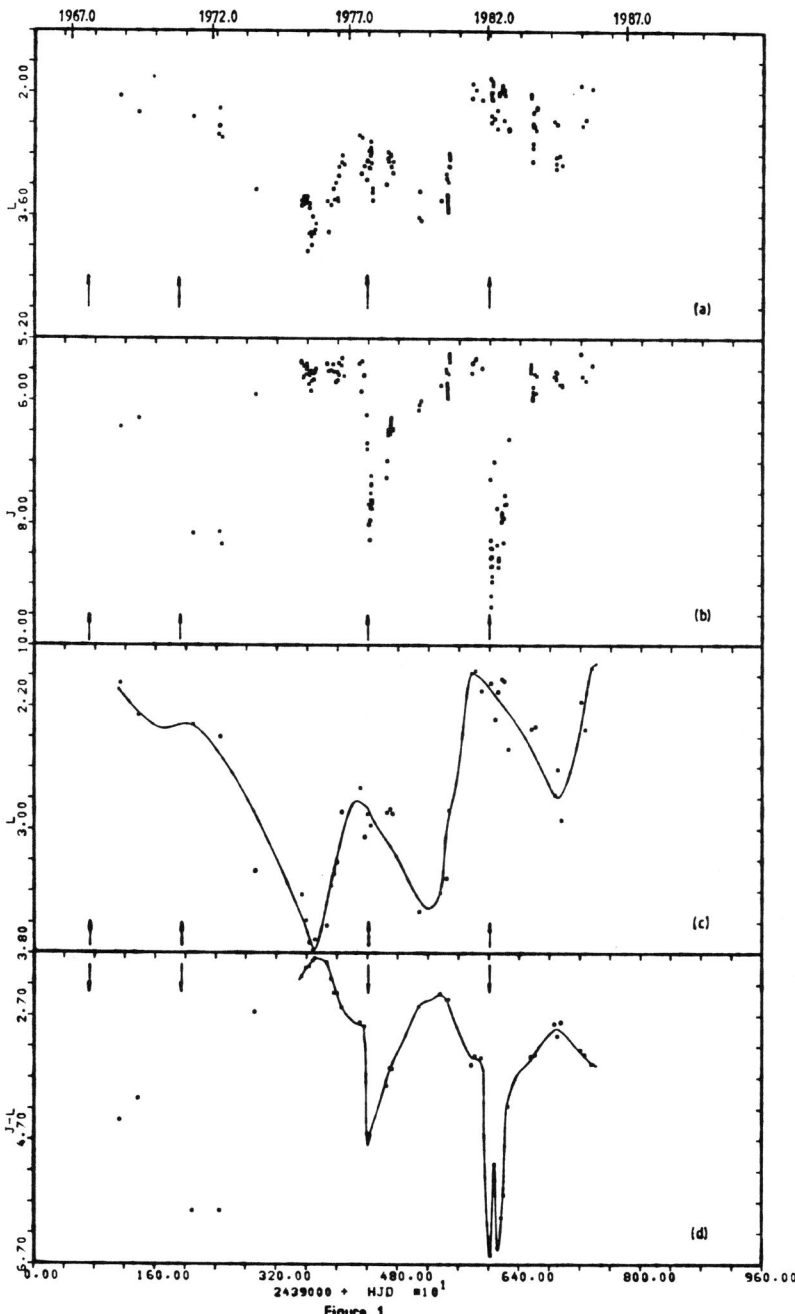

Figure 1 (a), (b) L, J light curves of RY Sgr.
(c), (d) L, (J-L) light curves binned into 38.57d intervals.
The vertical bars indicate times of onset of deep optical
minima. The continuous lines are freehand representations of
the long term variations.

5. CONCLUSION

With the long time scales involved in the RCB phenomenon, it is not easy to amass enough data to give a clearer picture of the mechanism involved. If the times of deep minima of RY Sgr are predictable, this will allow the star to be observed intensively before and during such an event. Spectra may give details of gas streaming motions and the onset of the chromosphere spectrum, while optical photometry may show whether the decline is related to a particular phase in the pulsation cycle.

6. ACKNOWLEDGEMENTS

The bulk of the data used here was obtained by many SAAO staff astronomers in pursuit of a general program to monitor RCB stars in the infrared. The first two points are due to Lee and Feast (1969) while two other early points come from Lee (1973).

REFERENCES

Feast, M.W., 1978. IAU Colloquium No.46, 246.
Kilkenny, D., 1982. Mon. Not. R. astr. Soc., **200**, 1019.
Lee, T.A. & Feast, M.W., 1969. Astrophys. J., **157**, L173.
Lee, T.A., 1973. Publ. astron. Soc. Pacific, **85**, 637.

RY SGR: PULSATION RELATED PHENOMENON

W. A. Lawson
Mount John University Observatory
Dept. of Physics, Univ. of Canterbury
Christchurch
New Zealand

The absorption line splitting reported by Cottrell and Lambert has been shown to be associated with the semi-regular pulsations. The event has a time scale of about 10 days (0.23 of a period) occuring about the radial velocity reversal at minimum radius. The velocity reversal is more rapid than indicated in an earlier radial velocity analysis by Alexander et al. There is evidence that the phase of the line splitting event is similar to the phase of the onset of the obscurational light declines.

1. INTRODUCTION

The low amplitude periodic variations of RY Sgr were first noted by Jacchia (1933). Jacchia found a 39 day period with an amplitude of 0.5 mag. in his own observations of RY Sgr, covering 1920 to 1932, and concluded that the star was pulsating. A photometric and spectroscopic survey of RY Sgr (Alexander et al 1972) confirmed that the star was a pulsating variable. A mean period of 38.6 days was found in the photometric and radial velocity curves.
    A series of high resolution spectra of RY Sgr, obtained by Cottrell and Lambert (1982), revealed absorption line splitting at certain phases. The line split spectra were obtained 80.8 days apart. The separation suggests a correlation between the line splitting and the 38.6 day pulsation period reported by Alexander et al. The relationship was not investigated due to a lack of data.

2. RESULTS

A continuing series of high resolution échelle spectra and accompanying photometry, initiated in 1983 (Lawson 1985), have been obtained with the two 61cm reflectors at Mount John University Observatory. An initial series of spectra, completed in June 1984, isolated the line splitting event and established an ephemeris to aid future observations. The spectral range chosen was 5000Å to 7000Å. From August to November 1984 (JD 2445924-6007) spectra and photometry were obtained as regularly as possible to determine the nature of the radial velocity variations and correlate the line splitting event. A plate showing line splitting was obtained on JD 2445971. The light and radial velocity curves are reproduced in Figure 1. The velocity reversal occurs as the star contracts to minimum radius and then re-expands.

Many strong absorption lines show splitting. Most of the split lines are C I and ionised species, for example; Ba II, Fe II, Si II and Ti II. Weak lines (10% to 15% central depth) do not appear split. These observations reflect comments made by Danziger (1963) about line split and line broadened spectra obtained by him in 1961. Danziger found many strong ionised lines were split and no obvious examples of split weak lines. He did not detect any split neutral lines. In contrast, Cottrell and Lambert (1982) found most, if not all, lines displayed splitting although the companion lines were often very weak. The ability to detect the weak companion lines is probably related to the superior signal-to-noise capability of the Reticon system employed by Cottrell and Lambert compared to the image intensifier/photographic plate system used in this programme.

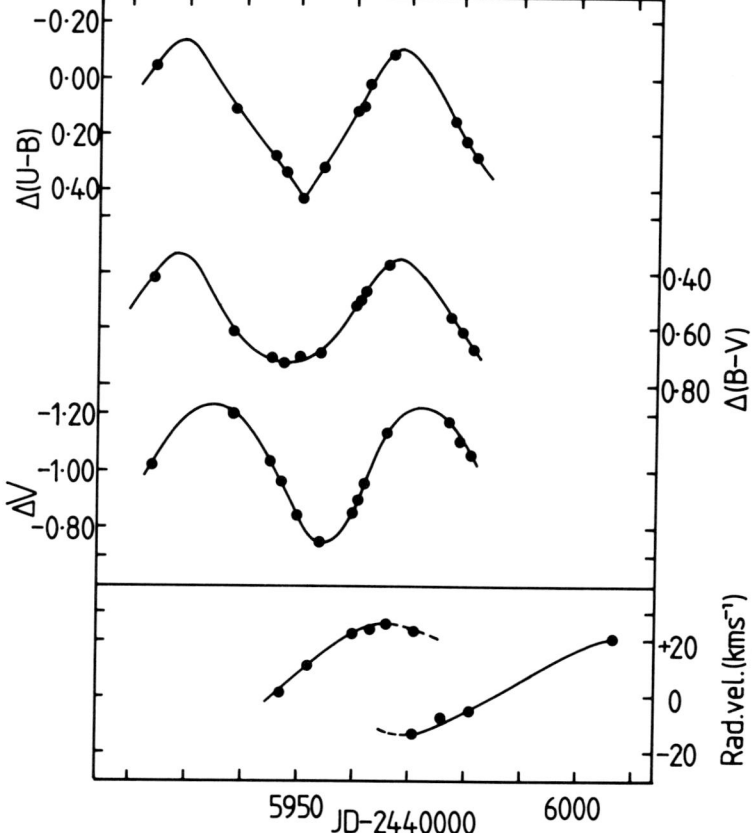

Figure 1. Photometric and radial velocity curves plotted against JD. The velocity split occurs at JD 2445971.

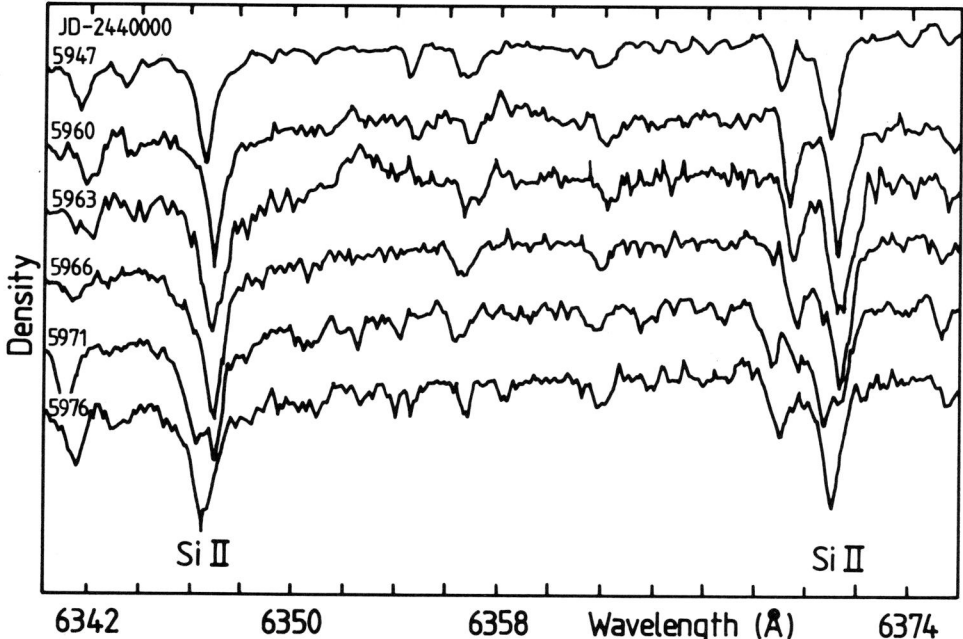

Figure 2. The evolution of the 6347Å and 6371Å Si II lines throughout the pulsation cycle.

The other notable variation seen in the absorption line spectra throughout the pulsation cycle is the general washing out of the spectrum approaching the line splitting event. This effect is especially noticeable in weak lines, many lines becoming almost indistinguishable from the continuum. The spectrum reverts to its normal form after the line splitting event and remains essentially unchcnged throughout the remainder of the pulsation cycle.

Spectra obtained 5 days prior, and 5 days after the line split spectra show little indication of the event other than minor asymmetric line profiles. This indicates the time scale of the event is about 10 days (0.23 of a period). The evolution of the 6347Å and 6371Å Si II lines throughout the pulsation cycle is illustrated in Figure 2.

The line splitting event has been interpreted as a shock wave appearing about the phase of minimum radius. This result is broadly consistent with Saio and Wheeler's (1985) models for $M < 1M_o$, $T_{eff} = 7000$ K pulsating R CrB stars.

3. DISCUSSION

The radial velocity reversal is more rapid than indicated in the radial velocity analysis of RY Sgr by Alexander et al (1972). Our data suggests the reversal occurs within 5 days (phase = 0.12) whereas the data of Alexander et al indicates the reversal occurs in a mean of 19 days (phase = 0.50). It is possible, because the low resolution spectra of Alexander et al failed to reveal line splitting, that the radial

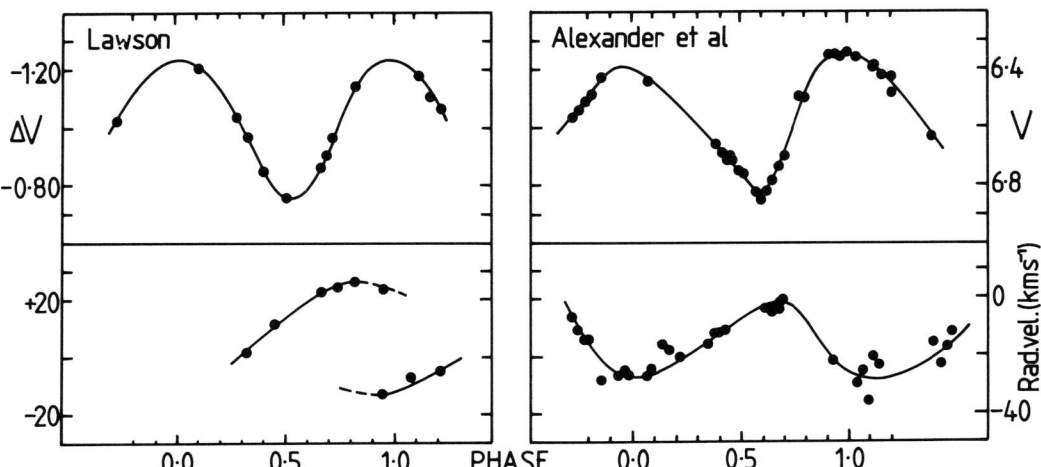

Figure 3. A comparison between the light and radial velocity curves of Alexander et al and our programme. The data of Alexander et al were obtained between JD 2440740 and JD 2440800.

velocity curve has been smoothed by averaging velocity measurements during the line splitting event. The light and radial velocity curves of Alexander et al and our programme are compared in Figure 3.

Pugach (1977) has found the beginnings of declines to obscurational minima are related to the phase of pulsation. The declines begin at, or just prior to, Vmax. Since the phase of Vmax is comparable to the phase of the line splitting event (see Figure 3), a correlation between the declines and the formation of the line splitting event appears likely.

4. CURRENT RESEARCH

Recent work has included continued photometric coverage and attempts to investigate the line splitting event in greater detail. Spectra obtained in June 1985 (JD 2446243-47) cover the decline of the event. Spectra covering the formation of the event are required to satisfactorily analyse the phenomenon.

REFERENCES

Alexander, J.B., Andrews, P.J., Catchpole, R.M., Feast, M.W.,
    Lloyd Evans, T., Menzies, J.W., Wisse, P.N.J. and Wisse, M.,
    1972, M.N.R.A.S., 158, 305.
Cottrell, P.L. and Lambert, D.L., 1982, Obs., 102, 149.
Danziger, I.J., 1963, thesis, Australian Nat. Univ.
Jacchia, L., 1933, P. Oss. Astr. U. Bol., 2, 173.
Lawson, W.A., 1985, thesis, Univ. of Canterbury.
Pugach, A.F., 1977, Info. Bull. Variable Stars, No. 1277.
Saio, H. and Wheeler, J.C., 1985, Ap. J., 295, 38.

DISCUSSION

WING: Your radial velocity curve for RY Sgr looks very similar to the one Abt obtained many years ago for the RV Tauri star U Mon, although I believe the splitting in U Mon was detected for a larger fraction of the cycle.

LAWSON: The line splitting in U Mon was detected for about half a period. Observations of shock waves in pulsating stars are not uncommon. W Virginis stars, for example, show line splitting. Saio and Wheeler have shown for R CrB stars that a shock is essential if the pulsation amplitude is to remain bound. Perhaps this is also true for other classes of pulsators.

N.K. RAO: Did I understand you said that there is line doubling at certain phases in R CrB?

LAWSON: There is perhaps an indication of line asymmetry in some R CrB spectra. More spectra are required.

LYNAS-GRAY: What is the maximum line-splitting observed in RY Sgr?

LAWSON: The degree of line splitting is 0.4 Å to 0.6 Å.

ANOMALOUS UV-EXTINCTION AND THE EFFECTIVE TEMPERATURE OF V348 SGR

D. Schönberner and U. Heber
Institut für Theoretische Physik und Sternwarte
der Universität Kiel
Olshausenstr. 40, 2300 Kiel, F.R.G.

Our idea of the evolutionary state of the hot R CrB-star V348 Sgr is seriously restricted by our ignorance of its effective temperature. A direct determination from photospheric lines is very difficult, owing to contamination by emission lines. Using Johnson photometry, Houziaux (1968) estimated a spectral type of B0 or B1. This result was, however, based on the assumption that V348 Sgr behaves photometrically as a main sequence star. V348 Sgr appears, however, to be an extreme helium star since hydrogen absorption lines are completely absent (Houziaux, 1968). This conclusion has been strengthened by a study of Heber et al. (1984) in which IUE-spectra (low resolution) of V348 Sgr are compared with those of well-known extreme helium stars. From the similarity between these spectra, it was concluded that V348 Sgr must be a helium supergiant with an effective temperature of about 16000 K, though its (dereddened) spectral flux distribution differs in a peculiar way from that of HD 124448 which has about the same temperature. It should, however, be noted that the dereddening was performed by eliminating the λ = 2200 Å dip with Seaton's law for the interstellar reddening. Obviously, the latter is not applicable when describing the extinction towards V348 Sgr.

Owing to the importance of a reliable temperature determination of this unique object, we decided to further examine the extinction problem. We assumed that the extinction to V348 Sgr is composed of two parts, namely a normal interstellar contribution which can be described by Seaton's parametrization, and a circumstellar contribution with different absorption properties in the UV. As a first preliminary choice for the latter, we selected an absorption law belonging to amorphous carbon grains (carbon "smoke") of r = 0.01 µm as measured by Stephens (1980). Figure 1 illustrates the extinction caused by these amorphous carbon grains as compared to the normal extinction law. In the optical region both are alike, but in the UV they differ considerably. The amorphous grains give more extinction in the near UV, but less for λ < 2500 Å. The peaked absorption occurs at $\lambda \approx$ 2500 Å, contrary to the normal law of extinction which peaks at $\lambda \approx$ 2200 Å. In this respect, it should be noted that Greenstein (1981) found the first evidence of circumstellar matter containing amorphous carbon when he studied the UV-spectrum of the central star of A 30. Also, Hecht et al. (1984) noted a peculiar extinction in UV-spectra of R CrB-stars and suggested glassy or amorphous carbon grains as being responsible.

For the interstellar extinction, we used the value found by Heber et al. (1984), E(B-V) = 0.45, which is necessary to remove the 2200 Å dip in the IUE-spectrograms. The additional contribution of the circumstellar carbon grains that is necessary to flatten the whole UV-spectrum was estimated by trial and error to E(B-V) = 0.15. The result is displayed in Figure 2 where the old fit of Heber et al. (1984) is also shown. With our choice of circumstellar extinction, the observations are now reasonably well matched by a helium model atmosphere (unblanketed) of 20000 K from the UV to the red spectral region. The fit, though, is not perfect since it appears that we somewhat overcorrected at $\lambda \approx 2500$ Å. The fit suggests a weaker bump for the carbon smoke extinction than that given by Stephen's (1980) measurements. Of course, we cannot expect to get final results with only one of different possible extinction curves for amorphous carbon grains (cf. Stephens, 1980). In the future we shall have to model the circumstellar extinction using better exposed UV-spectrograms of V348 Sgr.

We can, however, quite safely conclude that circumstellar extinction by amorphous carbon grains is important in interpreting UV-spectrograms of V348 Sgr, and its consideration leads to a substantially higher effective temperature of about 20000 K (cf. Fig. 2). The total reddening amounts to E(B-V) = 0.6, thus being close to the value of E(B-V) = 0.6 .. 0.7 which one gets by comparing Johnson photometry with theoretical colours of helium model atmospheres (Heber and Schönberner, 1981). Finally, we would like to emphasize that this reddening is only valid along the line of sight to the star. In fact, Dahari and Osterbrock (1984) determined a reddening up to E(B-V) $\approx$ 1.5 from emission lines. Since these lines originate in the shell, this large reddening must not be in contradiction to our result. Instead, it points to an inhomogeneous dust distribution around the star.

Dahari and Osterbrock (1984) identified a weak photospheric absorption at 4089 Å as being caused by Si IV. The very existence of Si IV in the photosphere of V348 Sgr would be in variance with our temperature estimate of only 20000 K. However, the Si IV line at 4116 Å is not reported by Dahari and Osterbrock, despite the fact that this line should be comparable in strength to the 4089 Å line. We therefore inspected the old tracings of Houziaux (private communication) and found no evidence of a photospheric absorption at 4116 Å. We concluded that the line at 4089 Å is obviously not due to Si IV, but instead to O II, λ = 4089.3 Å. The latter line is clearly evident in hotter stars, and a number of other strong O II lines have also been identified in V348 Sgr. Thus, the absorption spectrum of V348 Sgr, as far as is known, also appears to be consistent with an effective temperature of 20000 K.

REFERENCES

Dahari, O., Osterbrock, D.E.: 1984, Astrophys. J. **277**, 648
Greenstein, J.L.: 1981, Astrophys. J. **245**, 124
Heber, U., Schönberner, D.: 1981, Astron. Astrophys. **102**, 73
Heber, U., Heck, A., Houziaux, L., Manfroid, J., Schönberner, D.: 1984, Proc. 4th European IUE Conf., Rome, Italy, p. 367
Hecht, J.H., Holm, A.V., Donn, B., Wu, C.C.: 1984, Astrophys. J. **280**, 228
Houziaux, L.: 1968, Bull. Astron. Inst. Czechoslowakia **19**, 265
Seaton, M.J.: 1979, Mon. Not. Roy. Astron. **187**, 73
Stephens, J.R.: 1980, Astrophys. J. **237**, 450

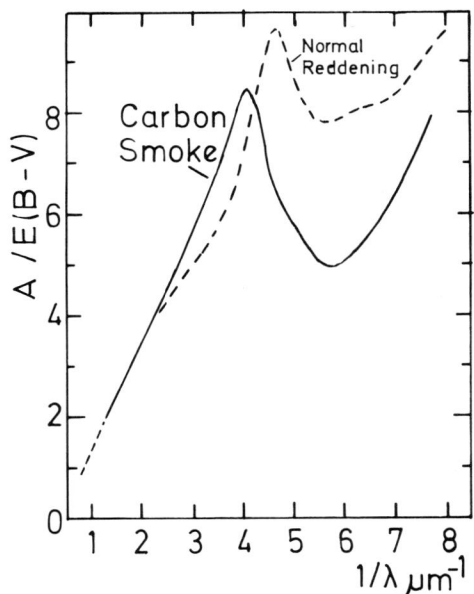

Fig. 1 The reddening law of carbon smoke consisting of amorphous carbon grains with r = 0.01 μm (Stephens, 1980). Figure adapted from Greenstein (1981).

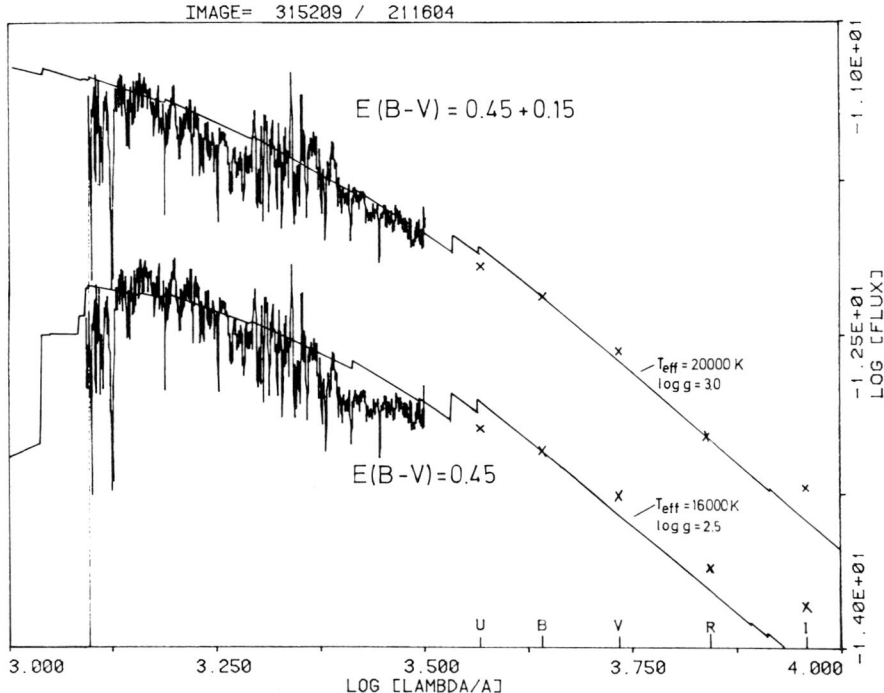

Fig. 2 Fit of dereddened spectra (SWP 15209, LWR 11604) of V348 Sgr with helium-carbon atmospheres for interstellar reddening E(B-V) = 0.45 (bottom), and for the same interstellar reddening plus a circumstellar contribution of E(B-V) = 0.15 with the law as shown in Fig. 1 (top).

ON THE MASS AND LUMINOSITY OF V348 SGR

D. Schönberner
Institut für Theoretische Physik und Sternwarte
der Universität Kiel
Olshausenstr. 40, 2300 Kiel, F.R.G.

In this short note, an attempt has been made to estimate mass and luminosity of the unique object, V348 Sgr, which appears to be a "Rosetta Stone" for our understanding of late stages of stellar evolution, as
i)   its photosphere is virtually hydrogen-free and carbon-rich as with an extreme helium star,
ii) it fades irregularly as does a R CrB-star and
iii) it is surrounded by a nebular shell similar to that surrounding Planetary Nebulae.

We know a great deal more about the above-mentioned objects than about V348 Sgr and this knowledge can be used to obtain estimates on mass and luminosity of the latter.

The UV-spectrum taken with the IUE-satellite reveals that V348 Sgr must be a supergiant as are other extreme helium stars (Heber et al., 1984). Together with its galactic coordinates ($l = 11°$, $b = -8°$) and its large radial velocity ($V_r \approx 160$ km/s), it is thus tempting to assume that it lies in the galactic bulge (see also Webster and Glass, 1974). Assuming therefore a distance $d = 9$ kpc, or $(V_o - M_o) = 14.8$, we get from the observations (with $V \approx 12$, $A_v \approx 2$) $V_o \approx 10$, $M_v \approx -4.8$. With BC $= -1.7$, as follows from $T_{eff} \approx 20000$ K (Schönberner and Heber, this volume), we arrive at $M_{bol} \approx -6.5$ or $\log L/L_\odot \approx 4.5$. This value is consistent with that normally assigned to extreme helium or R CrB-stars ($\log L/L_\odot \approx 4$).

Coming next to the mass, some analogies are helpful. It has already been mentioned that IUE-spectra of V348 Sgr show its similarity to B-type supergiants (Heber et al., 1984), which in turn have $\log(L/M) \approx 4$ (in solar units). Central stars of planetary nebulae also have $\log(L/M) \approx 4$ and, last but not least, extreme helium and R CrB-stars have $\log(L/M) = 4.1 \pm 0.5$ Assuming also for V348 Sgr a luminosity to mass ratio of this size, its mass must then be of the order of 1 $M_\odot$.

The above estimates of mass and luminosity for V348 Sgr are consistent with its observed properties and place it well into the category of peculiar low mass stars in a very advanced stage of their evolution. Further investigations of the properties of V348 Sgr will certainly also improve our knowledge of the origin and evolution of extreme helium and R CrB-stars.

REFERENCES

Heber, U., Heck, A., Houziaux, L., Manfroid, J., Schönberner, D.: 1984, Proc. 4th European IUE Conf., Rome, Italy, p. 367
Webster, B.L., Glass, I.S.: 1974, Mon. Not. Roy. Astron. Soc. **166**, 491

DISCUSSION

TUTUKOV: You said that there is something like a planetary nebula connected with the star. Could you give some parameters of this nebula, and some possible evolutionary scenarios?

SCHÖNBERNER: The nebular shell is of very low excitation since the star is cool. Probably only a very small part of the shell is ionized. I think it is premature to say anything about possible scenarios.

POTTASCH: Is anything known about the abundances in the nebula?

SCHÖNBERNER: Yes and no. The nebula is supposed to have normal abundances. We need better observations to clarify the situation concerning the nebular abundances.

LIEBERT: What is known about the abundances of the star V348 Sgr?

SCHÖNBERNER: Photospheric absorption lines of hydrogen are not detectable in the present observational material. The C II lines (especially in the UV) appear unusually strong. It is reasonable to assume that the photosphere is extremely hydrogen-deficient and somewhat carbon-rich, i.e. comparable in composition to the Extreme Helium Stars.

POTTASCH: The determination of the distance from the radial velocity is very uncertain. Planetary nebulae are known in similar directions with similar velocities and which are much closer.

SCHÖNBERNER: Agreed, it is only an estimate.

POTTASCH: Is there any other method of getting the distance?

SCHÖNBERNER: I don't think so.

FEAST: What is the absorption, is the interstellar absorption reasonable for the distance?

SCHÖNBERNER: The star is well out of the galactic plane and hence the interstellar absorption gives only a lower limit to the distance.

LYNAS-GRAY: Would the inclusion of amorphous carbon (circumstellar) absorption make any difference to the MV Sgr effective temperature determination?

SCHÖNBERNER: We will certainly have to investigate this in the future. However, if I remember correctly, in the case of MV Sgr there appeared no need for the introduction of an additional circumstellar absorption.

KILAMBI: Your estimate of 0.15 magnitude of circumstellar absorption is the result of fitting the energy distribution, or is there any other confirmation for that?

SCHÖNBERNER: Yes. This particular circumstellar extinction was necessary to match the UV energy distribution. I don't know of other confirmations.

FEAST: Observations of $\lambda 4430$ will also give limits on the interstellar absorption.

SCHÖNBERNER: Yes, but you need spectra with very good signal to noise ratio, which are presently not attainable.

THE LARGE MAGELLANIC CLOUD R CrB STAR - HV12842

D.H. Morgan and K. Nandy
Royal Observatory,
Edinburgh

N. Kameswara Rao
Indian Institute of Astrophysics
Bangalore 560034
INDIA

The star HV12842 is one of the 5 R CrB stars listed in the catalogue of variable stars in the LMC (Payne-Goposchkin 1971) and is located 4 deg. to the north of 30 Doradus. At this location it falls in the north-eastern part of one of two standard LMC fields used by the UK 1.2 m Schmidt Telescope. Since 1976 a series of I-plates (normally 90 minute exposures of hypersensitized Kodak IV-N emulsion through a Schott RG 715 filter) has been obtained on both the standard fields (LMC (N) and LMC (S)). This paper describes the behaviour of HV 12842 during the period 1976-1985 as it appears on the 38-I-plates of LMC(N) taken with the UK 1.2 m Schmidt Telescope. Earlier, the UBV magnitude at maximum light are given as V=13.65, B-V=0.51, U-B=-0.11 (Sherwood 1974) and the spectral type as F (Feast 1979).

The magnitudes of HV 12842 were estimated by eyeball comparison with a sequence of neighbouring stars for which magnitudes had been estimated through an eyeball comparison with the UBV photoelectric sequence No.VIII of Martin (1977). I-magnitudes in the Cousins photoelectric system were calculated for the stars of the photoelectric sequence using the (B-V) vs (V-I) colours of Cousins (1978) and an assumed reddening of E(B-V)=0.05 with a reddening relation E(V-I)/E(B-V)=1.25. The colours were checked by visual inspection of the available UKST objective prism plates available at the ROE.

These I-magnitudes were then transformed to UKST I-magnitudes using the transformation of Blair and Gilmore (1982). A new procedure has now been started using measurements made with the COSMOS measuring machine at the ROE (for details see MacGillivray and Stobie 1984). Machine magnitudes of all stars within 5 arcmin of HV 12842 and 5 arcmin of the Martin NO VIII sequence were obtained on plate 17385 (1981 Dec 5). This yielded a value of I = 13.4 for HV 12842 at maximum brightness. The photoelectric sequence and the HV 12842 regions suffer from very similar degrees of vignetting and there were no apparent density gradients across the 2 degrees between the regions.

The results are shown in Fig.1. Since the time difference between successive plates is sometimes very small compared with the total duration of the observations, points close together at maximum have been given slightly greater separation than is strictly accurate. The eyeball estimates are perfectly adequate to show the nature of the brightness variations of HV 12842.

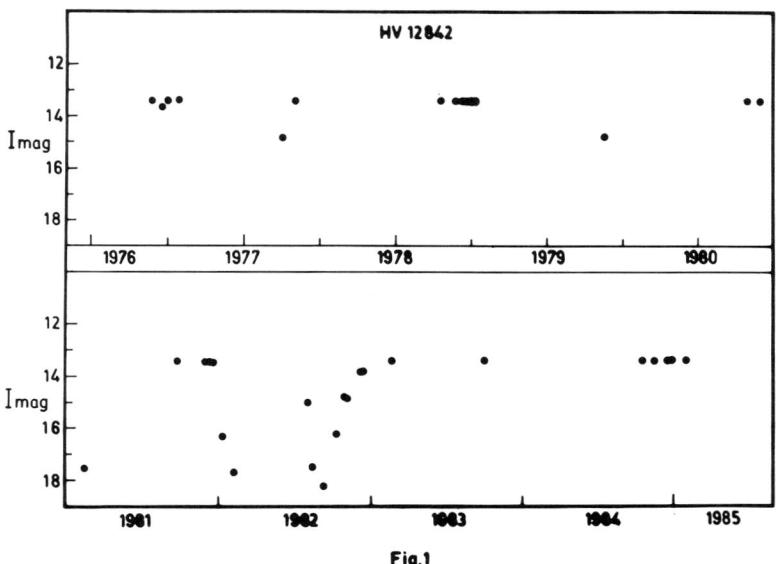

Fig. 1. I Magnitudes of HV 12842 obtained during the period 1976-1985.

Fig.1. shows HV 12842 to have the typical light curve of an R Cr B star with a constant magnitude at maximum brightness and a large number of complex minima. During the period concerned 5 minima **are** dectected Sep. 1977, Nov. 1979, Feb. 1981, Feb. 1982 and Sep. 1982 ;i.e. a minimum at least once in every 1.8 years similar to R CrB. The best monitored minimum is that of Aug-Dec 1982, though even then we cannot tell whether the minimum started from maximum light or when the star was brightening from an earlier minimum in Feb.1982. The recovery seems to be slower than the drop in light.

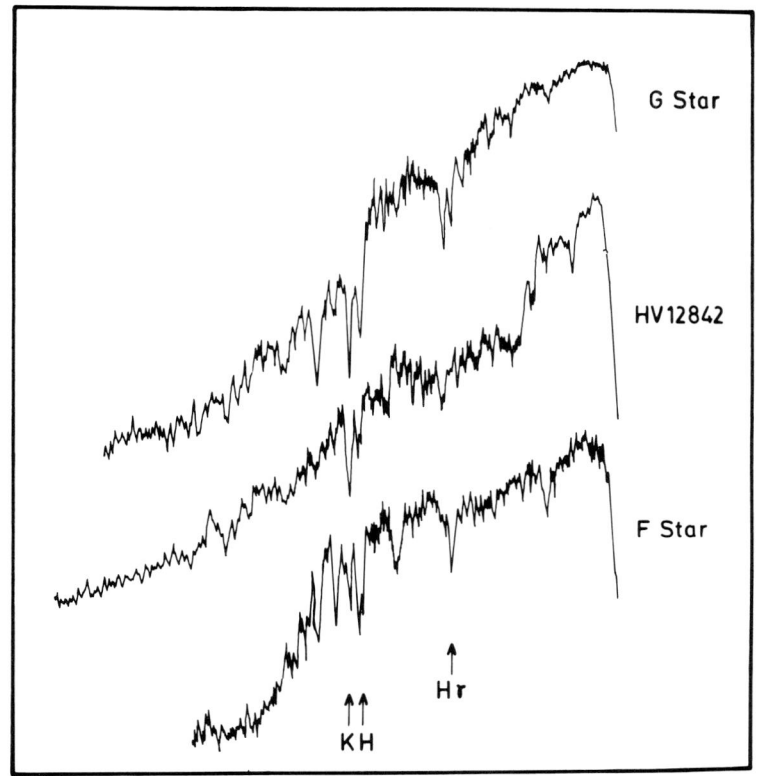

Fig. 2. Spectra of HV 12842 and two comparison stars of similar magnitudes.

Fig.2. shows a tracing of the spectrum of HV 12842 and some comparison stars of a similar magnitude and spectral class. These tracings were made from plate UJ 9015P, an UKST objective prism plate at 800 A/mm dispersion on hypersensitized Kodak IIIa-J emulsion exposed for 45 min. The plate was taken on 1984 Feb.3 when the star was at maximum. The plate was widened to give 80 micron images. For details of the prisms see Cannon et al. (1982). The spectrum of HV 12842 shows strong H and K lines and other weaker lines; but it does not show the drop in flux at 4000 A or strong hydrogen lines as would be normal in F-G stars. Assuming a spectral type of F5I with normal (V-I) colours and an LMC distance modulus at 18.5 we obtain, for HV 12842, V=13.9 and Mv = -4.6.

The reduction of COSMOS measure of HV 12842 in other colours from plates obtained during the minimum of Sep. 1982 has started and the results will be published later. However, these do indicate that particularly on the recovery part of the light curve the colors get redder than usual similar to other R CrB type stars.

REFERENCES

Cannon, R.D., Dawe, J.A., Morgan, D.H., Savage, A., and Smith, M.G., 1982, Proc. ASA., 4, 468.
Cousins, A.W.J., 1978: M.N.A.S.S.A., 37, 62.
Blair, M., and Gilmore, G., 1982, PASP, 94, 742.
Feast, M.W. 1979, in IAU Coll. 46 Changing trends in variable star research ed. F.M.Bateson, J. Smak, I.H. Ulrich, p.246.
MacGillivry, H.T., and Stobie, R.S., 1984, Vistas in Astronomy. 27, 433.
Martin, W.J., 1977, Memoirs R. astr. Soc., 83, 95.
Payne-Gapaschkin, C.E., 1971, Smithsonian No.13 Cont.
Sherwood, V.E., 1974, in IAU Symp. 67, p.147.

V HYDROGEN DEFICIENT BINARIES

HYDROGEN-POOR BINARY STARS

Mirek J. Plavec
Department of Astronomy, University of California
Los Angeles, California 90024
U. S. A.

ABSTRACT. Hydrogen-poor and helium-rich stars are easy to produce in interacting binaries. Thus they should be found among Population I binaries, in which a large-scale mass transfer has occurred between the components (possibly associated with mass loss from the system). For in such cases, those layers are now on the surface of the "loser" (and, most likely, also on the surface of the "gainer") that were subject to hydrogen burning and the associated mixing of processed material. Helium overabundance in these objects will be accompanied by an overabundance of nitrogen and underabundance of carbon, as a result of the CNO process. All the Algol-like semidetached binaries should be mild helium stars; so far this has been demonstrated only in β Lyrae, for the He/H ratio is not extreme in such cases. Extreme helium stars require a more complex process, with two stages of mass transfer and/or loss ("case BB"); υ Sagittarii and KS Persei seem to be good examples of this process. The optically invisible components of these two stars seem to have been detected with the IUE. Good model atmospheres do not exist yet, so caution must be exercised in interpreting the UV data.

1. ALGOL BINARIES AS MILDLY HYDROGEN-POOR STARS

Close binary star systems represent the simplest way to produce hydrogen-deficient and helium-rich stars. In every eclipsing binary system of the Algol type, both stars should show deficiency of hydrogen. Algol binaries are semi-detached systems: the mass-losing star ("loser") fills its critical Roche lobe and gas streams through the Lagrangian L_1 point toward the "gainer." The mass-transferring process begins when the more massive component fills all its available volume during its secular expansion. In many systems which we observe, this happened after the more massive star had left the main sequence and was crossing the Hertzsprung gap in the H-R diagram, expanding to become a red giant; this mode of mass transfer is known as "case B."

Soon after the onset of mass transfer, the process accelerates rapidly and the rate at which the loser loses mass may be quite high (on order of 10^{-4} $m_\odot$ per year for stars initially near 10 $m_\odot$. After the

loser becomes the less massive component, the process slows down
significantly, and it is at this "slow phase" that we observe the Algol
binaries. Algol, or β Persei, is the brightest object of this type, but
we know many others, since the eclipses of the hotter but smaller gainer
by the cooler but larger loser easily attract attention.

The initially more massive star filled its critical Roche lobe at
a time when all hydrogen had already been exhausted in the core, and
energy is now generated by hydrogen burning in a shell surrounding the
inert, helium-rich core. This shell acts as a barrier for mass loss,
which stops when just a bit of the star's envelope is left outside this
hydrogen-burning shell. Interior to that shell, the evolution of the
core proceeds undisturbed, as if nothing were happening to the star.
That is, the helium-rich core shrinks, becoming hotter and denser. In
stars initially less massive than 2.7 $m_\odot$, the contraction and heating of
the core is halted by electron degeneracy. When the core stops
contracting, there is also no longer any push on the envelope outside
the shell; rather, whatever has been left of it loses the support from
within and collapses on the shell. The star detaches itself from its
critical Roche lobe and the mass outflow ceases. The reduction in
radius is so drastic that the star virtually becomes a ready-made white
dwarf.

In stars initially more massive than 2.7 $m_\odot$ contraction is halted
by the ignition of helium, which now begins to be converted into carbon
by the triple-alpha process. This event reverses contraction in the
core into expansion, and by the so-called "mirror effect" (Kippenhahn
and Weigert 1967), the envelope begins to shrink. Again, the star
detaches itself from the Roche lobe and mass loss ends.

At the time of the termination of the mass transfer, the hydrogen
content at the surface of the star is about $X = 0.20$ (by mass) (Kriz
1969; De Greve, de Loore and van Dessel 1978), compared to the initial
(adopted) value of $X = 0.70$. Why should hydrogen be deficient at the
surface of the loser? The layers now exposed were initially part of the
convective core of the star. Hydrogen burning did not actually occur
there, but convective mixing maintained the same chemical composition
throughout the region, as long as it remained convective. But the
convective core shrinks with time, so that the hydrogen content in the
now-exposed layers is eventually stabilized.

If the hydrogen content by mass is down to about $X = 0.20$, then
the helium content is about $Y = 0.77$. Since the atomic masses are in
the ratio 4:1, we can say that at the end of mass transfer in case B, we
have a "half-and-half" mixture of hydrogen and helium, as far as the
number of atoms is concerned. For a long time, the loser has been
slowly sending this helium-enriched material toward the gainer.
Therefore, we can expect that the atmosphere of the gainer, too, will be
helium-enriched and, if there is any accretion disk around the gainer,
or any other kind of circumstellar material, that material, too, should
show hydrogen deficiency.

However, we cannot expect striking effects. After all, hydrogen
with its high continuous and line opacity still remains plentiful.
Klinglesmith (1971) published models with $X = 0.143$ and $Y = 0.857$, which
give us good insight. A comparison of these models with the more

detailed models calculated by Kurucz (1979) for atmospheres with normal composition shows that in the optical spectral region, deviations in the continuous flux are likely to be negligible. Larger deviations show in the far ultraviolet (in the region of the short-wavelength IUE "SWP" camera), but the picture is not clear for the following reason: Kurucz's models take into account line blanketing by thousands of lines, while Klinglesmith considered only line blanketing by hydrogen. The far ultraviolet region is rich on strong absorption lines even for temperatures 10,000 - 14,000°K considered by Klinglesmith, and the more transparent hydrogen-poor atmosphere produces deeper absorption lines than the normal atmosphere. This effect shows very strongly in Klinglesmith's models of the Balmer lines, which are strikingly deeper than in normal atmospheres. The same curious consequence of mild hydrogen deficiency also makes the Balmer jump deeper. These predictions cannot be tested on Algol systems, though. Many of them are surrounded by circumstellar matter which tends to fill in the Balmer absorptions to such a degree that we often observe emission lines instead. Similarly, a circumstellar hydrogen cloud often produces emission at the Balmer discontinuity, observed in the combined spectrum of the system as the so-called "(near)-ultraviolet excess."

It seems that the only way in which to detect the mild hydrogen deficiency in Algols is to study the metallic-line profiles at high dispersion and compare them with profiles calculated by means of spectrum synthesis on the basis of Klinglesmith's (or some more up-to-date) hydrogen-poor models. So far, this was succesfully done only in one case, by Balachandran et al. (1986) for the famous binary β Lyrae. There is another important abundance effect closely associated with depletion of hydrogen, namely a fairly drastic change of the abundance ratios C/N or even C/N/O as a by-product of the CNO bi-cycle. Most of carbon has been converted into nitrogen in the hydrogen-poor region, and this is precisely what the Texas people established for β Lyrae. They conclude that the Klinglesmith model with $X = 0.143$, $Y = 0.857$ (that is, by numbers of atoms, $N[H] = 0.4$, $N[He] = 0.6$)) is quite fitting. Considering the uncertainties involved at each step, the agreement with the prediction from the theory of mass transfer is very satisfactory. Surprisingly, it also tells us that β Lyrae is near the end of the mass-transfer episode rather than in its rapid phase. The currently adopted large inequality in masses, about 2 $m_\odot$ for the loser and 12 $m_\odot$ for the gainer, supports this conclusion.

Perhaps the fairly large strength of the circumstellar emission lines of He I in β Lyrae also indicates the overabundance of helium (see Figure 1). The circumstellar emission lines offer another method for exploring the abundance anomalies in Algols. I have discovered far-ultraviolet emission lines in at least five binaries apparently similar to β Lyrae (the W Serpentis stars) and in thirteen "ordinary" Algols (Plavec 1980; Plavec et al. 1984). Again, the best approach is to compare the relative strengths of the N V doublet at 1240 Å with the C IV doublet at 1550 Å. My spectra do show considerable differences between individual objects, and some show N V as stronger, contrary to what one expects for solar-like abundances.

2. BINARIES WITH EXTREME HELIUM STARS

While in β Lyrae and in the Algols the hydrogen deficiency is so mild that we must painstakingly look for it, two binary stars have long been known to show striking hydrogen deficiency in the spectra. These are υ Sagittarii (HD 181615/181616) and KS Persei (HD 30353). There seems to be no more than one hydrogen atom per 10,000 helium atoms in the atmospheres of the visible components in these systems. This practically complete depletion of hydrogen cannot be achieved in the mass transfer case B. Case C (in which the mass transfer starts when the more massive star is an expanding helium-buring giant) seems to be indicated by the long orbital periods (138 days for υ Sgr and 360 days for KS Persei). However, case C does not reduce X below 0.20 either (Lauterborn 1970), and does not therefore offer any good explanation.

I proposed a solution of this dilemma (Plavec 1973) in an invited talk which passed largely unnoticed. Paczynski (1971) studied the evolution of pure helium stars on the helium main sequence and concluded that they will move into the giant region of the H-R diagram provided their masses lie between about 1 and 2 m$_\odot$. At about the same time, Harmanec (1970) and Horn (1971) studied the final stages of case B for moderately massive stars. These stars shrink after the mass loss episode has ended and move close to the helium main sequence. Horn pointed out that mixing of the outermost layers can be expected after the hydrogen-burning shell dies out, since rapid contraction may cause rotational instability. This process will substantially reduce the abundance of hydrogen in the atmosphere, and whatever is left, will be passed over to the gainer since the loser may easily reach its critical Roche lobe for the second time. My suggestion was that υ Sagittarii and KS Persei are systems at this stage of evolution, which was subsequently called case BB by Delgado and Thomas (1981). The same model was independently rediscovered by Schönberner and Drilling (1983). Substantial progress in the theory of binary star evolution between 1973 and 1983 made it possible for them to make some more detailed comments. De Greve and de Loore (1977) found that case BB occurs for initial primaries in the range between about 8 and 15 m$_\odot$. Since the final mass after case B is given approximately as $m_{1f} = 0.04 \times m_{1i}^{1.62}$ (De Greve and de Loore 1977; for a more accurate formula, see De Greve 1982), these stars end up as helium stars with masses from just above the Chandrasekhar limit up to about 3.3 m$_\odot$. In the subsequent evolution, these helium stars expand again and cross the H-R diagram from left to right to become supergiants.

From various lines of evidence we know that υ Sagittarii and KS Persei are supergiants, and that their effective temperature cannot be far from 10,000°K (Schönberner and Drilling 1983, Drilling and Schönberner 1982). Helium stars with masses below 2 m$_\odot$ reach that region of the H-R diagram after they have exhausted helium in their cores and start burning helium in a shell. They expand to such large radii that they will fill the critical Roche lobe for the second time in most binary systems, and a new phase of mass transfer begins: this is the BB phase, at the end of which the helium-buring shell declines and the star shrinks again.

Our idea is that υ Sagittarii and KS Persei are observed just during the BB stage. This stage is short-lived, lasting on the order of 30,000 years, but the loser is a prominent supergiant, calling attention to itself by its anomalous chemical composition. Thus, in principle, the two peculiar binaries seem to be well explained. However, there are several constraints on the model which enable us to check on it. I will discuss here υ Sagittarii, with which I am more familiar. The star KS Persei (HD 30353) was thoroughly studied by Danzinger, Wallerstein and Böhm-Vitense (1967) and its ultraviolet spectrum was studied by Drilling and Schönberner (1982). KS Persei and υ Sagittarii show so many similarities that my "caveats" formulated in the following section will also apply to KS Persei.

3. CHECKING THE MODEL OF UPSILON SAGITTARII

Upsilon Sagittarii shows some signs that the mass loss is still going on, for example its Hα line is seen in emission in spite of the extreme hydrogen deficiency which must prevail in the circumstellar envelope as well. But the system is not hopelessly obscured by circumstellar matter, so we can assume that the loser is in a slower phase of mass loss. We can then conjecture that the mass of the loser will not be far from 1.5 $m_\odot$. Optically, the loser is the only visible component of the binary system. Its radial velocity curve derived by Wilson (1914) and by Seydel (1929) appears to be well determined (Eggen, Kron and Greenstein 1950), and gives K(1) = 49.1 km s^{-1} with an indication of a small eccentricity (e ≃ 0.05). The orbital period of the system is 137.96 days. This leads to a mass function f(m) = 1.693. With our estimate of m(1) = 1.5 $m_\odot$ within fairly small limits, the mass determination still requires the knowledge of inclination i. There were reports of eclipses, which would constrain this quantity very significantly, to values not very far from 90°. Unfortunately, there is no certainty.

Gaposchkin (1945) announced the detection of two minima of depths 0.15 mag and 0.08 mag respectively, from a survey of the Harvard patrol photographic plates. Because of the very small ranges, one must accept this report with caution. Eggen, Kron and Greenstein (1950) did observe the two minima photoelectrically, but with only one-half the amplitudes. A systematic coverage of υ Sagittarii in 1979 by J.J. Dobias, R.P.S. Stone and me failed to detect these eclipses, although small light fluctuations were seen, and very shallow eclipses could be obliterated by them. The deeper minimum, according to the two above-mentioned sources, occurs when a hotter but unseen component is to be eclipsed. This component would then be much smaller and optically fainter, so the shallowness of the eclipses does not surprise. Perhaps i = 80° is a good guess if the eclipses really exist; but if not, i may be much smaller. We must consider both alternatives.

If the system is eclipsing, then the mass of the gainer should be about 3.6 $m_\odot$ and the mass ratio would be 2.4:1 in its favor. The separation between the components would be about 190 $R_\odot$ and the mean radius of the loser would be near 60 $R_\odot$. We could then predict a radial velocity range of K(g) = 20 km s^{-1} for the unseen star. If the system does not eclipse and i = 60°, the mass of the gainer would be 4.6 m$_\odot$, and for i = 40° it would be 8.8 m$_\odot$. The system would measure well over 200 R$_\odot$ across, but the size of the loser would remain nearly the same on account of its decreasing share of the total mass of the system. Another consequence would be the diminished radial velocity range predicted for the gainer, from 20 km s^{-1} at i = 80° to 8 km s^{-1} at i = 40°.

If, in spite of the antics of the loser, the gainer still is a main-sequence star, which is quite possible, we can estimate its properties: we should expect a star between B6 and B2, depending on its mass. Such a star should be more easily detectable in the ultraviolet, and it appears that it had indeed been detected. The first report came from Duvignau, Friedjung, and Hack (1979), who found evidence for it with the Copernicus and S 2/68 satellites. The object was at the limit of detectability for Copernicus, but some lines were identified, including the N V resonance doublet at 1240 Å. This identification as well as the general appearance of the spectrum in the region 1159 - 1254 Å suggested the resemblance to B 0 supergiants. The S 2/68 observations in the region 1400 - 1600 Å are rather confusing but do suggest a fairly early spectral type, more likely of higher luminosity than not. The Copernicus observations suggested a radial velocity range for the gainer of 12 $\pm$ 8 km s^{-1}. This value would be best compatible with i $\simeq$ 50° and mass of the gainer m(g) = 6 m$_\odot$, but of course the uncertainty is too large. The observations failed to detect eclipses, which should be deeper in the ultraviolet.

Hack (1981) reports on her observations with the IUE: "The continuum clearly shows the presence of a hot companion and a reddening of about +0.30 [i.e., E(B-V) = 0.30 mag]; a very good fit of the observations is obtained with a composite spectrum formed by adding the flux of Alpha Cyg (A2 Ia) reddened for E(B-V) = 0.31 to the flux of Zeta Oph (O9.5 V, E(B-V)=0.31) reduced by a factor of 35. Hence, if the primary is an A-type supergiant, M$_V$ = -7 or -8, ... the companion should be an O9 dwarf which at λ 1500 has about the same luminosity as the primary, but in the visual is about 100 times less luminous than the primary." Contrary to that, Schönberner and Drilling (1983) have found that the combination of a B2 Ib supergiant with an A supergiant which is 5 mag brighter in V provides a much better fit both for the continuum and the absorption lines. But it creates a puzzle how a B2 Ib supergiant can be much less luminous that the A supergiant. Schönberner and Drilling attempt to circumvent this difficulty by assuming that the secondary is not in thermal equilibrium because it is accreting matter from the loser. However, a high rate of mass transfer is not supported by any observations. Moreover, if it existed, an accretion disk would surround the gainer and luminosity predictions would much depend both on dM/dt as well as on the inclination of the disk with respect to the line of sight; we would quite possibly obtain a distinctly non-stellar flux distribution as the disk could easily dominate over the star's radiation.

The above fitting attempts are about the best we can do, but it is necessary to realize that this approach cannot yield the correct picture. The optically observed star is extremely hydrogen-poor and so is most likely the companion. Ordinary stars with solar-like composition cannot be used to represent their respective flux distributions. I would like to demonstrate my point with the aid of Fig. 2. This figure shows the flux distribution of υ Sagittarii obtained at the end of May, 1980, from my observations made nearly simultaneously with the IUE and the Lick Observatory ITS scanner at the 3m Shane telescope. The two scans do not match too well near 3200 Å, and I left a gap there, since neither instrument gives a reliable response in the vicinity of this wavelength. The mismatch does not seriously affect the overall picture of flux distribution. The color excess used is E(B-V)=0.25, and was obtained by eliminating the 2200 Å interstellar bump. The uncertainty of this value is about 0.05 mag, but a much larger margin of error is unlikely, so the actual fluxes are fairly well represented by my diagram.

The optical spectrum can be very well matched by two different models: One is a Kurucz (1979) model with a normal composition (N(H) = 0.9, N(He) = 0.1), an effective temperature of 11,000°K, and log g = 2.5. The other model shown is Klinglesmith's model with hydrogen reduced to N(H) = 0.06, N(He) = 0.94, an effective temperature of 10,000°K, and log g = 2.5 again (Klinglesmith 1971). The surface gravity has a negligible effect, and the effective temperatures and color excesses could be adjusted to provide an even better match. The helium-rich model postulates deep Balmer lines, while none are actually seen at the low resolution of the scan (about 7 Å). It is true that one could argue that the hydrogen lines are filled in by emission which shows at Hα, but a more decisive cause is no doubt the virtual absence of hydrogen in the star. Apart from that, both models represent the Paschen continuum equally well and ...

... They equally dismally fail shortward of the Balmer discontinuity, but each of them in a different way. Both postulate a huge Balmer jump, but there is very little of it in the star. Thus, it is not surprising that the models offer no guidance between the Balmer jump and about 2500 Å. The hydrogen-poor model then roughly represents the observed flux down to about 1500 Å, beyond which wavelength one has to postulate a sharply rising flux from the other component. On the other hand, the 11,000°K model does not require any additional source at short wavelengths; in fact, there is a deficiency of flux there! One can certainly argue that this model is not valid for a hydrogen-depleted star, and I can only counter by saying that it is just as much or as little valid as the combinations of two standard spectra discussed above.

Line blocking is very serious in the ultraviolet in the Kurucz model, and will be even more serious in the hydrogen-depleted model, since the absorption lines will be deeper. Quite possibly the alleged supergiant character of the hotter spectrum is due just to this line blocking. I am almost tempted to speak of the "alleged supergiant character of the alleged hotter spectrum." For we really do not know what the shape of flux distribution is for either star.

A powerful blow to a simple interpretation has now come from an observation made by Polidan with the <u>Voyager</u> spectrometer, which is sensitive down to the Lyman limit. The recorded spectrum (Figure 3) does show the sharp rise in flux between 1450 Å and 1250 Å, but it also shows an equally abrupt decline at 1100 Å. In this respect, this is a unique spectrum, since no other object shows it. An early B star of any luminosity class maintains high flux levels around 1100 Å. You will notice that Polidan uses a smaller value of the color excess (0.12 mag), but the interstellar reddening curve seems to be rather flat there, so interstellar reddening cannot be responsible for the abrupt drop of flux at 1100 Å.

In short, the character of the secondary component (the gainer) is far from known. What is most urgently needed above all is <u>modern model stellar atmospheres, with line blocking taken into account, for stars with various degrees of hydrogen depletion, all the way to pure helium stars.</u>

4. EXTREME HELIUM BINARIES AND BINARY STAR EVOLUTION

Careful IUE observations are highly desirable in order to establish: 1) the existence or absence of eclipses; 2) the radial velocity curve of the gainer, if observable.

Establishing the gainer's mass -- directly from radial velocities velocities and/or indirectly from spectrophotometry -- is very important for checking on the general model outlined here, and on evolutionary calculations for interacting binaries. If the present mass of the loser is near 1.5 $m_\odot$, then its initial mass must have been 10 - 11 $m_\odot$ (De Greve 1982). There is a puzzling discrepancy in the remnant masses at the end of case B between de Loore and De Greve on the one side, and Delgado and Thomas (1981) on the other: according to the latter authors, an initial mass of 9 $m_\odot$ yields, after mode B of mass transfer, a helium star of 2.0 $m_\odot$, while De Grève and De Loore give 1.66 $m_\odot$ as the remnant mass for a star initially of 10 $m_\odot$. But this does not change my argument: If the initial mass was 9 - 11 $m_\odot$ and is now 1.5 $m_\odot$, then 7.5 - 9.5 $m_\odot$ of gas left the loser -- but only a fraction of that could have landed on the gainer, which according to our above estimates, should be about 3.6 - 6 $m_\odot$ now and almost certainly no more than 9 $m_\odot$, and naturally must have entered the mass transfer process with some decent mass of its own. The implication is that <u>the process seems to be more a process of mass loss from the binary system than mass transfer between the components</u>; and it may be that stellar wind plays a very significant role. There is currently a very extensive debate going on as to whether the mass transfer process is nearly conservative or not, and if the case of υ Sagittarii is well established, this would seem to be a powerful argument for strongly non-conservative processes.

We see that we still have a long way to go before we understand the helium-rich binaries to our satisfaction. But, if we improve our knowledge on these rather bizarre systems, we will also have a valuable test on the current theories of binary star evolution.

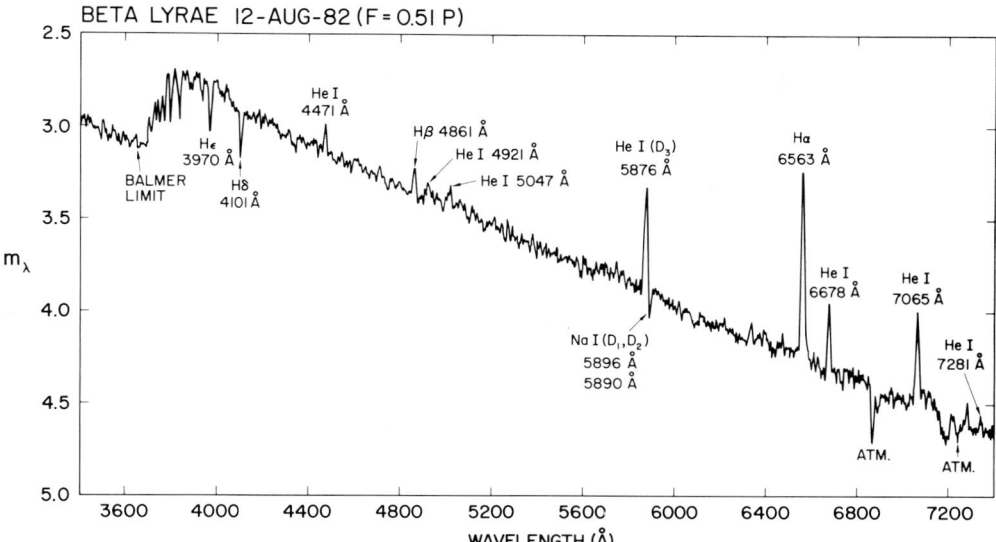

Fig. 1. - Optical scan of Beta Lyrae shows strong emission lines of HeI.

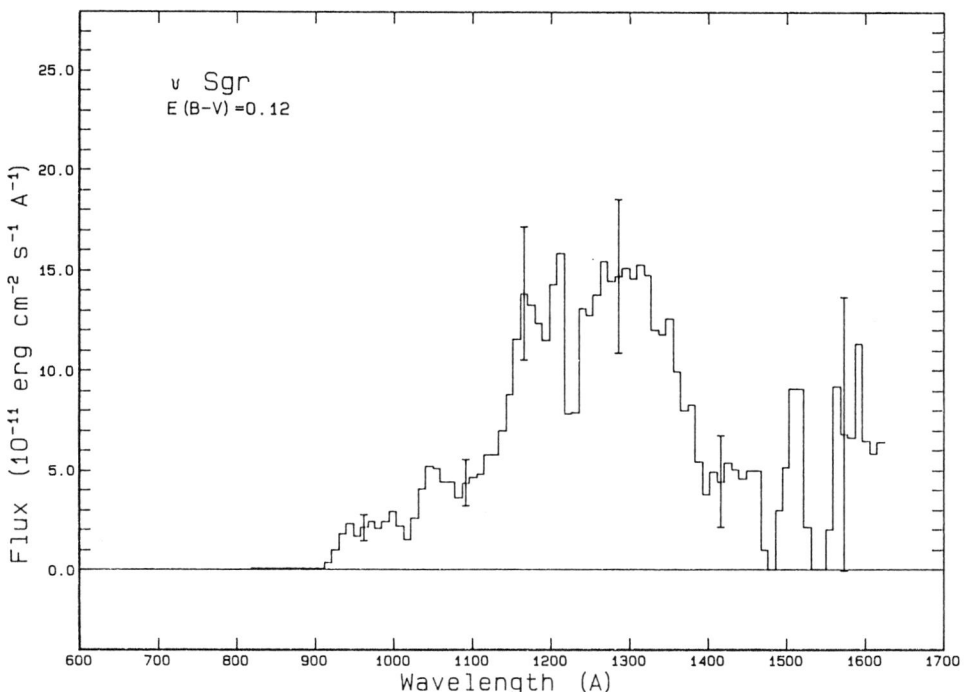

Fig. 3. Voyager scan of Upsilon Sagittarii shows no hot star flux.

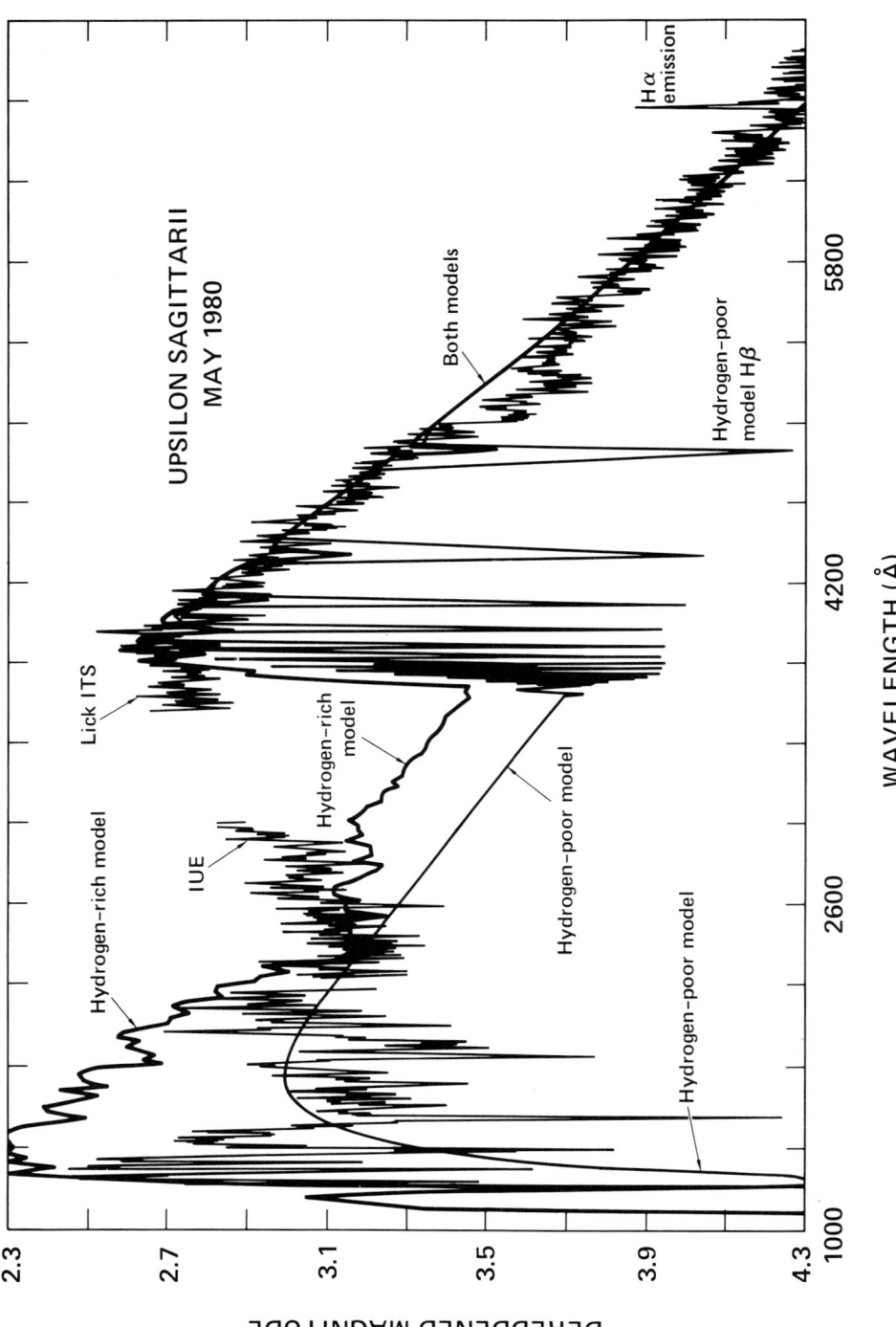

Fig. 2. - Flux distribution of Upsilon Sagittarii and two models.

REFERENCES

Balachandran, S., Lambert, D.L., Tomkin, J., and Parthsarathy, M.: 1986, Mon. Not. R.A.S., in press.
Danzinger, I.J., Wallerstein, G., and Bohm-Vitense, E.: 1967, Ap. J. 150, 239.
De Greve, J.P.: 1982, Ap. Sp. Sci. 84, 447.
De Greve, J.P. and De Loore, C.: 1976, Ap. Sp. Sci. 43, 35.
De Greve, J.P. and De Loore, C.: 1977, Ap. Sp. Sci. 50, 77.
De Greve, J.P., De Loore, C., and Van Dessel, E.L.: 1978, Ap. Sp. Sci. 53, 105.
Delgado, A.J. and Thomas, H.-C.: 1981, Astr. Ap. 96, 142.
Drilling, J.S. and Schönberner, D.: 1982, Astr. Ap. 113, L22.
Duvignau, H., Friedjung, M., and Hack, M.: 1979, Astr. Ap. 71, 310.
Eggen, O.J., Kron, G.E., and Greenstein, J.L.: 1950, P.A.S.P. 62, 171.
Gaposchkin, S.: 1945, A. J. 51, 109.
Hack, M.: 1981, in Photometric and Spectroscopic Binary Systems, ed. E. B. Carling and Z. Kopal (Dordrecht: Reidel), 453.
Harmanec, P.: 1970, Bull. Astr. Czech. 21, 113.
Horn, J.: 1971, Bull. Astr. Czech. 22, 37.
Kippenhahn, R. and Weigert, A.: 1967, Zs. Ap. 65, 251.
Klinglesmith, D.A.: 1971, Hydrogen Line Blanketed Model Stellar Atmospheres (Washington: NASA SP-3065).
Kriz, S.: 1969, in Mass Loss from Stars, ed. M. Hack (Dordrecht: Reidel), 257.
Kurucz, R.L.: 1979, Ap. J. Suppl. 40, 1.
Lauterborn, D.: 1970, Astr. Ap. 7, 150.
Paczynski, B.: 1971, Acta Astr. 21, 1.
Plavec, M.J.: 1973, in Extended Atmospheres etc., ed. A. H. Batten, (Dordrecht: Reidel), 216.
Plavec, M.J.: 1980, in Close Binary Stars: Observations and Interpretation, ed. M.J. Plavec, D.M. Popper, and R.K. Ulrich, (Dordrecht: Reidel), 251.
Plavec, M.J., Dobias, J.J., Etzel, P.B., and Weiland, J.L.: 1984, in Future of the Ultraviolet Astronomy, ed. J.M. Mead, R.D. Chapman, and Y. Kondo (NASA Conf. Publ. 2349), 420.
Schönberner, D. and Drilling, J.S.: 1983, Ap. J. 268, 225.
Seydel, F.L.: 1929, Pub. Amer. Astron. Soc. 6, 278.
Wilson, R.E.: 1914, Lick Obs. Bull. 8, 132.

DISCUSSION

N.K. RAO: We looked at the ANS observations of the Upsilon Sgr during 1974-75. Some of the observations show that there was an eclipse with a depth of about 0.1 mag in the ultraviolet which coincides with the phase of Hiltner's earlier optical observations. The companion is of the spectral type B8, but it is probably a disk rather than a star. We also investigated the IS reddening, the IS atomic lines and the polarization and found that M_V is around -4 rather than -7. This has already been published (J. Ap. Astr. **6**, 101, 1985).

PHOTOMETRIC AND SPECTROSCOPIC VARIABILITY OF THE HYDROGEN-DEFICIENT BINARY CPD-58°2721

[1]K.Morrison, [2]J.S.Drilling, [3]U.Heber, [1]P.W.Hill, and [1]C.S.Jeffery.

[1]University Observatory, Buchanan Gardens, St Andrews, Fife KY16 9LZ, Scotland.
[2]Department of Physics and Astronomy, Louisiana State University, Baton Rouge, LA 70803-4001, U.S.A.
[3]Institut für Theoretische Physik und Sternwarte der Universität, Olshausenstrasse 40, D-2300 Kiel, Federal Republic of Germany.

ABSTRACT. The hydrogen-deficient star CPD-58°2721 (=LSS1922) has been observed to show radial velocity variations with a range of 140 km s^{-1}. It shows light and colour variations with amplitudes of 0^{m}07 on a timescale of about 9 days. The absorption spectrum is nitrogen-rich and is seen to vary on a short timescale (4 days). CPD-58°2721 therefore appears to be a hydrogen-deficient binary similar to KS Per and $\vee$Sgr. The photometric variability is attributed to radial pulsation.

1. INTRODUCTION

CPD-58°2721 (=LSS1922) was found to be a hydrogen-deficient A star by Drilling (1980) with a spectrum very similar to that of $\vee$Sgr. Subsequent ultraviolet spectrophotometry by Schönberner et al (1982) yielded an effective temperature of 11100±500 K, very similar to that of $\vee$Sgr (T_{eff}=10500 K, Drilling et al. 1984a). Unlike $\vee$Sgr, but like KS Per, CPD-58°2721 does not appear to have an infrared excess (Drilling at al. 1984b).

The two known binaries ($\vee$Sgr and KS Per) do not show the large abundance of carbon that is observed in single extreme helium stars (e.g., Heber 1983). The low C:N ratio observed in the photospheres of the former implies that layers of the star depleted in hydrogen via CNO-cycle hydrogen-burning have been revealed. Schönberner and Drilling (1983) have shown that the hydrogen-deficient binaries are the product of case BB mass transfer (Delgado & Thomas, 1981).

Besides the spectroscopic similarity to $\vee$Sgr, no *direct* evidence has previously been found that CPD-58°2721 is a hydrogen-deficient binary. Here we report the discovery of variability in light, colour, radial velocity and spectrum.

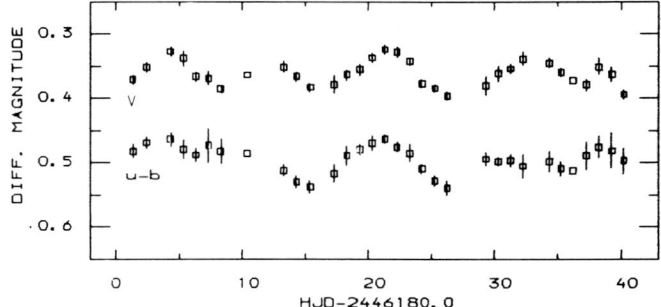

Figure 1. Differential Strömgren photometry of CPD-58°2721 obtained at SAAO in 1985.

2. PHOTOMETRY

Photometry of CPD-58°2721 was obtained during 6 weeks in 1985 May-June with the 0.5m telescope at SAAO. Observing techniques are described by Jeffery et al. (1985). Mean magnitudes and colours of the programme star and two comparison stars were:

	$\langle V \rangle$	$\langle b-y \rangle$	$\langle m_1 \rangle$	$\langle c_1 \rangle$
CPD-58°2721	10.349	0.574	-0.047	0.194
LSS 1915	9.990	0.338	-0.026	-0.034
HD93712	8.962	0.010	0.150	0.994

The differential Strömgren V and u-b curves are shown in Fig. 1. Both light and colour traces show variations of 0^m07 on a timescale of between 5 and 12 days with a mean period of about 9 days. Since the colour variations follow the light variations, they probably reflect effective temperature changes produced by radial pulsation.

3. RADIAL VELOCITIES

High resolution spectrograms have been obtained with the AAT, ESO 3.6m and CTIO 4m telescopes in the period 1982 July to 1985 October. A simple comparison of the Ca II H and K line-region shows large variable shifts of the stellar lines relative to the interstellar lines. Stellar radial velocities determined relative to the interstellar medium are given in Table I, where we find the amplitude of the velocity variation to be ~ 140 km s^{-1}. These velocity changes can only be produced if the star is a member of a binary system. The data of 1985 March and April indicate a period shorter than that of υ Sgr (138 days), but its exact value cannot yet be determined.

TABLE I.

Date	Relative Velocity (km s⁻¹)	Observer	
1982 Jul 2.4	0 ± 10	Hill	AAO
1983 May 30.1	-100 ± 8	Drilling	CTIO
1984 Apr 15.4	0 ± 10	Morrison	AAO
1985 Mar 13.6	-64 ± 11	Morrison	AAO
1985 Apr 6.1	+34 ± 5	Heber	ESO
1985 Apr 9.1	+38 ± 5	Heber	ESO
1985 Oct 7.4	-38 ± 5	Hunger	ESO

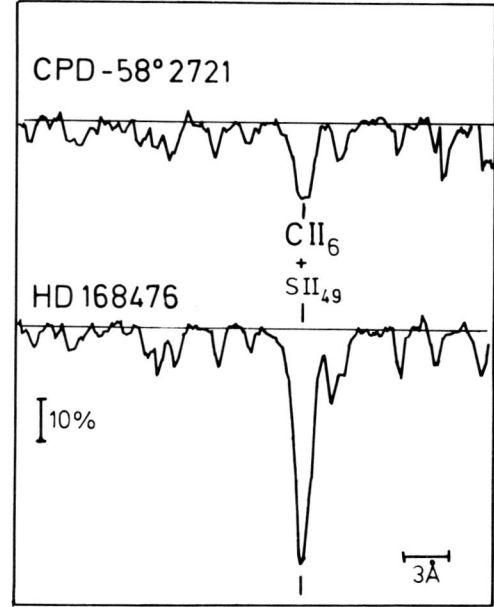

Figure 2. Comparison of the spectra of CPD-58°2721 (top) and the carbon-rich helium star HD168476 in the wavelength region of CII 4267Å.

4. SPECTRUM VARIATION

The spectrum of CPD-58°2721 shows strong lines of NII, while CII is weak. This is in contrast to the C-rich spectra of the extreme helium stars such as HD168476 (Fig. 2) but similar to the N-rich spectrum of KS Per (Wallerstein et al. 1967). Two echelle spectrograms obtained with CASPEC and the ESO 3.6m telescope on 1985 Apr 6 and 9 show marked changes in the strengths of low excitation metallic lines (Fig. 3). However there is no general weakening of the lines, but a clear dependence on the excitation potential.

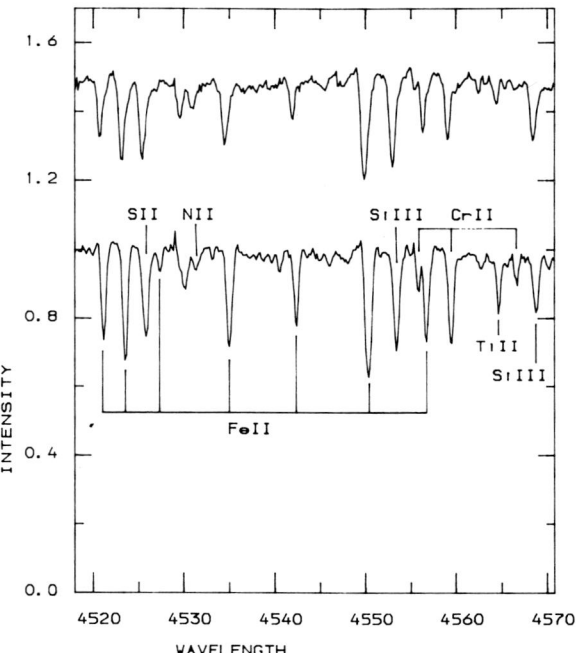

Figure 3. CPD-58°2721. Comparison of two CASPEC spectra taken on 1985 April 6 (top) and April 9 (bottom) in the wavelength region of SiIII (multiplet 1).

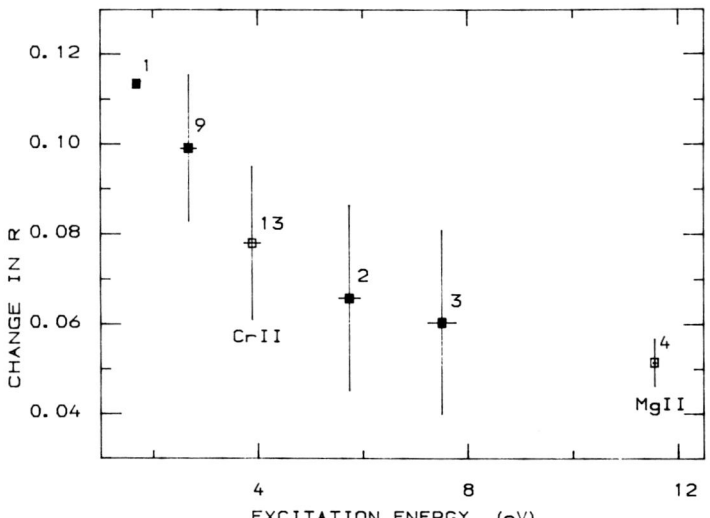

Figure 4. Changes in the residual intensities (R) of FeII lines (numbers given) between the two spectra of Fig. 3, shown as a function of excitation potential. Line-strength changes in MgII and CrII are also shown for comparison.

Lines with low excitation potentials show the largest changes, while those with high excitation potentials change less. This is demonstrated by the behaviour of the FeII lines shown in Fig. 4. Note also that the high excitation SiIII lines in Fig. 3 have almost identical equivalent widths in the two spectra. Hence we conclude that these spectrum changes are probably related to the effective temperature changes deduced from the colour variations.

The Hα and Hβ line profiles appear to be variable and consist of a photospheric absorption and an additional emission component (Fig. 5). The latter has a radial velocity curve different from the photospheric lines and at some phase, fills in the photospheric absorption line (see Fig. 5). Strong and variable Hα emission has been found in KS Per and υ Sgr and has been interpreted to be due to mass transfer through the Lagrangian point (Nariai 1967,1972). Our observation of variable Hα and Hβ emission suggests that mass transfer may also occur in CPD-58°2721.

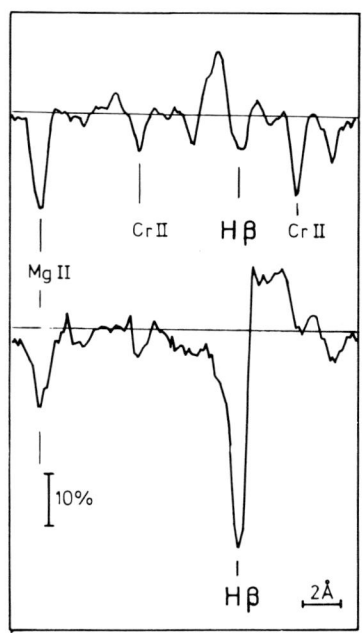

Figure 5. CPD-58°2721. Hβ profiles from two CASPEC spetra taken on 1985 April 9 (top) and October 7 (bottom)

5. DISCUSSION

We have shown that CPD-58°2721 is a single-lined spectroscopic binary which resembles the two other class members in (nearly) all observational respects. Its radial velocity amplitude ($\geq$140 km s^{-1}) is larger than that of KS per and υ Sgr ($\sim$100 km s^{-1}). All three objects have variable Balmer line emission indicative of mass transfer in the binary system.

The photometric variability of CPD-58°2721 may also be compared with the photometric variability observed in KS Per (Osawa et al. 1963) and in υ Sgr (Eggen et al. 1950). These variations have been variously attributed to pulsation and to eclipses. The observed variablility of the spectrum supports the interpretaion that CPD-58°2721 is pulsating.

Drilling (1986) reports that the absolute magnitude of CPD-58°2721 appears to match those of the single H-deficient stars. Jeffery et al. (1986) report a rough period-temperature sequence for variability in the single H-deficient stars. The mean period of CPD-58°2721 fits the Jeffery et al. sequence quite well and we therefore suggest that the pulsational properties of the H-deficient binaries are likely to be similar to those of the single H-deficient stars.

REFERENCES

Delgado,A.J. & Thomas,H.-C., 1981. Astr.Astrophys., 96,142.
Drilling,J.S., 1980. Astrophys.J., 242,L43.
Drilling,J.S., Schönberner,D., Heber,U. & Lynas-Gray,A.E., 1984a. Astrophys.J., 278,224.
Drilling,J.S., Landolt,A.U. & Schönberner,D., 1984b. Astrophys.J., 279,748.
Drilling,J.S., 1986. in 'Hydrogen-deficient stars and related objects', eds. K.Hunger, N.K.Rao, & D.Schönberner.
Eggen,O.J., Kron,G.E. & Greenstein,J.L., 1950. Publs.astr.Soc.Pacific., 62,171.
Heber,U., 1983. Astr.Astrophys., 118,39.
Jeffery,C.S., Skillen,I., Hill,P.W., Kilkenny,D., Malaney,R.A. & Morrison,K., 1985. Mon.Not.R.astr.Soc., 217,701.
Jeffery,C.S., Hill,P.W., & Morrison, K., 1986. in 'Hydrogen-deficient stars and related objects', eds. K.Hunger, N.K.Rao, & D.Schönberner.
Nariai,K., 1967. Publs.astr.Soc.Japan, 19, 564
Nariai,K., 1972. Publs.astr.Soc.Japan, 24, 495.
Osawa,K., Nishimura,S. & Nariai,K., 1963., Publs.astr.Soc.Japan., 15,313.
Schönberner,D., Drilling,J.S., Lynas-Gray,A.E. & Heber,U., 1982. in 'Advances in Ultraviolet Astronomy: Four Years of IUE Research' (NASA CP-2238), p.593.
Schönberner,D. & Drilling,J.S., 1983. Astrophys.J., 268,225.
Wallerstein,G., Greene,T.F. & Tomley,L.J., 1967. Astrophys.J., 150,245.

HYDROGEN DEFICIENCY IN ALGOL SECONDARIES

Praveen Nagar and K.D. Abhyankar
Centre of Advanced Study in Astronomy
Osmania University
Hyderabad - 500 007, INDIA

ABSTRACT. A study was carried out for 309 Algol systems to show that most of the secondary components of low mass-ratio systems are hydrogen deficient stars. This hydrogen deficiency is evident from their positions on a $\Delta \log L$ - $\Delta \log T$ plot, where Δ is the difference between the observed quantity, $\log L$ or $\log T_e$, and the value of the same quantity for a star of the same mass on the empirical Main Sequence. The secondary components are shown to occupy the region of mass losing helium dwarfs on this plot. A correlation has also been obtained between the excess of effective temperature and the mass ratio of the system.

1. INTRODUCTION

A method of distinguishing the mass-losing components of close binaries from the post and pre-main sequence objects was proposed by Abhyankar (1984). This method makes use of a transformation of variables ($\log L$, $\log T_e$) to ($\Delta \log L$, $\Delta \log T_e$), where the new set of variables is defined as follows:

$\Delta \log L$ = Observed $\log L$ - $\log L$ for a star of same mass on the empirical Main-Sequence,

and $\Delta \log T_e$ = Observed $\log T_e$ - $\log T_e$ for a star of same mass on the empirical Main-Sequence.

A plot of these two quantities, ($\Delta \log L$ vs. $\Delta \log T_e$), provides a powerful tool for distinguishing the stars in various stages of evolution. In Figure 1, which is taken from Abhyankar (1984), all the stars in pre-Main-Sequence contracting phase and the post-Main-Sequence expanding phase lie on the right hand side of the origin. The origin itself represents the empirical Main-Sequence. The track marked as I is for a rapidly mass accreting star of mass 1.5 M(Sun) in a binary according to Neo et al (1977) and the track marked as II represents the mass-losing component in a binary with initial masses of 1.8 and 0.7 solar masses; it is based on the calculations of Refsdal and Weigert (1969). The numbers along both the tracks indicate the mass of the star at various stages. The evolutionary tracks of helium white dwarfs in close binaries are marked as III and are based on the calculations of Webbink (1975). The helium Main-Sequence given by the dash-dot curve is due to Hansen et al (1972). Here track I shows that as the mass-gaining star heats up due to the infalling material it moves slightly to the right indicating a cooling of the surface due to the enlarged envelope which does not allow accretion of more mass.

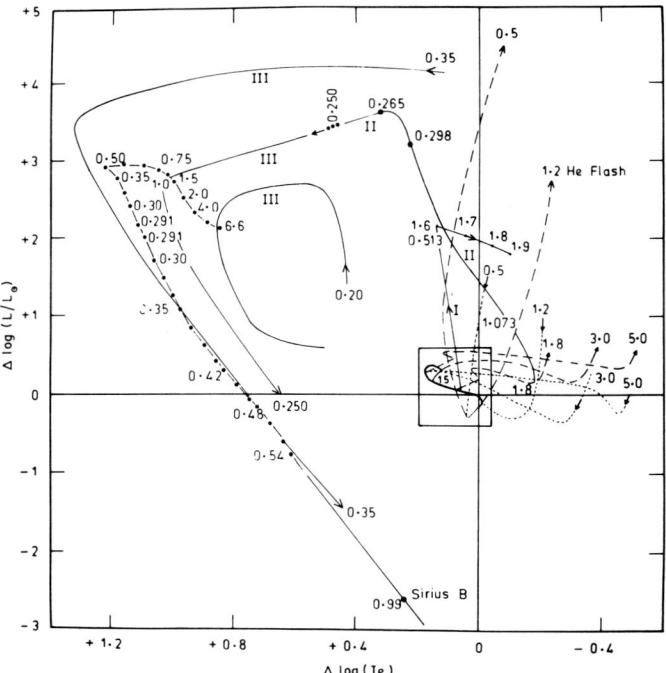

Figure 1. - A Δ log L - Δ log Te plot for the evolutionary tracks of various objects. The numbers along the tracks represent mass of the evolving star.

It is the track for the mass-losing component of the binary which is of the main importance to us here. As we can see in track II a mass-losing star gets hotter and more luminous compared to the star of same mass on the empirical Main-Sequence. As a consequence of the mass loss from the envelope the hotter layers are exposed which also have a lower hydrogen to helium ratio.

It was shown by Abhyankar (1984) that nearly 70 per cent of the secondary components of semi-attached systems with well determined absolute dimensions and mass lie in the region of mass-losing stars in which hydrogen to helium ratio is reduced considerably. But the number of stars in his sample was small, only 26 semi-detached systems if we exclude the few RS CVn systems. It was felt that in order to get a confirmation for his results a much larger sample is needed. In the next section we talk about the data used in the present study.

2. THE DATA AND RESULTS

A catalogue presented by Brancewicz and Dworak (1980) is the source of data for the present study. In this catalogue they have provided calculated values of the geometric and physical parameters for 1048 binaries. The intrinsic error in these calculated parameters is about 5 per cent. In the above mentioned catalogue 701 systems are classified as Algols.

Using another catalogue of classical Algol type systems by Budding (1984), we put these 701 systems in five different groups depending on their semi-detached status (SD) which varies from 0.1 to 0.9 in steps of 0.2. Since we are interested only in those systems where the secondary has evolved and has lost considerable mass from its envelope, we consider stars with SD = 0.9, 0.7 and 0.5 as true representatives of evolved Algols. Therefore, finally we are left with 309 systems that we have used here in the present study.

In Figures 2 and 3 the primary and secondary components of the above 309 systems are plotted in a $\Delta \log L - \Delta \log T_e$ plane. Systems with different SD status are shown with different symbols. Figure 2 shows the primary components to be concentrated around origin, which represents the unevolved Main-Sequence. But on the other hand the secondary components of the same systems have a marked luminosity and temperature excess as shown in Figure 3. It is also clear from that figure that the secondary components of the most Algol systems lie in the region occupied by the mass-losing stars and the helium white dwarfs. This indicates that these secondary components have lost a significant portion of their envelope and therefore are exposing their helium rich cores. It is almost certain that only a fraction of the mass lost by the secondary is transferred to the primary component, the rest of the mass is lost from the system thereby reducing the mass ratio q of the two stars.

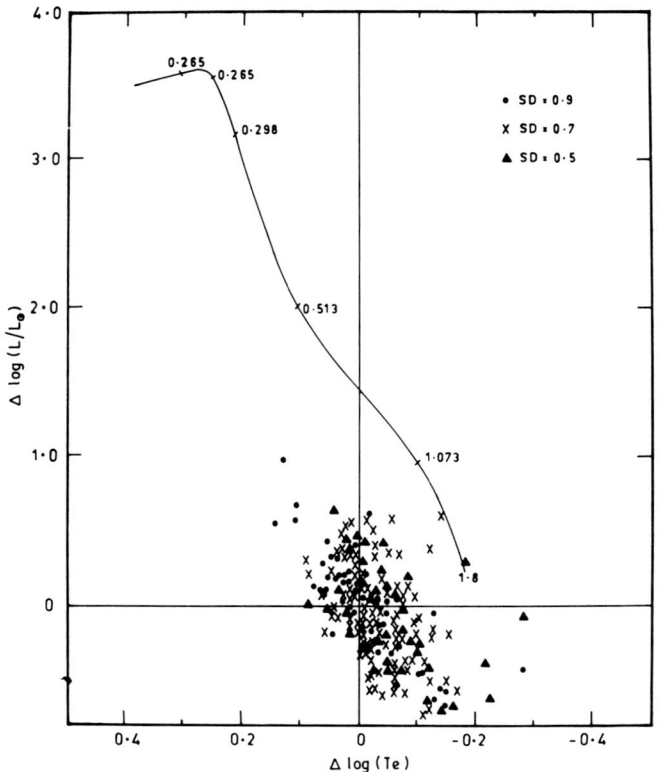

Figure 2. - The primary components of Algol systems plotted on $\Delta \log L - \Delta \log T_e$ plane. Stars with different semi-detached status are shown by different symbols. The solid curve is the evolutionary track of a mass losing star.

We have already seen in Figure 3 that in addition to the well known excess of luminosity we also have an excess of effective temperature ($\Delta \log T_e$) arising due to mass loss, therefore one would expect a correlation between $\Delta \log T_e$ and q.

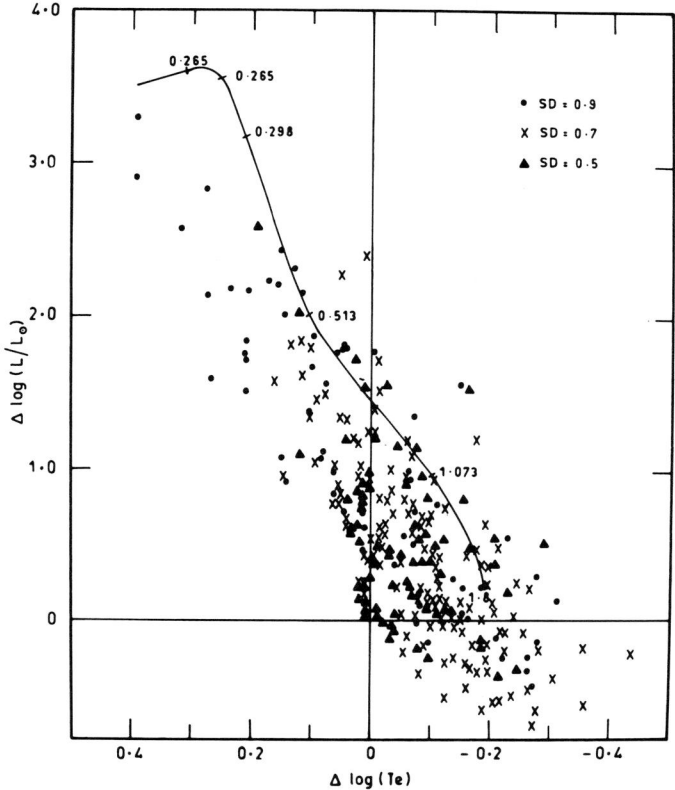

Figure 3. - The secondary components of Algol systems plotted on the $\Delta \log L$ - $\Delta \log Te$ plane. The different symbols and the solid curve have the same meaning as in Figure 2.

In Figure 4 we have plotted these quantities for all the Algols with SD = 0.9, .07 and 0.5. One can easily make out the trend that the temperature excess is less for the stars with high mass ratio and becomes more for the ones with lower values of q. In order to see this trend clearly we reduced the scatter by averaging the points lying in small intervals of q. The smooth curve obtained by joining these averaged points is also shown in Figure 4. It follows closely the trend expected from a mass losing component of the Algol System.

In Figure 5 we have plotted the luminosity excess against the mass ratio. This plot confirms the results of previous studies by Ziolkowskie (1969) and Giuricin et al (1983) that the systems with low mass ratios have large luminosity excesses for the secondary components.

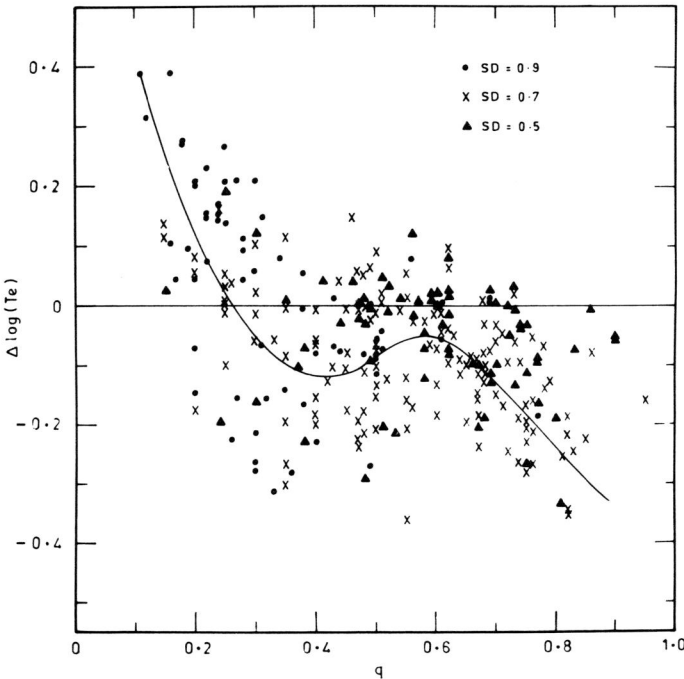

Figure 4. - A plot of temperature excess Δ log Te vs. mass ratio for Algol secondaries.

3. CONCLUSION

With the help of Figures 3, 4 and 5 we can conclude that the secondary of an Algol system with small mass ratio (q > 0.5) is likely to show temperature and luminosity excesses, and therefore can be classified as a hydrogen deficient star. It is clear from Figure 3 that good fraction of our sample occupies the region of mass losing stars and helium dwarfs. We suggest that spectroscopic observations should be made for these systems in order to do abundance analysis of the secondary components although these observations may not be easy to make due to the complications caused by the spectrum of the primary component.

Acknowledgements: The authors are thankful to the Department of Science and Technology, New Delhi for financial support in the form of a research grant.

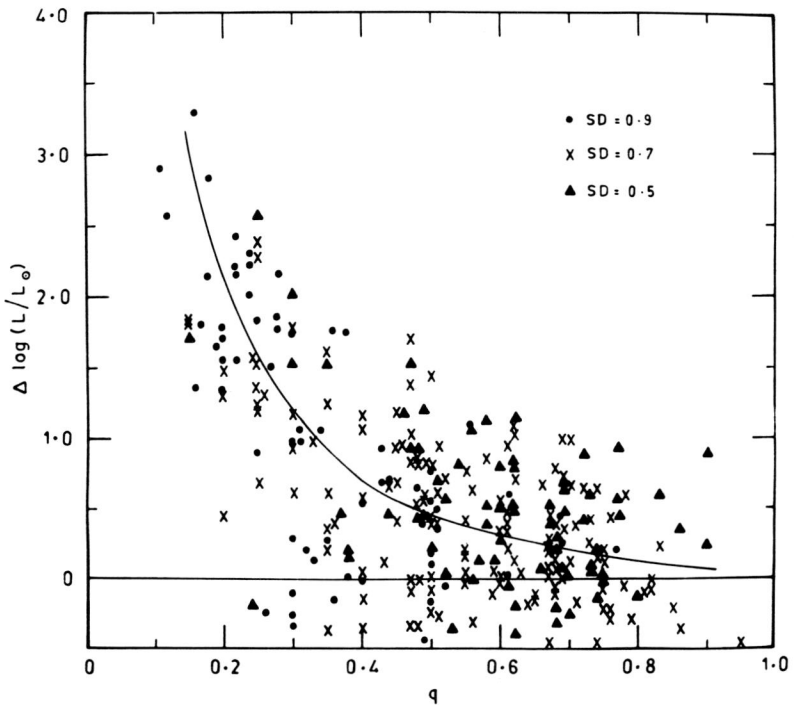

Figure 5. - A plot of the luminosity excess Δ log L vs mass ratio for Algol secondaries

REFERENCES

Abhyankar,K.D.: 1984, Astrophys. Sp. Sc. 99, 355.
Brancewicz,H.K. and Dworak,T.Z.: 1980, Acta. Astron. 30, 501.
Budding, E.: 1984, Review paper presented at the 3rd Asian-Pacific IAU meeting in Japan.
Giuricin,G., Mardirossian,F. and Mezzetti,M.: 1983, Astrophys.J. Suppl. 52, 35.
Hansen,C.J., Cox,J.P. and Herz,M.A.: 1972, Astron. Astrophys. 19, 144.
Neo,S., Miyaji,S. and Nomoto,K.: 1977, Pub. Astron. Soc. Japan 29, 249.
Refsdal,S. and Weigert,A.: 1969, Astron. Astrop. 1, 167.
Webbink,R.F.: 1975, Monthly Notices R.A.S. 171, 555.
Ziolkowski,J.: 1969, Astrophys. Sp. Sc. 3, 14.

VI   INTERMEDIATE HELIUM STARS

INTERMEDIATE HELIUM STARS:
ATMOSPHERIC PARAMETERS, OBLIQUE ROTATORS AND SHELLS

K. Hunger
Institut für Theoretische Physik und Sternwarte
der Universität Kiel
Olshausenstr. 40, 2300 Kiel, F.R.G.

1. INTRODUCTION

Intermediate helium stars are defined by

$$0.3 < n_{He}/n_{H} < 10$$

n_{He}/n_{H} being the number ratio of He over H. The upper bound which separates the intermediates from the extremes is well-defined (see Sect. 7), whereas the lower bound is rather soft.
   The first intermediate helium star discovered was Sigma Ori E (Greenstein and Wallerstein, 1956). Since then, 23 intermediates with $V \leqslant 11^m$ have been found (see the list of Walborn, 1983 and also the Annex), mostly through surveys such as that by McConnell et al. (1970) and McConnell (1972). (To this list, the newly discovered object SB 939, Langhans and Heber, 1985, should be added.) It is anticipated that a substantial fraction has as yet escaped detection. The spectral type centres around B2V while n_{He}/n_{H} is typically of the order of unity.
   Up to the mid-seventies, most of the work done on the intermediates was in photometry and spectral analysis of the photospheres. The results were reviewed by Hunger (1975) who gave a complete bibliography. In the following, mainly the period starting 1975 is reported. For the older literature the reader is referred to the above cited review. He is also referred to the reviews by Bolton (1983) and Hunger (1986).
   The previous results (Hunger, 1975) indicate that the intermediates do not form a uniform class but are divided into 2 distinct subclasses, one with $M \leqslant 2 \, M_\odot$, which may be linked to the extremes, and one with $M > 2 \, M_\odot$ which belongs to a young population, and which are often fast rotators with strong and variable magnetic fields and variable strengths of the absorption lines. In the former, He-enrichment would be the result of nuclear evolution while in the latter, He-enrichment may be brought about by diffusion (Osmer and Peterson, 1974, Vauclair, 1975). This subclass is considered to be the extension of the Ap-star sequence.
   The existence of a low-mass component has been questioned by

various authors (see the discussion in the paper by Walborn, 1983). In particular, the idea of a nuclear He-enrichment of a young star is hardly acceptable. Since the low masses are derived from low gravities, and these in turn from small equivalent widths, the calibration of (mostly ESO) photographic plates has also been questioned. As can be seen in Sect. 4, the disputed calibrations have been re-checked and found to be correct (Hunger, 1986). The small masses, hence, can only be ascribed to the distances being systematically underestimated. We concentrate on this important issue in Sect. 4. Clues as to whether we are dealing with nuclear enrichment or with diffusion will also come from the metal abundances (Sect. 3).

A major part of this review is devoted to the prototype Sigma Ori E as this is the best-studied object, both observationally and theoretically. The discussion covers atmospheric parameters, distance, mass (Sect. 4), He-variability, oblique rotator model (Sect. 5) and (partly) circumstellar material (Sect. 6).

The period 1978-1985 brought a wealth of data on the circumstellar matter around intermediates, mainly through IUE but also through ground-based IR, IRAS and VLA observations. For the variable component of the intermediates, the picture of a magnetic field modulated wind emerges which feeds circumstellar clouds. We have only briefly treated this interesting subject as it will be dealt with extensively in the review by Barker (this conference). Likewise, for the sake of space, we have not reported on the promising attempts to model magnetized stellar atmospheres (Madej, 1971; Carpenter, 1985) and also hot and windy magnetospheres (Nerney, 1980; Havnes, 1981; Havnes and Görtz, 1984; Nakajima, 1985).

2. STATISTICS, ROTATION

Among the intermediates listed by Walborn (1983), about one third (7 stars) are spectrum variables (Walborn, 1975; Bond and Levato, 1976; Pedersen and Thompson, 1977; Pedersen, 1979). Six of the seven stars have variable magnetic fields and one a static field (Landstreet and Borra, 1978; Borra and Landstreet, 1979; Borra et al. 1983;). Five have $v \sin i \approx 150$ kms^{-1}, with periods $P < 2d$, and three have $v \sin i \leqslant 30$ kms^{-1}, with $P \geqslant 9d$ (Walborn, 1983). All variables obey the $(P, v \sin i)$-relation for radii $R > 4 R_\odot$, i.e. all variables are considered as being oblique rotators. Four of them belong to the Orion aggregate, one to IC 2944. From these properties, the age τ is estimated to be: $10^6 a < \tau \leqslant 10^8 a$. The stars are listed in Table I.

TABLE I

List of He-variable and magnetic intermediates

	Period (days)	v sini (kms^{-1})
HD 37017	0.9^1	170^2
HD 37479	1.2	170
HD 37776	1.5	160
HD 58260	1.7	≤30
HD 64740	1.3$_3$	160
HD 184927	9.5	≤17
CPD-46°3093	variable ?^{4}	150
HD 96446	non-variable	≤30

1 Pedersen (1979)
2 Walborn (1983)
3 Bond and Levato (1976)
4 Groote et al. (1982)

The question of a bimodal distribution of the intermediates is also encountered in the statistics of rotational velocities (Walborn, 1983). From these statistics, the possibility that two classes of rotators exist, one with v sini < 80 kms^{-1} and one with v sini > 120 kms^{-1}, cannot be excluded. The statistics, however, depend largely on the way the data are presented. If a uniform binning of the data is used, then the histogram of rotational velocities flattens out and the distinction between slow and fast rotators is lost. In the same way, the differences between early main-sequence B stars and intermediates disappear. Furthermore, there is no correlation between v sini and galactic latitude so that, with this information, one has to conclude that the intermediates have essentially the same rotational velocities as main-sequence stars.

3. CNO ABUNDANCES

Walborn (1983) raised the question as to whether metal abundance anomalies can be seen in intermediates and whether they can clarify their evolutionary status. If helium enrichment were the result of nuclear burning, anomalies of the CNO elements would be present. Since no conspicuous anomalies could be found, Walborn concluded that all intermediates may be Population I main-sequence objects. One problem, however, at least with the helium variables, is that the metal lines are also variable (they vary in antiphase with the helium lines) (Hunger, 1974; Lester, 1979; Levato and Malaroda, 1979; Shore and Adelman, 1981; Walborn, 1982;). Analyses based on a few spectrograms may not be representative of the whole star.

The other problem is of a more general nature: abundance tables are often based on inhomogeneous sets of observations, on different model atmospheres, f-values and also on the way of presentation which renders direct comparison difficult. In addition to the analyses cited by Hunger (1975), the following abundance analyses have been published: HD 60344 (Kaufmann and Hunger, 1975), HD 64740 (Lester, 1976), HD 133518 (Gerlach, 1976), HD 120640 (as a result of the analysis now classified as a normal B star; Detz, 1977), HD 186205 (Lee and O'Brien, 1977) and CPD-46°3093 (Heber and Hunger, 1981; Groote et al., 1982).

A homogeneous table of abundances was published by Hunger (1975). It comprises 8 intermediates, 2 of which are helium variables and one (HD 144941) a border case between the extremes and the intermediates. If we leave out these 3 objects and adopt for hydrogen the normalization, $\log n_H = 12$, then we find a remarkable uniformity in the abundances of the (other) 5 stars: HD 168785, CPD-69°2698, HD 60344, HD 184927 and HD 96446. For the sake of space, we have quoted in Table II only the mean logarithmic abundances, plus the root mean square deviations. The latter proved to be smaller than the typically encountered ± 0.3 dex for the individual abundances, from which we conclude that there is practically only one single type of non-variable intermediate helium star. This statement also pertains to the helium abundances and to the metallicity Z. The range of effective temperatures and gravities is likewise very small, the differences, however, are probably real.

TABLE II

Logarithmic abundances (by numbers) of He, C, N and O, and the metallicity Z, for the average of 5 non-variable intermediate helium stars

	H	He	C	N	O
Intermediates	12.0	11.9 ± 0.05	8.4 ± 0.1	8.0 ± 0.2	8.4 ± 0.2
B-stars (Scholz, 1972)	12.0	10.95	8.6	7.9	8.9
Intermediates		Z	T_{eff}		log g
		0.01 ± 0.003	24600 K ± 2700 K		3.7 ± 0.3

The CNO abundances are essentially normal in the non-variable intermediates. The latter statement allows an important conclusion to be drawn, with regards to evolution. Even if only a very small fraction of helium is generated by nuclear processes, the very sensitive N/C ratio switches from < 1 to > 1 (Caughlan, 1965). Since N/C is found to be smaller than unity, it must be concluded that He enrichment in the non-variable intermediates cannot be the result of nuclear burning but must be due to a selective effect as with diffusion. This statement is not necessarily true for the variables as, according to analyses by Schacht (quoted by Hunger, 1975) and Lester (1976), nitrogen is significantly enhanced in HD 64740. There is also an example for a non-variable with nitrogen enhancement: HD 186205 (Lee and O'Brien, 1977).

Diffusion leading to He enrichment apparently acts only in a
narrow range of effective temperatures (22000 K < T_{eff} < 27000 K), and
in an even narrower range of gravities (3.4 < log g_{eff} < 4.0). According
to gravity, intermediates have already evolved away from the main
sequence or have not yet reached it. The metallicity (Z = 0.01) points
to the former interpretation.

Hunger (1975) quoted also the masses for the above-described
sample of non-variable intermediates. The masses are determined
spectroscopically (see Sect. 4) and, hence, depend quadratically on the
assumed distances. The mean mass proved to be 2.4 ± 1.5 $M_{\odot}$ which is
more than a factor of 3 smaller than the mass of a normal composition
star. Hence, if one wants to restore the normal B2-star mass (8 $M_{\odot}$),
one must assume distances which are a factor 1.8 larger than hitherto
determined.

Because of the importance of the mass problem (see also Odell,
1974), the atmospheric parameters and the distance of Sigma Orionis E
shall be discussed in some detail in the following section.

4. ATMOSPHERIC PARAMETER AND MASS OF Sigma Ori E

As the mass is of crucial importance when one wants to find out whether
helium-enrichment is due to diffusion, and hence only a surface effect,
or whether it is due to evolution, as is the case with the extreme
helium stars, we shall now discuss the mass of Sigma Ori E, the so far
best-studied example of an intermediate.

The masses of the intermediates are determined spectroscopically
as there are no binaries known in this class, except for HD 37017. The
principle of the method is as follows: spectral analysis yields T_{eff}
and gravity g while photometry yields apparent magnitudes. If the
distance is known and hence the absolute magnitudes, then from
$L = 4R^2 \pi \sigma T_{eff}^4$ the radius R can be determined, and from $g = GM/R^2$
finally the mass M. The 3 quantities that determine the mass, hence,
are T_{eff}, gravity and distance.

T_{eff} is determined from the IUE-flux plus V-magnitude (Remie and
Lamers, 1982). As most of the flux of a B star is carried in the IUE
band, T_{eff} is determined with a high precision, namely with ± 2.5%.
(This method is largely independent of gravity and chemical composition
in the parameter range under consideration.)

Once T_{eff} is known, gravity and helium content can be determined
from the profiles of the Balmer and helium lines. (This has to be done
simultaneously - see the paper on Sigma Ori E by Groote and Hunger,
1982.) Up to the present paper, only photographic spectrograms have
been used. Since gravity depends sensitively on the profiles, the
calibration of the photographic plates is important. The ESO
calibration, which yields small equivalent widths and hence
uncomfortably small gravities and masses, has been questioned (see
Walborn, 1983). To settle the dispute, ESO CASPEC spectrograms have
been taken from Sigma Ori E. (This spectrograph employs a linear CCD
detector.) The result is shown in Figure 1, where the profile of H_γ is

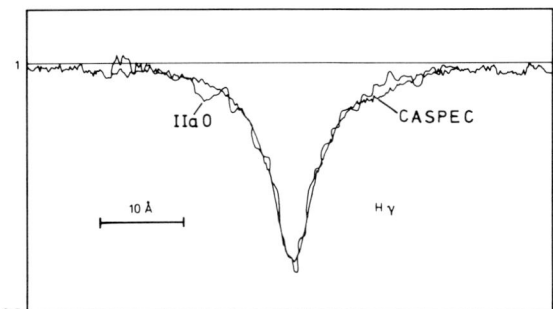

Fig. 1 The profiles of H_γ (Sigma Ori E) obtained photographically (IIa O) and with CASPEC.

reproduced. It fully confirms the photographic profile and, hence, clearly demonstrates the reliability of the ESO calibration. Gravity indeed comes out as small as quoted in the paper by Groote and Hunger (1982): log g = 3.85, again with good precision (Δ log g = $\pm$ 0.15), which means that Sigma Ori E is not on the main sequence but is evolving either to or away from it. (In the derivation of T_{eff} and log g, a slight inconsistency is introduced, in as much as for the continuous flux distribution (Kurucz, 1979) normal composition models are employed, while for the line profiles, He-rich models are used. The latter do not account for metal line blanketing. Consistent He-rich metal line blanketed models may lead to gravities which are larger by Δ log = 0.1.)

The only quantity left which could be blamed for the small mass of our prototype object is the distance. If one employs the dynamical parallax of Sigma Ori AB, d = 400 pc (Heintz, 1984), then with T_{eff} = 22500 K and log = 3.85 the mass M = 3.0 $M_\odot$ results, which is roughly a factor 3 too small for a normal composition near main- sequence star (8 $M_\odot$) but in agreement with the mass of a fully mixed star with equal amounts of hydrogen and helium (Röser, 1975) (see Fig. 2). To avoid this conclusion, one must postulate a distance of 700 pc which would simply mean that Sigma Ori E is not a physical member of the quintuplet system Sigma Orionis, but a background star. Whether this is true must be checked by a differential study of the UV interstellar lines and the lambda 2300 feature of the system Sigma Orionis.

5. HELIUM VARIABILITY AND THE OBLIQUE ROTATOR

Since a substantial fraction of the intermediates, i.e. those whose masses are well in excess of 2 $M_\odot$, are helium variables (see Sect. 2), we have to discuss the surface distribution of the chemical elements and possible implications on the outer structure.

The most direct access to the oblique rotator model is offered by the cap model of Mihalas (1973) (see also Hensler, 1979). In this

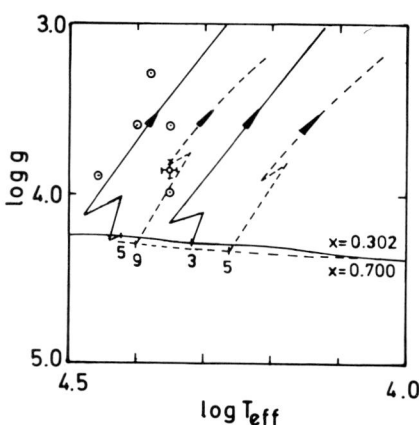

Fig. 2 (g, T_{eff})-diagram. Fully drawn: zero age main sequence and evolutionary tracks for mixed stars with X = 0.302. Dashed: the same as before, for normal composition stars. Circles correspond to non-variable intermediates, the error bars to Sigma Ori E.

simple approach, the helium-enriched surface area is approximated by one or two circular caps which are defined by the diameter, orientation and chemical composition. The 3 parameters can be extracted from the observed phase variations of the equivalent widths $W_\lambda(\phi)$ of the helium lines and their variable radial velocity shifts $\Delta v_R(\phi)$. The latter play an important role as roughly the same $W_\lambda(\phi)$ variations result for a small cap with a large amount of He as for a large cap with a medium amount of He. The radial velocities, however, will tell us which model is appropriate: the small and He-strong cap produces large amplitudes of Δv_R, while the large cap produces small amplitudes of Δv_R because the radial velocities are smeared out. The problem with the intermediates is that, so far, no radial velocity shifts have been observed, though efficient cross-correlation techniques have been employed. For Sigma Ori E the radial velocities are constant within 2 kms^{-1} while the equivalent width of He I 4471 varies by as much as a factor of 1.5 (Hunger, 1974; Bolton, 1974; Groote and Hunger, 1977).

A way out of the dilemma of the missing R.V. variations is saturation (Landstreet and Borra, 1978; Groote and Hunger, 1982). In the intermediates, both in the spectrum of the cap and in the disk, the strong He lines are saturated. Figure 3 shows the unrotated profiles of HeI 4471, for the number fractions of Helium ε_{He} = 0.90 (cap) and ε_{He} = 0.35 (disk). It appears that, due to a slight difference in the temperature structure, the helium "poor" disk produces a core that is even slightly deeper than that of the helium rich cap. From this example, one would expect the cores of strong He lines to be stationary while the wings of the strong lines and the cores of unsaturated (faint) He lines would, at least partially, reflect the motion of the helium-rich cap. The theory based on the method developed by Stoeckly and Mihalas (1973) and applied by Gruschinske (1982) to HeI 4471 yields, however, the surprising result that, when the cap is approaching the observer, the core is shifted to the red and, when the cap is receding, the core is shifted to the blue (Fig. 3). This means that, in the core, we see predominantly the helium "poor" disk while, in the wings, we see mainly the helium-rich cap. This strange behaviour is due to the combined effect of saturation and rotation: it can be shown by series expansion, that the intensity of a rotated line profile at a given $\Delta\lambda$ is the sum of the intensity of the unrotated profile at this wavelength plus $3/8 \times \lambda/c \times v\sin i$ times the wave length derivative of the unrotated profile. As the profile of the disk line is sharper than that of the cap line, and accordingly the derivative of the disk line larger than that of the cap line, the former dominates in the core of the rotated line. Consequently, the strong line cores exhibit an inverse shift while the weak lines are shifted directly; the medium-strong lines are practically unshifted. A cross-correlation method (Groote and Hunger, 1977) is hence bound to find near zero velocity amplitudes. This situation exists as long as the helium lines are core-saturated. In the helium-weak variable stars, with their unsaturated lines, radial velocity shifts indeed are observed.

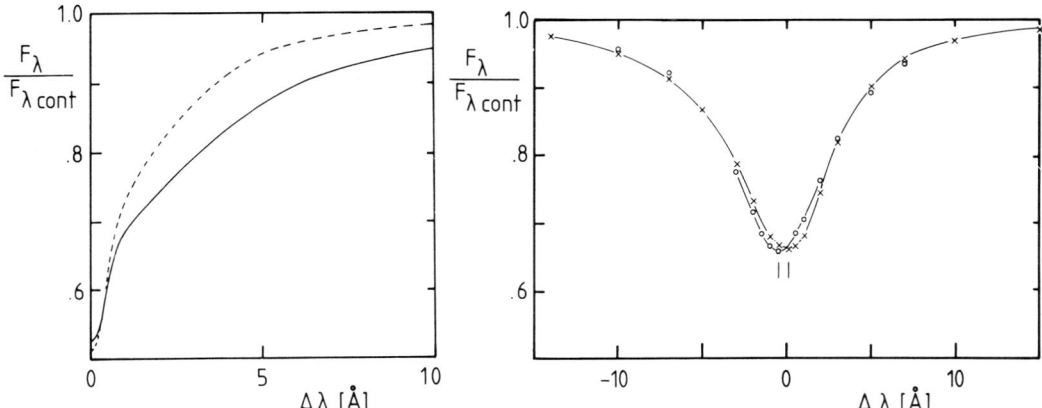

Fig. 3 Left panel: Unrotated profile of He 4471 (red wing only). He "poor": dashed, He rich: fully drawn. Right panel: The (symmetrisized) profile of He 4471, in two phases: xxx He cap is approaching, ooo cap is receding. The core is shifted inversely by 40 kms^{-1}.

For most of the strong variables, models with 1 or 2 He caps seem to be the rule. One example of a banded geometry is HD 37776. It represents the first case of a stellar magnetic quadrupole with current loops (Fig. 4) (Thompson and Landstreet, 1985). According to Groote and Kaufmann (1981) and Shore and Adelman (1981), helium is distributed in bands across the surface. It appears to be the only intermediate with variable Balmer absorption lines. (Recent CASPEC spectrograms, however, prove that also in Sigma Ori E Balmer lines are slightly variable — the variability apparently being correlated with the magnetic field.)

Surface enrichment in spots and bands is undoubtedly due to diffusion whereas the cause of the overall enrichment is still a matter of debate (see Sect. 4). The non-uniform surface distribution is probably the result of a complicated interplay of wind and magnetic braking, which acts differently on the different ions according to

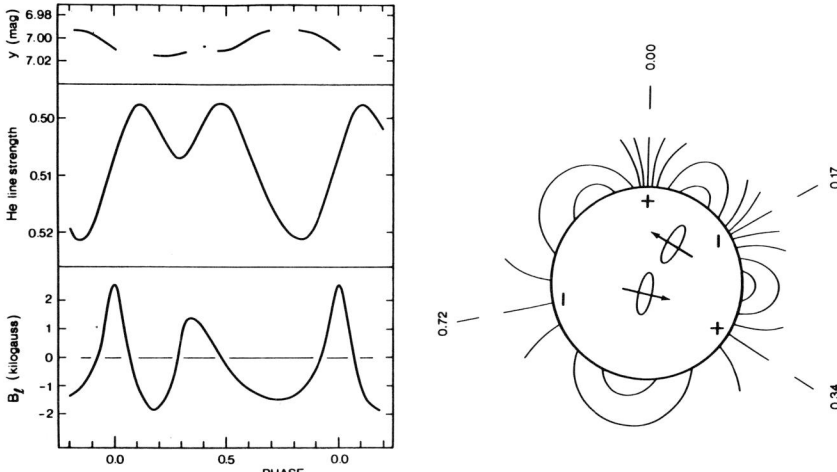

Fig. 4   Phase variation of the y-magnitude, He line strength and magnetic field of HD 37776. This variability is best explained by a banded helium distribution which is tied to a magnetic quadrupole field (right panel) (Thompson and Landstreet, 1985).

their charge to mass ratio (see, for instance, Shore, 1977). This not only leads to caps of helium but also to caps of the various metal ions which, as a rule, seem to vary in antiphase to helium (see Sect. 3).

Whatever lastly the theory for the horizontal surface abundance gradients may be, one must be aware of vertical abundance gradients which may be present in the atmosphere and which may influence the profiles (see also Bolton, 1983). For instance, He may float above a sea of hydrogen. This situation is opposite to what is known from White Dwarfs, where H floats above the DAs. However, from parameter studies of DA-model atmospheres and corresponding line formation calculations (Jordan, 1985), we can draw conclusions as to our helium variables: as long as the transition depth τ_{tr}, i.e. the depth where the helium content varies drastically with depth, is either very small or very large, normal profiles will result. However, when τ_{tr} is in the range $10^{-2} < \tau_{tr} < 1$, a temperature inversion may occur and hence strange profiles result.

So far, only one object has been found with unusual line profiles, the helium-weak variable HD 49333. The observed profiles of the strong helium lines in HD 49333 are far too shallow for any conceivable classical model and any reasonable v sini. Fig. 5 shows the example of He I 4471. According to Groote et al. (1985), the photosphere has a helium-poor surface layer atop a helium-normal bottom, the transition occurring at $\tau_{tr} = 0.4$. Because He is a poor absorber, the temperature stratification remains unaffected and hence no temperature inversion occurs, which explains why no emission is seen.

Fig. 5 He 4471 of HD 49333. The unusually shallow profile is formed in an atmosphere where the helium content varies drastically at $\tau_{4000} = 0.4$, from helium poor at the surface ($\varepsilon_{He} = 0.001$) to approximately helium normal at the bottom (... observed profile).

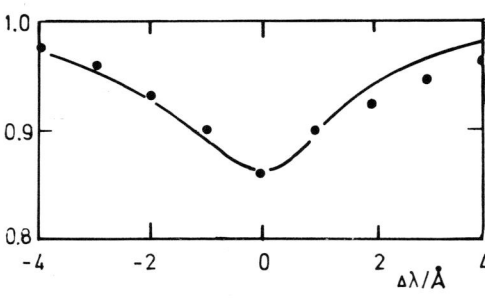

6. CIRCUMSTELLAR MATTER

Circumstellar matter seems to be present around all or most of the helium-strong variables. It may be the result of mass loss. Other hypotheses are that the clouds are a relict of protoplanetary matter, or that they are accreted from interstellar space. There have also been attempts to identify the helium-strong variables as Be-stars (Bolton, 1983; Harmanec, 1984). Mass loss is observed, for instance, in Sigma Ori E: $10^{-9}-10^{-10} M_\odot y^{-1}$ (Hamann, 1981), which makes the first hypothesis plausible. The circumstellar matter is responsible for all observed variability, e.g. in Sigma Ori E, except for the variable He and metal lines, and the magnetic field.

Since probably all of the helium-strong variables also have strong magnetic fields (see Sect. 2), the wind is modulated by the magnetic field: it is funnelled from the polar caps and stored in clouds near the magnetic equator, i.e. the circumstellar matter around the variable intermediates occurs mostly in the form of localized clouds. In Sigma Ori E, the clouds become manifest by strong absorption in the U-band (Hesser and Ugarte, 1976; Hesser et al., 1976, 1977), by H_α-emission (Walborn, 1974; Walborn and Hesser, 1976), by additional Balmer absorption lines (Groote and Hunger, 1976, 1977), by polarization (Kemp and Herman, 1977), by IR excess (Groote et al., 1980; Groote and Kaufmann, 1981; Groote and Hunger, 1982) and by CIV UV lines (Shore and Adelman, 1981). All features are variable, with a common period of 1.19080 days which is the period of rotation. (In the new CASPEC spectrograms, also H_β-emission is seen.)

From the 8 independent variables observed, a unique solution has been found for the models of the photosphere and shell of Sigma Ori E (Fig. 6) (Groote and Hunger, 1982). The angle between the axes of rotation and magnetic field is large, which may be typical for the variable intermediates. The clouds are located near the intersection of the magnetic and rotational equators. They radiate mainly ff-emission, with a temperature of 15000 K, which accounts for the infrared excess observed in the bands J, H, K and L (Fig. 7). The density is of the order of 10^{12} cm^{-3}, and the total mass $10^{-10} M_\odot$, which means that the clouds are replenished by the stellar wind within one year.

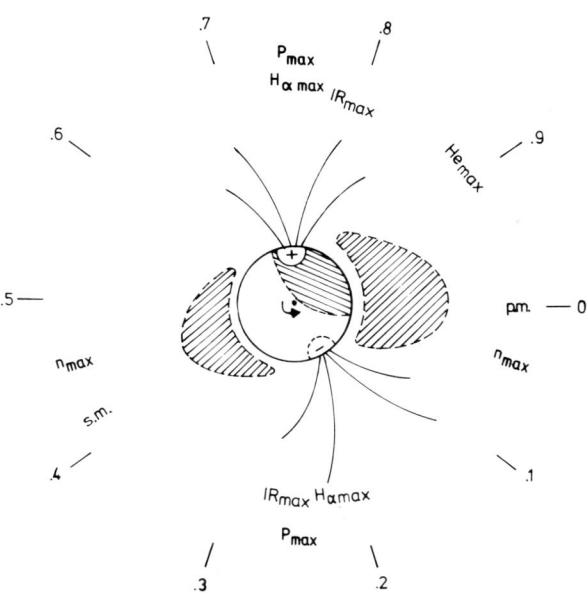

Fig. 6 Pole on model of Sigma Ori E. The line of sight rotates in the plane. The phases are given by numbers on the periphery. The phases of maximum H_α emission, IR emission, helium line strength, shell absorption and polarization, and minimum of light curves are indicated. The (cross-section of the) belt, helium cap and magnetic poles are shown.

The (variable) M-band excess has not been confirmed (Bonsack and Dyck, 1983; Odell and Lebofsky, 1984). If it were real, it would mean that there are corotating grain clouds, this being rather unlikely in view of the proximity of the hot star. Another explanation could be synchroton radiation from energetic electrons. The presence of energetic particles is considered to be typical in rotating magnetospheres (Havnes, 1981). In this case, Sigma Ori E would also be a strong radio-emitter. Radio-emission was searched for by Altenhoff and Wendker (1978) but the resolution of the 100 m Effelsberg telescope proved to be insufficient. Sigma Ori E was finally discovered with the VLA, at the wave lengths of 2 and 6 cm, by Drake et al. (1985). (This was a serendipitous discovery, as the literature on Sigma Ori E was apparently not known). IRAS also detected an excess of several magnitudes (Walker, 1985). From the 12 mμ/25 mμ ratio, a black body temperature of 200 K is determined which confirms the temperature of 270 K derived from Fig. 7.

The only other helium variable source found with VLA is the binary HD 37017. However, this source was too faint for detection with IRAS. This discovery once again opens up the question as to the M-band excesses of intermediates. It also makes the helium-strong variables interesting objects for observations in the far infrared.

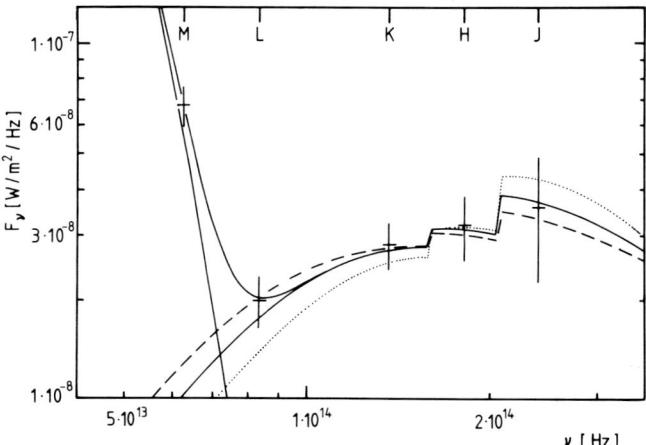

Fig. 7 The infrared excess of Sigma Ori E. Observations are marked by (long) crosses. The fully drawn curve (lower part of the diagram) corresponds to free-free emission with T = 15000 K. The M-band excess corresponds to a black-body temperature of 270 K.

CONCLUSIONS, PROBLEMS

The central problem of the intermediates is whether He-enrichment is a surface phenomenon or whether the entire envelope of an intermediate has been transformed by nuclear processes. An N/C ratio below unity definitely rules out the latter alternative. A ratio close to 0.4 is found in all non-variable intermediates, except for HD 186205 (which may be a candidate for variability). This ratio, hence, can be used to define the boundary between the intermediates and evolved extreme helium stars. With this definition the border case HD 144941, for instance, having n_{He}/n_H = 15 and N/C = 3, cannot be considered as intermediate.

The low gravities of the non-variable intermediates (log g = 3.7 $\pm$ 0.3), confirmed by recent CCD-spectrograms, mean that the intermediates are *not* on the main sequence. Whether this also implies that the intermediates have low masses, depends critically on the distances assumed. Distances published so far lead to masses which are by a factor of 2-3 too small for stars with normal composition. In view of the small N/C ratios observed, the distances must be regarded with scepsis.

A further problem not yet thoroughly discussed is: what is the distinction between the non-variables and variables among the intermediates, apart from the variability? Most of the variable features observed in Sigma Ori E, for instance, are caused by corotating clouds. Why do some intermediates have clouds and others not? Is it a matter of wind or accretion? The localization of the clouds, undoubtedly, is related to magnetic fields. Why, then, do some variables have strong magnetic fields - and these in turn are related to the non-uniform surface distribution of the chemical elements - and others not? The simple discrimination "rotation" apparently does not apply. Is it then the mass or the N/C ratio? From the foregoing, it is clear that we are still far away from completely understanding the intermediate helium stars.

ACKNOWLEDGMENTS

It is a pleasure to thank my former colleagues, Dr. D. Groote and Dr. J.P. Kaufmann for their collaboration on the intermediate helium star which, in part, led to this review. The Deutsche Forschungsgemeinschaft supported the travel to Mysore.

REFERENCES

Bolton, C.T.: 1974, Astrophys. J. **192**, L7
Bolton, C.T.: 1983, Harv. Obs. Bull. **7**, 241
Bond, H.E., Levato, H.: 1976, Publ. Astron. Soc. Pac. **88**, 905
Bonsack, W.K., Dyck, H.M.: 1983, Astron. Astrophys. **116**, 64
Borra, E.F., Landstreet, J.D.: 1977, Astrophys. J. **212**, 141
Borra, E.F., Landstreet, J.D.: 1979, Astrophys. J. **228**, 809
Carpenter, K.G.: 1985, Astrophys. J. **289**, 660
Caughlan, G.R.: 1965, Astrophys. J. **141**, 688
Detz, A.: 1977, Astron. Astrophys. Suppl. **28**, 403
Drake, S.A., Abbot, D.C., Bieging, Z.H., Churchwell, E., Linsky, J.L.:
    1984, in Radio Stars, ed. R. Hjellming and D. Gibson, New
    York: Plenum
Gerlach, M.: 1976, Dissertation, Technische Universität Berlin
Groote, D., Hunger, K.: 1976, Astron. Astrophys. **52**, 303
Groote, D., Hunger, K.: 1977, Astron. Astrophys. **56**, 129
Groote, D., Hunger, K., Schultz, G.V.: 1980, Astron. Astrophys. **83**, L5
Groote, D., Kaufmann, J.P.: 1981, Astron. Astrophys. **94**, L23
Groote, D., Kaufmann, J.P.: 1981, in **23rd** Liège Astrophys. Coll, p. 435
Groote, D., Kaufmann, J.P., Lange, A.: 1982, Astron. Astrophys. Suppl.
    50, 77
Groote, D., Kaufmann, J.P., Theil, U.: 1985, to be published
Gruschinske, J.: 1982, private communication
Hamann, W.R.: 1981, private communication
Harmanec, P.: 1984, Bull. Astron. Inst. Czechosl. **35**, 193
Havnes, O.: 1981, in **23rd** Liège Astrophys. Coll., p. 403
Havnes, O., Goertz, C.K.: 1984, Astron. Astrophys. **138**, 421
Heintz, W.D.: 1974, Astrophys. J. **79**, 397
Heber, U., Hunger, K.: 1981, Astron. Astrophys. **101**, 269
Hensler, G.: 1979, Astron. Astrophys. **74**, 284
Hesser, J.E., Ugarte, P.: 1976, IAU, Circ. No. 2911
Hesser, J.E., Walborn, N.R., Ugarte, P.: 1976, Nature **262**, 116
Hesser, J.E., Moreno, H., Ugarte, P.: 1977, Astrophys. J. **216**, L31

Hunger, K.: 1975, in Problems in Stellar Atmospheres and Envelopes, eds. B. Baschek, W.H. Kegel, G. Traving, Springer, New York, Heidelberg, Berlin, p. 57
Hunger, K.: 1986, unpublished
Hunger, K.: 1986, in IAU Coll. No. **90**.
Jordan, S.: 1985, private communication
Kaufmann, J.P., Hunger, K.: 1975, Astron. Astrophys. **38**, 351
Kemp, J.C., Herman, L.C.: 1977, Astrophys. J. **218**, 770
Kurucz, R.L.: 1979, Astrophys. J. Suppl. **40**, 1
Landstreet, J.D., Borra, E.F.: 1978, Astrophys. J. **224**, L5
Langhans, G., Heber, U.: 1985, to be published
Lee, P., O'Brien, A.O.: 1977, Astron. Astrophys. **60**, 259
Lester, J.B.: 1976, Astrophys. J. **210**, 153
Lester, J.B.: 1979, Astrophys. J. **233**, 644
Levato, H., Malaroda, S.: 1979, Publ. Astron. Soc. Pac. **91**, 789
McConnell, D.J.: 1972, Publ. Astron. Soc. Pac. **84**, 388
McConnell, D.J., Frye, R.L., Bidelman, W.P.: 1970, Publ. Astron. Soc. Pac. **83**, 730
Madej, J.: 1981, in **23rd** Liège Astrophys. Coll., p. 379
Mihalas, D.: 1973, Astrophys. J. **184**, 851
Nerney, S.: 1980, Astrophys. J. **242**, 723
Nakajima, R.: 1985, Astrophys. Space Sci. **116**, 285
Odell, A.: 1974, Astrophys. J. **194**, 645
Odell, A.P., Lebofsky, M.: 1984, Astron. Astrophys. **140**, 468
Osmer, P.S., Peterson, D.M.: Astrophys. J. **187**, 117
Pedersen, H.: 1979, Astron. Astrophys. J. Suppl. **35**, 313
Pedersen, H., Thompson, B.: 1977, Astron. Astrophys. J. Suppl. **30**, 11
Remie, H., Lamers, H.J.G.L.M.: 1982, Astron. Astrophys. **105**, 85
Röser, M.: 1975, Astron. Astrophys. **45**, 335
Shore, S.N.: 1977, Bull. Americ. Astron. Soc. **9**, 307
Shore, S.N., Adelman, S.: 1981, in **23rd** Liège Astrophys. Coll., p. 429
Stoeckly, T.R., Mihalas, D.: 1973, NCAR-TN/STR-**84**, 1
Thompson, I.B., Landstreet, J.D.: 1985, Astrophys. J. **289**, L9
Vauclair, S.: 1975, Astrophys. J. **45**, 233
Walborn, N.R.: 1974, Astrophys. J. **191**, L95
Walborn, N.R.: 1975, Publ. Astron. Soc. Pac. **87**, 613
Walborn, N.R.: 1982, Publ. Astron. Soc. Pac. **94**, 322
Walborn, N.R.: 1983, Astrophys. J. **268**, 195
Walborn, N.R.:, Hesser, J.E.: 1976, Astrophys. J. **205**, L87
Walker, H.: 1985, Astron. Astrophys. **152**, 58

DISCUSSION

HILL: When you look at the helium lines with phase, do you find that the inverse shift problem for the strong lines goes away when <u>all</u> the lines are weak?

HUNGER: Even in the phase when the disk is shown, the strong He lines are saturated.

VARDYA: You have talked about He at the top and with a H and He mixture below. How does this fractionation take place?

HUNGER: Fractionation takes place by diffusion in the presence of wind and magnetic field. You have to play with the two things, wind and magnetic field. When the wind is braked down by the magnetic field to a certain value, then you might have all the He ions transported upwards where they recombine to neutral helium. Thus, He can accumulate in the region of optical depth 0.3. This effect you get only for a definite velocity determined by the wind and also by the magnetic field. This led people to believe that He-enrichment is latitude-dependent, with respect to the magnetic equator. This model, however, leads to bands rather than spots.

MICHAUD: What do you believe to be the effective temperature range of the helium-rich stars?

HUNGER: The He-rich oblique rotators cluster around T_{eff} = 22000 K. The hottest known object is HD 37771 (T_{eff} = 27000 K).

MICHAUD: How well-known is that?

HUNGER: 27000 is by Kaufman, I think. Do you think it is too high?

MICHAUD: I am not saying that. It has an important effect on the separation mechanism.

KILAMBI: What is the observed variation in the strength of the magnetic field in these stars?

HUNGER: Of the order of ΔB = 5000 G typically.

KILAMBI: In your model, you have said that the He is concentrated in a spot. What would be the size estimates of these spots?

HUNGER: In σ Ori E, He in the spot is enriched up to He = 0.90. The spot covers some 30% of the surface.

DESHPANDE: If σ Ori E has a synchrotron component in its radiation, then polarimetric observations will help in proving this. Have any such observations been made?

HUNGER: Yes. Polarization measurements have been taken. They are difficult to perform as one has amplitudes of the order of only 0.1% which led us to assume that the clouds are not full but are truncated van Allen belts. The polarization, hence, is not connected with sychronoton radiation.

WEHLAU: In your model of the He spots, is the He distribution assumed to be uniform within the spots?

HUNGER: Yes.

GARRISON: You mentioned that you have a problem with the mass for σ Ori E. I notice that you have assumed 400 parsecs. However, I prefer 500 pc to the Ori group from the results of unpublished work I have done on the cluster. I believe this will help to relieve the mass problem.

HUNGER: That changes in the right direction. However, you need 600 pc in order to bring it up to the normal main sequence mass. Our estimate of 400 pc comes from Heintz which should be reliable.

GARRISON: The other comment is that I really like the models that you are coming up with because, for a long time, even at low dispersion, I think one has seen structures and strange profiles in the hydrogen and helium lines. The previous models have not really addressed this problem.

MICHAUD: Have other anomalies been observed?

HUNGER: Anomalies in metal abundances are hard to detect because weak lines are washed out by rotation.

MAGNETIC FIELDS AND WINDS OF THE INTERMEDIATE HELIUM STARS

Paul K. Barker
Astronomy Department
University of Western Ontario
London, Ontario N6A 3K7
Canada

ABSTRACT. The intermediate helium stars are exceedingly rare hot analogs of the classical Ap stars, and are the earliest type stars to possess observable global ordered magnetic fields. A recent discovery is the existence of stellar winds which have large scale magnetospheric structure embedded within them. The nature and geometry of the detected fields are summarized, and the modulation of the circumstellar material by the field is illustrated for two examples: the rapid rotator σ Ori E, and the slow rotator HD 184927. The complex variety of stellar wind phenomenology which may be encountered is displayed by a sample of ten helium strong stars. A few of these objects show Hα emission, and thus are the only known magnetic Be stars.

1. INTRODUCTION

This review is restricted to the intermediate helium stars (or helium strong stars) which occur in the region of the H-R diagram near B2V and show abnormally strong lines of neutral helium for that spectral type. Many of the helium strong stars are spectroscopic, photometric, and magnetic variables, with a single period common to all forms of variability. Thus these objects are hot analogs of the Ap stars. In the oblique rotator model for such stars, the magnetic/spectroscopic/photometric period is identified as the stellar rotation period. The spectral morphology and photospheric parameters are reviewed by Walborn (1983), Bolton (1983), and Hunger (1986).

The surface gravities (and hence evolutionary status) of the helium strong stars remain somewhat controversial (cf. Walborn 1983); careful analysis of accurate observed profiles is required to investigate this question, and it is hoped that the work of Hunger (1986) and Odell (1986a) may help settle the issue. Intuitively, one expects that the helium strong, helium weak, and classical Ap stars form a unified sequence in temperature along the main sequence; for all three groups, surface abundance anomalies result from diffusive processes in the outer layers of a magnetic star. For the most luminous objects in this sequence, radiation pressure and mass loss also play an important role in determining the equilibrium atmospheric structure; the stellar winds and magnetic fields of the helium strong stars are described here.

Ultraviolet observations of three rapidly rotating magnetic helium strong stars in Orion (Shore and Adelman 1981) suggested the existence of weakly variable mass-losing stellar winds. An extensive program of contemporaneous UV and optical spectroscopy, and magnetic field measurements, was stimulated by this discovery. In 1982, the Helium-Rich Magnetic Emission-line Star Working Group (Hermes) was formed by P.K. Barker, C.T. Bolton, D.N. Brown, J.D. Landstreet, and S.N. Shore. Phase resolved observations have now been obtained for a variety of helium strong stars which provide a wide-ranging sample in rotation rate and magnetic field strength, in order to investigate the interaction of magnetic, rotational, and radiative forces in the winds from these stars. This review is based largely on work published or in progress by the Hermes Working Group.

2. MAGNETIC FIELDS

The processes affecting the profile observed when a spectral line is formed in the presence of a magnetic field have been reviewed by Landstreet (1980, 1982). Usually the existence of a global stellar magnetic field which has a line of sight component (a mean longitudinal field) is inferred from the presence of a characteristic "S-wave" circular polarization profile across the spectral line. The mean stellar surface field can be measured only for a very few sharp-lined stars with strong fields; it is important to realize that the ratio of surface to longitudinal field, and even the very detection of a longitudinal field, depend extremely strongly upon the field geometry and its orientation relative to the line of sight at the time of observation. Field detection becomes progressively more difficult for successively higher order multipole components, and present techniques cannot detect any locally strong but globally complex magnetic fields. If any such disordered fields should exist, indirect inferences based on, for example, the presence of gyroresonance radiation (Underhill 1984) might become practical in the future. The photon counting problem is severe for present optical techniques: a 100 gauss longitudinal field typically produces a peak circular polarization of only 0.005%.

Among the OB stars in general, several searches with null results (with errors $\sim$500 gauss) are reviewed by Borra, Landstreet, and Mestel (1982). An additional 31 stars ranging from O4 to B8 have been observed by Barker et al. (1981, 1985, 1987) again with null results (with errors $\sim$100 gauss). Apart from Rigel (Severny 1970) only the helium spectrum variables possess detectable fields within this spectral range.

The helium strong stars observed in the search for magnetic fields and/or stellar winds are listed in Table 1, where the information is drawn from Barker et al. (1982), Borra and Landstreet (1979), Landstreet and Borra (1978), Thompson and Landstreet (1985), Walborn

(1983), and references therein. For HD 37017 the table gives a revised period (Landstreet 1986) based on additional magnetic observations; Bolton (1986) has provided a revised period for HD 184927 from analysis of optical spectral features. Seven of the nine stars observed magnetically show positive detections, with maximum mean longitudinal field typically ∼1-3 kilogauss. When the field is variable, the strength varies sinusoidally (in all but one case) on the independently determined stellar rotation period. The sinusoidal variation implies a field which is predominantly dipolar in geometry; examples are shown in Figure 1. For a dipole, the ratio of surface to mean longitudinal field must exceed ∼2.4 even at the most favorable field orientation, so these stars commonly have true surface fields of ∼10 kilogauss.

As for the Ap stars, the period-v sin i relation permits an estimate of the inclination i of the rotation axis to the line of sight, provided one has a reasonable estimate of the stellar radius. An estimate of the obliquity β of the magnetic axis to the rotation axis then follows from the ratio of magnetic extrema. Given the small

TABLE 1

Magnetic Fields and Rotation of the Helium Strong Stars

HD	B_ℓ Extrema gauss	V sin i km s^{-1}	Period days	IUE Images*
36485	?	80	?	5
37017	-350/-2170	170	0.901190 ± 0.00005	24
37479	+3100/-2300	170	1.190811 ± 0.00001	30
37776	+2500/-2150	160	1.53869 ± 0.00007	30
58260	+2200/ ?	<30	?	7
60344	<490	<30	?	2
64740	+400/-800	160	1.33016 ± 0.00016	23
96446	-1400/ ?	<30	?	8
133518	<250	<30	?	3
184927	+2250/ 0	<17	9.52793 ± 0.00078	14

* High dispersion SWP observations

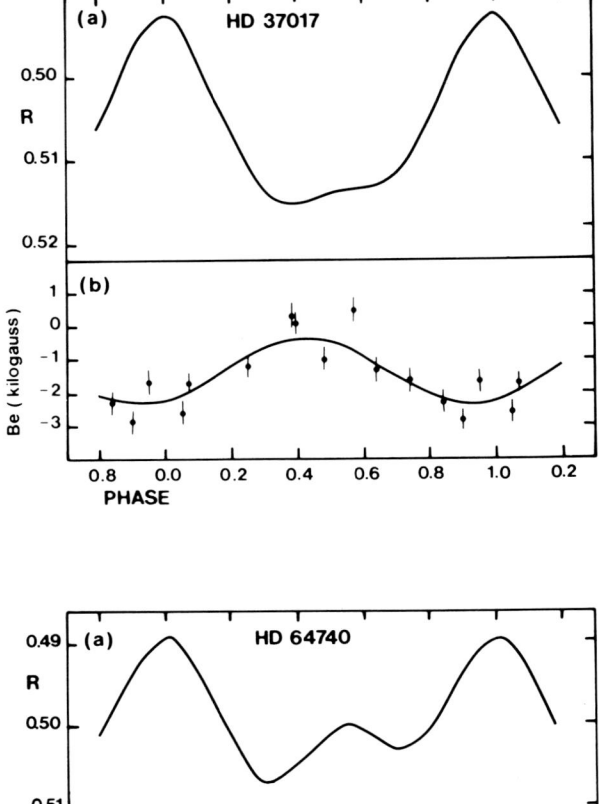

Figure 1. Typical sinusoidal variation of the mean longitudinal magnetic field on the stellar rotation period--indicating predominantly dipolar field geometry--for the helium strong stars HD 37017 (rotation period 0.90 day) and HD 64740 (rotation period 1.33 day). Also shown are the photometric helium line strength variations (a smaller value of the index R indicates stronger helium lines). Notice that the helium maxima occur at times of greatest exposure of the magnetic polar regions. From Borra and Landstreet 1979.

sample size, there is no clear indication of any preference for high or
low obliquity among the helium strong stars. The stars HD 58260 and
HD 96446 have apparently constant fields over many years; it is not
known whether the magnetic and rotational axes are parallel, or the
stars are simply viewed rotationally pole-on.

Figure 2 shows the unique non-sinusoidal magnetic field variation
of HD 37776. This cannot result from a dipolar field distribution, and
Thompson and Landstreet (1985) argued that for this object the quadrupolar field component is dominant--the first such case discovered.
Interestingly, in HD 37776 both maximum and minimum helium line
strength occur at phases when the mean longitudinal magnetic field is
close to zero--quite unlike the stars with strong dipolar fields, for
which helium maxima occur at or very near phases of magnetic extrema.

3. THE RAPID ROTATOR HD 37479 (σ Ori E)

This most extensively studied of the helium strong stars has come to be
regarded as the prototype, even though every object in Table 1 is
unique in some way. An analysis of optical and infrared behavior is

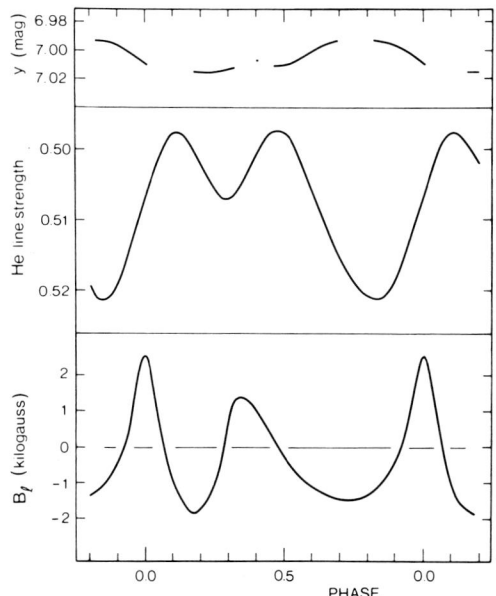

Figure 2. The unique mean longitudinal magnetic field variation of
HD 37776 (rotation period 1.54 day), indicating a predominantly
quadrupolar field geometry. The photometric helium line strength index
and Strömgren y magnitude variations are also shown. In sharp contrast
to the helium strong stars with dipolar fields, both maximum and
minimum helium line strength occur at phases when the mean longitudinal
magnetic field is close to zero. From Thompson and Landstreet 1985.

given by Groote and Hunger (1982), while magnetic observations are presented by Landstreet and Borra (1978). Possible interpretations of the data are discussed in both papers, which include the complete bibliography. Figure 3 summarizes the optical and magnetic variations on the stellar rotation period.

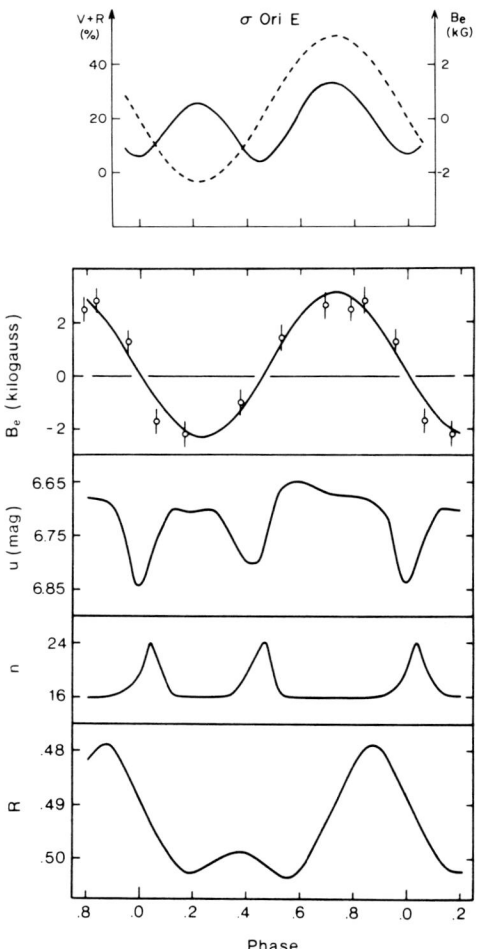

Figure 3. The best studied helium strong star HD 37479 (rotation period 1.19 day). The bottom panel (from Landstreet and Borra 1978) shows the variations in mean longitudinal magnetic field, photometric helium line strength index R, Strömgren u magnitude, and number n of the highest visible Balmer shell line. In the top panel (from Nakajima 1981) the solid line shows the corresponding variation of the Hα emission strength V+R (the sum of the red and violet emission peak intensities) expressed as a percentage of the local continuum level. The dashed line indicates the mean longitudinal magnetic field.

Figure 4 presents selected profiles of the C IV λλ1548, 50 resonance doublet observed with IUE; the rotational phases corresponding to those in Figure 3 are marked. The profiles are asymmetric, with a persistent shortward absorption extending to -600 km/s, indicating the existence of a stellar wind; however, there is no evidence for a fully developed P Cygni profile with longward emission above the continuum. As the star rotates, the absorption cores vary in depth by a factor of ∼2, but the variation is confined to wavelengths near line center: there is no detectable variation in the high velocity portion of the wind. Further, since it is not known whether all of the observed C IV arises in the wind--some contribution from photospheric C IV could be present--there is not even any unambiguous evidence for variability in the wind. Perhaps only the photospheric C IV, if any, varies as the star rotates. The nature and amplitude of the variations appear to be stable and repeatable during two epochs of observation separated by two years. The Si IV λλ1394, 1402 resonance doublet displays variability analogous to that in C IV, but of much lower amplitude because here the photospheric contribution certainly dominates. Detailed examination of other UV wavelength regions is in progress; Shore and Adelman (1981) found some evidence for variability in other species, especially C II λλ1334, 5.

HD 37479 is also a radio source: Drake et al. (1984) measured fluxes of ∼3.5 mJy at 6 cm and ∼3 mJy at 2 cm with the VLA. The spectral index of -0.1 suggests that a non-thermal mechanism such as gyroresonance emission is probably the source of radiation. As an aside, Drake et al. also detected a flux of 1.8 mJy at 2 cm from HD 37017, but HD 37776, HD 58260, and nine other Bp and Ap stars, were not detected, with 3σ upper limits of <0.5 mJy at 2 cm.

Groote and Hunger (1982) have developed the model shown in Figure 5 to interpret the available data (excluding UV observations). Here only the Hα and UV observations will be emphasized. HD 37479 shows weak double-peaked Hα emission which varies in strength and V/R ratio on the rotation period (Figure 3); thus, σ Ori E is a magnetic Be star. The highly ionized species such as C IV cannot arise from the same circumstellar regions as the Hα emission, so the (unsolved) problem is to determine the geometrical configuration of the stellar wind and the Hα emitting volume (as is the case for the classical Be stars).

First, Groote and Hunger have proposed the two clouds shown in Figure 5, to explain the variable Hα emission and Balmer shell absorption lines. However, high quality observations by Bolton (1985) show clearly that the V and R emission peaks rise and fall in intensity as the star rotates, but remain fixed in velocity at -500 and +500 km/s relative to line center, respectively. If the Hα emission region is trapped by, and forced to corotate with, the magnetic field, then this material must be at least two stellar radii above the photosphere to appear at this velocity. It is not obvious how to reconcile Groote and Hunger's model with these observations, but nevertheless there does appear to be some kind of trapped Hα magnetosphere within the wind. A similar situation exists for the Hα emission in HD 37017 (Odell 1986b).

Second, superposed on Figure 5 are the phases of maximum C IV absorption, obtained by Barker et al. (1986) from the entire IUE data

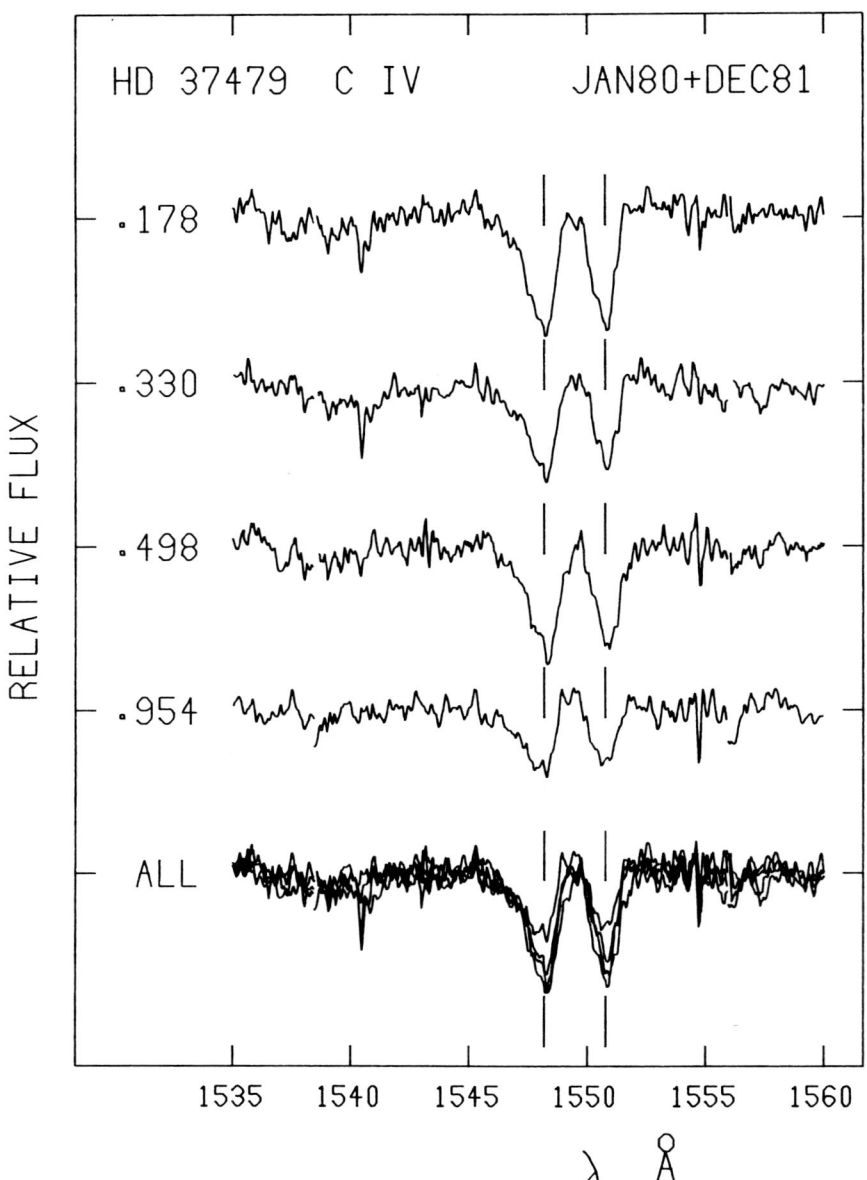

Figure 4. The asymmetric C IV resonance doublet in HD 37479, indicating the existence of a stellar wind. The vertical dashed lines mark the component rest wavelengths; the persistent shortward absorption extends to -600 km/s. Only a few representative profiles are shown, with the rotational phases (see Figure 3) marked; the four displayed phases are shown superposed at the bottom. The vertical spacing between horizontal tick marks for adjacent spectra equals the local continuum level. From Barker et al. 1986.

set (average spacing 0.05 in phase). One maximum coincides with maximum exposure of the negative magnetic pole. Thus one interpretation could be that the stellar wind emerges along the open field lines above this pole, and is thereby collimated into a jet or cone. However, the other C IV maximum is shifted by ~0.2 in phase relative to the positive pole--and furthermore leads this pole in rotation! This cannot be realistic. In fact, examination of Figure 3 shows that both C IV maxima coincide with the phases of helium minima. While there is not yet any explanation for this, the photospheric C IV abundance could be related to the photospheric helium abundance, and be dependent upon the local surface temperature and gradient of magnetic field lines. This strengthens the suggestion that the C IV variation is possibly purely photospheric in origin, and hence that there is no evidence for stellar

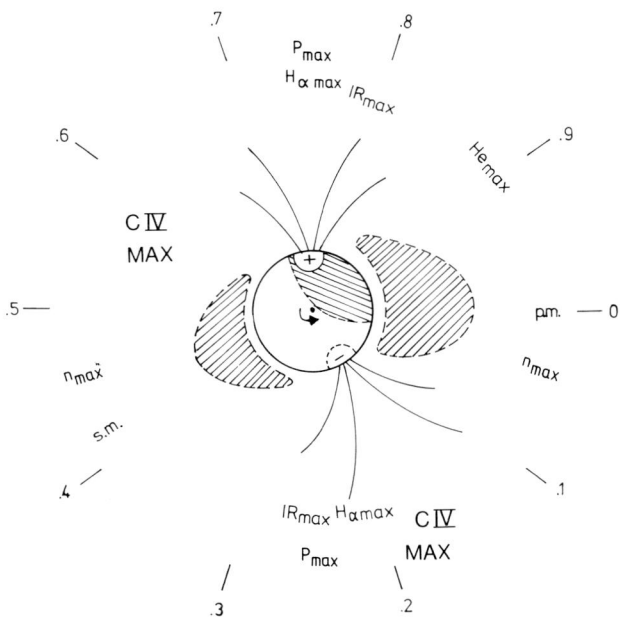

Figure 5. A perfect example of international non-collaboration. This model for σ Ori E (Groote and Hunger 1982) shows (viewed rotationally pole-on) the phases of maxima in Hα emission, IR emission, helium line strength, shell absorption, and linear polarization. The primary and secondary light curve minima are marked by p.m. and s.m. respectively. The projected orientation of the magnetic polar regions is also shown. The shaded region on the star marks the location of the helium strong spot, while the shaded circumstellar regions show the Hα emission clouds. Superposed on this diagram are the phases of maximum C IV absorption, obtained independently by Barker et al. 1986. These phases coincide with the times of helium minima, and do not correspond to the magnetic field phasing (see Figure 3).

wind variability in this star. Another explanation might be that the
C IV ions are stratified in extended atmospheric lobes above the regions
of minimum helium abundance, forming a superionized magnetosphere
embedded within the general stellar wind. In this scenario, the magnetospheric density and geometry might be arranged so that, as observed,
there is no detectable overt C IV emission at any phase.

4. THE SLOW ROTATOR HD 184927

This recently discovered helium strong star was studied optically by
Levato and Malaroda (1979) and references therein. A magnetic field
was found by Barker et al. (1982); the variations in strength of the
field and optical features are shown in Figure 6. The lines of He I
vary in phase with the magnetic field, but lines of H, Si III, and N II
vary in antiphase; thus there is a helium rich region near the positive magnetic pole.

The observations of C IV in Figure 7 provide a dramatic contrast
to those of σ Ori E. The doublet is highly variable on the stellar
rotation period. When the positive magnetic pole is closest to the
subsolar point, each component of the doublet displays strong longward
emission, whereas when the magnetic field is close to zero, C IV is
strongly in absorption. Exactly as for HD 37479, maximum C IV absorption coincides with the phase of helium minimum. The steep shortward
edge to the longward emission occurs essentially at the rest wavelength. At all phases, there is an extended shortward absorption which
reaches -600 km/s from line center, and reveals the presence of a
stellar wind. Exactly as for HD 37479, this extended shortward absorption does not vary as the star rotates. The Si IV doublet shows very
asymmetric profiles rising steeply on the longward side, with an extended shortward absorption. Again, the amplitude of variations is
much less than at C IV because of the dominant photospheric contribution. Other UV wavelength regions have not yet been examined for
variability.

Interpretation of these observations is less complete than for
σ Ori E. As argued for that star, there may be some photospheric (or
atmospheric lobe) contribution to C IV, explaining the coincident C IV
absorption maximum and helium minimum. The high velocity wind far from
the star does not vary. The C IV seen in emission must be circumstellar in origin, but its greatest strength occurs at low velocity,
and when the positive magnetic pole is closest to the line of sight.
It is not easy to see whether the C IV emission arises in the wind
close to the star, or perhaps in a magnetosphere trapped by horizontal
field lines. In any case, the C IV emission modulation shows that the
wind/magnetosphere is strongly controlled by the field. Further, the
amplitude of the modulation implies that a significant portion of the
C IV emitting material is highly compact and close to the star.

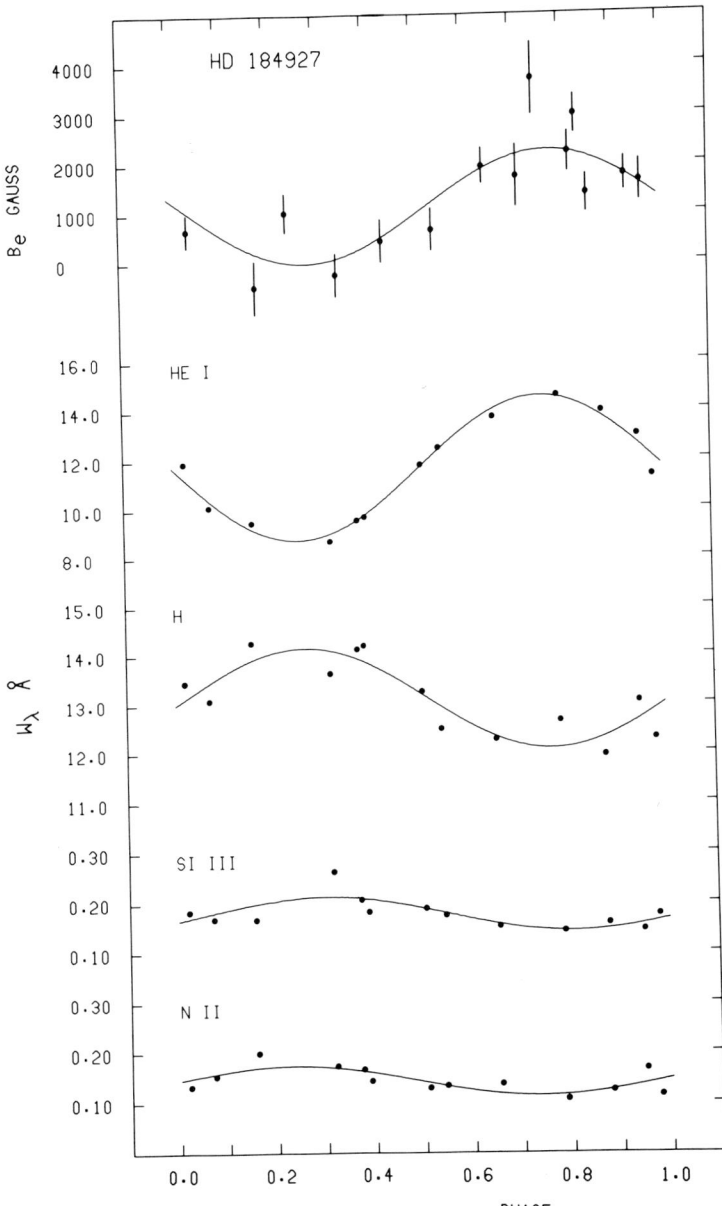

Figure 6. The mean longitudinal magnetic field and equivalent widths of optical features in the slow rotator HD 184927, phased on the ephemeris JD 2,444,796.0 + 9.536E. The smooth curves show least-squares sine wave fits to the data. From Barker et al. 1982.

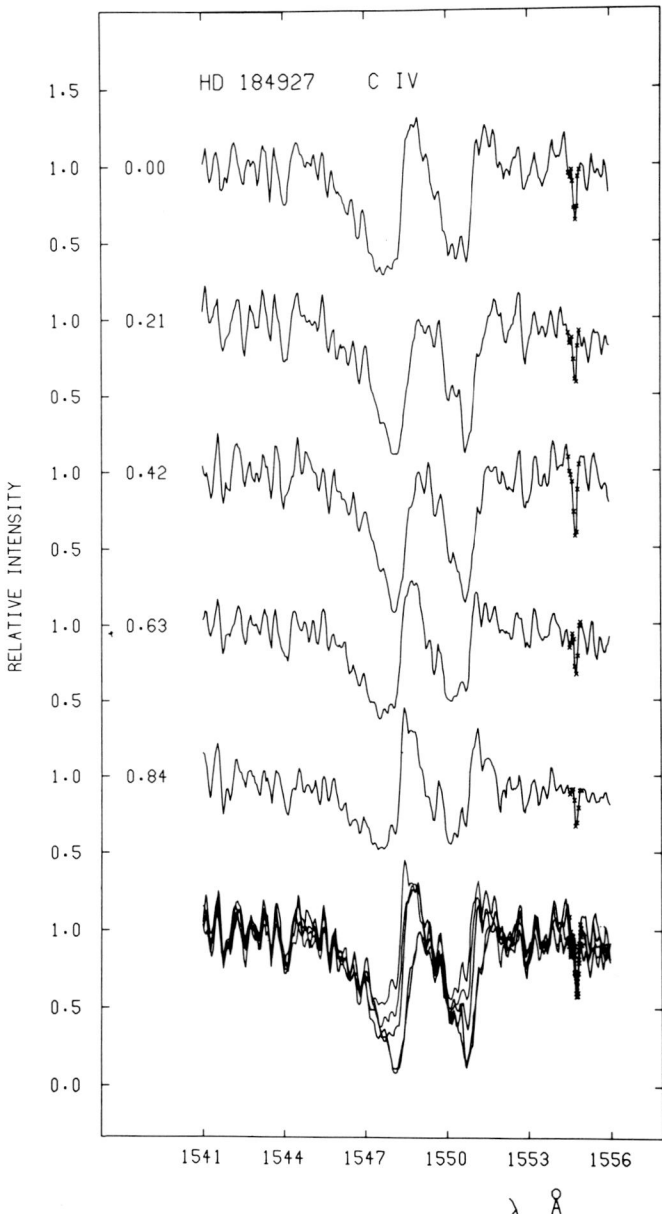

Figure 7. The top five profiles show the C IV resonance doublet in HD 184927 at the marked rotational phases; the same five spectra are superposed at the bottom. From Barker et al. 1982.

5. STELLAR WIND PHENOMENOLOGY

Figures 8 and 9 display the characteristic range of variation in C IV for nine of the stars listed in Table 1. HD 120640 in Figure 9 is now known not to be helium strong. HD 36485 profiles are not shown, but in fact are not much different from those for HD 120640, and are not variable. There may not even be any C IV present: the three sharp absorptions near the C IV doublet rest wavelengths probably result from Fe III (the same three features are seen in HD 58260 and HD 96446 superposed on the C IV emission). The variety in the C IV profile shape and in the amplitude of any variability for this sample of stars is overwhelming, if not dismaying. The most common feature is the presence of extended shortward absorption wings, indicating mass loss in a stellar wind, for those stars with C IV absorption.

The only cases of dominant C IV emission occur for stars in which the line of sight to the observer roughly corresponds to the magnetic axis: that is, at the maximum field phase of HD 184927, and also for HD 58260 and HD 96446, both of which have large and apparently constant fields. The lack of magnetic and spectral variations in HD 58260 and HD 96446 may arise either from their being viewed rotationally pole-on, or from intrinsic slow rotation; HD 184927 is a known slow rotator. These three stars are similar in that the C IV emission has a pronounced asymmetry, which is not true of the C IV emission in HD 64740, the star with next strongest emission. Walborn (1974) has reported weak broad emission wings to Hα in the rapid rotators HD 37017, HD 37479, and HD 64740--but not in HD 58260 or HD 96446. The present spectroscopic data show little or no Hα emission in HD 184927. Thus one arrives at the first taxonomic conclusion regarding this sample of helium strong stars: the strongest and most asymmetric C IV emission occurs in the stars which are either intrinsically the slowest rotators, or which are observed pole-on, whereas Hα emission is strongest in the most rapid rotators. The lack of significant Hα emission in HD 37776 may be related to the unique quadrupolar field geometry for this star.

A comparison of these C IV profiles with the stellar wind profiles seen in OB and Be stars in general is intriguing. The superionized resonance doublets in luminous OB stars typically display a broad absorption trough (with or without longward emission) which often has superposed narrow absorption components shifted to shortward wavelengths within the trough. Among Be stars there is sometimes no trough, only the shifted narrow components (SNCs). These features can show minor irregular variations in the luminous OB stars, and dramatic irregular variations in the Be stars, but not in normal low luminosity B stars; the present observational status is reviewed by Henrichs (1984, 1986). The origin of the SNCs is not understood for any of these stars, but it has been suggested that they may result from ejection of discrete "blobs" of material into the ambient stellar wind (which produces the general trough). The blobs in turn may result from fluctuations in the mode and amplitude of non-radial pulsation (recently discovered for many of these stars). The relevant points here are: first, there do not appear to be any irregular fluctuations in the wind

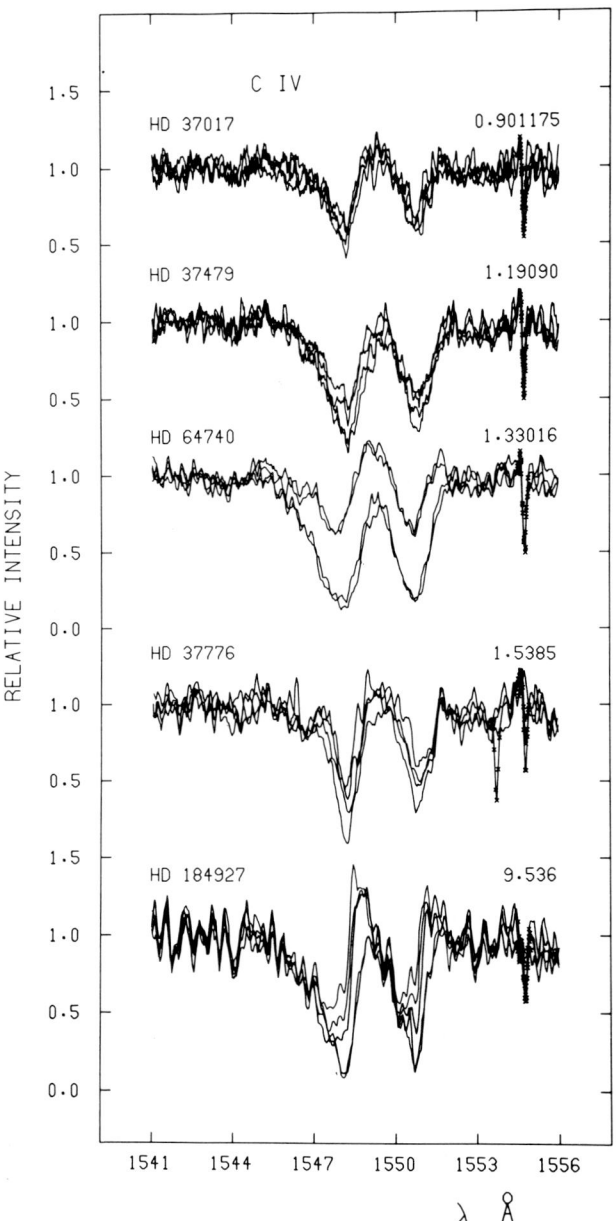

Figure 8. Here and in Figure 9, IUE spectra are superposed to show the characteristic variation of C IV in the helium strong stars. This figure shows the stars with known photometric period; the period in days is marked for each object. From Barker et al. 1982.

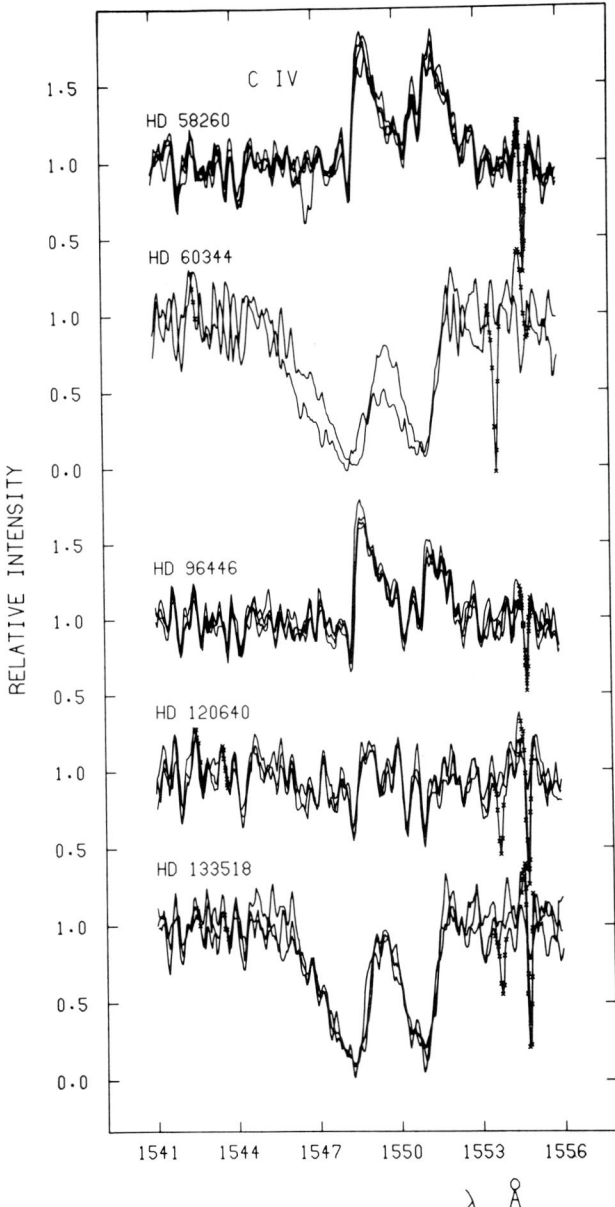

Figure 9. As in Figure 8, for the stars of unknown rotation period. HD 120640 is now known not to be a helium strong star, while HD 36485 (recently discovered as helium strong) is described in the text. From Barker et al. 1982.

profiles of helium strong (or helium weak) stars; second, no helium peculiar star has ever been observed with SNCs present; third, non-radial pulsation has not yet been observed among the helium peculiar stars (but has anyone looked?).

If future work confirms these points, the different phenomenological behavior for helium peculiar vs. OB and Be stars might be explained by supposing that a strong global magnetic field acts to suppress non-radial pulsation, and hence the ejection of blobs and production of irregular profile variability and SNCs. Or, perhaps the helium peculiar stars have quiescent winds simply because they are nothing but normal main sequence B stars that happen to have magnetic fields.

6. HELIUM WEAK STARS

Only slightly out of place at this Colloquium, these stars seem to be intermediate objects between the helium strong and Ap stars: helium weak stars are generally analogous to the Ap stars in regard to spectrum variations and presence of magnetic fields, and occur on the main sequence at spectral types between those of helium strong and Ap stars. Brown, Shore, and Sonneborn (1985) discovered weakly asymmetric C IV absorption in HD 21699, which varies on the magnetic and rotational period. This was interpreted as evidence for collimated jets emerging from above the magnetic poles. Interestingly, the only helium weak stars to show this behavior are the "sn" stars HD 21699, α Scl, and 36 Lyn (with sharp hydrogen and metallic lines but broad diffuse helium lines), in a survey conducted by Shore and Brown (1986). In this pleasing sequence from early B to A type stars, the Ap stars themselves presumably lack the C IV morphology shown by the helium peculiar stars only because they are insufficiently luminous to possess radiatively driven winds--although they may possess corotating magnetically driven winds (Rakos 1981).

7. FUTURE DIRECTIONS

In summary, the helium strong stars have winds and magnetospheres whose structure is modulated by the global stellar magnetic field. Despite the complex phenomenology, one might hope that the varietal sample available will ultimately permit elucidation of the interacting effects due to rotation rate, magnetic geometry, radiative and diffusive forces, and stellar orientation.

Some aspects of these factors have been addressed theoretically. The role of diffusion is reviewed by Bolton (1983) and Michaud (1986). Nakajima (1981) developed a model for the $H\alpha$ emission in which circumstellar gas--constrained to move only along the field lines--is trapped in a corotating magnetosphere whose inner and outer boundaries are determined by the balance between centrifugal and gravitational forces. This qualitatively predicts that the rapid rotators should have stronger $H\alpha$ emission than the slow rotators, as observed, but the model does not include radiative forces. Peterson and Theys (1981) considered the extension above the photosphere of early B star atmospheres with strong horizontal magnetic fields. Limber (1974), Saito (1974), and Nerney (1980) constructed wind models driven entirely by magnetically enforced corotation. Shore (1978) attempted to model the balance between radiative and diffusive forces in magnetic B stars. Barker (1982) and Friend and MacGregor (1984) constructed magnetic wind models incorporating the effects of line radiation pressure.

Unfortunately, no model yet includes the effects of all the forces known or suspected to be influential among the helium strong stars. Not only must all these forces be treated simultaneously, but in addition, fully three-dimensional models must eventually be sought if one hopes to approach the complex realities of Figures 8 and 9. Such calculations have not yet been attempted. Progress will be slow, but--based on the present qualitative intuitive physical scenario outlined here--one hopes it will be substantial.

ACKNOWLEDGEMENTS

I particularly wish to thank Dr. K. Hunger for the invitation to attend IAU Colloquium 87 in Mysore, and especially my friends and collaborators Drs. C.T. Bolton, D.N. Brown, J.D. Landstreet, and S.N. Shore of the Hermes Working Group for sharing their physical insight during numerous discussions. Hermes thanks the IUE Observatory staff for their friendly and efficient operation of the spacecraft, and Dr. I.B. Thompson for his contribution to some of the magnetic observations. This work was supported by the Natural Sciences and Engineering Research Council of Canada through grants to Drs. J.M. Marlborough and J.D. Landstreet.

REFERENCES

Barker, P.K. 1982, in IAU Symposium 98, Be Stars, ed. M. Jaschek and H.-G. Groth (Dordrecht: Reidel), p.485.
Barker, P.K., Bolton, C.T., Brown, D.N., Landstreet, J.D., and Shore, S.N. 1986, in preparation.
Barker, P.K., Brown, D.N., Bolton, C.T., and Landstreet, J.D. 1982, NASA CP-2238, p.589.
Barker, P.K., Brown, D.N., and Marlborough, J.M. 1987, in preparation.
Barker, P.K., Landstreet, J.D., Marlborough, J.M., and Thompson, I.B. 1985, Ap. J., 288, 741.
Barker, P.K., Landstreet, J.D., Marlborough, J.M., Thompson, I., and Maza, J. 1981, Ap. J., 250, 300.
Bolton, C.T. 1983, Workshop on Rapid Variability of Early Type Stars, Hvar Obs. Bull., 7, No. 1.
Bolton, C.T. 1985, private communication.
Bolton, C.T. 1986, private communication.
Borra, E.F., and Landstreet, J.D. 1979, Ap. J., 228, 809.
Borra, E.F., Landstreet, J.D., and Mestel, L. 1982, Ann. Rev. Astron. Astrophys., 20, 191.

Brown, D.N., Shore, S.N., and Sonneborn, G. 1985, A. J., 90, 1354.
Drake, S.A., Abbott, D.C., Bieging, J.H., Churchwell, E., and Linsky, J.L. 1984, in Radio Stars, ed. R. Hjellming and D. Gibson (New York: Plenum).
Friend, D.B., and MacGregor, K.B. 1984, Ap. J., 282, 591.
Groote, D., and Hunger, K. 1982, Astr. Ap., 116, 64.
Henrichs, H.F. 1984, ESA SP-218, p. 43.
Henrichs, H.F. 1986, in O, Of, and Wolf-Rayet Stars, ed. P.S. Conti and A.B. Underhill, in press.
Hunger, K. 1986, this publication.
Landstreet, J.D. 1980, A. J., 85, 611.
Landstreet, J.D. 1982, Ap. J., 258, 639.
Landstreet, J.D. 1986, private communication.
Landstreet, J.D., and Borra, E.F. 1978, Ap. J. (Letters), 224, L5.
Levato, H., and Malaroda, S. 1979, Pub. A. S. P., 91, 789.
Limber, D.N. 1974, Ap. J., 192, 429.
Michaud, G. 1986, this publication.
Nakajima, R. 1981, Science Reports of the Tohoku University, 8th Series, 2, No. 3, p. 130.
Nerney, S. 1980, Ap. J., 242, 723.
Odell, A.P. 1986a, this publication.
Odell, A.P. 1986b, this publication.
Peterson, D.M., and Theys, J.C. 1981, Ap. J., 244, 947.
Rakos, K.D. 1981, NASA CP-2171, p. 167.
Saito, M. 1974, Pub. Astr. Soc. Japan, 26, 103.
Severny, A. 1970, Ap. J. (Letters), 159, L73.
Shore, S.N. 1978, Ph. D. Thesis, University of Toronto.
Shore, S.N., and Adelman, S.J. 1981, 23rd Liège Astrophys. Coll., p.429.
Shore, S.N., and Brown, D.N. 1986, in preparation.
Thompson, I.B., and Landstreet, J.D. 1985, Ap. J. (Letters), 289, L9.
Underhill, A.B. 1984, Ap. J., 276, 583.
Walborn, N.R. 1974, Ap. J. (Letters), 191, L95.
Walborn, N.R. 1983, Ap. J., 268, 195.

DISCUSSION

BALASUBRAMANIAM: In measuring the magnetic fields, what criteria do you use to eliminate lines that are temperature sensitive?

BARKER: None. For the early B stars, one is limited to lines of helium and hydrogen. The hydrogen lines are stronger and steeper, and produce a stronger polarization signal; generally only H_β is observed. In any case, there is no evidence for any coronal or prominence-like fine structure in these stars.

BALASUBRAMANIAM: Again: are the magnetic fields derived from two different lines?

BARKER: Only one line is observed in each star because the photon counting problem is unbelievably severe, it takes half a night on the 3.6 meter telescope to observe one star at one line.

ODELL: Is it possible to get incorrect field measurements if the line profile is varying with a timescale of 1 day?

BARKER: Yes, very easily, unless one uses a large enough telescope that the entire sequence of polarization observations can be completed in a time much less than the rotation period. This of course is true whether or not the observed profile has any weak emission component.

HUNGER: The R-index of Pedersen, if really not O.K. can only get messed up because the filter in between $\lambda\,4009$ Å and $\lambda\,4020$ Å does not measure the true continuum, because these two He I lines may overlap.

MICHAUD: Is the same phasing between He and H observed for all He rich stars and, is it the same for the He weak stars?

BARKER: As shown in Figure 1, the helium strong stars with dipolar fields have helium maxima when the magnetic poles are closest to the subsolar point. The phasing is different for HD 37776 with its quadrupolar field. For the helium weak star 3 Sco, the phasing is also different even though this does have a dipolar field.

HUNGER: As to the anti-correlation of CIV and H_α emission, HD 60344 has a low mass (1.3 $M_\odot$), and hence may belong to the low mass subgroup if that really exists. HD 133518 is probably not He-rich.

BARKER: I believe Walborn in his classification paper did still include HD 133518.

LIEBERT: I may be committing an unpardonable sin: The unwritten rule whenever a talk on magnetic stars is presented, is never to ask basic questions, like where does the magnetic field come from. Since the Orion OB group is close and relatively easy to study, why is σ Ori E so unique? The more global question to Dr. Hunger what fraction of B stars relly have He strong magnetic behaviour?

HUNGER: Roughly one third.

GARRISON: There is an interesting, tight cluster of about 12 stars surrounding σ Ori E. It includes also a helium-weak star and a mild Ap star. The main sequence is very tight, about the width of the line drawn, according to some unpublished MK work I did on it. If you plot the He-weak star according to its color, it falls about 0.7 magnitude below the main sequence, whereas its spectrum is cooler and above the main sequence - the truth is probably somewhere in between.

HILL: This question is for Dr. Garrison. In this small cluster containing the Orion complex, do you find any other stars with the same T_{eff} as σ Ori E but not with its peculiarities?

BARKER: Not exactly the same, but there are stars bracketing it on the main sequence.

GURM: Are there any X-ray studies from these objects?

BARKER: It has been suggested (Groote, Kaufmann, and Hunger, Astron. & Astrophys., 1978, 63, L9) that HD 64740 may be coincident with the X-ray SOURCE 4U 0750-49. I do not believe that any helium strong star has ever been deliberately observed at X-ray wavelength.

GURM: There seems to be coronal behaviour as there is a strong collimated wind from the magnetic pole. It forces us to think that there is a sun like phenomenon. However, differences could arise because some dust may be present.

BARKER: Many helium strong stars show IR excesses, but for HD 37479, Groote and Hunger argued that emission from dust is unlikely.

HELIUM-RICH STELLAR ATMOSPHERE MODELS FOR B STARS

ANDREW P. ODELL*
Institute for Astronomy
University of Vienna
Tuerkenschanzstrasse 17
A-1180 Vienna, Austria

STEPHEN A. VOELS
Joint Institute for
 Laboratory Astrophysics
University of Colorado
Boulder, CO 80309 USA

ABSTRACT. The hydrogen and helium line equivalent widths from stellar atmosphere models for early B stars are presented. The models, which range in helium number fraction from 0.10 to 0.85, result from the non-LTE statistical equilibrium code of Mihalas.

I. RESULTS

Model atmospheres were computed with the non-LTE code of Mihalas (1972), with modifications listed by Abbott and Hummer (1985). The code assumes a plane-parallel atmosphere in radiative and hydrostatic equilibrium, allows for only hydrogen and helium composition, and does not include backwarming by a wind. The inaccuracies due to inclusion of only two bound levels of helium in non-LTE become more serious with higher helium abundance.

Our initial purpose for computing these models was to compare them to line profiles of intermediate helium rich stars, but there are several other applications as more distant B stars are observed with improved instrumentation. They will be useful in connection with the general study of helium abundances and gradients within the Galaxy, the chemical evolution of the Galaxy, and the relative rates of change of helium and heavy elements, as well as similar studies of B stars in other nearby galaxies.

Tables I through III list the equivalent widths of the five hydrogen lines and 14 helium lines for which profiles have been computed, as well as the temperature, gravity and helium number fraction of the atmospheres. In addition, for each of these lines, the intensity as a function of wavelength and zenith angle is available for computing line profiles which are rotationally broadened and/or arise from spotted surfaces. Odell will provide these results on computer tape, if contacted at the address below.

We are grateful to Dr. D. C. Abbott for useful discussions and to Dr. D. Mihalas for the use of his code. The National Science Foundation provided partial support for the project through grant AST84-17656 to Northern Arizona University. The National Center for Atmospheric Research (NCAR) provided the computing facilities.

II. REFERENCES
Abbott, D. C., and Hummer, D. G. 1985, Ap. J. $\underline{294}$, 286.
Mihalas, D. 1972, Non-LTE Atmospheres for B and O Stars (NCAR-TN/STR-76).

---

*on leave of absence from Northern Arizona University, Flagstaff, AZ 86011.

TABLE I

Hydrogen Line Equivalent Widths

Teff	Log g	NHE/NTOT	ALPHA	BETA	GAMMA	DELTA	EPSILON
24000	3.85	0.35	3.42	4.48	4.64	4.84	5.18
24000	3.85	0.60	3.25	4.18	4.31	4.50	4.82
24000	4.30	0.35	4.25	5.66	5.91	6.19	6.67
24000	4.30	0.60	4.07	5.33	5.55	5.81	6.25
22500	3.40	0.35	2.92	3.81	3.90	4.04	4.29
22500	3.40	0.60	2.84	3.61	3.68	3.82	4.07
22500	3.85	0.10	3.57	4.81	4.99	5.19	5.55
22500	3.85	0.35	3.76	4.98	5.16	5.37	5.76
22500	3.85	0.60	3.73	4.84	4.98	5.19	5.57
22500	3.85	0.85	3.28	4.08	4.16	4.31	4.60
22500	4.30	0.10	4.46	6.07	6.35	6.64	7.15
22500	4.30	0.35	4.72	6.34	6.61	6.92	7.46
22500	4.30	0.60	4.79	6.31	6.55	6.83	7.36
22500	4.30	0.85	4.51	5.76	5.91	6.13	6.56
20000	3.85	0.35	4.25	5.69	5.87	6.11	6.55
20000	3.85	0.60	4.51	5.91	6.06	6.29	6.75
20000	4.30	0.35	5.33	7.22	7.52	7.85	8.45
20000	4.30	0.60	5.71	7.62	7.89	8.21	8.85
18000	3.85	0.10	4.71	6.49	6.73	6.99	7.47
18000	3.85	0.35	5.41	7.41	7.68	7.97	8.53
18000	3.85	0.60	6.03	8.23	8.50	8.82	9.45
18000	3.85	0.85	6.61	8.88	9.13	9.44	10.13
18000	4.30	0.10	5.79	8.04	8.41	8.77	9.41
18000	4.30	0.35	6.56	9.09	9.52	9.92	10.66
18000	4.30	0.60	7.24	10.02	10.48	10.91	11.79

The equivalent widths are total widths, to continuum at ± 30 Å.

NHE/NTOT refers to the helium number fraction NHE/(NH+NHE).

TABLE II

Blue Helium I Line Equivalent Widths

Teff	Log g	NHE/NTOT	3889	4026	4121	4387	4438	4471	4713
24000	3.85	0.35	0.43	2.31	0.37	1.53	0.23	1.97	0.32
24000	3.85	0.60	0.59	3.17	0.51	2.06	0.33	2.65	0.44
24000	4.30	0.35	0.57	3.01	0.48	1.90	0.29	2.52	0.42
24000	4.30	0.60	0.77	3.97	0.65	2.50	0.39	3.28	0.55
22500	3.40	0.35	0.34	1.85	0.31	1.27	0.22	1.60	0.27
22500	3.40	0.60	0.49	2.64	0.43	1.75	0.30	2.24	0.37
22500	3.85	0.10	0.28	1.29	0.22	0.85	0.14	1.17	0.28
22500	3.85	0.35	0.48	2.53	0.40	1.62	0.26	2.14	0.35
22500	3.85	0.60	0.67	3.49	0.56	2.21	0.35	2.90	0.48
22500	3.85	0.85	0.88	4.46	0.73	2.80	0.44	3.68	0.63
22500	4.30	0.10	0.36	1.72	0.27	1.05	0.15	1.53	0.27
22500	4.30	0.35	0.62	3.20	0.51	1.97	0.30	2.67	0.44
22500	4.30	0.60	0.85	4.28	0.70	2.63	0.40	3.51	0.59
22500	4.30	0.85	1.10	5.38	0.91	3.31	0.52	4.39	0.77
20000	3.85	0.35	0.53	2.69	0.43	1.66	0.26	2.26	0.38
20000	3.85	0.60	0.74	3.75	0.60	2.29	0.36	3.09	0.51
20000	4.30	0.35	0.66	3.27	0.51	1.93	0.28	2.71	0.45
20000	4.30	0.60	0.89	4.37	0.71	2.60	0.39	3.56	0.61
18000	3.85	0.10	0.33	1.31	0.21	0.71	0.10	1.15	0.22
18000	3.85	0.35	0.55	2.47	0.38	1.36	0.20	2.02	0.35
18000	3.85	0.60	0.74	3.39	0.54	1.87	0.27	2.71	0.46
18000	3.85	0.85	1.00	4.55	0.76	2.55	0.38	3.60	0.63
18000	4.30	0.10	0.36	1.49	0.22	0.75	0.10	1.27	0.22
18000	4.30	0.35	0.60	2.71	0.41	1.42	0.19	2.19	0.36
18000	4.30	0.60	0.79	3.60	0.57	1.91	0.26	2.85	0.48

TABLE III

Red Helium I Line Equivalent Widths

Teff	Log g	NHE/NTOT	4921	5015	5047	5876	6678	7065	7281
24000	3.85	0.35	1.30	0.45	0.27	0.78	0.91	0.17	0.32
24000	3.85	0.60	1.69	0.58	0.35	1.05	1.17	0.26	0.38
24000	4.30	0.35	1.56	0.55	0.31	1.02	1.07	0.27	0.35
24000	4.30	0.60	1.98	0.70	0.41	1.34	1.36	0.36	0.42
22500	3.40	0.35	1.11	0.39	0.25	0.63	0.79	0.10	0.29
22500	3.40	0.60	1.48	0.51	0.31	0.86	1.01	0.17	0.34
22500	3.85	0.10	0.79	0.31	0.17	0.55	0.61	0.17	0.24
22500	3.85	0.35	1.37	0.48	0.28	0.85	0.94	0.21	0.32
22500	3.85	0.60	1.79	0.63	0.37	1.15	1.21	0.29	0.39
22500	3.85	0.85	2.18	0.79	0.47	1.49	1.49	0.40	0.46
22500	4.30	0.10	0.94	0.36	0.18	0.69	0.69	0.24	0.25
22500	4.30	0.35	1.60	0.57	0.32	1.09	1.08	0.31	0.35
22500	4.30	0.60	2.05	0.74	0.43	1.43	1.38	0.40	0.42
22500	4.30	0.85	2.48	0.93	0.56	1.80	1.70	0.52	0.52
20000	3.85	0.35	1.38	0.50	0.28	0.91	0.92	0.26	0.31
20000	3.85	0.60	1.81	0.66	0.37	1.23	1.19	0.34	0.38
20000	4.30	0.35	1.55	0.57	0.30	1.10	1.01	0.35	0.33
20000	4.30	0.60	1.99	0.74	0.41	1.43	1.29	0.43	0.40
18000	3.85	0.10	0.64	0.27	0.12	0.59	0.48	0.26	0.17
18000	3.85	0.35	1.08	0.42	0.20	0.85	0.68	0.33	0.23
18000	3.85	0.60	1.42	0.54	0.27	1.07	0.85	0.39	0.27
18000	3.85	0.85	1.84	0.71	0.39	1.38	1.10	0.47	0.34
18000	4.30	0.10	0.65	0.26	0.11	0.61	0.45	0.25	0.14
18000	4.30	0.35	1.11	0.42	0.20	0.89	0.67	0.34	0.20
18000	4.30	0.60	1.43	0.54	0.27	1.11	0.83	0.40	0.24

ANALYSIS OF THE HELIUM STRONG STAR HD 37017

ANDREW P. ODELL*[2]
Institute for Astronomy
University of Vienna
Tuerkenschanzstrasse 17
A-1180 Vienna, Austria

ABSTRACT. High resolution and signal-to-noise spectra (0.22 A; S/N $\geq$ 50) of the H-alpha and He I 6678 A lines have been obtained over most of the cycle of the intermediate helium strong star HD 37017 with the Coude spectrograph and CCD camera at KPNO. These spectra are modeled with the non-LTE code of Mihalas to derive properties of the star. Both of the observed lines show small equivalent width variations on the previously known timescale of 0.901 days. The profiles can be simulated by a model with temperature 20,000K, log g=3.4, and helium number fraction 0.25, and which has a helium-rich spot of radius $40°$ and helium number fraction of 0.60 on the equator. Other possible models are discussed, as well as a model of the weak emission on the wings of H-alpha.

1. INTRODUCTION

The intermediate helium strong stars are early B stars, normally associated with very young groups of stars, which show abnormally strong lines of He I that seem to vary in strength with a timescale of about one day. They also exhibit variable hydrogen line profiles, both in absorption and emission, and variable magnetic fields. With these features in mind, I undertook a study at KPNO to obtain high quality spectra of the H-α line and a helium line to quantitatively interpret the behavior of several of these stars. At that time, only red-sensitive CCD detectors were available, and so the He I line at 6678 A was chosen. This line is formed in non-LTE conditions in the stellar atmosphere, and so models using this physics must be employed in any interpretation.

The star HD 37017 in the belt region of Orion is the star with the least extreme helium enrichment of those in Orion, and so is the first chosen here for analysis. The only detailed atmospheric study to date

---

* on leave of absence from Northern Arizona University, Flagstaff, AZ.

[2] Visiting Astronomer, Kitt Peak National Observatory, operated by the Association of Universities for Research in Astronomy, Inc., under contract with the National Science Foundation.

was made by Lester (1972), using photographic spectra, and with no accounting for possible variability on short timescale. He concludes that the temperature is 21,000K, log g is 4.4, number fraction of helium is 0.19, and the mass and radius are small for the effective temperature. Blaauw and van Albada (1963) solved for a binary orbit with a period of 18.65 days. Pedersen (1979) showed that the He I line at 4026 A varies in strength with period of 0.901175 days.

This paper summarizes an analysis of the H-α and the He I 6678 lines in HD 37017 in terms of non-LTE model atmospheres enriched above normal in helium atoms. Heasley and Wolff (1983) have analyzed the H-α profile of normal B stars, and found that the cores of such lines are in agreement with non-LTE atmospheres but the wings near the line core are too deep in the observed profiles (the discrepancy not exceeding 3%). Heasley, Wolff, and Timothy (1982) compared the profiles of the He I lines in several normal B stars to non-LTE models and concluded that the He I 6678 line is somewhat too strong in the observed profiles compared to models which mimic the star in the blue helium and hydrogen lines. This discrepancy is most serious for stars of higher temperature and lower surface gravity than the star under discussion.

2. THE SPECTRA

A total of 13 usable spectra of HD 37017 were obtained on three nights during January 1982 with the Coude spectrograph and feed telescope of KPNO using a Fairchild CCD camera. The configuration covered wavelengths 6540 A to 6690 A with a dispersion of 0.22 A/pixel. A signal-to-noise over 50 could be achieved in ten minutes for the helium stars in Orion.

Figure 1 shows the spectrum at five phases throughout the 0.901 day cycle, each scan being the average of two or three individual spectra. In the left panel is the H-α profile, and in the right panel is the Helium I 6678 A profile.

Typically the spectra show a rounded profile of the helium line, with equivalent width about 0.8 A and rotation of about 130 km/sec. They also show a sometimes-rounded, sometimes-pointed H-α line, with emission appearing on the red or the blue wing. The C II lines which appear on the red wing indicate a temperature of about 20,000K, even though other researchers have suggested a somewhat higher value.

3. THE ANALYSIS

Since only two lines were observed, the following strategy was adopted for the analysis. I assume a temperature of 20,000K based on the results of Lester (1972) and the visibility of the C II lines. The helium line strength implies a range of gravities and compositions which are almost independent of the assumed temperature. The hydrogen line strength implies a similar requirement, but one which is more sensitive to the assumed temperature and is not sensitive to the gravity. Thus for each spectrum, at the assumed temperature, there is only one gravity and composition for which the model has the observed line strengths.

ANALYSIS OF THE HELIUM STRONG STAR HD 37017

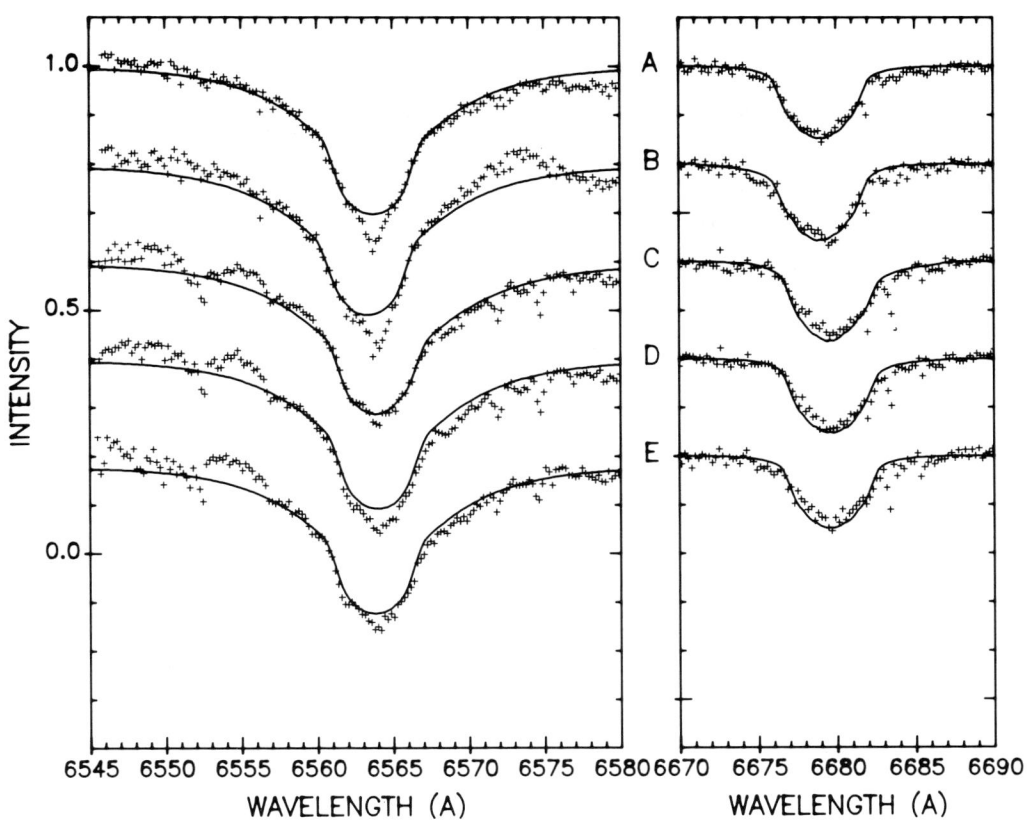

FIGURE 1--Spectra of HD 37017 taken at five phases during its cycle. In the left panel is the H-α profile, and in the right is He I 6678A.

Profile	Phase	W (H-α)	W (He I)	Comments
A	0.06	3.05	0.72	shows sharp H-α core
B	0.31	2.85	0.81	sharp H-α core, red emission
C	0.64	3.15	0.84	blue emission on H-α
D	0.72	3.35	0.76	blue emission on H-α
E	0.86	3.70	0.75	blue emission on H-α

The solid line through each profile is the prediction of a non-LTE model atmosphere which has a temperature 20,000K, log g=3.4, NHE/NTOT (helium number fraction)=0.25, rotational velocity 130 km/sec. The model, which is viewed equator-on, has a helium rich spot on the equator 40 degrees in radius and with NHE/NTOT 0.60.

The phase convention is the same as Pedersen (1979) where P = .901175 day and phase zero is at HJD 2442777.5. This means that the helium lines reach maximum strength at phase 0.5; this is when the spot is viewed on the center of the model's disk.

Models of helium rich stellar atmospheres were generated for several temperatures, surface gravities, and helium number fractions with the non-LTE code written by Mihalas. These atmospheres, described in Odell and Voels (1986), use simple geometry (plane-parallel) with no vertical or horizontal variation of composition. The variablility of line strength observed in the intermediate helium strong stars has been interpreted as due to regions of helium enhancement on the surface. Here it is assumed that the region inside the spot differs from its surroundings only in helium content, and that the intensity of light coming from any point on the star's surface is that from an atmosphere with the local composition, *i.e.*, the temperature and surface gravity are assumed the same inside the spot as outside it.

It was found that intensities at a certain zenith angle and wavelength can be interpolated linearly in log g and NHE/NTOT to about 1% in the range of characteristics for which atmospheres were computed. Thus for any assumption of stellar characteristics, line profiles can be computed from the model atmospheres. From the line profile of He I 4026 A, the R-index defined by Pedersen (1977) can be predicted. This index is contaminated in the sense that the He I line at 4009 A affected the blue continuum band. Unfortunately, this other line was not included in the computation of the model atmospheres, and so the He I 4387 A line (from the same series) was used to correct for this effect. In all of the models reported here, this correction was small and amounted to less than three percent.

4. RESULTS OF THE MODEL

Figure 2 shows the equivalent width of the helium line as it varies with phase. Both the He I 4026 A line strength parameter R of Pedersen (1979) and the He I 6678 A equivalent width show a constant value for the first and last quarter of the cycle, which I interpret as the helium rich spot being on the far side of the stellar disk. From the line profile taken during this phase (spectrum A in Figure 1), the normal stellar surface can be understood as having surface gravity log g=3.4 and NHE/NTOT=0.25. The solid lines through the two profiles in spectrum A are the prediction of this model. Although the helium line agrees well with the model, there is a sharp absorption feature at the center of H-α which is discrepant from the model. On the far red wing of H-α (between 6575 and 6580 A), there is a slight depression of the observed intensities below the predicted profile; this is due to the C II lines, which are not included in the model atmosphere calculation.

I had thought that variations in the profile of the helium line would uniquely define the geometry on the stellar surface. Figure 2 shows an increase in both the R-index and the He I 6678 A line equivalent width as the spot of helium-rich material comes onto the visible face of the star. The spotted model reveals that almost the only effect of the spot appearing on the approaching limb of the star is to increase the line strengths. The changes in the profile which are expected even from this extreme case (spot on the equator) are too small to be noted in the 6678 A line profile.

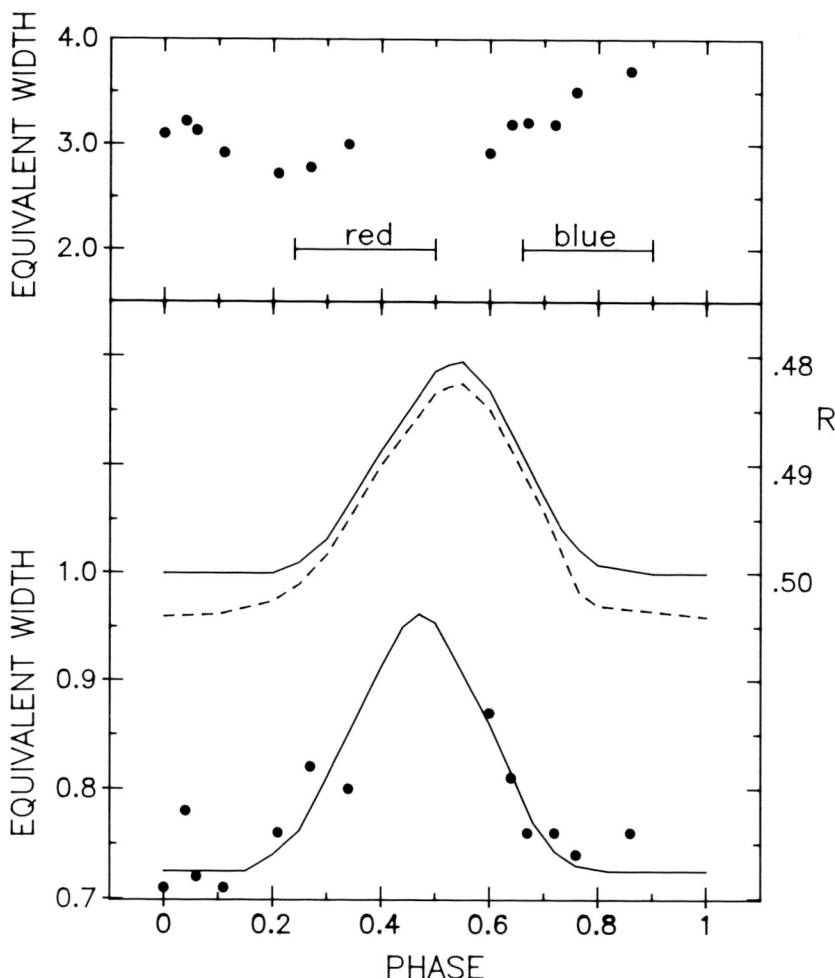

FIGURE 2--The top panel presents H-α equivalent width, as a function of phase (ephemeris same as figure 1). Also shown in the top panel is the range of phases where emission is seen on the red and blue wings of H-α.

The bottom panel shows the R-value (He I 4026 line strength observed by Pedersen (1979), shown dashed), and He I 6678A equivalent width. The predictions of the model stellar atmosphere, described in the figure 1 caption, are shown in solid lines.

Spectrum B in Figure 1 (phase 0.31) shows the helium line somewhat stronger due to the appearance of the spot on the approaching limb. The hydrogen line still exhibits the sharp core, and also has an emission feature on the red wing, at about 6573 A, or a velocity of +500 km/sec from the absorption center and about 5 A (250 km/sec) wide. In the individual spectra it persists until phase 0.50 before disappearing.

During the last part of the rotation cycle, as the spot passes toward the receeding limb, the helium lines return to their initial strength, as shown in Figure 2. The hydrogen line loses most or all of the central absorption feature, but in spectrum C, D, and E there can be seen an emission feature on the blue wing, at 6553 A, or -500 km/sec from the absorption core, about 250 km/sec wide. The emission is seen between phase 0.64 and 0.92.

Groote and Hunger (1982) point out that for spotted models of Sigma Orionis E, the behavior of a helium line depends on its strength. In the weaker lines, increasing the helium abundance increases the depth of all points across the line, while in the stronger lines, the central core of the line actually gets shallower, even though the line becomes stronger as a whole. According to this definition, the He I 6678 A line is intermediate in strength and should show no large variation of profile as the spot becomes visible on the surface.

The lack of line profile variation in the observations means that other models would also yield acceptable fits. It is not necessary to place the spot on the equator, as was done here. If it were closer to one pole, the line profiles would vary even less. The data do not require a spot--the banded model proposed by Bolton (1984) could also be made to fit the changes of the helium lines.

A more interesting case would be that geometry suggested by the magnetic field variations observed by Borra and Landstreet (1979). If the stellar radius were 4 $R_\odot$ as they take it to be, then the rotation speed and period imply an inclination about 42°. The magnetic field behavior then requires an obliquity of magnetic pole to rotational pole of 39°. A model with a normal helium abundance everywhere but in a spot centered on the magnetic pole cannot reproduce all of the helium line characteristics. Even though this geometry would naturally explain why the other pole does not exhibit a spot (that pole is never on the visible disk), the spot would have to be too big and/or too helium rich to produce such small variations in the helium line strength.

The hydrogen line does not behave in a simple fashion compared to the models. For any temperature and gravity, the hydrogen lines increase strength with increasing helium content. This is due to a more rapid decline of continuum opacity than line opacity with increase in helium abundance. Figure 2, however, shows that the hydrogen line has maximum equivalent width at phase 0.00 rather than at helium line maximum. The hydrogen line strength being out of phase with the helium spot cannot be understood in terms of any changes of the model stellar atmosphere.

The photospheric hydrogen line is certainly modified by both an excess absorption near the line core and the emission which appears on each wing. It is possible that this interference causes the hydrogen strength anomalies, but a more likely explanation is the pervasive telluric water vapor absorption and its erratic effect on the H-α line.

5. THE CIRCUMSTELLAR MATERIAL

The behavior of the H-α emission and absorption puts very severe restrictions on the model invoked to explain them. The emission is coupled to the rotation period, presumably forced to rotate with the stellar surface through the magnetic field. In order to have the proper velocity compared to the photosphere (i.e., about three times the rotation velocity), the emitting gas must be at least two stellar radii above the surface, and further out if it is at high latitude.

However, the velocity of the gas is observed to be always the same, appearing first at high velocity, as can be seen in the blue wing of the H-α line in spectra C, D, and E of Figure 1. If the material were to appear from behind the stellar disk, the component of the velocity in the line of sight of the earth would be only 150 km/sec -- far from that which is observed. If the gas were close to the stellar surface, so that it would be moving directly toward the earth when it became visible, it would never achieve a velocity shift larger than 200 km/sec.

Another feature which is seen in the individual spectra is that the red emission is stronger (about 0.25 A in equivalent width) than the emission which appears later on the blue wing (which is never more than 0.13 A in equivalent width). This is only possible if the emitting region which is seen receding is not the same as the one seen later to approach, or if the gas is optically thick and not homogeneous.

Nakajima (1981) has suggested a magnetospheric model in which gas left over after star formation is trapped at the points where the magnetic and rotational equators intersect. Qualitatively, this model is appealing, but the observations require a more complicated geometry. The profile shown in that paper for an obliquity of 90° between magnetic and rotational poles bears remarkable similarity to the features seen in HD 37017. However, the gas density must be at least 100 times less than the critical density in Nakajima's model. Further, the time dependent behavior of his model will be different from the observations, and the second junction point of magnetic and rotation equator must be empty for this star.

A model which qualitatively fits the appearance and disappearance of the spectral features requires that the gaseous region strech about one-third of the way around the star. Thus the sharp absorption core appears on H-α when the cloud is superposed on the star's disk between phase 0.0 and 0.35. The gas only contributes a noticeable red emission feature when it is seen at elongation and receding, from phase 0.25 to 0.50. The gas is first seen approaching earth at elongation when the phase is 0.65 and remains visible until phase 0.90.

Whether this distribution of gas will quantitatively predict the line profiles which are observed, and whether this model is compatible with the magnetosphere of Nakajima remains to be demonstrated. The model also leaves unexplained the asymmetry of the gas distribution on the two sides of the star.

6. CONCLUSIONS

1. HD 37017 is only moderately enriched in helium over normal B stars with helium number fraction about 0.25. The star is rotating at about 130 km/sec, projected to the line of sight.

2. The surface gravity is low, about 3.4 for log g. This implies a mass much smaller than that of typical B stars and suggests that the star is helium-rich throughout.

3. The variation of the line profiles during what is presumably one rotation is minimal; the details of the variable line strengths over the cycle cannot be explained by a model having normal composition and a spot that is always visible.

4. The generally accepted explanation of the source of the emission is mass trapped by the magnetic field, which is carried by that field as the star rotates. In HD 37017, the behavior of H-α can be understood in terms of a thin stream of gas about two stellar radii from the surface which passes across the face of the star between phase 0.00 and 0.35. The gas is so tenuous that it can only be seen in emission when it is at the elongation points.

ACKNOWLEDGMENTS

I would like to thank Dr. Werner W. Weiss for useful discussion, John Handy for computer programming help and Syl Rice for editing assistance. There were several individuals at KPNO whose help in data acquisition is much appreciated. This research was partly supported by National Science Foundation through grant AST84-17656 to Northern Arizona University.

REFERENCES

Blaauw, A. and van Albada, T. S. 1963, Ap. J. 137, 791.
Bolton, C. T. 1984, Proceedings of the Hvar Workshop on Rapid Variabililty of Early Type Stars.
Borra, E. F. and Landstreet, J. D. 1979, Ap. J. 228, 809.
Groote, D. and Hunger, K. 1982, Astron. Astrophys. 116, 64.
Heasley, J. N. and Wolff, S. C. 1983, Ap. J. 269, 634.
Heasley, J. N., Wolff, S. C. and Timothy, J. G. 1982, Ap. J. 262, 663.
Lester, J. B. 1972, Ap. J. 178, 743.
Nakajima, R. 1981, The Science Reports of the Tôhoku Univ. Series 8, Vol. 2, No. 3, p. 130.
Odell, A. P. and Voels, S. A. 1986, included in these proceedings.
Pedersen, H., and Thomsen, B. 1977, Astron. Astrophys. Suppl. 30, 11.
Pedersen, H. 1979, Astron. Astrophys. Suppl. 35, 313.

SB 939 - A NEW INTERMEDIATE HELIUM STAR AT HIGH GALACTIC LATITUDES

G. Langhans and U. Heber
Institut für Theoretische Physik und Sternwarte
der Universität Kiel
Olshausenstr. 40, 2300 Kiel, F.R.G.

ABSTRACT. SB 939 was analyzed spectroscopically for the basic atmospheric parameters T_{eff}, log g and helium content. It was shown to be an intermediate helium star (helium abundance = 22% by number) situated unusually far away from star-forming regions.

1. INTRODUCTION

Slettebak and Brundage (1971) carried out an objective prism survey near the southern galactic pole in order to search for faint blue stars. Star no. 939 (CD-40° 15910) in their list was noted to have a peculiar spectrum. Subsequently, Graham and Slettebak (1973) observed the star once again at higher resolution (39 Å/mm) and assigned a spectral type of B2V, but realized that the He I lines appeared abnormally strong. In order to clarify its nature, we observed SB 939 at high resolution with the ESO-Cassegrain Echelle Spectrograph at the 3.6 m telescope and carried out a spectroscopic analysis in order to determine its atmospheric parameters.

2. EFFECTIVE TEMPERATURE, GRAVITY AND HELIUM ABUNDANCE

The effective temperature is derived from the total flux. UV-fluxes are taken from IUE-observations, visual from Stroemgren colours and unobservable parts of the spectrum from adapted model fluxes (Kurucz, 1979). The model flux distribution with T_{eff} = 17400 K fits the observations very well (see Fig.1). Surface gravity and helium abundance are determined by fitting theoretical line-profiles to the observed Balmer- and He I lines. An excellent match is reached for log g = 3.8 and $y = n_{He}/(n_H + n_{He})$ = 0.22 (number fraction) (see Fig.2+3). This clearly demonstrates the overabundance of helium in the photosphere of SB 939 - it can be regarded as an intermediate helium star. The helium enrichment is only moderate and smaller than in most of the known helium stars. Its radial and projected rotational velocities are very small (v_{rad} = -27 ± 6 km/s, V_{rot} * sini = 0).

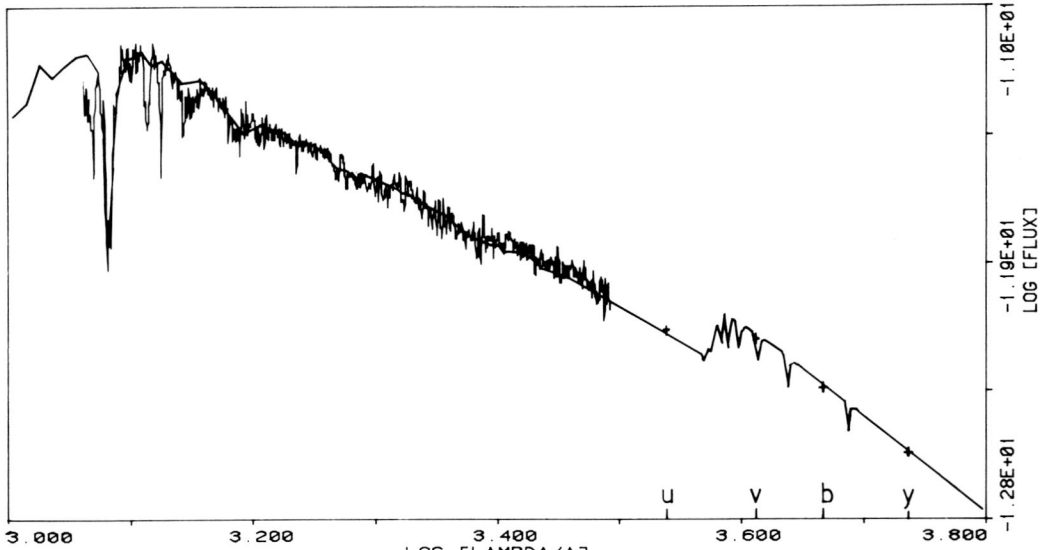

Fig. 1: Energy distribution of SB 939 compared to the flux distribution calculated from the final model (T_{eff}=17400 K).

3. DISCUSSION

All previously known intermediate helium stars are strongly confined to the galactic plane. Walborn (1983) gave a complete list of 20 intermediate helium stars brighter than 10th magnitude. They all have $|b^{II}| \leq 19°$ - consequently, SB 939 (V = 10.3, b^{II} = - 74°) is the first intermediate helium star discovered at high galactic latitudes. We can derive its mass only by comparison with evolutionary calculations (Hejlesen, 1980) since its distance is not known. If the helium anomaly is confined to the surface, a mass of 6.3 solar masses will be derived. Therefore, its distance above the galactic plane (z) is very large: z = 2.7 kpc. Hence, SB 939 is rather a massive star which lies far away from the star-forming regions (i.e. in the galactic plane). SB 939 possibly had been ejected from the galactic plane as a runaway B star (Blaauw, 1961) and subsequently slowed down to its present low velocity.

REFERENCES

Blaauw, A.: 1961, Bull. Astron. Inst. Neth. **15**, 265
Graham, J. A. and Slettebak, A.: 1973, Astron. J. **78**, 295
Hejlesen, P.M.: 1980, Astron: Astrophys. Suppl. **39**, 347
Kurucz, R. L.: 1979, Astrophys. J., Suppl. **40**, 1
Slettebak, A., Brundage, R.K.: 1971, Astron. J. **76**, 338
Walborn, N. R.: 1983, Astrophys. J. **268**, 195

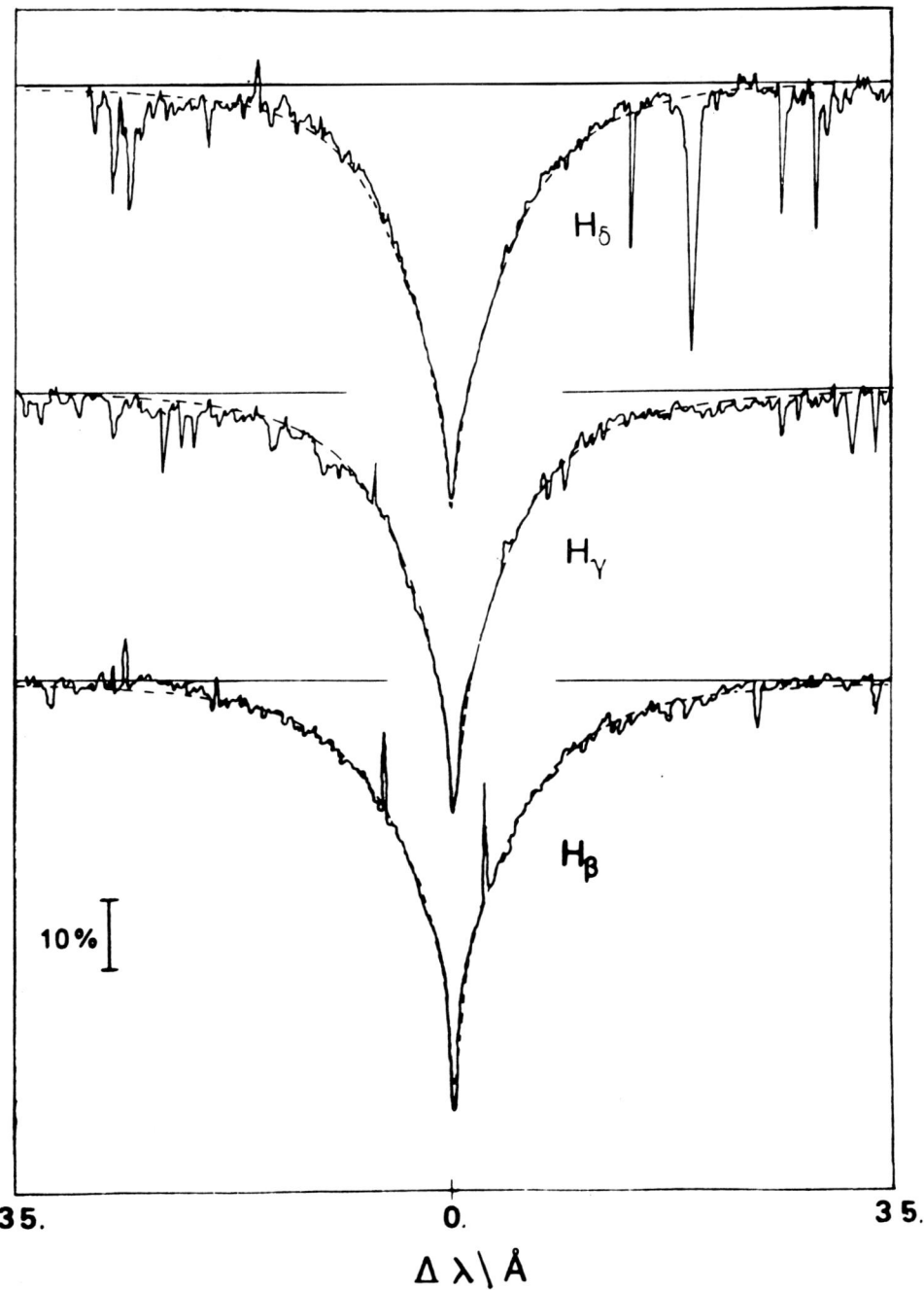

Fig. 2 Comparison of observed Balmer line profiles with theoretical profiles calculated from the final model. 10% continuum height is marked by a vertical bar.

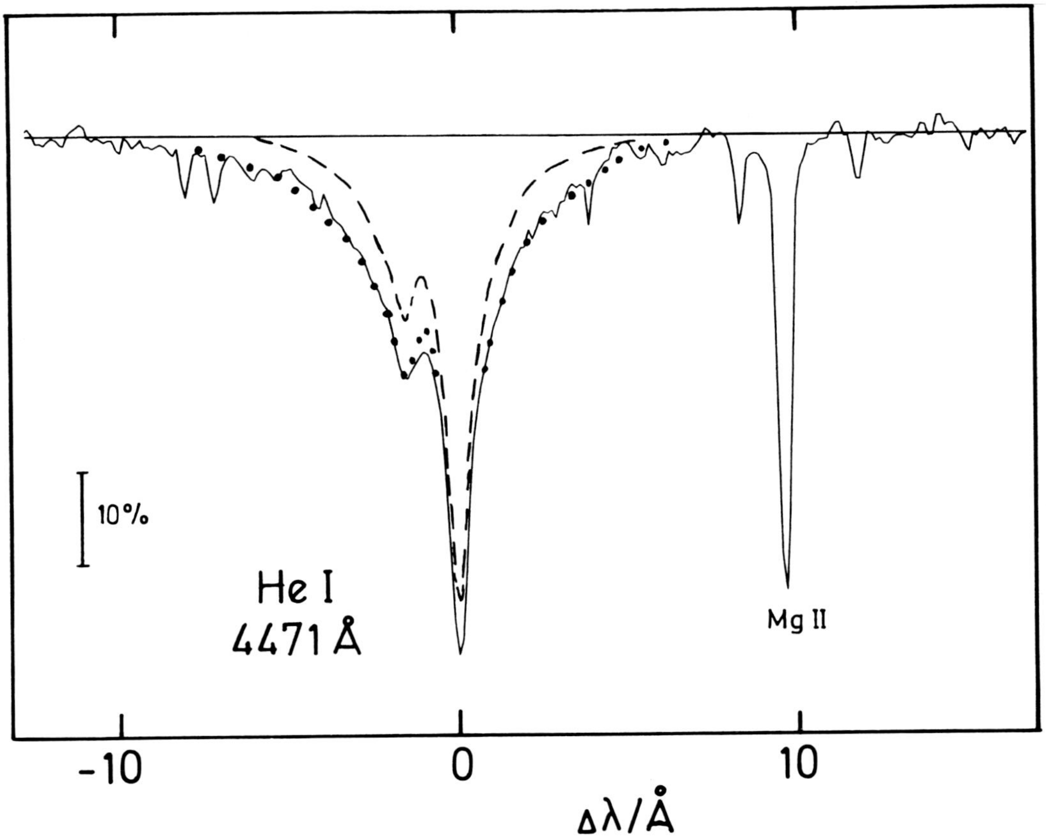

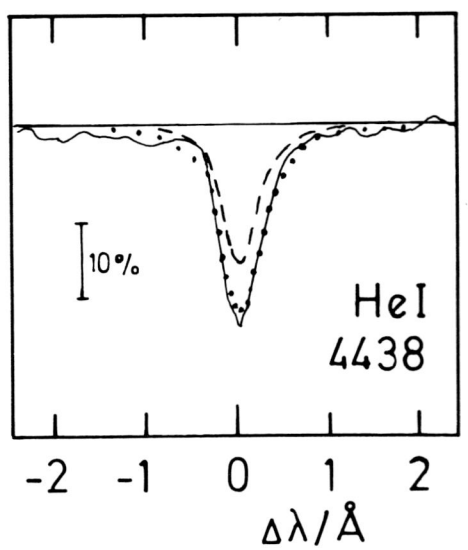

Fig. 3

Comparison of observed He I line profiles to theoretical ones:
top: λ 4471 Å, the strongest helium line observed.
right: λ 4438 Å, the weakest helium line observed.
Theoretical profiles calculated assuming normal He abundance (y=0.1) are dashed. Dotted profiles are calculated with y=0.22.

SPECTRAL VARIATIONS OF THE RAPIDLY OSCILLATING Ap STAR HD 60435

J.M. Matthews, Dept. of Astronomy, U. of Western Ontario
R.W. Slawson, Las Campanas Observatory, U. of Toronto
W.H. Wehlau, Dept. of Astronomy, U. of Western Ontario

ABSTRACT. The peculiar F star HD 60435 exhibits broadband light oscillations of low amplitude with periods between 15 and 4 minutes. Photographic spectra obtained in January/February 1985 using the University of Toronto 0.6 m telescope on Las Campanas, Chile, reveal that the star also undergoes slow spectral variations like those observed in many Ap stars. The data further suggest that Sr II maximum may coincide with maximum amplitude of light oscillations. According to the oblique pulsator model (Kurtz 1982) for such variables, this implies that Sr II is concentrated near at least one of the magnetic poles of HD 60435.

I. INTRODUCTION

Rapid photometric variations have been convincingly detected in 11 cool peculiar A-F stars to date, mostly by Kurtz (e.g. 1982, 1984). The observed periods range from about 15 to 4 minutes; typical amplitudes are less than $0^m.01$ in Johnson B or V. The oscillation amplitudes are modulated, usually with timescales of days. For stars whose magnetic field variations have been determined, these modulation periods are identical to the magnetic periods, and are phased such that maximum oscillation amplitude occurs during peak field strength.

Kurtz (1982) has accounted for most of the observed properties of the rapid variations through his oblique pulsator hypothesis, which itself is an extension of the oblique rotator model. (That model, in which Ap stars possess dipole magnetic fields inclined to their rotation axes, has been very successful in explaining their long-term magnetic, spectroscopic, and photometric variations.) According to this picture, a rapidly oscillating Ap star is an oblique rotator which is also pulsating nonradially, but with the pulsation poles aligned with the magnetic axis instead of the rotation axis. The periods are indeed consistent with high-overtone nonradial p-modes, and the model readily explains the observed relation between the apparent magnetic field strength and oscillation amplitude (as well as other empirical features, such as fine structure in the Fourier spectra).

HD 60435 is an F0p star (V = 9.0) which has one of the richest oscillation spectra yet analysed among the rapidly oscillating Ap stars. Kurtz (1984) first detected periods near 12 and 6 minutes in this star. Coordinated multi-site observations by Matthews et al. (1985) confirmed the presence of the "12-min" oscillations, and uncovered additional periods near 15 and 4 min. Additional photometry obtained a year later (Matthews et al. 1985a) showed these oscillations to be part of a pattern of frequencies spaced by roughly 55 µHz - and half that value - which might be expected for a spectrum of overtones of p-modes with even and odd degree (ℓ = 2, and 1,3).

The amplitude of the "12-min" oscillations varies with a timescale of about 8 days. Fourier spectra of the light curves also show fine splitting of peaks (predicted by the oblique pulsator model) which can account for beating with that period. Long-term photometric monitoring of HD 60435 by Matthews et al. (1985, 1985a) has yielded a light curve with a period of approximately 7.7 days (presumed to be the rotation and magnetic period of the star). All of these results are consistent with the oblique pulsator hypothesis.

A sample light curve of the rapid oscillations of HD 60435, and the corresponding Fourier amplitude spectrum, are shown in Figure 1.

In the absence of magnetic field measurements of this relatively faint star to refine tests of the oblique pulsator interpretation, spectroscopic data can supply another independent determination of the star's rotation, and perhaps some indirect information about the magnetic field geometry. As a first step, we decided to obtain several low-dispersion photographic spectra of HD 60435, in conjunction with simultaneous or contiguous rapid photometry. These would determine:
    a) if the star is a spectrum variable, and if so, to what degree;
In the event of sufficiently strong variations...
    b) a rough estimate of the variation period (or timescale) for comparison with previous results; and
    c) any correlations between the line strength variations and the modulation of the rapid oscillations.

II. OBSERVATIONS

A dozen spectra were exposed using the U. of Toronto 0.6 m telescope and classification spectrograph (Las Campanas, Chile) during 31 January to 11 February 1985, by RWS and JMM, at a dispersion of 67 Å/mm. (The spectra were recorded on baked IIa-O plates; a typical exposure was about two hours long.

Coincident with these observations, rapid photometry of HD 60435 was performed by JMM, using the U of T telescope and the 0.9 m telescope of the Cerro Tololo Inter-American Observatory. (These measurements were part of a larger campaign undertaken by Matthews et al. (1985a).) The observations consisted of continuous 20-second integrations through a B filter. No comparison star was employed, but on nights of stable sky transparency, the star's rapid coherent oscillations can be distinguished from the slower random variations in the atmospheric aerosol extinction (see Figure 1).

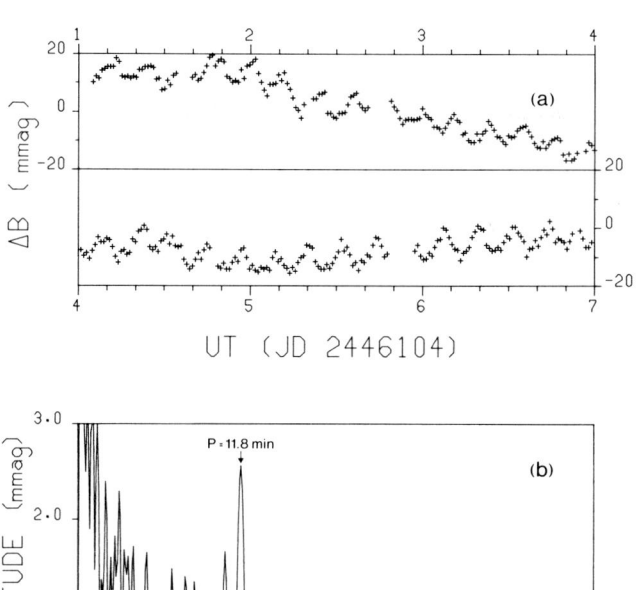

FIGURE 1. a) A light curve of HD 60435. Each point represents a 60-s integration through a B filter, using the CTIO 0.9 m telescope. b) A Fourier spectrum of the data in a). The power present at low frequencies is attributed to slow incoherent changes in sky transparency.

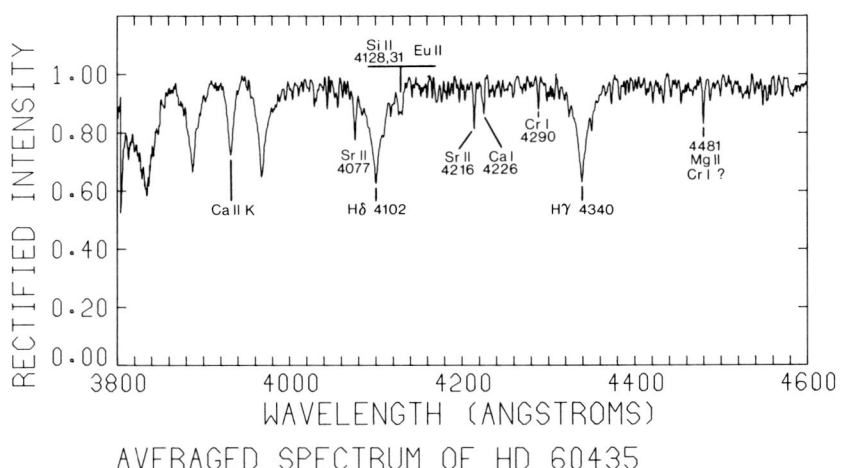

FIGURE 2. A plot of the average of the 12 digitized photographic spectra of HD 60435.

III. RESULTS

The photographic spectra were digitized using the PDS microdensitometer of the U of T. A tracing of the average of all 12 spectra is presented in Figure 2. The appearance of this spectrum agrees with an F0p(SrCr) classification for HD 60435. A few of the strongest lines are labelled. (Many lines which are apparent upon visual inspection of the plates appear as 'noise' in the continuum of the tracing.)

To display variations in line strengths, each individual spectrum was divided by the average spectrum of Figure 2. In the divided spectra, lines weaker than average will appear as 'bumps' in the continuum, while stronger lines are seen as depressions. The divided spectra are plotted in Figure 3. On some spectra (e.g. 3 Feb), there are indications that the photometric properties of the plate are non-uniform, and hence, the line strength information from those is suspect. However, on others (e.g. 31 Jan, 9 Feb), there is little doubt that the line strength changes are genuine. The constancy of the Hγ and Hδ lines was taken as an indicator of the reliability of the scan features.

Beside each spectrum in Figure 3 is an estimate of the B amplitude of the "12-min" oscillation of HD 60435, derived from the amplitude spectrum of the light curve from each night of observation (e.g. see Figure 1b for the night of 8 Feb).

To show the relation between the oscillation and spectroscopic variation even more clearly, the values of $R_\lambda/R_c - 1$ were determined for the Sr II λ4215 line in the ratioed spectra shown in Figure 3. Figure 4 is a plot of these values as a function of the time of observation. The values of ΔB are also plotted there. The lines drawn through the points are meant only to suggest the occurence of variations, and to indicate a correlation between the spectral and photometric variability.

A few limited conclusions may be drawn from the data in Figures 3 and 4:

a) HD 60435 is a spectrum variable, although a rather mild one.

b) A comparison of the relative line strengths and oscillation amplitudes suggests that Sr II is strongest at times of maximum photometric amplitude. On nights when Sr II λ4215 is noticeably stronger than average, the measured B amplitudes were $\gtrsim 2.0$ mmag. On 8 Feb, when the amplitude reached its peak value (2.5 mmag) over the 12-day interval, the Sr II λ4077 line was also stronger than average. That line shows no marked deviation from average on any other night.

c) Variations of the λ4481 line are also apparent. In addition to the Mg II doublet, there is also a Cr I line at this wavelength; however, the observed variation shows no obvious correlation with the photometric modulation.

In the context of the oblique pulsator model, result b) implies that Sr II may be concentrated at one magnetic pole of HD 60435.

Unfortunately, the S/N ratio of these spectra is too low to permit any more specific conclusions based on these data. The preliminary findings are encouraging enough to prompt further spectroscopic investigations of HD 60435, at higher resolution and S/N. Two of the authors (JMM and WHW) are currently planning such an observing programme.

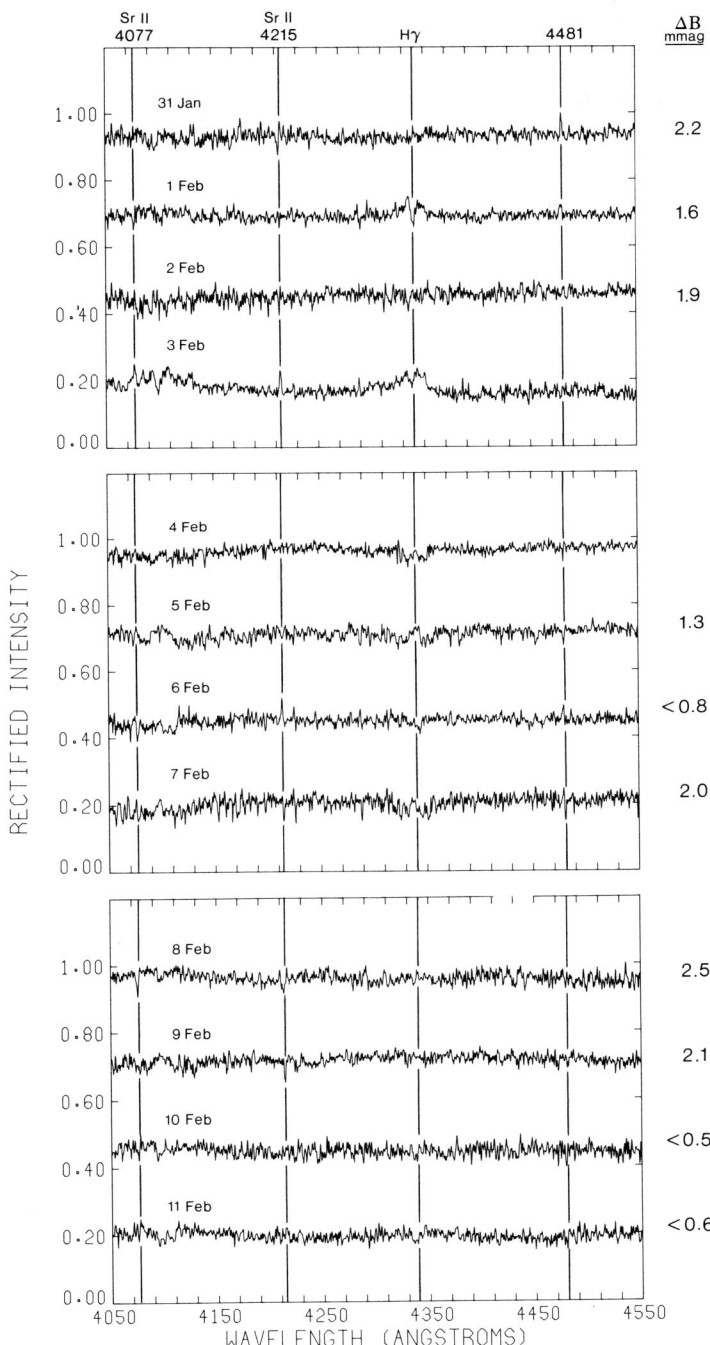

FIGURE 3. The 12 individual spectra, divided by the average spectrum of Figure 4. On the right are the corresponding B amplitudes of the "12-min" oscillation measured on those nights.

ACKNOWLEDGMENTS. We would like to thank Ron Lyons (U. of Toronto) and David Bohlender (U. of Western Ontario) for their generous assistance in the PDS processing of the plates, and Mira Rasche for her help in preparing the figures for publication. This project was funded by a grant from the Natural Sciences and Engineering Research Council of Canada.

REFERENCES

Kurtz, D.W. 1982. M.N.R.A.S. 200, 807.
-----. 1984. M.N.R.A.S. 209, 841.
Matthews, J.M., Kurtz, D.W., and Wehlau, W.H. 1986. Ap. J. 300, 348.
-----. 1986a. in preparation.

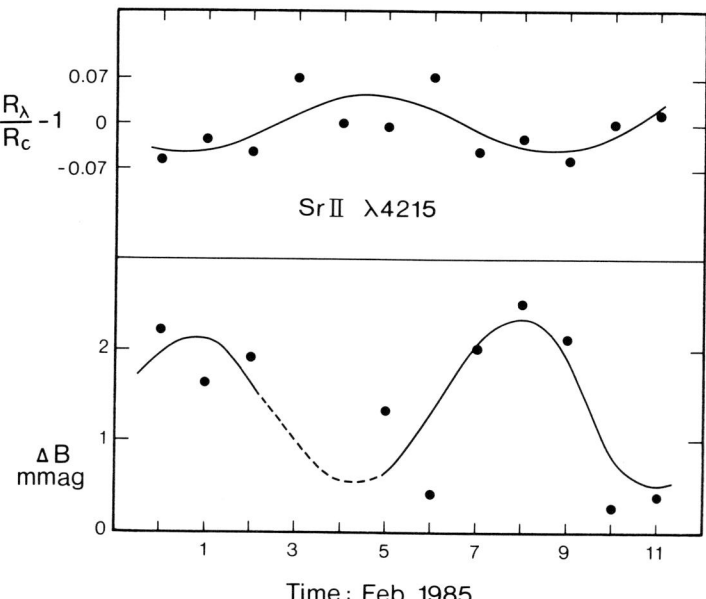

FIGURE 4. The variations of the Sr II λ4215 line strength and of ΔB (the amplitude of the light variability) plotted against time. The curves are fits by eye to both sets of data.

DISCUSSION

RAMADURAI: Is there any limit to the magnetic field strength at all? You said there are no observations yet, but is it possible to put some limits?

WEHLAU: No; there is no information at all and we have made tentative arrangements to see whether there is a magnetic field or not. It is a 9th magnitude star. The measurement requires observing time on a large telescope in the southern hemisphere.

RAMADURAI: How strong are the abundance anomalies in this star? Is it in the same range as for Ap stars?

WEHLAU: It is clearly an Ap star with Sr, Cr anomaly. However, the spectra show it to be only a mild spectrum variable, so that higher resolution spectra are required to study the spectrum variability.

G.S.D. BABU: Since this magnetic field is not yet observed, do you think the application of Kurtz's model for explaining the rapid oscillation is justified?

WEHLAU: Yes, I think that is quite correct. We really do need to get magnetic information in order to see what is going on. HD 60435 shows the features typical of the rapid oscillator. The oblique pulsator model appears to explain these features for other rapid oscillators, so, it seems appropriate to use it for this star as well; still it would be very desirable to measure the magnetic field and to test the applicability of the oblique pulsator model.

G.S.D. BABU: Is any spectrophotometry done for this star?

WEHLAU: No.

VII   RELATED OBJECTS

HELIUM RICH SUBDWARF O STARS AND CENTRAL STARS OF PLANETARY NEBULAE[*]

R. H. Méndez[1,2] and C. H. Miguel
Instituto de Astronomía y Física del Espacio, C.C. 67,
1428 Buenos Aires, Argentina

U. Heber
Institut für Theoretische Physik und Sternwarte der
Universität Kiel, Olshausenstrasse 40,
D-2300 Kiel, Federal Republic of Germany

R.P. Kudritzki
Institut für Astronomie und Astrophysik der Universität
München, Scheinerstrasse 1,
D-8000 München, Federal Republic of Germany

1. INTRODUCTION

In this review we will discuss the hottest subluminous H-deficient stars, namely those with T_{eff} > 30000 K. In the absence of reliable distance determinations for hot subluminous stars, the best way to discuss their properties and evolutionary status is to find their positions on the log g - log T_{eff} diagram. In the last few years, after extensive computational work, first in Kiel and more recently also in Munich, it has become possible to obtain log g and T_{eff}, together with the surface He abundance, directly by fitting the observed H and He absorption line profiles with theoretical profiles obtained from non-LTE model atmospheres and associated line formation codes. The non-LTE models are plane-parallel, in hydrostatic and radiative equilibrium, and the atmosphere is assumed to consist of H and He only. A recent paper by Groth et al. (1985) gives most of the references on the application of this non-LTE model atmosphere approach to the study of all kinds of hot subluminous stars.

Let us call y = N(He) / (N(He)+N(H)). Figure 1 shows the positions in the log g - log T_{eff} diagram of 51 subluminous stars. They have widely different values of y. Objects known to be members of close binary systems have been omitted. In Section 2 we will discuss the He-rich sdO stars, 0.1 < y < 0.8, and the extreme He-rich sdO stars, defined as showing no trace of H absorption lines in their spectra (y≃1). Section 3 will be devoted to H-deficient central stars of planetary nebulae (CSPN). Let us remark that, up to now, only one such CSPN has been positioned on the log g - log T_{eff} diagram: K1-27 (Méndez et al. 1985). All the other CSPN represented in Figure 1 are H-rich ($0.01 < y < 0.13$).

2. HE-RICH AND EXTREME HE-RICH sdO STARS

Before considering the He-rich sdO's, it is convenient to mention some recent work on the He-poor sdB's and sdOB's (Heber et al. 1984, Heber 1985). These stars appear now to be convincingly explained as evolving away from the extended horizontal branch (EHB) and towards the white dwarf configuration (an EHB star is defined to have a ratio q (core mass/total mass) so high that H-burning becomes negligible). Their low He abundances are attributed to gravitational settling. Their heavier-element abundances are not yet well understood.

The He-rich sdO stars (filled circles in Figure 1) pose two problems: how to produce a high photospheric He abundance, and how to keep it high against gravitational settling, since some H is left in their photospheres. Let us consider first the second problem. Groth et al. (1985) have studied the role of convection at photospheric and subphotospheric levels. They conclude that photospheric convection helps to explain the persistence of He in the photospheres of the observed He-rich sdO stars only if they had a high initial photospheric He abundance. This has an important evolutionary implication: these stars cannot have reached their present positions in the log g - log T_{eff} diagram following evolutionary tracks similar to those of EHB stars; because once a star becomes He-poor, the photospheric convection zone cannot develop, and the star is bound to remain He-poor (indeed, the hottest He-poor sdOB's in Figure 1 are well within the "convective" region).

Therefore, now it seems possible to predict what horizontal branch stars should become He-rich sdO's. First, they must have q so high that they either cannot reach the top of the asymptotic giant branch (AGB), or, if they do, and eject a planetary nebula, their subsequent evolution is so slow that the nebula is dispersed before the star becomes hot enough to ionize it (according to Schönberner (1983), this would be the case if the stellar mass after nebula ejection is smaller than 0.55 solar masses). Second, they must have q not so high as to become an EHB star, with an inert H-rich envelope. It would be necessary to specify more precisely, from the interplay between diffusion and mass loss, in which regions of the log g - log T_{eff} diagram is gravitational settling expected to dominate, and to confirm, by means of more detailed evolutionary calculations, if it is really possible to obtain evolutionary tracks leading to the filled circles in Figure 1 while avoiding the "He-sinking" regions.

Of course, a complete picture should also explain how does the high initial photospheric He abundance originate. At the present time the problem is not solved, and we can only make guesses.

Consider a star, initially somewhat more massive than the Sun, on the red giant branch, and assume (Hunger and Kudritzki 1981) that it suffers a severe mass loss, before and/or as a consequence of the core

He flash. We do not want it to lose all its H-rich envelope, because a star starting from the He main sequence would not explain most of the points in Figure 1. If a small, inert H-rich envelope remains, then we have an EHB star, whose subsequent evolution we expect to produce sdB's and sdOB's, all He-poor. It would seem that the only alternative is to start with a more normal horizontal branch star, burning He in the core and H in a shell. In this case we expect a normal initial photospheric He abundance, and the only hope seems to be additional mass loss and/or mixing during the subsequent double-shell-burning phase. Groth et al. (1985), following Paczynski (1971), favor the idea of mixing, apparently the less unlikely choice, speculating that it could arise as a consequence of the He shell flashes known to occur in double shell configurations.

The properties of extreme He-rich sdO's (plus signs in Figure 1) add some confusion to the picture. They are spread all over the log g - log T_{eff} diagram; in particular, some of them are placed among the CSPN

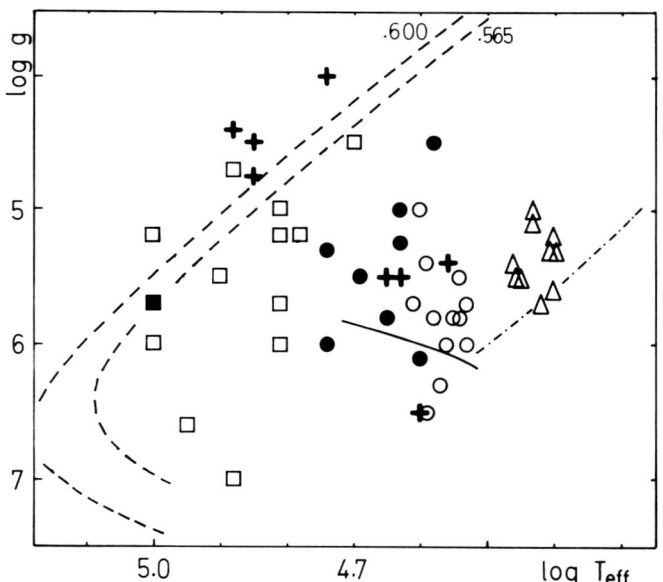

Figure 1. The positions of 51 subluminous stars in the log g - log T_{eff} diagram. The triangles are the extremely He-poor sdB stars, y<0.015, all with temperatures below 30000 K. The open circles are He-poor sdOB stars, y<0.09. Filled circles indicate the He-rich sdO stars, 0.1<y<0.8. The plus signs are extreme He-rich sdO stars, showing no trace of H absorption lines in their spectra. The open squares are H-rich CSPN, 0.01<y<0.13. The filled square is the H-deficient central star of K1-27. The full line is the He main sequence (Paczynski 1971). The dashed - dotted line is an EHB, extrapolated from Caloi et al. (1978). The dashed lines are theoretical evolutionary tracks for stars of 0.6 and 0.565 solar masses, descending from the asymptotic giant branch (Schönberner 1981, 1983).

(see Husfeld, Heber and Drilling, these Proceedings), but no nebula has been detected around them. One can speculate that these objects are really following post-AGB evolutionary tracks but somehow, in spite of their comparatively large masses, manage to evolve so slowly that the nebulae have disappeared long ago, or that it may be possible for evolutionary tracks corresponding to high q values to intersect the post-AGB tracks, or that it may be possible for a star to reach the top of the AGB without ejecting a planetary nebula.

3. H-DEFICIENT CSPN

3.1. Introductory notes on the spectral classification of CSPN

Some parts of the current scheme of spectral classification for CSPN (see e.g. Smith and Aller 1969, Lutz 1978) need revision. Objects having very different spectra are grouped together under the "O VI" label (Méndez and Niemela 1982, Heap 1982). The "continuous" objects are not continuous (Kudritzki et al. 1981a, b) and will have to be given other spectral types when better spectrograms become available. All this produces a certain amount of unnecessary confusion. The easiest way to overcome these problems is to redefine the main groups so that they become as homogeneous as possible.

Table I gives a preliminary revised spectral classification scheme (Méndez, in preparation). There are four main factors affecting the spectral characteristics of a CSPN: surface abundances, mass loss rate, T_{eff} and log g. The spectral types in Table I have been arranged in such a way that (as far as we can tell) the photospheric H-abundance decreases towards the left, the mass loss rate decreases towards the bottom, and surface gravity increases towards the bottom ("hg" stands for high gravity). The parentheses are used to indicate the chemical element whose lines predominate in the visual and photographic spectrum. For example, the central star of K1-27 shows only He II absorptions, and is therefore called O(He) (or perhaps hgO(He); the use of the label hg in this case is not yet clearly defined), while the central star of Longmore 4 shows very strong C IV absorptions, and is therefore called O(C). Parentheses are also used to describe peculiarities (e.g. " N strong ", see Table II).

We strongly emphasize that the arrangement of spectral types in Table I is schematic, particularly concerning the photospheric H abundance. Suitable model atmospheres for many of these stars are lacking, and therefore a precise determination of T_{eff}, log g and surface abundances is not yet possible. However, from what we know about very hot stars, it seems reasonable to call a CSPN "extremely H-deficient" if its spectrum does not show H lines.

3.2. A list of extreme H-deficient CSPN

To produce a list of extreme H-deficient CSPN is a delicate task. Most CSPN are faint, and many are embedded in nebulae of very high surface brightness. An efficient sky-light-suppressor, a spectral resolution of

TABLE I

REVISED SPECTRAL CLASSIFICATION SCHEME FOR CSPN (PRELIMINARY)

H-deficient		H-rich
WC	WN	
A30-78		WR-Of
		Ofp, Of
O(C), O(He)		O(H)
hgO(C), hgO(He)		hgO(H)

about 0.5 Å and a good signal-to-noise ratio are sometimes necessary to detect stellar H or He lines masked by strong nebular emissions. Therefore, in many cases, the available spectral descriptions are not good enough to be sure about the absence of stellar H lines. We have listed in Table II only those objects about which we are sure, and we have added a supplement with slightly less certain objects, for which we are not able to provide a more precise classification. In Table II we have also added references on nebular abundances, which will be discussed later.

We remark that in Table II we have included the star Sanduleak 3, following the suggestion by Barlow and Hummer (1982) that it is the remnant central star of a planetary nebula which is no longer visible. This suggestion has been confirmed by van der Hucht et al. (1985), who discovered IR emission from a circumstellar dust shell, with a temperature within the range of dust temperatures found to be common in planetary nebulae.

We would like to comment on three objects we have not included in Table II: He 2-99 (309-4°1), He 2-113 (321+3°1) and M1-67 (50+3°1).

We start with the central star of He 2-99, classified WC9 by Smith and Aller (1969). On calibrated image-tube spectrograms obtained with the 4-m telescope at the Cerro Tololo Inter-American Observatory (CTIO), we have found that the spectrum of this star, from 3700 to 6600 Å, is very similar to that of the Population I WC9 star HD157451. The full widths of emission lines are similar (with one outstanding exception: the C III emission at 5696 Å is narrower in He 2-99), and also the equivalent widths are similar, several lines being slightly stronger in the spectrum of He 2-99. In a recent paper on WC9 stars by Torres and Conti (1984) we find that most of the emission lines of HD157451 are stronger than those of HD164270. Finally, Smith and Aller (1971) have shown that the lines of HD164270 are stronger and wider than the lines of the WC9 CSPN BD+30°3639. Clearly, if we do not want to question the

TABLE II. A LIST OF EXTREME H-DEFICIENT CSPN

Object	PK designation	Spectral type	Refs. on spectral type	Refs. on nebular abunds.
K1-27	286 -29°1	O(He)	1	
K1-16	94 +27°1	hgO(C)	2	
Longmore 3	258 -15°1	O(C)	1	
Longmore 4	274 +9°1	O(C)	1	
NGC 246	118 -74°1	O(C)	3,4	3
Abell 30	208 +33°1	A30-78 (N strong)	4	13
Abell 78	81 -14°1	A30-78 (N strong)	4	13
NGC 5189	307 -3°1	WC 2	5	
NGC 2452	243 -1°1	WC 3	5,6	15
NGC 2867	278 -5°1	WC 3	5	16
NGC 6905	61 -9°1	WC 3	5,6	15
NGC 7026	89 +0°1	WC 3	5,6	15
Sanduleak 3		WC 3	7	
NGC 5315	309 -4°2	WC 4	5	14
NGC 6751	29 -5°1	WC 4 (N strong)	8	15
IC 1747	130 +1°1	WC 4	5,6	15
NGC 40	120 +9°1	WC 8	6,9	15
BD+30°3639	64 +5°1	WC 9	6,9	14
SwSt 1	1 -6°2	WC 9	5,6	17
M4-18	146 +7°1	WC10	10	18
CPD-56°8032	332 -9°1	WC10	10	
V348 Sgr	11 -7°1	WC10	10	

Supplement: slightly less certain objects

Object	PK designation	Spectral type	Refs. on spectral type	Refs. on nebular abunds.
NGC 1501	144 +6°1	WC early	4	
NGC 2371-2	189 +19°1	WC early	4	15
NGC 6578	10 -1°1	WC early	11	11
IC 2003	161 -14°1	WC early	4	14,15
M3-30	17 -4°1	WC early	12	12

References:

1 Méndez et al. 1985
2 Grauer and Bond 1984
3 Heap 1975
4 Heap 1982
5 Méndez and Niemela 1982
6 Aller 1977
7 Barlow et al. 1980
8 This paper
9 Smith and Aller 1969
10 Webster and Glass 1974
11 Kaler 1985
12 Kaler and Shaw 1984
13 Jacoby and Ford 1983
14 Torres-Peimbert and Peimbert 1977
15 Aller and Czyzak 1983
16 Aller et al. 1981
17 Flower et al. 1984
18 Goodrich and Dahari 1985

TABLE III

A LIST OF CSPN IN WHOSE SPECTRA H LINES HAVE BEEN CLEARLY SEEN

*Longmore 1	*LSS 1362	*Longmore 13	*NGC 6629
*NGC 1360	*NGC 3242	*He 2-151	Abell 46
*NGC 1535	*NGC 4361	*He 2-162	*NGC 6720
*Abell 7	IC 3568	Abell 39	*Abell 51
*IC 418	*Longmore 8	NGC 6210	NGC 6804
IC 2149	*MyCn 18	*He 2-182	NGC 6826
*Abell 15	*Abell 36	*H 2-1	NGC 6853
*NGC 2392	*He 2-108	*IC 4637	*NGC 6891
*He 2-5	*NGC 5882	Sa 4-1	NGC 7008
*EGB 5	*LSE 125	*Tc 1	*NGC 7009
*Abell 31	*Sp 1	*M1-26	NGC 7094
*IC 2448	NGC 6058	Abell 43	*NGC 7293
*Abell 33			

Note: we have suitable spectrograms of all those objects marked with an asterisk. The others are taken from the following references: Aller 1968, Greenstein and Minkowski 1964, Heap 1977, Lutz 1977, Sanduleak 1983.

widely accepted statement that subluminous WC9 stars have weaker and narrower lines, we have to consider the central star of He 2-99 as a massive, Population I star. Until the situation is clarified, it seems prudent to stop counting He 2-99 among the planetary nebulae.

The central star of He 2-113 has been classified WC10 by Webster and Glass (1974). We have not included it in Table II because they say there may be some H at or near the surface of the star.

For a similar reason we have not included the WN8 central star of M1-67 (now accepted again as a planetary nebula, see van der Hucht et al 1985): although very probably H-deficient, it appears to show some H in its spectrum (Aller 1977).

In conclusion, Table II lists 27 extreme H-deficient CSPN. It may be interesting to compare this number to the amount of H-rich CSPN, defined as those showing in their spectra clear evidence of H lines. From spectrograms at our disposal, and the information available in the literature, we have counted 49 confirmed H-rich CSPN, which are listed in Table III. Of course, it would be wrong to conclude that 35% of all CSPN are extremely H-deficient, because there is a huge selection effect favoring the detection of CSPN with WC spectra. Therefore, 35% appears to be a quite solid upper limit.

3.3. WC early CSPN

The spectral types we have adopted for these stars are based on the classification criteria defined by Méndez and Niemela (1982). These

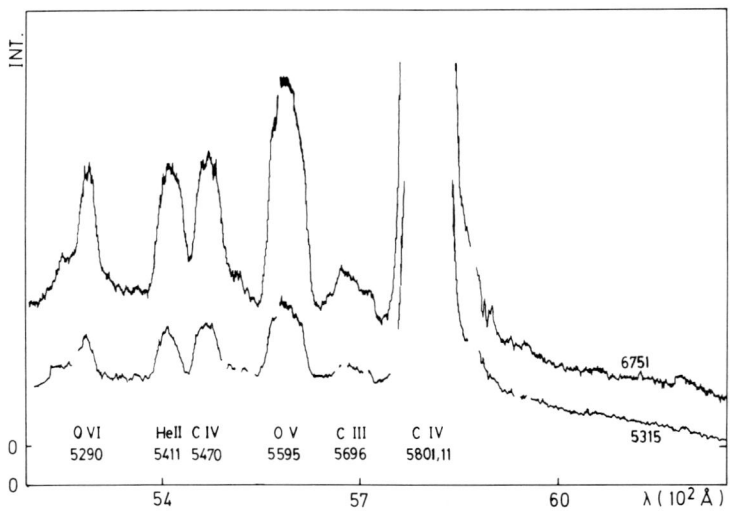

Figure 2. Intensity tracings of the WC4 central stars of NGC 6751 and NGC 5315. The nebular emission lines have been omitted. The levels of zero intensity are indicated for each object.

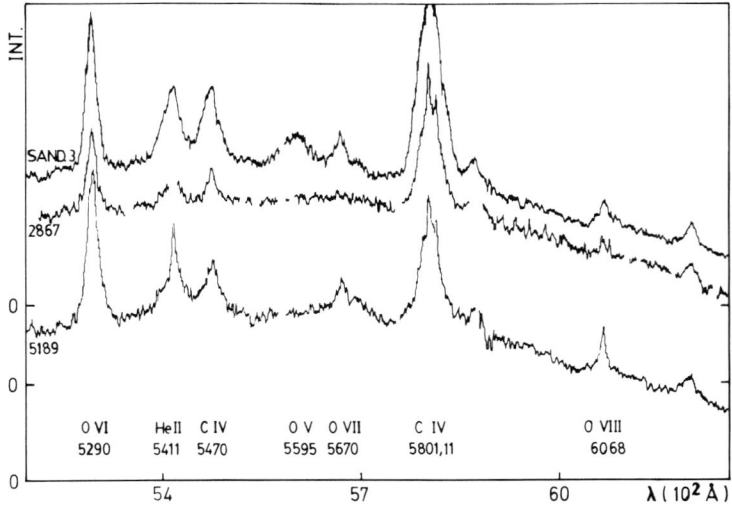

Figure 3. The same as Figure 2, for Sanduleak 3 (WC3) and the central stars of NGC 2867 (WC3) and NGC 5189 (WC2).

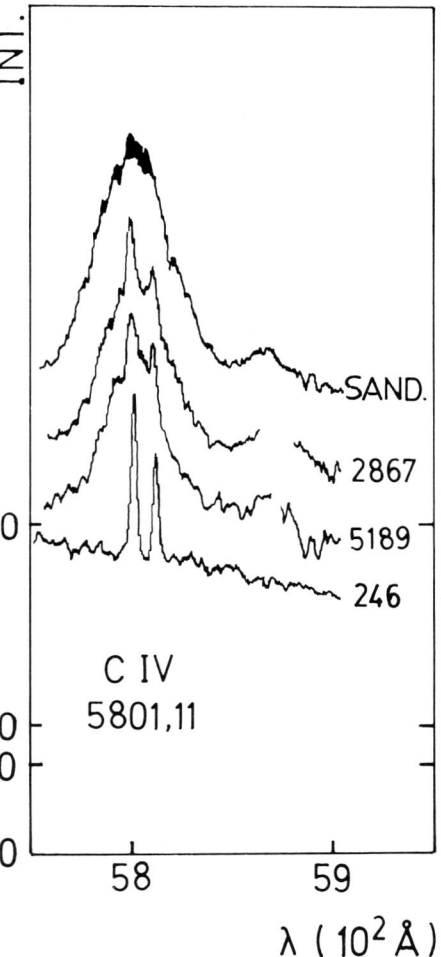

Figure 4. The C IV doublet in the spectra of Sanduleak 3 and the central stars of NGC 2867, 5189 and 246. The levels of zero intensity are indicated for each object.

criteria provide a smooth connection with the classification system used in the Sixth Catalogue of galactic WR stars (van der Hucht et al. 1981).

Figures 2,3 and 4 show intensity tracings obtained from calibrated image-tube spectrograms taken with the CTIO 4-m telescope. Figure 2 shows that the central stars of NGC 5315 and NGC 6751 are WC4, not WC6 as repeatedly misclassified in the literature; notice the weakness of the C III feature at 5695 Å compared to O V 5595 and O VI 5290. It is important to remark that we have not found any example of spectral types WC5, 6 or 7 among CSPN. Méndez and Niemela (1982) show in their Figure 3 the very different distributions of WC subtypes for CSPN and for Population I WC stars. Since no selection effect is expected to be working against the discovery of WC5, 6 or 7 CSPN, we have to conclude that this striking deficiency is real.

Figure 3 shows the spectrum of the earliest known WC CSPN (NGC

5189). Notice the weakness of the O V feature at 5595 Å and the strength of O VIII 6068. Another interesting detail is the complex profile of C IV 5801-11, also visible in the spectrum of NGC 2867. Figure 4 shows these C IV doublet profiles in more detail, compared to the C IV profile in the spectrum of the O(C) central star of NGC 246. Apparently, in the cases of NGC 2867 and 5189 the wind is becoming optically thin, and we are beginning to see "photospheric" components shining through it. We can reasonably expect that, when the stellar wind dissipates further, the central stars of NGC 2867 and 5189 will become objects very similar to the central star of NGC 246.

3.4. O(C) and O(He) CSPN

Figure 5 shows intensity tracings of the central star of NGC 246, again obtained from calibrated image-tube spectrograms taken with the CTIO 4-m telescope. Notice the O VI absorption at 5278 Å. It is interesting

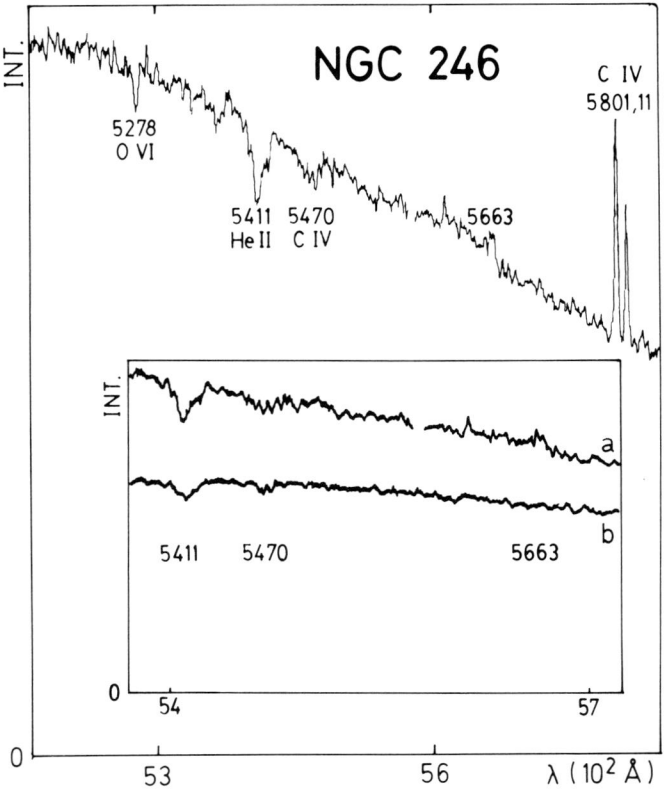

Figure 5. Intensity tracings of the O(C) central star of NGC 246. The insert shows the region from 5400 to 5700 Å in more detail. The levels of zero intensity are indicated in each case. The spectrogram labeled "a", which is reproduced twice, was taken on January 8, 1979. The spectrogram labeled "b" was taken with the same spectrograph and emulsion on June 30, 1982.

to note that in the spectrum of this star, from 3700 to 6600 Å, there is no evidence of O VII or O VIII lines, with the possible exception of a weak emission at 5663 Å, which in spite of its somewhat discrepant wavelength might be attributed to O VII. If one believes in the close connection between NGC 5189 and NGC 246, then the conclusion is that the very high ionization features are lost with the wind. We further note that the emission feature at 5663 Å, undoubtedly present in 1979, was no longer there in 1982 (see the insert in Figure 5).

Now we turn our attention to the hgO(C) and O(C) central stars of K1-16 (Grauer and Bond 1984), Longmore 3 and Longmore 4 (Méndez et al. 1985). Their spectra are dominated by C IV and He II absorptions. The central star of K1-16 does not show O VI emissions at 3811, 3834 Å; we tentatively use the absence of these emissions as the criterion to apply the label "hg" to an O(C) CSPN. In the case of Longmore 3 and 4 we do not have the necessary information; they might also be hgO(C). A reliable determination of T_{eff}, log g and surface abundances for O(C) objects appears to be within reach in the very near future (Husfeld 1986).

Grauer and Bond (1984) have discovered that the central star of K1-16 sometimes pulsates, and have stressed the spectroscopic and photometric similarities with the previously known hot pulsator PG1159-035. Subsequently, Bond et al. (1984) reported the discovery of two additional hot pulsators, again with similar spectroscopic and photometric characteristics. These four objects appear to define a new pulsational instability strip at the hot edge of the HR diagram, the cause of the pulsation being very probably the cyclical ionization of C and O (Starrfield et al. 1984). The obvious inference is that other O(C) CSPN may also be pulsating, for example Longmore 3 or 4. As far as we know, this possibility has not been checked yet.

Sion et al. (1985) have reported the discovery of strong O VI absorptions in the spectra of several members of the "PG1159" class, confirming the plausibility of the proposed pulsation mechanism, and have suggested an evolutionary link connecting the WC early CSPN with the PG1159 stars, the central star of K1-16 being considered a transition object, because of its still visible nebula and of its longer period and presumably lower surface gravity than the PG1159 pulsators.

The existence of these hot pulsators opens the possibility of detecting period changes produced by stellar evolution in very short times (a few years). There is no doubt that these stars will receive a lot of attention in the near future.

A final comment in this section concerns the O(He) central star of K1-27 (Méndez et al. 1985), whose existence demonstrates that not all the H-deficient CSPN show strong carbon lines. What is the reason for the difference? At the present time we can offer no answer.

3.5. Do all H-deficient CSPN belong to a single evolutionary sequence?

Whatever their surface abundances, CSPN are expected to follow post-AGB evolutionary tracks leading from low to high surface temperatures and from low to high surface gravities. The mass loss rate should decrease sooner or later along any of these tracks. Based on these expectations,

TABLE IV. H-DEFICIENT CSPN: COMPLEMENTARY DATA

Object	Spectral type	Neb. exc. class	Stellar temp. (10^3 K)	V_∞ (Km/s)	Neb. exp. vel. (Km/s)	References
K1-27	O(He)		100			0
K1-16	hgO(C)		>80	8500		1
NGC 246	O(C)	10	>85	>3200	39	2,1,3,4
Abell 30	A30-78 (N st)		>72	4900	40	5,6,4
Abell 78	A30-78 (N st)		>69	5000		1,6
NGC 5189	WC 2	7		3800	37	7,6,4
NGC 2452	WC 3	7	97			8,9
NGC 2867	WC 3	7	91			10,9
NGC 6905	WC 3	7	62		44	8,9,4
NGC 7026	WC 3	6	58		37	8,9,4
NGC 5315	WC 4	6	50			11,9
NGC 6751	WC 4 (N st)	4	61		40	8,9,4
IC 1747	WC 4	6			28	8,4
NGC 40	WC 8	2	35	2600	29	8,9,6,4
BD+30°3639	WC 9	1	32	2000	23	10,9,12,4
SwSt 1	WC 9	3	33	2000	13	13,14,6,15
M4-18	WC10	1	22		<15	16,4
CPD-56°8032	WC10	1				17
V348 Sgr	WC10	1	23			17,16

References:

0 Méndez et al. 1985
1 Kaler and Feibelman 1985
2 Heap 1975
3 Heap 1982
4 Sabbadin 1984
5 Kaler 1983
6 Cerruti-Sola and Perinotto 1985
7 Johnson 1976
8 Aller and Czyzak 1979
9 Preite-Martinez and Pottasch 1983
10 Aller 1965
11 Martin 1981
12 Kaler et al. 1985
13 Acker 1975
14 Flower et al. 1984
15 Acker 1976
16 Goodrich and Dahari 1985
17 Webster and Glass 1974

it is natural to propose an evolutionary sequence for H-deficient CSPN, starting with the WC late objects, then WC early, A30-78, O and hgO.

Now we ask if the observational evidence supports this picture. For the discussion of this subject we have tried to use only the most reliable information, and to keep the interpretation of the observations to a minimum. For that reason we have avoided the color temperatures of hot CSPN estimated from IUE data, which we consider too unreliable, and we have not used any of the distance determinations found in the literature. We have selected the following observational characteristics: the nebular excitation class (expected to have some relation to the stellar effective temperature), the stellar T_{eff} itself, or in its defect a temperature derived from careful studies of nebular spectra, the terminal velocity of the stellar wind (expected to increase along the evolutionary tracks, see Heap 1982), and the nebular expansion velocity. The relevant information is collected in Table IV. In Figures 6 to 9 we have plotted these nebular and stellar characteristics as functions of the spectral type.

In all four cases we find a satisfactory correlation; the observational evidence is consistent with the proposed evolutionary sequence. Besides, in Sections 3.3 and 3.4 we have seen additional reasons to suggest that WC early CSPN will become O(C) objects like NGC 246, then hgO(C) objects like K1-16, then PG1159 stars, presumably ending as non-DA white dwarfs.

However, a few problems remain. Can we really incorporate A30 and A78 into the sequence? They have unique properties. First, their central stars show moderately strong nitrogen lines in the photographic and visual regions of the spectrum (a characteristic they share with NGC 6751). Second, their central stars have managed to eject H-deficient material, which is now seen in the inner regions of these two planetary nebulae. In contrast, all the other H-deficient CSPN appear to be surrounded by essentially normal H-rich nebulae (including NGC 6751). The references are listed in Table II. At this point it is necessary to remark that in many cases there might be H-deficient inner regions around H-deficient CSPN; very careful observations would be necessary to reject this possibility. The very high angular resolution to be obtained with the Space Telescope would be very useful for such studies. However, at the present time we do not have observational evidence of the existence of more such H-deficient inner regions, and A30 and A78 remain unique among the well known H-deficient CSPN. Therefore, their connection with the other objects is not clear.

Another problem is posed by the absence of carbon lines in the spectrum of the central star of K1-27, as mentioned at the end of Section 3.4.

Finally, if the WC late and WC early objects belong to the same sequence, why are there no WC5,6 and 7 CSPN? Looking at the stellar temperatures listed in Table IV, it would seem that there is a deficit of H-deficient CSPN between 35000 and 50000 K. This is not expected from evolutionary considerations.

On the other hand, the alternative of different sequences appears somewhat artificial. We would need non-WC predecessors for the WC early, and non-WC followers for the WC late. In connection with this, it would be very important to check if there are O(He) or O(C) CSPN with surface temperatures below 60000 K. No such objects are currently known to exist.

In summary, there may be more than one way to produce a H-deficient CSPN, and/or there are many details of post-AGB evolution that we do not understand.

3.6. On the masses of H-deficient CSPN

In this section we want to discuss the available evidence on possible

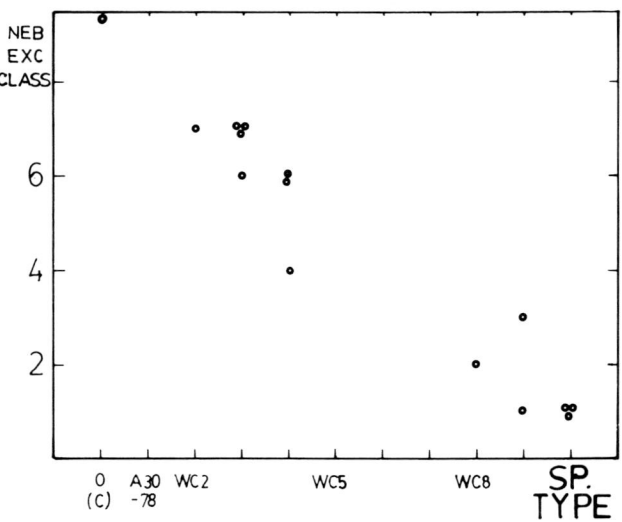

Figure 6. Nebular excitation class plotted as a function of spectral type.

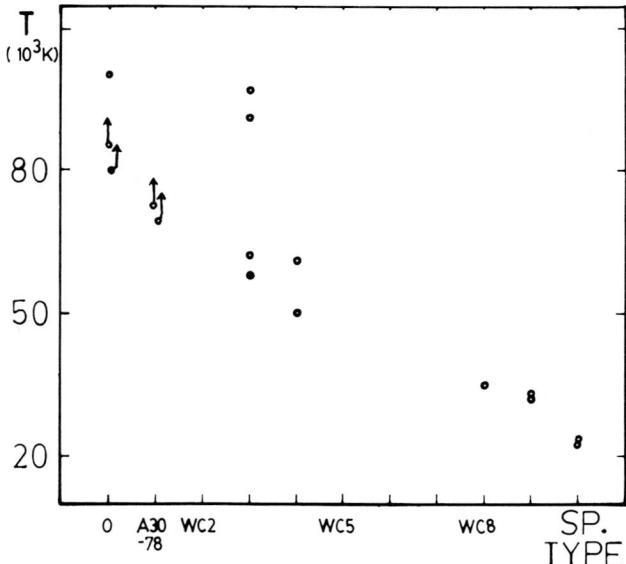

Figure 7. Stellar surface temperature plotted as a function of spectral type.

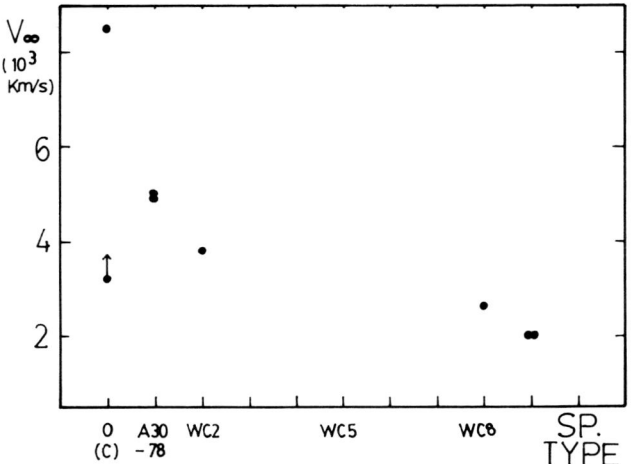

Figure 8. Terminal velocity of stellar wind plotted as a function of spectral type.

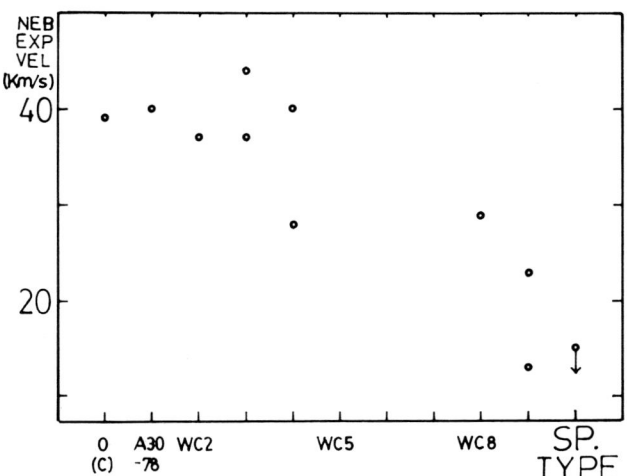

Figure 9. Nebular expansion velocity plotted as a function of spectral type. The trend is similar to that found by Phillips (1984).

differences between the masses of H-deficient and H-rich CSPN or of their ancestors.

Greig (1971,1972) states that WC objects are associated with what he calls "B" nebulae, which would appear to belong to an intermediate Population I having more massive ancestors than "non-B" nebulae. Peimbert and Torres-Peimbert (1983) state that "Type I" planetary nebulae, defined as having nebular abundances $N(He) / N(H) > 0.125$ and/or $\log (N/O) > -0.3$, frequently belong to the "B" class defined by Greig, and give additional arguments to suggest that Type I PN are associated with the most massive ancestors. Heap (1982, 1983), based on plots of absolute visual and ultraviolet magnitudes as functions of the nebular radius, suggests that WC CSPN may be more massive than O-type CSPN (meaning by "O-type" the stars we have called O(H) and hgO(H) in Table I).

On the other hand, Phillips (1984), on diagrams very similar to those used by Heap, finds no clear difference between the masses of WC, Of, O and sdO CSPN, and finds them less massive than CSPN classified as WR-Of. Perhaps this helps to understand why we are not inclined to give much weight to available distance determinations.

Let us study more carefully the relation between surface abundances of CSPN and nebular classes defined on the basis of morphological or nebular abundance determinations. We divide CSPN in two groups: the 27 H-deficient CSPN listed in Table II, and the 49 H-rich CSPN listed in Table III. Looking now for those objects that have been assigned a morphological class (see, e.g., Sabbadin 1984), we find among H-deficient CSPN 9 B and 4 non-B objects, and, among the H-rich CSPN, 6 B and 16 non-B objects. There would seem to exist a tendency for H-deficient CSPN to be associated with B nebulae and for H-rich CSPN to be associated with non-B nebulae.

In view of this, we should expect to find most of the H-deficient CSPN associated with Type I nebulae, as defined by Peimbert and Torres-Peimbert (1983). We have what seem to be reliable determinations of the required nebular abundances for 12 of the H-deficient CSPN listed in Table II. We find 4 of them associated to Type I nebulae (NGC 2371, 2452, 5315 and 6751), while 8 are definitely not (NGC 40, 2867, 6578, 6905, 7026; IC 1747, 2003; BD+30°3639). The prediction has failed.

From the evidence we have discussed, we conclude that there is no strong reason to associate the surface H-deficiency of CSPN with more massive CSPN or more massive progenitors.

We further note that the available evidence does not support the suggestion that WC early CSPN might be more massive than WC late CSPN (Peimbert 1985).

Unfortunately, the $\log g - \log T_{eff}$ diagram is still of very limited usefulness in the present discussion. Only one H-deficient CSPN (K1-27) is plotted in Figure 1, where it appears among the H-rich CSPN, again speaking against a significant difference in mass. However, we prefer to end this section by saying that, at the present time, it is not possible to make any solid statement on the subject.

3.7. Observational constraints

In its present state, the theory of stellar structure and evolution is not able to explain in detail the variety of characteristics observed in stars approaching the white dwarf configuration. A completely satisfactory description will probably have to wait until better theories for mass loss and convective mixing are developed. Normally, the best way for making progress is a semi-empirical approach, in which observational constraints play an essential role. With this in mind, we summarize those few observational constraints which appear to be well established from what has been discussed about CSPN:

(a) The lack of separation in Figure 1 between CSPN and extreme sdO stars not known to be surrounded by planetary nebulae. A careful search for faint nebulosities or dust shells around these sdO stars might yield positive results, as it did in the case of Sanduleak 3 (van der Hucht et al. 1985), perhaps leading to the elimination or modification of this constraint.
(b) Less than 35% of all CSPN are H-deficient.
(c) The absence of WC5, 6 and 7 CSPN.
(d) The existence of an evolutionary link connecting WC early, O(C), hgO(C) and PG1159 objects.
(e) The variety of surface abundances which is beginning to become apparent among H-deficient CSPN.
(f) Surface H-deficiency appears to be much more common than ejection of H-deficient material. However, as mentioned in Section 3.5, studies on the possible existence of nebular He abundance gradients or discontinuities near the H-deficient CSPN are lacking, and future work might change this conclusion.

ACKNOWLEDGEMENTS

We would like to thank the Directors and staffs of the Cerro Tololo Inter-American and European Southern Observatories for their hospitality. RHM is grateful to the IAU and the Local Organizing Committee of this Colloquium for financial support.

REFERENCES

Acker, A. 1975, Astron.Astrophys., 40, 415.
Acker, A. 1976, Ph.D.Thesis, Univ.L.Pasteur.
Aller, L.H. 1965, in Landolt-Bornstein, Group VI, Vol. 1, ed. H.H.Voigt (New York: Springer), p. 566.
Aller, L.H. 1968, IAU Symp. 34, p. 339.
Aller, L.H. 1977, R.A.S.C.Journal, 71, 67.
Aller, L.H. and Czyzak, S.J. 1979, Astrophys.Sp.Sci., 62, 397.
Aller, L.H. and Czyzak, S.J. 1983, Astrophys.J.Suppl., 51, 211.
Aller, L.H., Keyes, C.D., Ross, J.E. and O'Mara, B.J. 1981, M.N.R.A.S., 197, 647.

Barlow, M.J., Blades, J.C. and Hummer, D.G. 1980, Astrophys.J., 241, L27.
Barlow, M.J. and Hummer, D.G. 1982, IAU Symp. 99, p. 387.
Bond, H.E., Grauer, A.D., Green, R.F. and Liebert, J.W. 1984, Astrophys. J., 279, 751.
Caloi, V., Castellani, V. and Tornambe, A. 1978, Astron.Astrophys. Suppl., 33, 169.
Cerruti-Sola, M. and Perinotto, M. 1985, Astrophys.J., 291, 237.
Flower, D.R., Goharji, A. and Cohen, M. 1984, M.N.R.A.S., 206, 293.
Goodrich, R.W. and Dahari, O. 1985, Astrophys.J., 289, 342.
Grauer, A.D. and Bond, H.E. 1984, Astrophys.J., 277, 211.
Greenstein, J.L. and Minkowski, R. 1964, Astrophys.J., 140, 1601.
Greig, W.E. 1971, Astron.Astrophys., 10, 161.
Greig, W.E. 1972, Astron.Astrophys., 18, 70.
Groth, H.G., Kudritzki, R.P. and Heber, U. 1985, Astron.Astrophys., 152, 107.
Heap, S.R. 1975, Astrophys.J., 196, 195.
Heap, S.R. 1977, Astrophys.J., 215, 609.
Heap, S.R. 1982, IAU Symp. 99, p. 423.
Heap, S.R. 1983, IAU Symp. 103, p. 375.
Heber, U. 1985, 'The atmospheres of subluminous B stars II', Astron. Astrophys., in press.
Heber, U., Hunger, K., Jonas, G. and Kudritzki, R.P. 1984, Astron. Astrophys., 130, 119.
van der Hucht, K.A., Conti, P.S., Lundstrom, I. and Stenholm, B. 1981, Sp.Sci.Reviews, 28, 227.
van der Hucht, K.A., Jurriens, T.A., Olnon, F.M., The, P.S., Wesselius, P.R. and Williams, P.M. 1985, Astron.Astrophys., 145, L13.
Hunger, K. and Kudritzki, R.P. 1981, The ESO Messenger, No. 24, p.7.
Husfeld, D. 1986, Ph.D.Thesis, Munich Univ., Institut f. Astron. und Astrophysik.
Jacoby, G.H. and Ford, H.C. 1983, Astrophys.J., 266, 298.
Johnson, H.M. 1976, Astrophys.J., 208, 127.
Kaler, J.B. 1983, Astrophys.J., 271, 188.
Kaler, J.B. 1985, Astrophys.J., 290, 531.
Kaler, J.B. and Feibelman, W.A. 1985, Astrophys.J., 297, 724.
Kaler, J.B., Jing-Er, M. and Pottasch, S.R. 1985, Astrophys.J., 288, 305.
Kaler, J.B. and Shaw, R.A. 1984, Astrophys.J., 278, 195.
Kudritzki, R.P., Méndez, R.H. and Simon, K.P. 1981a, Astron.Astrophys., 99, L15.
Kudritzki, R.P., Simon, K.P. and Méndez, R.H. 1981b, The ESO Messenger, No. 26, p. 7.
Lutz, J.H. 1977, Astrophys.J., 211, 469.
Lutz, J.H. 1978, IAU Symp. 76, p. 185.
Martin, W. 1981, Astron.Astrophys., 98, 328.
Méndez, R.H., Kudritzki, R.P. and Simon, K.P. 1985, Astron.Astrophys., 142, 289.
Méndez, R.H. and Niemela, V.S. 1982, IAU Symp. 99, p. 457.
Paczynski, B. 1971, Acta Astron., 21, 1.
Peimbert, M. 1985, Rev.Mex.Astron.Astrof., 10, 125, see also p. 133.
Peimbert, M. and Torres-Peimbert, S. 1983, IAU Symp. 103, p. 233.
Phillips, J.P. 1984, Astron.Astrophys., 137, 92.

Preite-Martinez, A. and Pottasch, S.R. 1983, Astron.Astrophys., 126, 31.
Sabbadin, F. 1984, Astron.Astrophys.Suppl., 58, 273.
Sanduleak, N. 1983, Pub.A.S.P., 95, 619.
Schönberner, D. 1981, Astron.Astrophys., 103, 119.
Schönberner, D. 1983, Astrophys.J., 272, 708.
Sion, E.M., Liebert, J. and Starrfield, S.G. 1985, Astrophys.J., 292, 471.
Smith, L.F. and Aller, L.H. 1969, Astrophys.J., 157, 1245.
Smith, L.F. and Aller, L.H. 1971, Astrophys.J., 164, 275.
Starrfield, S., Cox, A.N., Kidman, R.B. and Pesnell, W.D. 1984, Astrophys.J., 281, 800.
Torres, A.V. and Conti, P.S. 1984, Astrophys.J., 280, 181.
Torres-Peimbert, S. and Peimbert, M. 1977, Rev.Mex.Astron.Astrof., 2, 181.
Webster, B.L. and Glass, I.S. 1974, M.N.R.A.S., 166, 491.

[1] Visiting Astronomer, Cerro Tololo Inter-American Observatory, operated by the Association of Universities for Research in Astronomy, Inc., under contract with the U.S. National Science Foundation.

[2] Member of the Carrera del Investigador Científico, CONICET, Argentina.

*Based partly on observations collected at the European Southern Observatory, La Silla, Chile.

DISCUSSION

FEAST: Is it known whether or not there are PN in the Magellanic clouds which fit the correlations you have shown?

MENDEZ: No. I don't think so.

LYNAS-GRAY: What is the helium abundance of the hydrogen deficient central star of K 1-27.

MENDEZ: The helium abundance (by number) is 0.6 $\pm$ 0.3 (Mendez et al. 1985, Astron. Astrophys. 142, 289). This result is to be regarded as preliminary, since hotter models and pure He models are not yet available.

SCHÖNBERNER: Is the nebular abundance known for K 1-27?

MENDEZ: We don't know it for this object. We know the nebular abundance for several WC central stars. In several cases the nebular abundance is normal. Nobody has checked whether there is a change in He abundance as you go near to the star. For that we need much better angular resolution. I would suggest space telescope should study this problem.

LIEBERT: Could you compare the likely atmospheric parameter for K 1-16 with K 1-27? I would have expected based on our models for PG 1159-035 that $T_e \approx 100000$ K and log g $\geq$ 6. K 1-16 should fall very close to K 1-27. But the spectra have some rather interesting differences in the absorption lines. In particular, you see He II 4540 of the Pickering series, which is strong and easily defined. In K 1-16, you don't see that. Normally this would be interpreted as a gravity effect. I just wondered if you had given any thought to fitting the absorption lines in K 1-16 with just helium models.

MENDEZ: I would not dare to, because you see all He II lines are severely contaminated with C IV lines. You would not know what you are doing. We need non-LTE model atmospheres for mixtures of He and C, which are not yet available.

DRILLING: I wanted to mention that I have looked for photometric variability in the LS objects and found none to show pulsations.

MENDEZ: Perhaps they are cooler, you would expect pulsation at about 100000 K more or less.

UV- AND VISUAL SPECTROSCOPY OF NINE EXTREMELY HELIUM RICH SUBLUMINOUS-
O-STARS

U.Heber[1], J.S. Drilling[2], D. Husfeld[3]

[1] Institut f. theor. Physik u. Sternwarte, Kiel, F.R.G.
[2] Louisiana State University, Baton Rouge, USA
[3] Universitätssternwarte München, F.R.G.

ABSTRACT. Nine helium rich sdO stars are found to show no trace of hydrogen on high resolution visual spectra. Effective temperatures derived from UV fluxes range from 42500 K to 80000 K. A dichotomy with respect to the C/N ratio is found which is reminiscent of the OBN and OBC stars near the main sequence. It is estimated that about 20% of the sdO stars are extremely helium rich. This fraction compares nicely with those of the (helium rich) DB white dwarfs (20%) and the helium rich central stars of planetary nebulae (less than 35%).

1. INTRODUCTION

Hydrogen deficiency is an outstanding feature of most of the sdO stars. Previous spectroscopic analyses revealed that the helium to hydrogen ratio (by number) is typically close to unity (see Hunger et al., 1981). However, a few sdO stars are known which are extremely hydrogen deficient and do not show any trace of hydrogen in their spectra.
    The sdO stars are immediate progenitors of the white dwarfs. Since about 20% of the white dwarfs are also helium rich (spectral type DB), it would be interesting to compare the properties of these two groups of stars.
    Presented here are ultraviolet and visual spectra obtained with the IUE Satellite and the ESO Cassegrain Echelle spectrograph (CASPEC). We concentrated on low galactic latitude sdO's (listed in Table I), mainly from the complete and homogenous survey of Drilling (1983).

2. OBSERVATIONS

High resolution visual spectra of nine helium rich sdO stars have been obtained at La Silla using the ESO-CASPEC attached to the 3.6m telescope. These spectra covered the spectral range from 3900 Å to 4800 Å and were reduced as described by Heber, Jonas and Drilling (these proceedings). A spectral resolution of 0.25Å was achieved.

Table I. V-magnitudes and galactic coordinates of the programmme stars

star	V	l^{II}	b^{II}	reference
LS IV+10°009	11.99	56	-19	Drilling (1983)
UV 0904-02	11.5	227	21	Berger and Fringant (1981)
UV 0832-01	12.0	232	28	Berger and Fringant (1981)
CPD-31°1701	10.54	246	-5	Garrison and Hiltner (1973)
LSS 1274	12.4	277	-5	this paper
LSE 153	11.35	314	15	Drilling (1983)
JL 9	13.37	323	-27	this paper
LSE 259	12.6	332	-8	Drilling (1983)
LSE 263	11.8	345	-23	Drilling (1983)

Most of the programme stars have been observed with the IUE satellite in the low and high resolution mode.

3. EFFECTIVE TEMPERATURES AND INTERSTELLAR REDDENING

Absolutely calibrated IUE low resolution spectra are well suited to derive effective temperatures of hot stars since the main flux is carried in the UV. For the large temperatures in question, however, the flux maxima shift towards unobservably small wavelengths and the IUE fluxes become less sensitive to T_{eff}. Moreover, since the programme stars lie at low galactic latitudes, interstellar extinction is non-negligible. The "UV-colour index" R, which is defined as the ratio of the integrated SWP fluxes to the integrated LWR fluxes has been calibrated empirically in terms of $T_{eff}(R)$ for hot subdwarfs (Schönberner and Drilling, 1984). R is shown to be almost insensitive to interstellar reddening. Schönberner and Drilling (1984) derived effective temperatures of twelve hot subdwarfs (T_{eff} > 60000 K) using their $T_{eff}(R)$ calibration. Subsequently, NLTE analyses of visual high resolution spectra for some of these sdO stars have been carried out (Husfeld et al., these proceedings and unpublished work by the authors). These spectroscopic results agreed within 27% with the UV-effective temperatures, in the worst case. Therefore, the R index allows the effective temperatures of our programme stars to be determined to an accuracy of some 20%. Since four programme stars have already been analyzed by means of NLTE analyses (CPD-31°1701, Giddings, 1980; LSE 153, LSE 259 and LSE 263, Husfeld et al., these proceedings), we can use these (completely independent) results to recalibrate R in terms of T_{eff} for extremely helium rich sdOs. (LSE 259 was neglected in the calibration since it is known to show P-Cygni profiles produced by a stellar wind, which affects the R index.) The resulting effective temperatures are given in Table II. According to their R indices, JL 9 is the hottest and CPD-31°1701 is the coolest star in our sample, while four stars have nearly identical T_{eff}s of some 50000 K. Interstellar reddening is determined by removing the λ 2200 Å bump.

TABLE II. R indices, effective temperatures and interstelar reddening

star	R	T_{eff}	E(B-V)
CPD-31°1701	2.40	42500[a]	0.00
LS IV+10°009	2.52	50000	0.07
LSS 1274	2.55	52000	0.15
UV 0832-01	2.55	52000	0.02
UV 0904-02	2.63	59000	0.02
LSE 263	2.66	70000[b]	0.10
LSE 153	2.85	70000[b]	0.09
JL 9	2.88	80000	0.07

[a] NLTE analysis, Giddings (1980)
[b] NLTE analysis, Husfeld et al., these proceedings

4. THE HELIUM LINE SPECTRUM

The CASPEC spectra are dominated by strong Stark broadened He II lines. Hydrogen is not detected, since the He II line profiles in question (λ 4340 Å and λ 4100 Å) do not show any distortion by corresponding hydrogen Balmer lines. Pronounced He I line spectra are present in five stars (CPD-31°1701, LSS 1274, LS IV+10°009, UV 0832-01, UV 0904-02), the He I lines being strongest for CPD-31°1701 and weakest for UV0904-02 among these five stars. Since the He I line strength is expected to decrease with increasing T_{eff}, this confirms the temperature sequence derived from the UV fluxes (see Table II).

For the hotter stars, He I lines are either very weak (only He I, λ 4471 Å is detected in LSE 153 and LSE 263, $W_\lambda \approx$ 100mÅ) or absent (LSE 259 and JL 9) which confirms the high effective temperatures deduced from the UV measurements.

5. CARBON AND NITROGEN LINE SPECTRA

The C/N ratios can give important clues as to the origin of the hydrogen deficiency. The "cool" stars ($T_{eff} \leq$ 60000 K) display C III, C IV, N III and N IV lines while the hotter stars display only the higher ionized species C IV, N IV and N V.

5.1 The "cool" stars

LS IV+10°009, LSS 1274, UV 0832-01 and UV 0904-02 show strong carbon and nitrogen line spectra, whereas CPD-31°1701 is carbon weak lined. This can be seen from Figure 1 where the spectrum of UV 0832-01 is compared to that of CPD-31°1701. The most striking difference to be noted is the strength of C III and C IV lines which are strong in the former but weak or even absent in the latter. This difference becomes even more apparent when we compare the C IV resonance lines in the UV (see Figure 2).

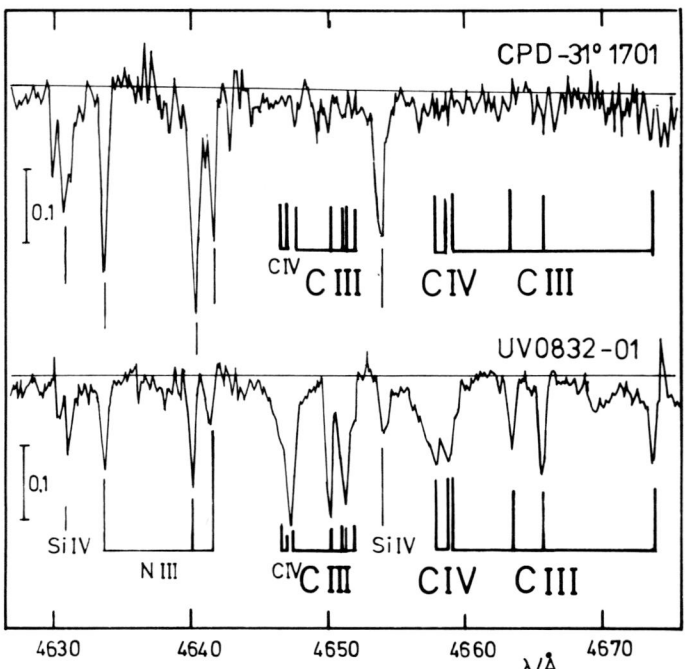

Figure 1: Comparison of the CASPEC spectra of CPD-31°1701 (top) and UV 0832-01 (bottom). 10% continuum height is marked by a vertical bar.

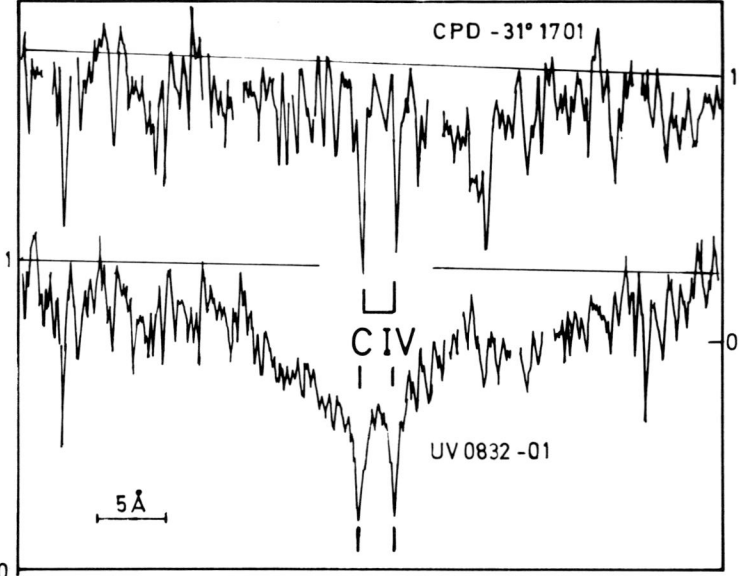

Figure 2: Comparison of C IV resonance lines of CPD-31°1701 (top) and UV 0832-01 (bottom).

5.2. The hotter stars

Nitrogen lines are present in the hotter stars in similar strength. As in the case of the "cool" stars, the carbon line spectra vary considerably. LSE 153 and LSE 259 are carbon strong lined while LSE 263 and JL 9 are carbon weak lined.

5.3. The C/N dichotomy

In conclusion, the high resolution spectra reveal strong evidence for a dichotomy of the group of extremely helium rich sdO stars with respect to the C/N ratio. Six of our programme stars have large C/N ratios while the other three must have much smaller C/N. This is reminiscent of the chemically peculiar OB stars near the main sequence (OBN and OBC stars).

Among the "cool" stars, the well known Garrison and Hiltner star CPD-31°1701 (the brightest star in our sample) is the only carbon weak lined star and, therefore, is probably not an archetypal extremely helium rich sdO but appears to be an exceptional case.

6. FREQUENCY OF HELIUM RICH SDO STARS

The sdO stars as well as the central stars of planetary nebulae (CPN) are immediate progenitors of the white dwarfs. About one fifth of the white dwarfs are helium rich (Liebert, these proceedings). Among the CPNs, less than 35% are helium rich (Méndez et al., these proceedings). It would, therefore, be interesting to know what fraction of the sdO stars is extremely helium rich. Two complete and homogeneous surveys have been studied in order to answer this question:
(i) The Slettebak-Brundage (1971) survey at the south galactic pole contains five sdOs out of which only one is extremely helium rich (Heber, 1986).
(ii) The Drilling (1983) survey at low galactic latitudes contains twelve sdOs out of which three stars are extremely helium rich.
Since there is no obvious bias against the extremely helium rich subdwarfs, we conclude that extreme hydrogen deficiency occurs as frequently among subluminous O stars as among the white dwarfs and the CPNs.

REFERENCES:

Berger, J., Fringant, A.-M.: 1981, Astron. Astrophys. **85**, 367
Drilling, J.S.: 1983, Astrophys. J. Letters **270**, L13
Garrison, R. F., Hiltner, W. A.: 1973, Astrophys. J. Letters **179**, L117
Giddings, J.R.: 1980, thesis, UCL, London
Heber, U.: 1986, Astron. Astrophys. **155**, 33
Hunger, K., Gruschinske, J., Kudritzki, R.P., Simon, K.P.: 1981, Astron. Astrophys. **95**, 244
Schönberner, D., Drilling, J.S.: 1984, Astrophys. J. **278**, 702
Slettebak, A., Brundage, R.K.: 1971, Astron. J. **76**, 338

DISCUSSION

LYNAS-GRAY: Did you use the standard Kudritzki's non-LTE code or did you include CNO levels following Husfeld et al.?

HEBER: I did not use any model atmospheres analysis here.

LYNAS-GRAY: You must have used some menas for the temperature analysis.

HEBER: The temperature determinations are purely empirical. We calibrated the R-indes in terms of T_{eff} by using CPD $-31°1701$ and LSE 153/LSE 263 as calibration standards. These stars have been analyzed from line profiles by means of NLTE model atmosphere techniques (Giddings, 1980, Husfeld, 1985).

NLTE-ANALYSIS OF THREE EXTREMELY HELIUM-RICH O-TYPE SUBDWARFS

D. Husfeld[1], U. Heber[2], J.S. Drilling[3]
[1] Universitätssternwarte, München, F.R.G.
[2] Inst. f. theor. Physik u. Sternwarte, Kiel, F.R.G.
[3] Louisiana State University, Baton Rouge, U.S.A.

ABSTRACT. Three extremely helium-rich sdO stars (LSE 153, LSE 259 and LSE 263) were analyzed spectroscopically by means of detailed NLTE model atmospheres. These stars are very hot, with effective temperatures ranging from 70000 to 75000 K and gravities between log g = 4.4 and 4.9. Upper limits for the hydrogen abundance were also derived. The evolutionary status of the sdO stars is discussed and it is concluded that they evolve from the asymptotic giant branch towards the white dwarf stage. A possible evolutionary link between these hot stars and the extreme helium stars of spectral type B is discussed.

1. INTRODUCTION

Drilling (1983) discovered 12 hot sdO stars at low galactic latitudes. Three of them (LSE 153, 259 and 263) were classified as helium-rich from low resolution spectra. Subsequently, high resolution spectra were taken which revealed no trace of hydrogen (Heber et al., 1986). Presented here is a NLTE analysis of the visual high resolution spectra described by Heber et al. (1986).

2. MODEL ATMOSPHERES

NLTE model atmospheres were constructed using the complete linearization approach as described by Mihalas (1972). The statistical equilibrium equations (for hydrogen and helium) and the radiative transfer equations can be solved simultaneously with the constraints of hydrostatic and radiative equilibrium. Plane-parallel geometry is assumed. The computer code has been newly written (Husfeld, 1986) and is an extended version of the code described by Kudritzki (1976). It allows any chemical mixture of hydrogen and helium in the atmosphere (metals are not included in the calculations). The hydrogen model atom used consists of 16 levels for H I and one level for H II; 16 levels of He I, 32 levels of He II and one level of He III are considered for the helium model atom. All these levels contribute to the continuous opacity. The lowest 5 levels of H I and He I, respectively, and the lowest 10 levels of He II are treated in statistical equilibrium, whereas the others are held in LTE, relative to the respective continuum state. Radiative bound-bound rates are neglected in the model atmosphere calculations.

Line formation calculations were subsequently carried out using more detailed model atoms for hydrogen and helium in the statistical equilibrium calculations. At this stage the radiative bound-bound rates were taken into account and, using the best broadening theories available, accurate line profiles for H I, He I and He II were calculated (for details see Kudritzki and Simon, 1978).

3. ANALYSIS

A large grid of NLTE model atmospheres and emergent line profiles was computed for a helium abundance $\varepsilon_{He} = n_{He}/(n_H + n_{He}) = 0.99$ (number fraction). The grid covered the following parameter range: 30000 K $\leq T_{eff} \leq$ 140000 K and $3.5 \leq \log g \leq 7.0$.

The above grid allows the construction of "fit curves" which represent lines of constant equivalent width in the $(g - T_{eff})$-plane. By plotting the "fit curves" for the observed equivalent widths of several lines, their intersection in the $(g - T_{eff})$-plane defines a first estimate for the model parameters. We used the follow-ing lines: He II, λ 4200 Å, λ 4542 Å, λ 4686 Å and He I, λ 4471 Å. Note that no He I, λ 4471 Å can be measured in the spectrum of LSE 259. An upper limit of 50 mÅ was assumed for the equivalent width. Figure 1 displays the fit diagram for LSE 153 as an example. Comparing the observed and the theoretically predicted line profiles, additional models with slightly modified temperature and gravity were calculated until a satisfactory match for all lines was achieved. Results are given in Table 1.

The effective temperatures of LSE 153 and LSE 263 can be determined very precisely (error margin 2.2% and 3.5%, respectively; see Table I). This is due to the fact that He I λ 4471Å is measurable in both spectra. The strength of this line is very sensitive to effective temperature. No He I line is present in the spectrum of LSE 259 and T_{eff} and log g have to be determined exclusively from He II lines, these being less sensitive to T_{eff}. Thus, the error range for LSE 259 is larger. Figure 2 displays the line profile fit for LSE 153.

To check the influence of the chemical composition, further models were computed with increased hydrogen abundance ($\varepsilon_H = 0.10 \ldots 0.20$). This increase always left the profiles of the unblended He II and He I lines — and therefore the adopted parameters T_{eff} and log g — unchanged, but distorted the He II λ 4339 Å and λ 4100 Å line profiles. Since the observed profiles of these lines appear symmetric, upper limits to the hydrogen abundance and, consequently, lower limits to the helium abundance were derived (see Table I).

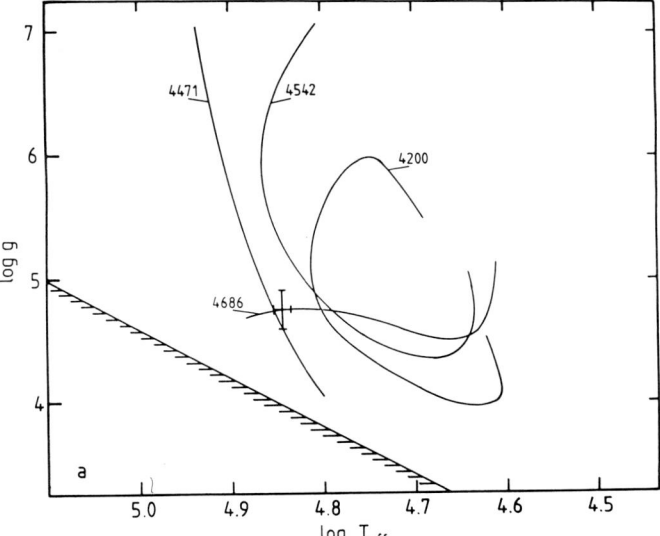

Fig. 1:

Fit curves for the observed equivalent widths (LSE 153). Also plotted is the finally adopted location together with the error limit. The hatched line represents the Eddington limit.

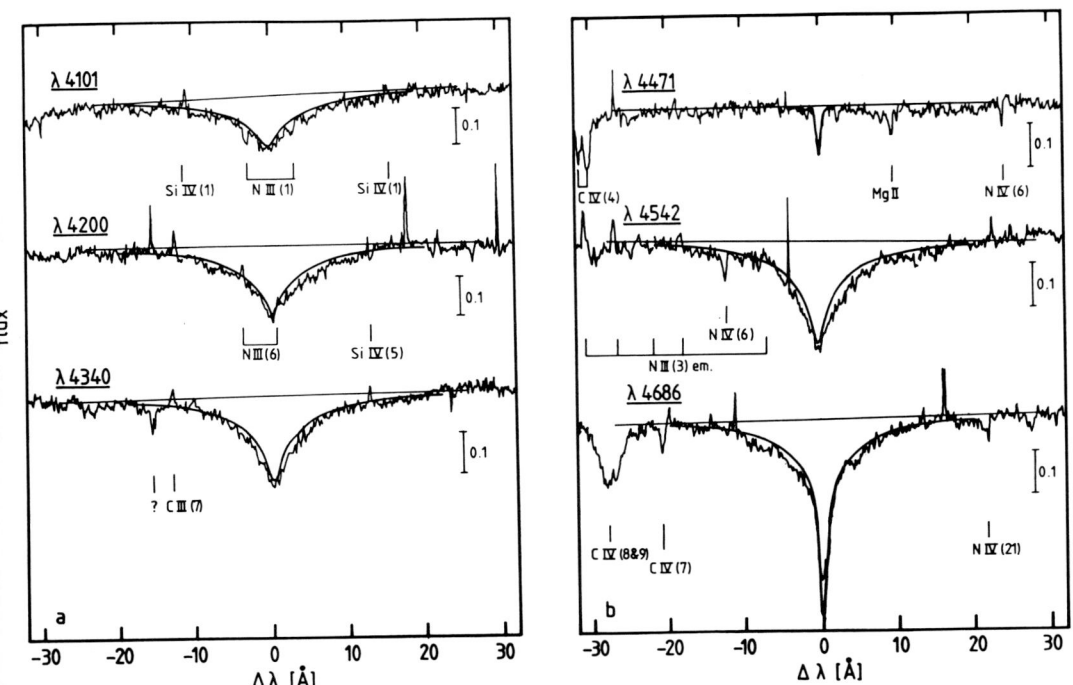

Fig. 2:

Comparison of the observed line profiles with the theoretically predicted profiles of the final model of LSE 153. The bars on the right represent 10% continuum height.

TABLE I: Atmospheric parameters of the program stars

Error limits indicate the parameter region in which an equally good fit to the line profiles can be achieved.

Object	T_{eff} [K]	log g [cgs]	ϵ_{He}
LSE 153	70 000 $\pm$ 1 500	4.75 $\pm$ .15	$\geq$ .90
LSE 263	70 000 $\pm$ 2 500	4.9 $\pm$ .2	$\geq$ .90
LSE 259	75 000 $^{+10\,000}_{-5\,000}$	4.4 $\pm$ .2	$\geq$ .95

4. DISCUSSION

In Figure 3 the atmospheric parameters for LSE 153, LSE 259 and LSE 263 are compared to those of other sdOs (filled and open circles), to the central stars of planetary nebulae (CPN; squares), and to extreme helium stars of spectral type B (triangles). All stars have been analyzed previously using model atmosphere techniques. As can be seen from Figure 3, LSE 153, LSE 259 and LSE 263 do not lie in the region in the (g - T_{eff})-plane where the other sdOs can be found (Hunger et al., 1981), with the notable exception of BD+37°442. The latter is also extremely helium-rich (Giddings, 1980). Instead, these stars lie close to some CPNs in Figure 3.

Also shown in Figure 3 is an evolutionary track for a helium star of 0.7 $M_\odot$ (Schönberner, 1977). The latter starts at the tip of the asymptotic giant branch (AGB) and leads through the observed region of both the extreme helium stars of spectral type B and the four extremely helium-rich sdOs, as can be seen from Figure 3. This suggests an evolutionary link between these two groups; thus, the helium-rich sdOs would be the immediate successors of the extreme helium stars of spectral type B. If this were the case, these sdOs would need to have metal abundances similar to those of their progenitors. The carbon abundance is of particular interest since the extreme helium stars are known to be carbon-rich. LSE 153, LSE 259 and BD+37°442 are indeed carbon-rich but LSE 263 is not and, thus, the latter cannot have evolved from a B-type extreme helium star.

Detailed abundance analyses are under way in order to test the possible relationship to the B-type extreme helium stars.

ACKNOWLEDGEMENT. This research was supported in part by grants Hu 39/21-1 and Ku 474/9/2 of the Deutsche Forschungsgemeinschaft and by NSF Grants AST 8018766, AST 8514574 and INT 8219240 and by NASA Grant NAG 5-71.

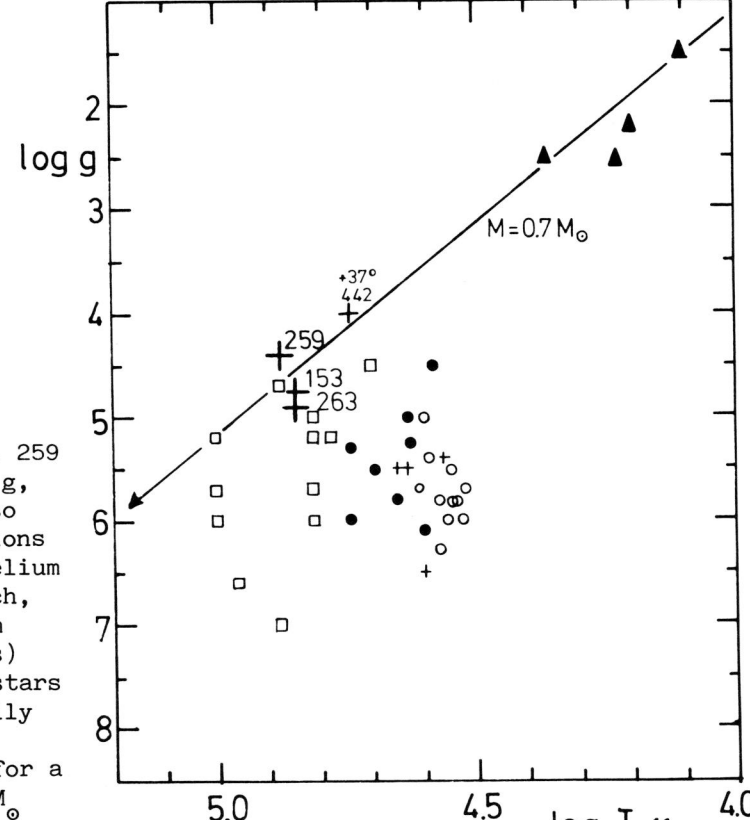

Fig. 3:

Position of LSE 153, 259 and 263 in the (log g, log T_{eff})-plane. Also shown are the positions of sdO stars (o = helium poor, ● = helium rich, + = extremely helium rich), CPNs (squares) and extreme helium stars (triangles). The fully drawn line is an evolutionary track for a helium star of 0.7 $M_\odot$ (Schönberner, 1977).

REFERENCES

Drilling, J.S.: 1983, Astrophys. J. **270**, L 13
Giddings, J.R.: 1980, PhD Thesis, UCL, London
Heber, U., Hunger, K., Drilling, J.S., Husfeld, D.: 1986, In Hydrogen deficient stars and related objects, eds. K. Hunger, N.K. Rao, D. Schönberner, Reidel
Hunger, K., Gruschinske, J., Kudritzki, R.P., Simon, K.P.: 1981, Astron. Astrophys. **107**, 313
Husfeld, D.: 1986, PhD Thesis, München
Kudritzki, R.P.: 1976, Astron. Astrophys. **52**, 11
Kudritzki, R.P., Simon, K.P.: 1978, Astron. Astrophys. **70**, 653
Mihalas, D.: 1972, NCAR Technical Note NCAR-TN/STR-76, Boulder/Colorado
Schönberner, D.: 1977, Astron. Astrophys. **57**, 437

HYDROGEN DEFICIENT PLANETARY NEBULAE: PRELIMINARY RESULTS

S.R. Pottasch[1], A. Mampaso[2], A. Manchado[2], J. Menzies[3]
[1]University of Groningen, [2]Instituto de Astrofisica de Canarias,
[3]South African Astronomical Observatory

ABSTRACT. New spectra of A78 and A58 at different positions in the nebulae are presented. An abundance gradient is found in A78, extending quite close to the center. Similarly the nebulous knot near the center of A58 has considerably higher heavy element abundances than the outer regions of this nebula. The ionization state is considerably lower in A58 than A78. In A78 most of the neon is in the form of Ne^{+3} and Ne^{+4}, indicating that the standard ionization correction factor as used by Jacoby and Ford, is substantially in error. Finally, the very high infrared excesses found in this nebulae are discussed.

I. Introduction

A little more than five years ago, Jacoby (1979) reported that the material near the center of the large planetary nebulae A78 and A30 was relatively much brighter in the [O III] line λ5007 than in the Hα line. This suggested that the central regions of these nebulae may be hydrogen deficient, although this possibility was not specifically mentioned by Jacoby. The spectroscopic measurements, first of Hazard et al (1980) of A30, then of Jacoby and Ford (1983) of both A78 and A30, confirmed that the central regions of these nebulae are indeed hydrogen deficient. The Balmer lines of hydrogen were not detected at all by these authors, although the recombination lines of both He^+ and He^{++} are easily seen.
This leads to the conclusion that He is a factor of at least 5 times more abundant than H, which is a factor 50 different that the normal H:He ratio in nebulae. Jacoby and Ford further report that the ratio of He to O, N and Ne is consistent with the hypothesis that the abundances are the same as in other nebulae with the exception that hydrogen has been converted to helium.

We have begun an investigation of the abundances in A78. The purpose originally was to determine if the outer regions of the nebula had normal chemical composition. The preliminary results are presented in section II, where it is shown that even in the central regions of this nebula a strong abundance gradient exists, and that the outer regions are still somewhat hydrogen deficient.

In section III we discuss the unusually large far infrared excess found in these nebulae. It is shown that this excess is at least two orders of magnitude greater than the value found in normal nebulae. While the origin of this larger excess is not understood, it may be related, either directly or indirectly, to the abundances anomoly,

Finally attention is directed to another nebula which also has an unusually large infrared excess: A58. Spectra are presented of parts of the outer regions of this nebula as well as the knot near the center, which may have been ejected less than 60 years ago. A preliminary analysis of the spectra show that the outer regions are underabundant in many elements, but that the knot has considerably higher abundances.

II. ABUNDANCE IN A78.

The spectra described here were taken with the IPCS on the 2.5m telescope at La Palma. A long slit (1" wide) was used which extended across the entire nebula in a NE to SW direction. It was intended that the slit be centered at the position measured by Jacoby and Ford (1983) i.e. midway between the central star and the visual companion located 10" to the NW. A photograph taken of the slit shows it to be about 2" to 3" further away

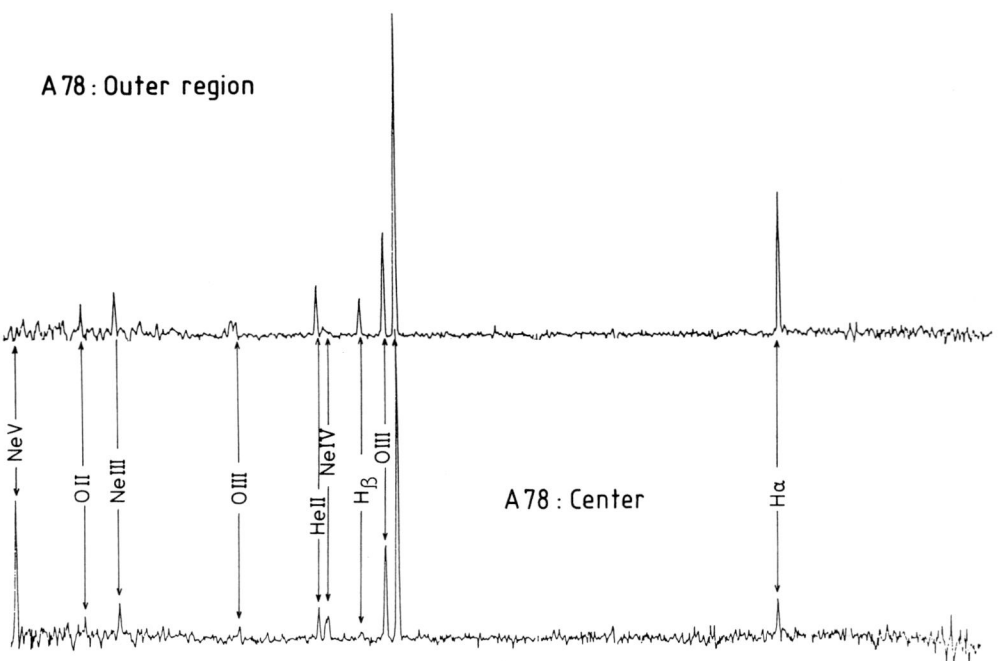

Fig. 1. The spectrum of A78. The above spectrum (a) shows a region 30" SW of the center in the outer nebula, while the lower spectrum (b) is a region near the center. Note the weakness of Hβ near the center.

from the central star however. This is probably 1" to the north of the edge of the 3″2 aperture used by Jacoby and Ford. Thus we have not measured precisely the same region.

The reason for comparing the relative position of the two measurements so closely is that our spectrum is substantially different than that measured by Jacoby and Ford. We can see the Balmer lines clearly, as shown in Fig. 1b, while they could not, indicating that these lines were much weaker on their spectrum. Furthermore they report seeing the [N II] line at $\lambda 6584$ Å which we see only weakly if at all. We both see the [Ne IV] line (two pairs the strongest of which is at $\lambda 4725$ Å) in considerable strenth. This line is much weaker in most nebulae and it is very unusual to see it so strongly.

In Fig. 1a we show the spectrum of the nebula at a position far from the central star, in the somewhat brighter region about 30" SW of the center. As can be seen, the spectrum is much different at this outer position. The Balmer lines are much stronger here. Furthermore the strong [Ne V] line at $\lambda 3425$ Å seen near the center has completely disappeared here. The [Ne IV] lines discussed above are also much weaker and they are now blended with, or perhaps replaced by, [Ar IV] lines which are usually stronger in most nebulae. The [O II] lines near $\lambda 3727$ Å clearly seen in the outer regions.

To obtain an approximation to the abundances we have performed a preliminary calibration of the spectra. The electron density in both the central and outer region is low enough so that its precise value need not be known. The electron temperature is obtained from the [O III] lines. The resultant abundances are given in Table 1.

TABLE 1: CHEMICAL COMPOSITION A78 : X/H

ELEMENT	OUTER REGION	CENTRAL REGIONS PRESENT	JACOBY FORD
HELIUM	0.21	1.2	> 6.0
OXYGEN	1×10^{-3}	1.4×10^{-2}	2×10^{-1}
NEON	2.5×10^{-4}	2.8×10^{-3}	1×10^{-1}

As can be seen from the table, there is a large abundance difference between the central position we have measured and that measured by Jacoby and Ford, which was to be expected because of the presence of the Balmer lines in our spectra. (It must be noted that we have used the measurements of Jacoby and Ford but not their derived abundances. This is because they correct for the missing ionization states in neon and oxygen (other than Ne^+ and O^{+2}) using the ratio of H^+ to He^{++} as a measure of the ionization of these elements. For neon however we have measured 3 ionization stages and can show that the traditional correction for missing ionization must be seriously in error.)

III. INFRARED EXCESS OF A30 AND A78.

Fig.2 is a plot of the total infrared flux emitted by almost 100 planetary nebulae vs. the 6cm. radio continuum emission of that nebula. A

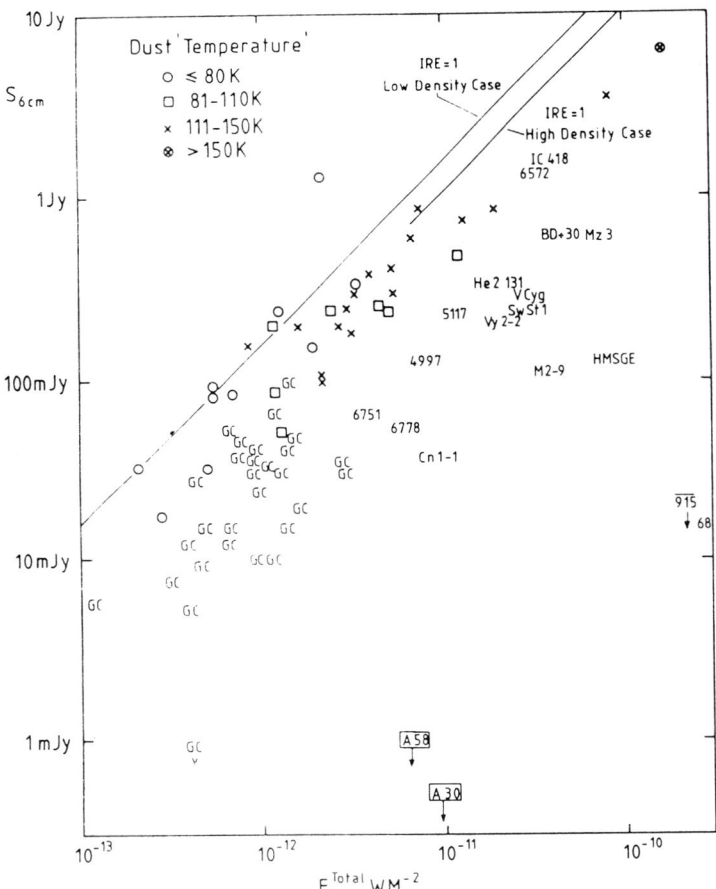

Fig. 2. The total infrared flux is plotted against the 6cm radio emission of many planetary and protoplanetary nebulae. Note the rather different positions of A30 and A58.

similar plot has been made and discussed by Pottasch et al. (1984). A30 is also plotted on the diagram using the IRAS far infrared measurements and an upper limit to the radio flux density as measured in Westerbork (Zijlstra, 1985). A78 is not plotted because its far infrared flux has never been measured (IRAS missed that part of the sky), but its position must be similar to A30. The upper limit to the radio flux densities is similar to A30 (Zijlstra, 1985) while the infrared flux shortward of 20 μm is also very similar (Cohen and Barlow, 1974). Thus the infrared radiation is at least two orders of magnitude higher than what normally would be expected from a nebula with such low radio continuum emission.

This difference cannot be due to an absence of hydrogen ions in the central regions, because helium ions will produce radio emission as well. But the reason for the discrepancy does not concern us here. What is important is that this property can be used in attempting to identify similar systems. A58 is one such system.

IV. ABUNDANCES IN A58

The central object in this nebula was first known as V605 Aql and became visible in 1917, reaching a magnitude of about 10 in 1919 (Lundmark, 1921). It faded thereafter and was last observed in 1923. The nebula is in Abell's (1966) list and is clearly a symmetric planetary. The expansion velocity of the 44" x 33" nebula is less than 40 km s^{-1} (Kamaswara Rao, unpublished) and the nebulae must therefore have existed long before the outburst. At present there is a star-like object near the center, whose spectrum (see below) shows it to be a nebula. There is a very faint star about 1" north of the central nebula (Seiter, 1984) which may be what remains of V605 Aql. There is no other obvious candidate for the central star of this nebula. The faint star is probably not hot enough (see below) to ionize the nebula.

We have taken spectra of the outer nebula and the central knot with the 1.9 m telescope in S.Africa (2" x 6" slit) and La Palma (long slit 1" wide). Here only the results of the former observation are discussed. These spectra are shown in Fig. 3a (outer region of the nebula in the bright SW region) and Fig. 3b (central knot). Both spectra are clearly nebular emission of medium excitation, but they are quite different. Especially the oxygen lines, [O III] and also [O I] are considerably stronger in the central knot than in the outer nebula.

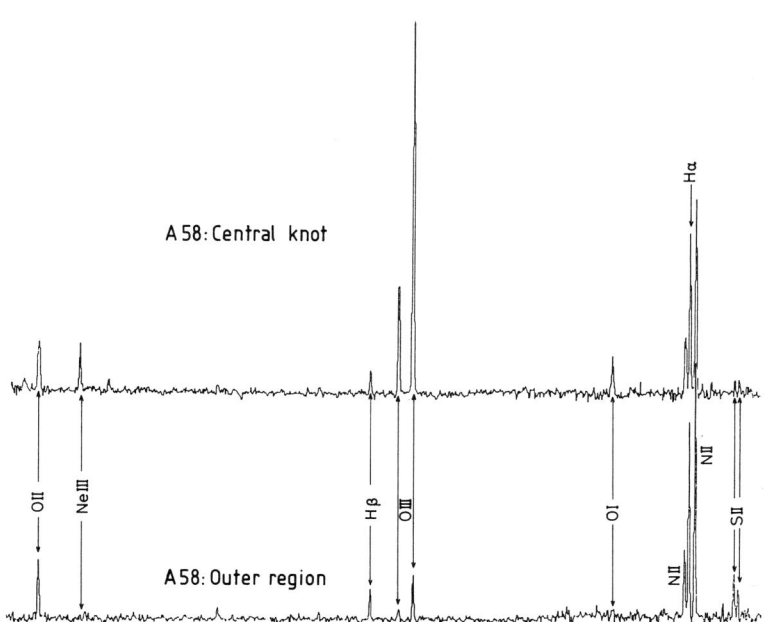

Fig. 3. (a) Spectrum of the central knot of A58 and (b) that of a bright patch in the outer nebula. Relative to Hβ, the O III and Ne III lines are extremely strong in the knot.

A preliminary analysis of this data, which will be discussed in detail in a future paper, gives abundances shown in Table 2. It may be seen from the table that the abundances in the outer parts of the nebula

TABLE 2: ABUNDANCES IN A58

ELEMENT	OUTER REGION	CENTRAL KNOT
HELIUM	0.081	0.22
OXYGEN	6×10^{-5}	1.3×10^{-3}
NITROGEN	6×10^{-5}	4×10^{-5}
NEON	4×10^{-5}	5×10^{-4}

are slightly lower than solar, with oxygen even somewhat lower. In the central knot, however, the abundances are an order of magnitude higher. Since it is likely that the central knot was recently ejected, probably at the time of the 'outburst' in 1917, it seems that this is also a case of relatively hydrogen poor material being ejected late in the nebular evolution. The hydrogen poorness is not as extreme as in A78 and A30, however.

ACKNOWLEDGEMENT: The Isaac Newton telescope is operated by the RGO in the Observatorio del Roque de los Muchachos of the Instituto de Astrofisica de Canarias.

REFERENCES

Abell, G.O. 1966, Astrophys. J. 144, 259

Cohen, N., Barlow, M.J. 1974, Astrophys. J. 193, 401

Hazard, C., Terlevich, R., Morton, D.C., Sargent, W.L.W., Ferland, G. 1980, Nature 285, 463

Jacoby, G.H. 1979, Publ. Astron. Soc. Pac. 91, 754
Jacoby, G.H., Ford, H.C. 1983, Astrophys. J. 266, 298

Lundmark, K. 1921, Publ. Astron, Soc. Pac. 33, 314

Pottasch, S.R., Baud, B., Beintema, D., Emerson, J., Habing, H.J., Harris, S., Hauck, J., Jennings, R., Marsden, P. 1984, Astron. astrophys. 138, 10

Seiter. W.C. 1985, Mitleitungen Astron. Ges. 63, 181 and 187.

DISCUSSION

HUNGER: As to the interpretation of your abundances, when you go for A78 from the central part to the outer part, then all the abundances are roughly increasing by a factor of 10. What is the implication of that?

POTTASCH: The abundances seem to be increasing from the more central parts, as Jacoby and Ford have measured, to the position a second or so of an arc away, that we have measured. The simplest interpretation is that this star is emitting increasingly hydrogen poor material as time goes on. This second of arc difference between the two positions means that the steep abundance gradient corresponds to a very short time scale for the change of composition of the ejected material. A second of arc corresponds to 100 yrs or few 100 yrs in time.

LIEBERT: Can you say anything about the relative masses of the outer nebulae and inner knots?

POTTASCH: I think Jacoby tried to do that. In A78 the mass within 10" of the exciting star is probably close to 10^{-3} M_o while the total mass is of the order of 10^{-1} M_o for the nebula. The knot in A58 probably has a relatively smaller mass because of its smaller size. Because both the density and the distance are poorly determined, these values should only be taken as indications.

FEAST: What would that mean to the bolometric luminosity?

POTTASCH: I would place it in the order of 3×10^3 L_o, with large uncertainties.

FEAST: Is it just possible that these central stars are evolving to R CrB stars? I mean their luminosities do not rule it out.

SCHÖNBERNER: I do not think so. The central star is evolving towards a white dwarf.

RAO: For Abell 58, the distance estimate by Ford is 3.5 kpc. From the reddening estimates, Van den Berg also gives about 4 kpc. I think after correcting for the reddening, the luminosity of the central star could be, when it was brightest around $M_v = -5$.

RAO: A30 has an IR disc around it. Is there such an evidence found for A58 and A78?

POTTASCH: Well it is not clear with A58 where the IR is coming from. IRAS observations have shown the strong infrared excess. There is no spatial information. I would not be surprised if the emitting region was much smaller than the nebula itself. A78 was not measured by IRAS. The indication in the near IR is that it is similar to A30 although these measurements have not been made in such detail. The IR may come from a region smaller than the total nebula.

MENDEZ: I want to know whether there is any evidence for expansion velocities in the central parts of A58.

POTTASCH: I do not think there is any direct measurement of A58, except for limits placed by the line width. I think, A58 does not look totally different from normal planetary nebulae, in the sense that the line width is certainly less than 40 kms^{-1}. A78 also does not have any radial velocity above normal for Planetary nebula.

THE ORIGIN AND EVOLUTION OF HELIUM-RICH WHITE DWARFS

James Liebert
Steward Observatory
University of Arizona
Tucson, Arizona 85721
USA

ABSTRACT. White dwarfs with helium-rich atmospheres constitute about one fifth of the white dwarfs hotter than 12,000 K. They appear to have a mass distribution similar to the hydrogen atmosphere (DA) stars, and are similar in other properties. However, the temperature distribution exhibits a deficiency of DB/DO stars in the interval 25,000-45,000 K, which implies evolution in the dominant surface composition as the stars cool. The hottest group of transition DO white dwarfs are the pulsating objects of the PG1159 class. The central star of K1-16 is a related object, as may be the newly discovered very hot star H1504+65, which shows no detected surface features of either hydrogen or helium.

1. INTRODUCTION: THE HELIUM WHITE DWARF SEQUENCE

The stellar parameters of the helium rich degenerates may be important in specifying the final boundary conditions for the prior post main sequence evolution of at least some of the hydrogen poor stars discussed at this meeting. The mass distribution of these remnants specifies how much envelope mass loss must have occurred; the rotation rates specify the angular momentum loss. Definite determination of the interior compositions, along with the mass distribution, would fix the extent of post main sequence nuclear evolution quite concretely. The stellar kinematics are a valuable clue to the progenitor population and mass range, especially for the hotter stars which have only recently become degenerate.

White dwarfs offer some diversity in the distribution of the atmospheric chemical compositions. The most basic division is that about four fifths of the hot white dwarfs retain hydrogen-rich surfaces, while one fifth have helium-dominated, and very hydrogen-poor, surfaces. There are no cases known in which the atmosphere is dominated by heavier species. Unfortunately, the usefulness of the atmospheric compositions of white dwarfs in the study of prior evolutionary states is much less clear than for the other stellar properties: Surface compositions can be quickly and completely changed due to the effects of gravitational and thermal diffusion, selective radiative acceleration processes, convective mixing, and accretion from the interstellar medium. Quite strong but undetected magnetic fields could play an important role; fields are detected at strengths above 10^6 gauss in only about 1% of the known white dwarfs. It is even possible that none of the helium rich degenerates bear a particular relationship to any of the higher luminosity stars with hydrogen-poor atmospheres! I shall, however, explore some evidence for more optimistic possibilities, suggesting that there is some relationship.

Over most of the observed range in temperature, the white dwarfs divide into the two sequences of hydrogen and helium-rich atmospheres. Over at least the temperature span of 5,000 K to 80,000 K, the hydrogen lines appear in the former, and they are classified DA. The helium-rich cases display much more diverse subgroups of spectra: These include DO stars above 45,000 K showing He II lines, cooler DB stars showing only He I lines, and still cooler objects below 10,000 K which are too cool to show any optical lines from their dominant atmospheric constituent. The cool helium atmosphere objects, in particular, have very transparent atmospheres -- i.e. we view to very high gas densities and pressures -- which makes contaminant elements detectable at very low abundances. Thus, the cool objects usually show evidence for carbon, dredged up in small amounts from the bottoms of their convective envelopes, and often detectable only in the ultraviolet (DQ or DC stars). A minority of the cool helium stars show traces of accreted metals sometimes accompanied by hydrogen (the DZ stars). The stars below 5,000 K have surfaces too cool to reveal the dominant atmospheric constituent, be it hydrogen or helium, and most of these stars are classified DC (featureless). A temperature map for these basic sequences is provided in Figure 1, which will be discussed more fully in Section 3.

In Section 2, the known parameters of this multi-faceted helium atmosphere sequence are compared with those for the dominant DA stars. Then, the possible evolutionary scenarios for cooling degenerates which form part of the sequence are outlined (Section 3). Some very hot, pulsating, precursor stars are discussed in Section 4, as well as possible origins of the helium-rich degenerate sequence.

2. PARAMETERS OF THE HELIUM ATMOSPHERE WHITE DWARFS

2.1. The Mass Distribution and Interior Composition

The physics describing a helium dominated atmosphere--particularly the line broadening theory and the opacities--are less accurately known than for corresponding stars with hydrogen rich compositions. In particular, the helium atmospheres are generally more transparent at a given temperature, so that gas pressures are much higher, convection in the outer envelope more prevalent, and the abundances of trace elements more important. These problems are most severe for the DC/DQ/DZ stars below about 10,000 K. Conversely, the hot DO stars must be observed primarily longward of the peak of the energy distributions, and of the strongest absorption edges and lines. The numerous DB white dwarfs constitute the best sample for determining surface gravities and masses using stellar atmospheres techniques. Several of the He I lines exhibit considerable sensitivity to surface gravity; the optical colors, though covering only weak bound-free continua, exhibit some sensitivity. Little gravity sensitivity is offered by observations at IUE ultraviolet wavelengths.

Early atmospheric analyses suggested that DB white dwarfs might have significantly lower surface gravities and smaller masses than do the DA stars, a result which was attractive to those concerned about how the chemical purity of a DB atmosphere is maintained (Wesemael 1979; Alcock 1979). The masses derived for DBs ($< 0.4 M_\odot$) were so much smaller as to imply a separate origin for these stars, as might be possible only from close binary evolution (Nather, Robinson and Stover 1979).

The most recent model grids include improved helium line profiles, blanketing and envelope convection. Some attention has been focussed on the effects of trace (and perhaps undetected) abundances of hydrogen and heavier elements. Shipman (1979) has consistently argued that the mean surface gravities and masses for DB stars were similar to the DA stars. The synthetic spectra from the new model grid of Wickramasinghe and Reid (1983) is also consistent with the stars having surface gravities near $\log g = 8.0$. Certainly the most comprehensive analysis is that of Oke, Weidemann and Koester (1984), who find that both the optical colors and the He I line profiles are consistent with the following result: The DB white dwarfs are distributed in a narrow mass range comparable with the DA stars, at $M \sim 0.55 +/- 0.10$ $M_\odot$. These authors therefore argue that the progenitors for the DB and DA white dwarfs may be the same. There is every reason, then, to expect that most helium atmospheres surround carbon-oxygen cores, the natural end for the asymptotic giant branch evolution of stars with original masses up to approximately eight solar masses.

2.2 Rotation

There is little evidence for rapid rotation among DB and DO white dwarfs, although little quantitative analysis comparable to that done for the DA stars is currently available. Wickramasinghe and Reid (1983) derive projected rotation velocities of < 135 km s-1 for several DB stars showing sharp line cores. Unpublished MMT echelle spectra covering the He I 5876A line obtained by the author and R. Green show that several more cool DB stars exhibit sharp cores not unlike the H-alpha cores found by Greenstein et al. (1977) and Pilachowski and Milkey (1984) for DA stars. The implication is that these DB stars rotate even more slowly than the Wickramsinghe and Reid (1983) limits. Likewise, two hot DO stars observed at high dispersion with the IUE Observatory--PG1034+001 (Sion, Liebert and Wesemael 1985) and KPD0005+5106 (Downes, Liebert and Margon 1985, and work in preparation)-- show sharp ultraviolet features attributed to their photospheric velocities. On balance, there is little evidence that the hot helium rich white dwarfs retained much angular momentum from earlier evolutionary states. Like the mass determinations of Oke, Weidemann and Koester (1984), these results are inconsistent with an origin for most DB stars involving close binary evolution of the type proposed by Nather, Robinson and Stover (1979).

2.3. Population Type

There is little evidence that the DO/DB/DQ/DZ white dwarfs are part of a different kinematical population (Sion and Liebert 1977), although total space motions (which require radial velocities) are available only for a few dozen stars. There is little evidence that the fractions of helium atmosphere white dwarfs found in young galactic clusters or having halo space motions are drastically different. These findings provide further support for an origin of these stars primarily from the old disk population.

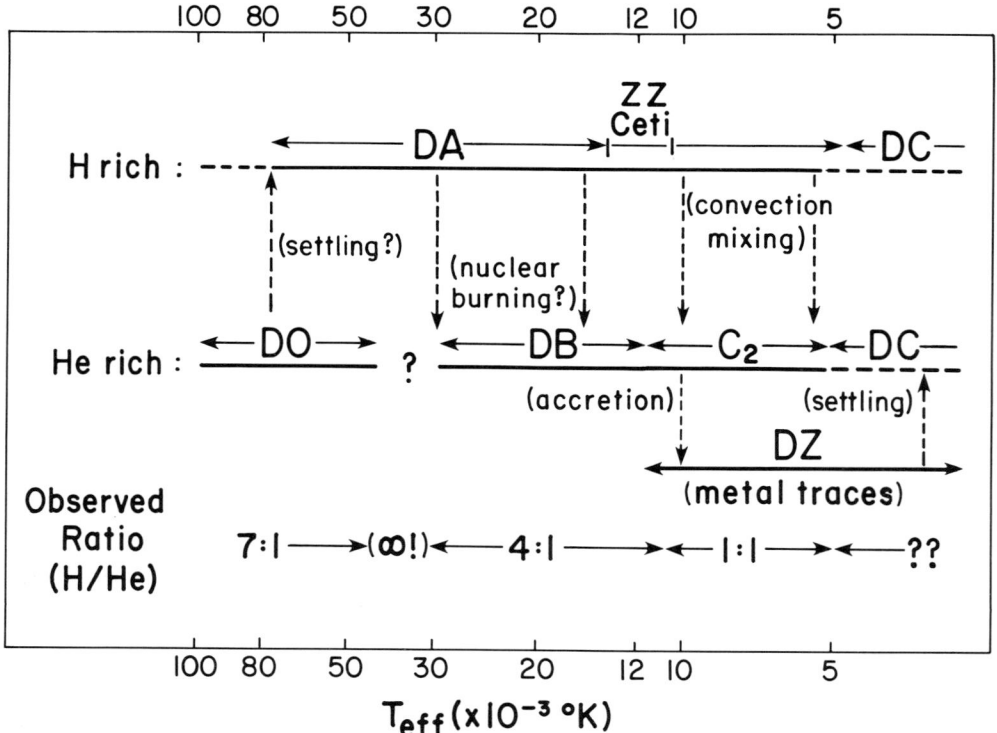

Figure 1. Spectral types of white dwarfs in the hydrogen and helium atmosphere sequences, as a function of temperature. Evolution in observed ratios is indicated at the bottom. Vertical arrows indicate processes (labelled) which might change atmospheric compositions.

2.4. Pulsations

Two groups of helium-rich degenerate stars (or pre-degenerate stars) constitute separate classes of pulsating variables. These are the DBV stars, whose prototype is GD358, with temperatures near 25,000 K and the "PG1159" or DOV variables. The latter are discussed in Section 4. The likely origin of the DB pulsational instability strip is linked to He II ionization (Winget et al. 1982). The temperature range is appropriately higher than that for the pulsating DA (ZZ Ceti) variables, although the exact temperature appropriate to the best analyzed case, GD358, is controversial (Koester et al. 1984; Liebert et al. 1986, and these proceedings). Indeed, the temperature scale for hot DB stars is important in establishing the extent of a deficiency of DB stars above about 25,000 K, as discussed in Section 3. Conceivably, the pulsation strip may be a clue in unravelling the evolutionary scenario.

3. EVOLUTION OF HELIUM ATMOSPHERE DEGENERATES

3.1. The Peculiar Temperature Distribution

The temperature or luminosity distributions of the helium rich white dwarfs differ significantly from those predicted from cooling theory and the assumptions of (1) a constant white dwarf formation rate for the last 10^9 years and (2) no change in the dominant atmospheric species as the white dwarfs cool. There is little evidence that these assumptions are invalid from the analysis of the luminosity function of DA white dwarfs (Greenstein 1979; Fleming, Liebert and Green 1986). For the helium-rich stars, the situation is quite different.

The temperature distributions of white dwarfs having hydrogen and helium-rich atmospheres, drawn from the complete sample of the Palomar Green Survey (Green, Schmidt and Liebert 1986) are shown as Figure 2. Only stars with temperatures at or above 12,000 K are shown. Temperatures for the DO stars are taken from Wesemael, Green and Liebert (1985), for the hot DBs from Liebert et al. (1986) and for the cooler DB stars from Shipman, Liebert and Green (1986). The data are not corrected for volume completeness. However, the shape of the discovery function may be estimated by comparing the distribution of DA stars with the luminosity function of Fleming et al. (1986). Moreover, the additional DB/DO stars analyzed in Oke, Weideman and Koester (1984) and one star from Koester, Liebert and Hege (1979) have been added to the helium star distribution. The PG1159-035 objects, borderline white dwarfs at log g ~ 7 and T_{eff} ~ 10^5 K, are plotted in the figure. These have few or no counterparts among the DA white dwarfs at the hot end, presumably because the latter have thick hydrogen envelopes and reach the white dwarf radii at somewhat lower temperatures.

The principal peculiarity in comparing the two histograms is the deficiency of helium stars in the interval log T_e ≈ 4.40-4.65, which corresponds to about 26,000-45,000 K. Admittedly, the temperature scale for DB stars is uncertain for T_{eff} > 20,000 K (Liebert et al. 1986, and this conference); had we adopted the optical scale, even fewer stars would have been assigned temperatures as high as 25,000 K. It is curious that the DA distribution shows a marginal bulge centered near log $T_e \sim 4.5$, but an assessment of the statistical significance of this is beyond the scope of this summary paper. Such an effect did not show up in the luminosity function with its coarser magnitude binning (Fleming et al. 1986).

In general, the ratio of DA to non-DA white dwarfs decreases from the higher temperatures to lower temperatures (Sion 1984). In the well defined Palomar Green sample, the ratio is 7.1 +/- 2.9 in the 40-80,000 K interval, may be even higher in the 20-40,000 K range (because of the deficiency of DB stars), but is only 3.4 +/- 0.7 over the 12-20,000 K interval (Fleming et al. 1986). At temperatures below ~ 10,000 K, there is now overwhelming evidence that the fraction of helium-rich white dwarfs increases towards something like a 1:1 ratio (Sion 1979, Wehrse and Liebert 1980, Sion 1984, Greenstein 1986).

3.2. The Evolution in Surface Abundances of Hot White Dwarfs

How can one account for the strange behavior in the DB-DO histogram and in the changes of the volume-corrected ratios of DA/non-DA white dwarfs with temperature? If most white dwarfs belong to the old disk population, with total ages generally >> 10^9 years, it is unlikely that the local white dwarf birthrate would show drastic fluctuations in the last billion years or so. It is therefore very likely that the observed distributions with temperature require changes in the dominant surface abundances of hot white dwarfs, due to effects such as those listed in the Introduction. Since the helium sequence is modest in number compared with the hydrogen sequence (at the higher temperatures), the need for such evolution would not be readily apparent in analyses of the latter alone.

A number of recent theoretical investigations indicate possible ways in which the required surface abundance evolution may proceed. In Figure 1, a working hypothesis is outlined. The ratio of DA to DO stars in the 40-80,000 K range is about 7. However, some of the DO stars show substantial trace abundances of hydrogen (e.g. the prototype HZ21), and could become DA stars with outer hydrogen layers of quite small mass. This would result in an initial increase in the DA/non-DA ratio.

In the 20-30,000 K range, the idea proposed by Michaud, Fontaine and Charland (1984) just might work: If diffusion tails of the carbon core and hydrogen envelope are able to cross in a sufficiently thin helium layer at high enough temperature, stable hydrogen burning via the CNO cycle (whose rate depends only linearly on the trace hydrogen abundance) might eat away the hydrogen surface layer. Indeed, for this mechanism to work efficiently enough requires a rather optimistic treatment of the known physics. Moreover, evolutionary models which also include a treatment of element diffusion (Iben and MacDonald 1985) do not predict enough burning for the outer hydrogen layer to disappear. Note, however, that the numbers require only a modest fraction of the DA stars to be converted to DB stars in this high temperature range.

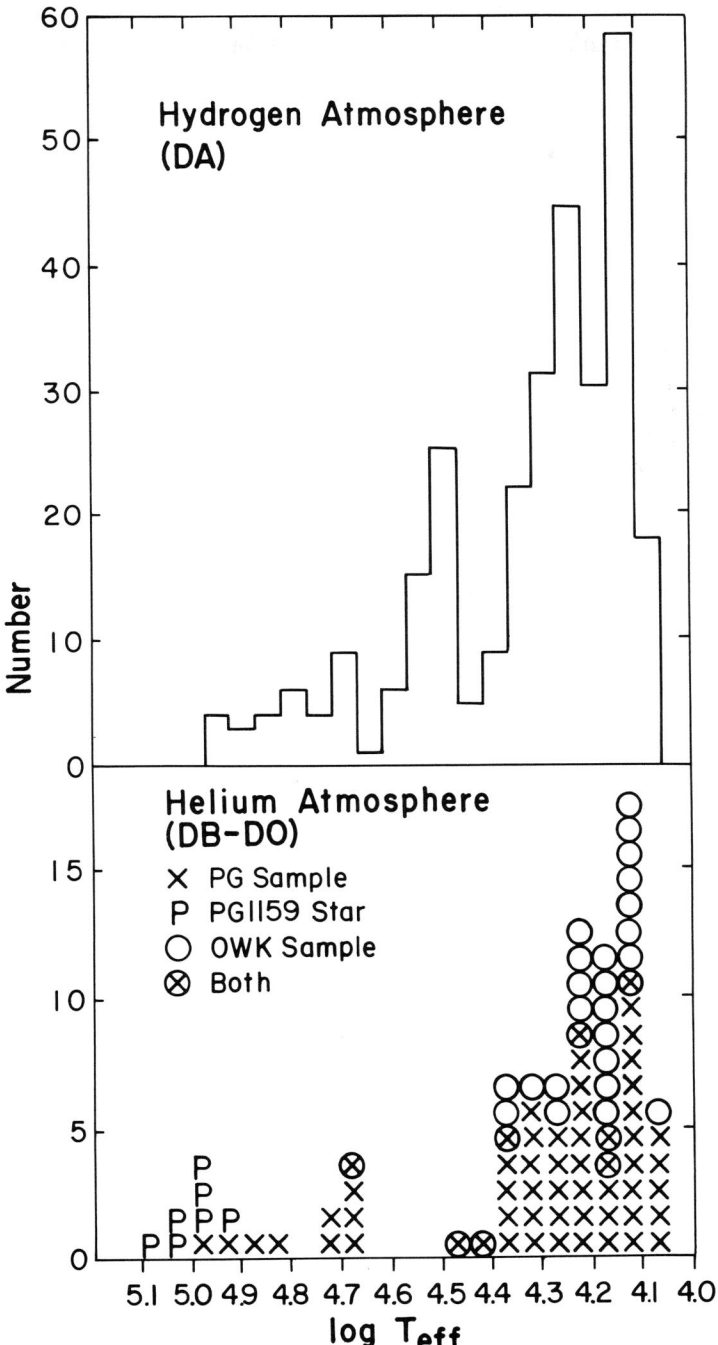

Figure 2. Histogram of temperatures for DA and DB-DO stars, from the Palomar Green Survey (PG) or previously known (OWK).

At lower temperatures -- possibly even as high as the 20-30,000 K range -- the onset of convective mixing can mix a hydrogen surface layer into a more massive helium envelope. Such mixing is believed to be the reason for the ~ 1:1 ratio at T_{eff} < 10,000 K, and implies that something like half of the hotter DA stars suffer this fate. Yet this mechanism is also invoked to cause the red temperature edge of the ZZ Ceti instability strip, which applies to most (or all) DA stars in the 10-13,000 K range (Winget and Fontaine 1982). Moreover, convective mixing should not work if the hydrogen layer mass remains higher than 10^{-7} $M_\odot$ as the stars reach these low temperatures.

Fortunately, the explanations for trace elements in the cooler DQ and DZ stars are on somewhat more solid ground. The traces of carbon appear routinely in helium-rich atmospheres below 10,000 K due to the dredge-up by the deepening outer convective envelope of the diffusion tail of the carbon core (Koester, Weidemann and Zeidler-K.T. 1982; Wegner and Yackovich 1984; Fontaine et al. 1984; Pelletier et al. 1986). For the DZ stars, the mechanism is accretion from the interstellar medium, although the messy interplay of accretion from a clumpy medium and diffusion in a convective envelope remains to be really understood (c.f. Alcock and Illarionov 1980; Wesemael and Truran 1982; Shipman and Greenstein 1983). A large fraction of these objects seem to show traces of hydrogen as well (Liebert, Wehrse and Green 1986), although this is certainly compatible with an accretion hypothesis.

The DBA stars are objects above 12,000 K which show traces of hydrogen. These seem to constitute more than 10% of the DB stars in the 12-18,000 K range (Shipman, Liebert and Green 1986). A survey of the H-alpha region with a precision (CCD) detector would likely increase the fraction further. If accretion were the cause of the DBA phenomenon, one would expect that the more extreme cases might show metals as well. Likewise, the hottest stars showing strong metals -- GD401 and GD40 -- do not show hydrogen features (c.f. Shipman and Greenstein 1983). An alternative explanation for the DBA stars is that these objects have recently mixed their outer hydrogen layers into helium envelopes, possibly evolving from DA to DB below 30,000 K. Perhaps the strange DAB object GD323 -- whose energy distribution and line spectrum have not been reconciled to any atmospheric model with a mixed H, He composition -- may provide a clue to an evolutionary answer (see Liebert et al. 1984). However, modest hydrogen pollution of helium-rich atmospheres is a characteristic of stars as hot as HZ21 (~ 50,000 K).

4. THE ORIGIN OF HELIUM RICH WHITE DWARFS

4.1. Precursor Stars: Helium-Rich PNNs and SdOs

Stars which feed the helium white dwarf sequence must include both some helium-rich planetary nebulae nuclei (PNNs) and some subdwarf O stars. Most sdO stars are actually hydrogen-rich, but with detectable helium. Moreover, it is also possible that many extremely He-rich sdOs and

PNNs retain enough envelope hydrogen that gravitational settling could turn them into DA atmospheres. Conversely, gravitational and thermal diffusion might also be overcome by selective radiative acceleration processes, which might preferentially expel hydrogen from such heliumrich envelopes, leaving an essentially pure helium surface as the remnant becomes a white dwarf.

The helium-rich PNNs constitute some 30% of the particular sample of nuclei for which photospheric abundance results are available (Mendez 1986, this conference). This is a somewhat higher fraction of helium enriched objects than for the hot (DO/DA) white dwarfs, but scarcely enough different for serious concern. The PNN selection bias favors analysis of central stars with low surface brightness nebulae; they should also tend to be nearby and low luminosity objects. Since the evolution time scale for the dying stellar core decreases with a high power of the core mass -- while the envelope dispersal time scale presumably is less sensitive to mass -- those PNNs with analyzable photospheres may have preferentially lower masses.

By this kind of reasoning, the subdwarf O stars may simply be cores of modestly lower mass than the PNNs which are observed after dissipation of the nebula (Heber et al. 1984). Of course, many PNNs have sdO spectral types, but the numbers of isolated sdO stars (over 200 in the PG Survey) exceed this type of PNN to comparable limiting stellar magnitudes.

The above ideas imply that most stars entering the helium-rich white dwarf channel may have lower core masses than those entering the DA sequence, but calculations show that the expected difference is small (Schönberner 1983) and is well within the uncertainties of the masses reviewed in Section 2. Moreover, there are other indications that some helium-rich sdOs and PNNs might have higher than average masses: First, Heber (1986, this conference) discusses several very luminous helium-rich sdOs which fit evolutionary tracks at higher than 0.6 $M_\odot$. Secondly, as discussed in Sections 4.2-3, the very hottest prewhite dwarf stars (at least those outside of high surface brightness PNs) appear to be hydrogen-poor stars like PG1159-035 and H1504+65. At the point where dying stars are descending in the H-R diagram to the white dwarf sequence, the hottest stars at a given luminosity and for a given composition should be the most massive. Finally, it now appears that a large fraction of planetary nebulae in the Magellanic Clouds -- which should logically include a higher fraction of more massive progenitors and more massive PNNs -- have nebulae which are enriched in helium (Boroson and Liebert 1986).

4.2. The Pulsating Stars: PG1159-035 and K1-16

Immediate precursors, which overlap in the H-R diagram both the hottest PNNs and sdOs, are the pulsating variable stars whose prototype is PG1159-035. We have previously described these (Wesemael, Green and Liebert 1985) as defining the high temperature end of the DO degenerate sequence; their surface gravities ($\log g \sim 7$) suggest that they are borderline white dwarfs using the definition of Greenstein and Sargent (1974). Indeed, I would not quarrel with calling them high gravity sdO stars. However, they are considerably hotter than nearly all previously analyzed sdO stars. At $T_{eff} \sim 10^5$ K, they are among the hottest field stars known, comparable to the very hottest PNNs whose photospheres have been analyzed. Their spectra show broad absorption lines (often with narrow emission cores) of He II, C IV and O VI, with no trace of hydrogen and indications that the corresponding N V transitions are weak or absent. At such high temperatures, however, it is difficult to establish to what degree that hydrogen is absent.

Some of the stars defining this spectroscopic class are pulsating variables (McGraw et al. 1979; Bond et al. 1984); the variations are complicated, non-radial pulsation modes, believed to be driven in ionization zones of the CNO species (Starrfield et al. 1984). The pulsation behavior is of interest to stellar pulsation theorists, and is of great potential value in measuring the rate of interior evolution in stars which should be evolving rapidly (Winget, Hansen and Van Horn 1983). Indeed, a period change in the prototype (PG1159-035) has recently been reported (Winget et al. 1985), though its interpretation is complicated by the possibility of rotation (Kawaler et al. 1985).

The planetary nebula K1-16 has a central star which is also (1) pulsating with complicated modes, (2) very hot, and (3) shows absorption lines with emission reversals of the same He II, C IV and O VI ions seen in the PG1159 stars (Grauer and Bond 1984; Sion, Liebert and Starrfield 1985; Grauer et al. 1986). Possible differences may be noted, however. K1-16 has somewhat sharper absorption lines, i.e. may be more luminous; and its typical periodicities are longer, which may be a consequence of higher luminosity. This behavior is of course exactly what is expected for a roughly vertical instability strip at the left edge of the H-R diagram. K1-16 also supports a link between the isolated PG1159 variables and that somewhat heterogeneous class of PNNs called the O VI stars (Sion, Liebert, and Starrfield 1985). The latter may exhibit strong Wolf-Rayet-like winds, which may drive the remaining hydrogen from the stars. K1-16, in particular, shows evidence for a subtle but high velocity wind (Kaler and Feibelman 1985). Moreover, there are even several "field" stars -- not associated with PNs but likely to be evolved, low mass objects -- which show very strong C IV and O VI emission (Sanduleak 1971, Barlow and Hummer 1982, Downes 1984).

4.3. H1504+65

An extremely hot, hydrogen-poor star has recently been discovered by the HEAO-1 X-ray satellite; H1504+65 shows some similarities to the PG1159 stars (Nousek et al. 1986). The ratio of its soft (EXOSAT) X-ray to optical fluxes is over 600 times that for PG1159-035. The energy distribution from 1000-6000Å is also steeper. The spectrum shows O VI and C IV features, but no trace of He II! The widths of the O VI absorption suggests a high surface gravity. There is as yet no conclusive evidence as to whether or not H1504+65 pulsates.

The absence of He II lines indicates that, if the atmosphere were helium dominated, the temperature must substantially exceed 150,000 K, the highest value for which Wesemael (1981) pure helium models are

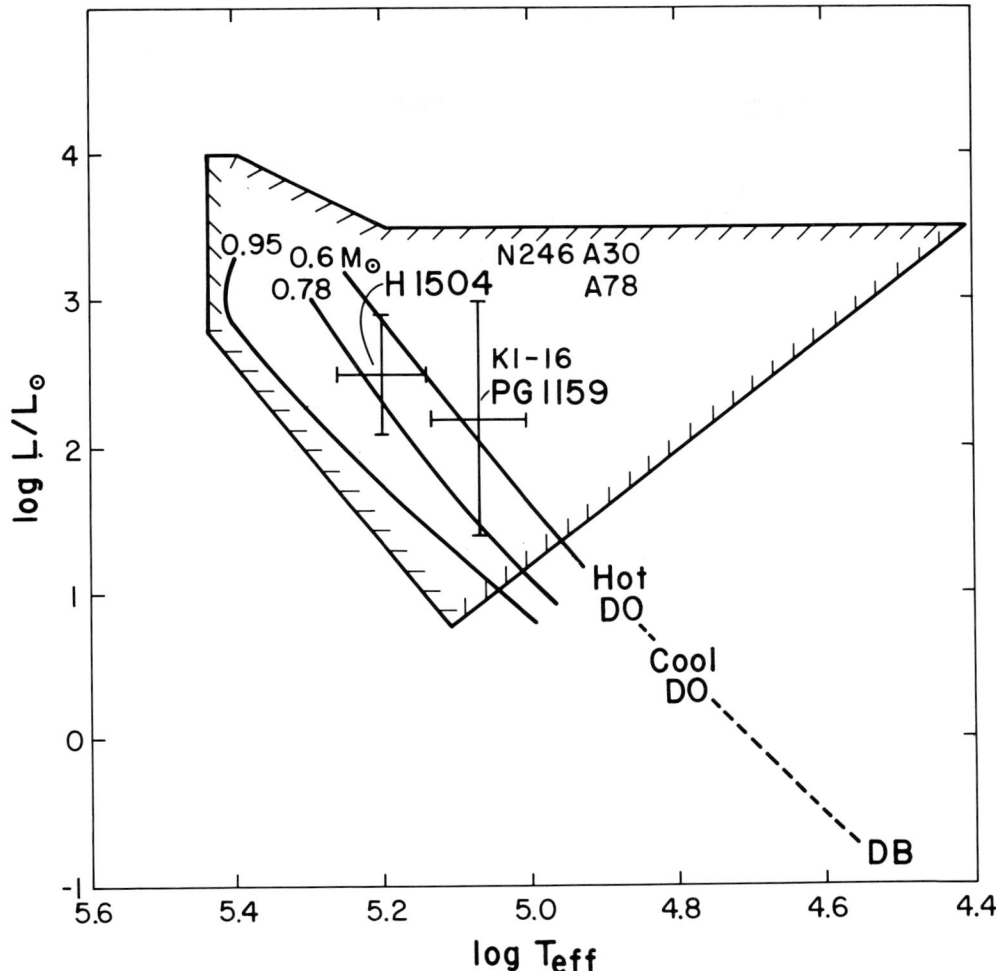

Figure 3. H-R Diagram for the hot helium white dwarfs, and some of their likely progenitors, as discussed in Section 4. Also shown are a cross-hatched region occupied by PNNs and evolutionary models (curves).

available. An alternative and still untested hypothesis is that the surface of H1504+65 is also helium poor, and presumably dominated by carbon and oxygen. This seems incompatible with stellar evolution models of the AGB and post-AGB phases, which predict that a helium layer of thickness $M > 10^{-3} M_\odot$ will be retained. Moreover, there are no known carbon/oxygen dominated surfaces among the > 1,000 known white dwarfs of cooler temperature. One might at least presume that the outer envelope retains enough helium for settling to later result in a helium-rich (DO) surface. On balance, Nousek et al. estimate the surface temperature as 160,000 +/- 30,000 K.

In Figure 3, adapted from Nousek et al., the stars H1504+65 and PG1159-035 are plotted with best current estimates of their error bars, and with theoretical evolution tracks taken from Kawaler et al. (1985). The hot, helium-rich white dwarf spectral groups (DO, DB), into which these stars presumably evolve, are also shown. Several O VI PNNs are indicated, although temperature/luminosity determinations of the photospheres are lacking. K1-16 is shown as more luminous than PG1159. It appears from the nebular analyses that the nuclei of NGC 246, Abell 30 and Abell 78 -- which do not pulsate (Grauer et al. 1986) -- may be higher in luminosity and/or lower in temperature than the instability strip. The large hatched region in the figure is approximately that occupied by the known PNNs with reasonably determined values given by Pottasch (1984).

It appears to be a reasonable conclusion that objects like PG1159 and H1504 have higher masses than those for most helium-rich sdO stars. Given the large uncertainties in the derived parameters, however, it is not clear that H1504+65 and the stars of the PG1159 class necessarily fit stellar evolution tracks of higher mass than the mean (~ 0.6 $M_\odot$) for DA white dwarfs. The point was made earlier (see also Fleming et al. 1986) that their temperatures do appear to be hotter than those for the hottest DA stars or exposed PNNs near log g ~ 7. However, if the DA stars retain thick (> $10^{-4} M_\odot$) hydrogen envelopes, their radii would be significantly larger (and surface temperatures cooler) for quite similar masses. Likewise, the more massive and hotter of the DA progenitors may remain buried in PN envelopes at lower luminosities. Certainly, a few of these -- see Pottasch (1984, Fig. XI-1) -- probably harbor PNNs well in excess of 100,000 K.

4.4. A Word about Origins

Three partly-overlapping groups of progenitor stars to the helium white dwarf sequence have been identified in the preceding discussion -- the helium-rich sdO stars, the helium-rich PNNs and the hotter, high gravity field stars like PG1159-035 and H1504+65. It is beyond the scope of this presentation to explore in detail how these stars lost their hydrogen surfaces in prior evolution. The cooler subdwarf progenitors could actually include both sdOs having had similar AGB evolution to the PNNS and also objects (sdBs, sdOBs) which evolved from an extended horizontal branch directly towards the white dwarf state (Groth, Kudritzki and Heber 1986). Both groups could become DO/DB white dwarfs with masses modestly lower than typical DA values.

For the apparently more massive progenitors, the appearance of oxygen and carbon features suggests a common origin for the PG1159 field stars and at least some PNNs of the O VI group. For these, it is tempting to consider an origin involving a late helium shell flash (c.f. Schönberner 1979, Iben and Renzini 1983, Iben et al. 1983). Such a hypothesis offers an explanation as to how a relatively massive star might have already shed its principal (hydrogen-rich) PN in an earlier episode. The problem is that the likelihood of a late shell flash should normally be higher for lower mass cores which take considerably longer to cool in the post-AGB phase. A remaining wildcard is the possible link between the helium shell flash timing and the mechanism which controls the final (superwind?) PN ejection. Thus, it remains to be worked out what fraction of helium-rich white dwarfs might owe their origin to a late helium shell flash, and what their mass distribution is.

I wish to thank Al Grauer, Harry Shipman, Francois Wesemael, John Nousek, Richard Green and other colleagues for some collaborative results presented at the meeting in advance of publication dates. In addition to discussions with numerous colleagues at this conference, I wish to acknowledge the preprints and valuable discussions with Rolf Kudritzki, Dirk Husfeld and Detlev Koester. This work was supported by the National Science Foundation grant AST 85-14778.

REFERENCES

Alcock, C. 1979, in White Dwarfs and Variable Degenerate Stars, Proc. IAU Coll. No. 53, eds. H. M. Van Horn and V. Weidemann, Univ. of Rochester Press, New York, p. 202.
Alcock, C., and Illarionov, A. 1980, Ap. J., 235, 541.
Barlow, M. J., and Hummer, D. G. 1982, in Wolf Rayet Stars: Observations, Physics, Evolution, Proc. IAU Symp. No. 99, eds. C. W. H. de Loore, and A. J. Willis, D. Reidel Publishing Co., Dordrecht, p. 387.
Bond, H. E., Grauer, A. D., Green, R. F., and Liebert, J. 1984, Ap. J., 279, 751.
Boroson, T. A., and Liebert, J. 1986, preprint.
Downes, R. A. 1984, Publ. A.S.P., 96, 807.
Downes, R. A., Liebert, J., and Margon, B. 1985, Ap. J., 290, 321.
Fleming, T., Liebert, J., and Green, R. F. 1986, Ap. J., in press.
Fontaine, G., Villeneuve, B., Wesemael, F. and Wegner, G. 1984, Ap. J., Letters, 277, L51.
Grauer, A. D., and Bond, H. E. 1984, Ap. J., 277, 211.
Grauer, A. D., Liebert, J., Fleming, T., Green, R. F., and Bond, H. E. 1986, preprint.
Green, R. F., Schmidt, M., and Liebert, J. 1986, Ap. J. Suppl., in press.
Greenstein, J. L. 1979, Ap. J., 233, 239.
Greenstein, J. L. 1986, Ap. J., in press.
Greenstein, J. L., Boksenberg, A., Carswell, R., and Shortridge, K. 1977, Ap. J., 212, 186.

Greenstein, J. L., and Sargent, A. 1984, Ap. J. Suppl., **28**, 157.
Groth, H. G., Kudritzki, R. P., and Heber, U. 1986, Astr. Ap., in press.
Heber, U., Hunger, K., Jonas, G., and Kudritzki, R. P. 1984, Astr. Ap., **130**, 119.
Iben, I., and Renzini, A. 1983, Ann. Rev. Ast. Ap., **21**, 271.
Iben, I., Kaler, J. B., Truran, J. W. and Renzini, A. 1983, Ap. J., **264**, 605.
Iben, I., and MacDonald, J. 1985, Ap. J., **296**, 540.
Kaler, J. B., and Feibelman, W. A. 1985, Ap. J., **297**, 724.
Kawaler, S. D., Hansen, C. J., and Winget, D. E. 1985, Ap. J., **295**, 547.
Koester, D., Liebert, J., and Hege, E. K. 1979, Astr. Ap., **71**, 163.
Koester, D., Vauclair, G., Dolez, N., Oke, J. B., Greenstein, J. L., and Weidemann, V. 1985, Astr. Ap., **149**, 423.
Koester, D., Weidemann, V., and Zeitler-K.T., E. M. 1982, Astr. Ap., **116**, 147.
Liebert, J., Wehrse, R., and Green, R. F. 1986, preprint.
Liebert, J., Wesemael, F., Sion, E. M., and Wegner, G. 1984, Ap. J., **277**, 692.
Liebert, J., Wesemael, F., Hansen, C. J., Fontaine, G., Shipman, H. L., Sion, E. M., Winget, D. E., and Green, R. F. 1986, Ap. J., in press.
McGraw, J. T., Starrfield, S. G., Liebert, J., and Green, R. F. 1979, in IAU Coll. No. 53, White Dwarfs and Variable Degenerate Stars, eds. H. M. Van Horn and V. Weidemann, Univ. of Rochester Press, Rochester, N.Y., p. 377.
Michaud, G., Fontaine, G. and Charland, Y. 1984, Ap. J., **280**, 247.
Nather, R. E., Robinson, E. L., and Stover, R. J. 1979, in IAU Coll. No. 53, White Dwarfs and Variable Degenerate Stars, eds. H. M. Van Horn and V. Weidemann, Univ. of Rochester Press, Rochester, N.Y., p. 453.
Nousek, J. A., Shipman, H. L., Holberg, J. B., Liebert, J., Pravdo, S. H., White, N. E., and Giommi, P. 1986, Ap. J., in press.
Oke, J. B., Weidemann, V., and Koester, D. 1984, Ap. J., **281**, 276.
Pelletier, C., Fontaine, G., Wesemael, F., and Michaud, G. 1986, Ap. J., in press.
Pilachowski, C. A., and Milkey, R. W. 1984, Publ.A.S.P., **96**, 821.
Pottasch, S. R. 1984, in Planetary Nebulae, Vol. 107, Astrophysics and Space Science Library, D. Reidel Publishing Co., Dordrecht, Figure IX-1, p. 218.
Sanduleak, N. 1971, Ap. J. Letters, **164**, L71.
Schönberner, D. 1979, Astr. Ap., **79**, 108.
Schönberner, D. 1983, Ap. J., **272**, 708.
Shipman, H. L. 1979, Ap. J., **228**, 240.
Shipman, H. L., and Greenstein, J. L. 1983, Ap. J., **266**, 761.
Shipman, H. L., Liebert, J., and Green, R. F. 1986, in preparation.
Sion, E. M. 1979, in White Dwarfs and Variable Degenerate Stars, Proc. IAU Coll. No. 53, eds. V. Weidemann and H.M. Van Horn, Univ. of Rochester Press, New York, p. 245.
Sion, E. M. 1984, Ap. J., **282**, 612.
Sion, E. M., and Liebert, J. 1977, Ap. J., **213**, 468.
Sion, E. M., Liebert, J., and Starrfield, S. G. 1985, Ap. J., **292**, 471.

Sion, E. M., Liebert, J., and Wesemael, F. 1985, Ap. J., **292**, 477.
Starrfield, S., Cox, A., Kidman, R. B., and Pesnell, W. D. 1984, Ap. J., **281**, 800.
Wegner, G., and Yackovich, F. H. 1984, Ap. J., **284**, 257.
Wehrse, R., and Liebert, J. 1980, Astr. Ap., **83**, 184.
Wesemael, F. 1979, Astr. Ap., **72**, 104.
Wesemael, F. 1981, Ap. J. Suppl., **45**, 177.
Wesemael, F., Green, R. F., and Liebert, J. 1985, Ap. J. Suppl., **58**, 379.
Wesemael, F., and Truran, J. 1982, Ap. J., **260**, 807.
Wickramasinghe, D. T., and Reid, N. 1983, M.N.R.A.S., **203**, 887.
Winget, D. E. and Fontaine, G. 1982, in Pulsations in Classical and Cataclysmic Variables, eds. J.P. Cox and C.J. Hansen, Boulder: JILA, p. 46.
Winget, D. E., Hansen, C. J., and Van Horn, H. 1983, Nature, **303**, 781.
Winget, D. E., Kepler, S. O., Robinson, E. L., Nather, R. E., and O'Donoghue, D. 1985, Ap. J., **292**, 606.
Winget, D. E., Robinson, E. L., Nather, R. E., and Fontaine, G. 1982, Ap. J. Letters, **262**, L11.

DISCUSSION

MENDEZ: I would like to make two comments: First one about Iben and Renzini's idea of explaining H-deficient stars as "born again" AGB stars. As discussed by Schönberner (1983, Ap.J. 272, 708), while the "fading times" of post-AGB objects increase with decreasing stellar mass, the interpulse times go through a maximum at about 0.57 $M_\odot$ and then start decreasing with decreasing mass. Therefore, the chance for a given post-AGB star to suffer a suitable He shell flash increases substantially for smaller masses, and we would expect non-DA white dwarfs to have smaller masses than DA white dwarfs. The fact that the mean masses of DA and non-DA white dwarfs are turning out to be quite similar, might then be used to argue that the "born again" mechanism is not the dominant one for the production of H-deficient white dwarfs.

LIEBERT: Quite right.

MENDEZ: I would like to point out that the gap you find in the temperatur distribution on non-dA white dwarfs appears to happen at the same temperature as the gap in the distribution of H-deficient CSPN. I do not know what this could possibly mean, but there it is.

T.M.K. MARAR: According to you there are more than half a dozen DO white dwarfs. The temperature range from 50,000° to 100,000°. Simply from the thermal emission they should all have been seen as soft X-ray sources.

LIEBERT: No. Hot helium white dwarfs should be heavily blanketed at soft X-ray wavelengths. One 12th mag star, PG 1034+001, at T_e = 80,000°K (Wesemael, Green and Liebert, 1985, Ap.J. Suppl.) was not detected in a pointed Einstein observation. The rest of the DO white dwarfs were not targeted (except for PG 1159).

T.M.K. MARAR: Is the survey complete with respect to all these DO white dwarfs and their regions of observations by Einstein?

LIEBERT: No... Two answers... The PG survey covered one quarter of the sky at galactic latitude above 30° and it is complete to B = 16 for stars bluer than U-B of -0.3 which is a blue color criterion. The Einstein satellite in survey mode covered only about 1% of the sky. In a field of 30 arcmin or so that is accessible we can find serendipitous X-ray sources. It is worth remarking that only about 2 hot white dwarfs that were genuinely new were found in this way, but in the Einstein survey mode which I actually worked to fair extent, I expected to find more white dwarfs, but did not.

T.M.K. MARAR: The mass distribution you mentioned applies only to single white dwarfs, I suppose (not to the ones in interacting binaries).

LIEBERT: Yes. It applies only to non interacting binary and single white dwarfs.

HUNGER: In cool degenerates, pulsation depends not only on g and T_{eff}, but sensitively on the composition. Do you think that pulsations in the hot pulsators are also influenced by the chemistry?

LIEBERT: Certainly. DA stars pulsate in the 10-13,000° K range, DB stars near 25,000° K. The indications are that the hydrogen envelope contains the driving for the former, and the He II ionization drives the latter. So we have actually looked at a few H rich stars such as the central stars. They are as dead as a door nail. They do not show propensity for pulsations. The exact chemical composition of the PG 1159 atmospheres is not yet determined, but it appears that they are helium rich, and carbon and oxygen rich. Starfield, Cox and collaborators attribute the driving by the ionization of oxygen and possibly carbon.

N.K. RAO: Regarding the born again red giant scenario there is some evidence that R CrB stars might be undergoing or passing through such a stage. In such a case one would expect to see the remnants of the first giant stage to be visible like the presence of planetary nebulae and cool dust shell at large distance. Both these are seen around R CrB.

LIEBERT: Some concern about the Iben et al scenario is also the optical depth effects. Whether the photosphere is something really big remains to be seen. So this sort of star looks like a red giant.

HILL: How many DB stars in the ragne $20-30 \times 10^3$ K are pulsating and, what is the type of pulsation?

LIEBERT: Our paper (Liebert et al; this proceedings) lists all of them, as we made an attempt to determine the temperature from IUE. Liebert, Winget and collaborators have discovered four such DB white dwarfs, including the prototype GD 358. The pulsations are non-radial overtones and may be caused by the helium ionization. There is a controversy about the temperature between the Kiel group and us, as to where the high temperature end really is, we would say around 28,000 K down to about 24,000 K being the cool end. According to the Kiel group it goes from 24,000 K to may be only 23,000 or 22,000 K. They have not analysed cooler pulsating stars. Hence it is possible that all the stars which are within the temperatur range are unstable like the pulsators, without proving that's the case. It is approximately the case with DA stars as they come through the instability strip of 10,000 to 12,000 K.

KILAMBI: You said that some of these white dwarfs are accreting matter. What kind of accretion rate are you expecting?

LIEBERT: The main problem has been how <u>little</u> they accrete, particularly the DB stars with nearly hydrogen-free atmospheres. There is also evidence that hydrogen is selectively excluded in the only cases where there is direct, observational evidence that accretion occurs, namely in the DZ white dwarfs with metallic features (in a He-dominated atmospheres). In the normal ISM the rates appear to be negligibly low; the DZ stars are the result (apparently) of chance encounters with interstellar clouds. As the metals diffuse downwards, the spectral type reverts to DC.

KILAMBI: You said between 30,000 K to 50,000 K some of these DB stars could be pulsating. Can you give a rough estimate of the pulsation period for these stars?

DISCUSSION

LIEBERT: The pulsating DB stars show complicated, non-radial pulsations with periods of several to tens of minutes. I believe they will be discussed further by Dr. Saio.

FEAST: What is the emission line you are measuring to give you the luminosity scale of the Planetary Nebulae in the Clouds.

LIEBERT: The luminosities are derived from $\lambda 5007$. The only line effectively to be used is H. Jacoby used [OIII]; however, using only [OIII] means that one would miss the lower excitation nebulae.

TEMPERATURES FOR HOT AND PULSATING HELIUM-RICH (DB) WHITE DWARFS OBTAINED WITH THE IUE OBSERVATORY

J. Liebert,[1] F. Wesemael,[2] C.J. Hansen,[3]
G. Fontaine,[2] H.L. Shipman,[4] E.M. Sion,[5]
D.E. Winget,[6] and R.F. Green [7]

ABSTRACT. Ultraviolet energy distributions are analyzed for several hot, helium atmosphere DB white dwarfs, including the four known pulsating stars which define an empirical DB instability strip. Temperatures are derived exclusively from fits to the ultraviolet energy distributions. The blue edge of the empirical DB instability strip lies at 30,000 ± 4,000 K, and the red edge lies near 24,000 ± 2,000 K. The hottest DB star -- and the only known one hotter than the instability strip -- is PG0112+104 at or above 30,000 K. This leaves no known helium-atmosphere degenerate stars in the interval $30,000 \leq T_e \leq 45,000$ K.

1. INTRODUCTION

The DB white dwarfs are those with helium-rich atmospheres that are hot enough for neutral helium lines to be seen in their spectra, but not hot enough for He II lines to be seen. Several important investigations have produced estimates of temperatures and other parameters for most of these stars, but the estimates remain very uncertain for those with $T_e > 18,000$ K. The determination of more accurate temperatures for the hotter DB stars is now important for two reasons: (1) Winget and collaborators have recently found several pulsating stars among the hottest known DB's, and it is important to establish the existence and the boundaries of the presumed instability strip for

---

[1] Steward Observatory, University of Arizona.
[2] Département de Physique, Université de Montréal.
[3] Department of Astrophysical, Planetary, and Atmospheric Sciences and JILA, University of Colorado.
[4] Department of Physics, University of Delaware.
[5] Department of Astronomy, Villanova University.
[6] Department of Astronomy and McDonald Observatory, University of Texas at Austin.
[7] Kitt Peak National Observatory, NOAO.

this new kind of variable star; (2) While the DB stars should range from about 11,000 K to about 40,000 K, above which He II lines will appear, there is currently no reliable determination of a DB having $T_e >$ 30,000 K. At the same time, Wesemael et al. (1985) have analyzed nineteen of the hotter DO stars, and the coolest of these has $T_e \approx$ 45,000 K. This leaves a gap at 30,000 < T_e <45,000 K in which no helium-atmosphere degenerate star is currently known.

The IUE observatory can play an important role in improving the temperature determinations for the hot DB stars. The IUE cameras cover a wavelength interval ($\lambda\lambda$1200-3000) which includes or is very near the Planckian peak; thus improved temperatures may be estimated from spectrophotometry of only modest quality. In the last few years, we have attempted to identify and observe with IUE the hot DB stars in the Palomar-Green Survey, as indicated from optical data. These include all four known pulsating DB stars. We report here the preliminary results of our ultraviolet spectrophotometry of these objects.

2. IUE ENERGY DISTRIBUTIONS AND EFFECTIVE TEMPERATURES

Low-dispersion IUE observations of several newly-observed PG objects were obtained in 1982-1984. Blanketed model atmosphere grids useful for comparison with these observations are available from Wesemael (1981), Koester (1980), and Wickramasinghe (1983). We have combined here the hotter Wesemael set with the cooler Wickramasinghe set (the W grid) for comparison with the observations and the predictions of the Koester models (the K grid). Two sets of IUE effective temperatures were derived for each observed star, one from the W and one from the K grid. The effective temperatures determined for our newly-observed objects will be given elsewhere.

In Figure 1 we illustrate the fitting of two of the hottest stars in our sample. GD358 is the first discovered pulsating DB star (Winget et al. 1982), while PG0112+104 was found to be hotter than GD358 based on optical spectrophotometry (Oke et al. 1984, hereafter OWK), and non-variable. Consistent fits for the SWP fluxes for both stars and both model sets are possible at temperatures of 29-30,000 K for PG0112+104 and 27-28000 K for the pulsating star GD358. In each case, the lower temperature is assigned from the K grid, while the W fit is 1,000 K higher. However, while there is good agreement between the K and W model fits to the IUE data for both hot stars implying that they differ in T_e by < 2,000 K, it is noteworthy that the temperature estimates using optical data (and the K grid) differ by a greater amount. OWK assign PG0112+104 a temperature of 28,900 K, in nice agreement with the IUE fits, and suggesting that this is the hottest known DB star. Yet the same authors (see Koester et al.

1985) favor an optically-derived temperature of 24,000 K for GD 358. The latter seems to be in sharp disagreement with the SWP region fits for both W and K models.

If we assume that an instability strip exists for DB stars (Winget et al. 1982, 1983) analogous to the well-defined temperature region of the ZZ Ceti (DA) variable stars, then these two stars may bracket the high temperature boundary near 27-28,000 K. Alternatively, since GD358 has the bluest energy distribution at IUE wavelengths of the four known pulsating stars, the high temperature boundary could be as low as 24,000 K if we use the optically-determined temperature.

In Figure 2, two somewhat cooler stars are plotted in the same way for comparison with the two grids. PG1654+160, which pulsates (Winget al. et 1984), appears to fit about 25-26,000 K, although the noisy LWR points appear too high (indicating a cooler temperature) for both model sets. The non-pulsating PG0853+164 may be assigned a fit near 22,000 K for both W and K curves. If there is a well-defined lower temperature limit to a pulsational instability strip, our results for GD190, 1654+160 and 1115+158 suggest that it is near 24-25,000 K, using IUE fluxes.

3. CONCLUSIONS

The temperature determinations reported herein from IUE data generally support the expectation of Winget and Fontaine (1982) and Winget et al. (1983) that an instability strip exists for pulsating DB stars. Realistically, the high temperature boundary is $30,000 \pm 4,000$ K, using IUE temperature determinations. The low temperature boundary is most likely within the interval $24,000 \pm 2,000$ K. Of course we are a long way from establishing whether all stars within the instability region actually pulsate, but the ordering of temperatures made from using the IUE energy distributions (regardless of the values of temperatures assigned) is consistent with this hypothesis. The empirical instability strip seems to lie close to that predicted by Fontaine et al. (1984) using ML3 convection (26-29,000 K).

Following this investigation, there remains only one normal DB star, PG0112+104, with (1) a temperature likely to be above the high temperature boundary for pulsational instability, and (2) an effective temperature near 30,000 K or above. For this conclusion we emphasize that the OWK optical determination is in excellent agreement. The new IUE results strengthen the hypothesis of a temperature gap at 30-45,000K, between the several dozen known and well-studied DB stars and the $\sim$ 20 hotter DO stars. The statistical significance of this result is discussed in Wesemael et al. (1985).

This work was supported by the National Science Foundation, by NASA, and by the NSERC Canada.

REFERENCES

Fontaine, G., Tassoul, M., and Wesemael, F. 1984, in Proc. 25th Liege Astrophysical Colloquium: Theoretical Problems in Stellar Stability and Oscillations, eds. A. Noels and M. Gabriel (Liège: Université de Liège), p. 328.
Koester, D. 1980, Astron. Astrophys. Suppl., 39, 401.
Koester, D., Vauclair, G., Dolez, N., Oke, J.B., Greenstein, J. L., and Weidemann, V., 1985, Astr. Ap., 149, 423.
Oke, J.B., Weidemann, V., and Koester, D. 1984, Ap. J., 281, 276.
Wesemael, F. 1981, Ap. J. Suppl., 45, 177.
Wesemael, F., Green, R.F., and Liebert, J. 1985, Ap. J. Suppl., 58, 379.
Wickramasinghe, D.T. 1983, M.N.R.A.S., 203, 903.
Winget, D.E., and Fontaine, G. 1982, in Pulsations in Classical and Cataclysmic Variable Stars, ed. J. P. Cox and C. J. Hansen (Boulder: University of Colorado), p. 142.
Winget, D.E., Robinson, E.L., Nather, R.E., and Balachandran, S. 1984, Ap. J. (Letters), 279, L15.
Winget, D.E., Robinson, E.L., Nather, R.E., and Fontaine, G. 1982, Ap. J. (Letters), 262, L11.
Winget, D.E., Van Horn, H.M., Tassoul, M., Hansen, C.J., and Fontaine, G. 1983, Ap. J. (Letters), 268, L33.

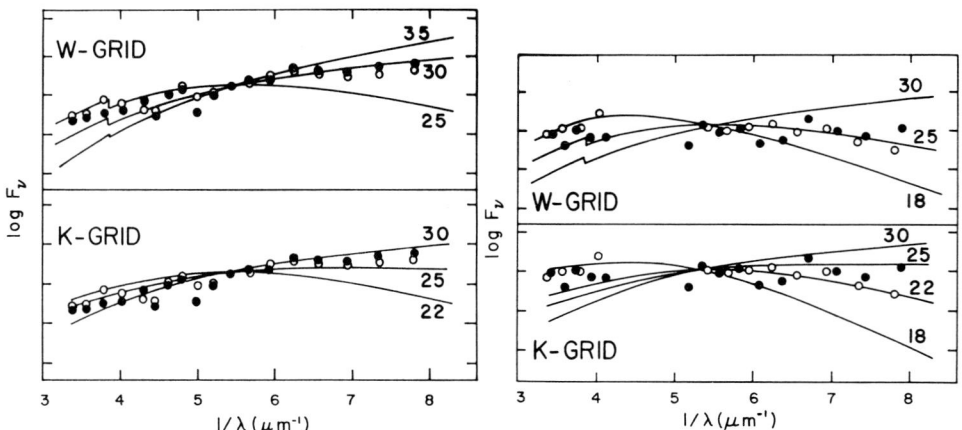

Fig. 1 - IUE energy distributions for PG0112+104 (filled circles) and GD358 (open circles), together with model fluxes from the W and K grids. Temperatures are in units of 10^3 K.

Fig. 2 - Same as Fig. 1, but for PG1654+160 (filled circles) and PG0853+164 (open circles).

LINE BAND PROFILES IN THE SPECTRA OF COOL MAGNETIC HELIUM-RICH WHITE DWARFS

I. Bues
Remeis Sternwarte Bamberg
Institut für Astronomie
der Universität Erlangen
Nürnberg, F.R.G.

For white dwarfs with effective temperatures smaller than 12000 K, the percentage of objects with a helium-rich atmosphere increases compared to the hydrogen-rich sequence. The carbon abundance, which can be determined from line and band strengths (Bues, 1973; Koester et al., 1982), varies by more than a factor of 1000 within this class. Moreover, for the subclass of white dwarfs with strong magnetic fields, the abundance ratio of H/He differs from that of the DB and DA sequences. The hot star Feige 7, analyzed by Liebert et al. (1977), shows lines of hydrogen and helium at a comparable strength for a moderately strong field of 10^3 Tesla. If there is any chance of finding white dwarfs which are descendants of hot, non-degenerate helium stars with rotation and magnetic fields, then it should be within these objects of mixed composition.

The most thoroughly investigated cool white dwarf of mixed composition is G 99-37 with T_{eff} around 6000 K (Bues, 1973; Liebert, 1976), with a magnetic field of moderate strength. Figure 1 shows a new spectrum taken the ESO 1.52 m telescope. It also shows the strongest bands of CH found in a white dwarf. In addition, our spectra, taken in 1984 (October) and 1985 (April), display H_β, several C I lines and variable branches of the Swan band of C_2. Figure 2 shows the v = 0 transition for 5 spectra taken on April 24th, 1985. From these spectra we derive a period of $\sim$ 4 hours.

For the analysis of the variable lines and bands, we calculated only a single flux-constant model atmosphere with a helium-rich composition, He/H = 1000, log g = 8 and T_{eff} = 6000 K. For the CH band at 430 nm the desired field strength did not vary by more than 10%, whereas for the C I lines and, especially, the C_2 branches, a factor of two is more likely. From CH, the value of 800 Tesla is derived (Rupprecht, 1983). The average field strength is nearly the same as for Feige 7 although the hydrogen abundance is smaller in G 99-37.

For the other two stars, EG 250 and EG 374, we used published spectra and polarization measurements (Landstreet, 1982). The small

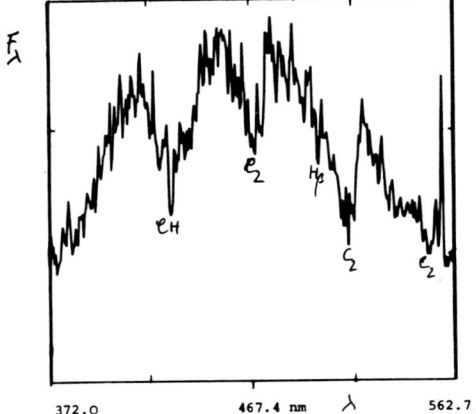

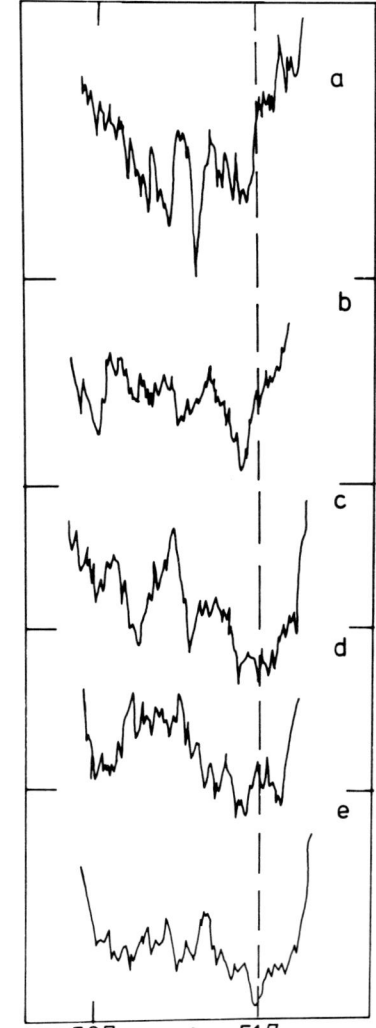

Fig. 1: Observed Flux for G 99-37

Fig. 2: Observed variation of the C_2 transition (Spectra taken with 114 Å/mm, ESO 1.52m)

field strength for one particular phase of EG 250, and the variation of polarization at the wavelengths of the CH and C_2 bands (Fig. 3), seem to confirm the previously derived abundance ratio. The same is true for the other phases with larger field strengths. In contrast, the energy distribution of EG 374 is in satisfactory agreement with theory (Fig. 4), whereas for the polarization an additional (unknown) absorber has to be assumed.

Concerning the relationship between non-degenerate helium stars and our three white dwarfs, we conclude that these two classes are probably not linked by evolution as neither masses nor magnetic fields fit.

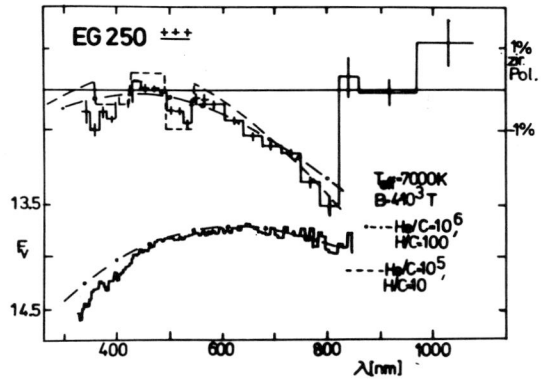

Fig. 3: Flux and circular polarization for one phase of EG 250 (model parameters are indicated, log g = 8)

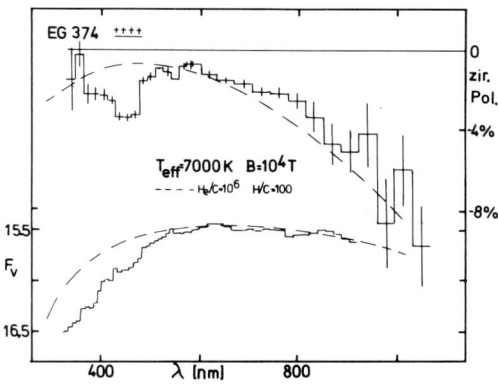

Fig. 4: Flux and circular polarization for EG 374

REFERENCES

Bues, I.: 1973, Astron. Astrophys. **28**, 181
Koester, D., Weidemann, V., Zeidler-K.T., E.-M.: 1982, Astron. Astrophys. **116**, 147
Landstreet, J.D.: 1982, private communication
Liebert, J.: 1976, P.A.S.P. **88**, 490
Liebert, J., Angel, J.R.P., Stockman, H.S., Spinrad, H., Beaver, E.A.: 1977, Astrophys. J. **214**, 457
Rupprecht, G.: 1983, Ph.D. Thesis, Universität Erlangen-Nürnberg

DISCUSSION

VARDYA: Do we see any molecules other than C_2 and CH? Like CN or CO?

BUES: No. There are definitely no traces of CN or CO in the atmospheres of cool white dwarfs with C_2 bands. We calculated the band strengths of CO in the UV region, but the molecule has not been observed. This yields limits on the abundance for O.

MICHAUD: You do not expect to see O or N since the C layer will be on top of N and O and is the most likely to leave traces in the atmosphere.

BUES: This is true for your diffusion mixture of layered envelopes. For an old (5×10^9 years) star the magnetic field should not interfere. Yet we have to keep in mind that the hydrogen abundance is larger than for normal cooled-down DB stars.

DETECTION OF AN EXTREMELY ACTIVE STATE OF AM CANUM VENATICORUM

K.R.Narayanan Kutty, T.M.K.Marar, V.N.Padmini
S.Seetha, K.Kasturirangan and U.R.Rao
ISRO Satellite Centre,
Airport Road,
BANGALORE 560 017, INDIA.

J.C.Bhattacharyya, S.Mohin and K.Jayakumar
Indian Institute of Astrophysics
BANGALORE 560 034, INDIA.

ABSTRACT: We report fast photometric observations on AM CANUM VENATICORUM (AM CVn) the ultra short period, hydrogen deficient variable. We have detected on 24th February, 1985 an intense flare of (Δm)peak ≈ 0.34 in white light lasting over 200s. Following this flare we observe an enhanced double humped structure lasting for 1051s which is the dominant periodicity exhibited by AM CVn. We have also detected the 525s and 1051s periods. In addition, we report flickerings, lasting typically 1-2 minutes, that are characteristic of cataclysmic variables.

1. INTRODUCTION

AM Canum Venaticorum (HZ29) is one of the fastest known white dwarf variables currently thought to consist of a degenerate semi-detached binary system. During its chequered career in the astronomical literature it has been considered as a white dwarf, a hot subdwarf, a magnetic rotator, a system of pair of subdwarfs, a triple system and even as a quasi stellar object.

First noted by Malmquist (1936) this system exhibits broad and probably double HeI absorption lines and is conspicuous by the total absence of hydrogen lines. The He spectrum is found to be too shallow and too wide (Greenstein and Matthews, 1957) which initially led to the conclusion that it may be a DB white dwarf. Smak (1967) who made the first detailed photometric observations detected a double humped structure with a maximum variability of 0.05 mag. He determined a period of 1054s and was the first to propose a binary theory for the system. Later Ostriker and Hesser (1968) improved on the period determination (1051.118s) and also detected a 525.53 s period. They found that a model of a close pair of white dwarfs will necessitate much shorter periods for the assumed masses and therefore proposed a model consisting of hot subdwarfs or a magnetic white dwarf. However, during later observations no circular polarisation has been detected in the system and the 525s period is not always present. Krzeminski (1972) found that a single period did not fit all data sets and determined two periods namely 1051.043s and 1051.056s. The model he put forward was that of a triple system.

Warner and Robinson (1972) conducted fast photometric observations with an integration time of 1 s. They refined the period value as 1051.0505s but could not detect any significant rate of change of this period. They reported in addition, several flickerings with periodicities ranging from 20s to 150s. Their observations supported the binary system model which was later expounded by Faulkner et al (FFW,1972). Spectroscopic data obtained by Robinson and Faulkner (1975) showed no doubling of helium lines and also confirmed absence of radial velocity variation. However the white dwarf binary model seems to best fit the peculiar profile and strength of the spectrum. Patterson et al., (1979)determined a period increase on a time scale of 10^5 years with a mass transfer of $3(\pm1.5)\times10^{-7}$ $M_\odot$/year which is much higher than that predicted by the FFW model. They determined a period of 1051.212 ± 0.015s and a coherent periodicity of 26.3s.

More recently, Solheim et al., have reported that the 1051s period is infact slowly decreasing and hence concluded that this period may represent the rotation period of the accreting white dwarf in the binary system. Elsworth et al., (1982) had observed a possible outburst from the system, the like of which has not been reported earlier. The outburst (flare) showed a 30% (~0.28 mag) increase in intensity in white light which lasted a total duration of about 50s.

Keeping in view all these enigmatic and variable characteristics of this object we decided to monitor it in the course of our regular program of optical observations of X-ray emitting cataclysmic variables. Our main objective was to look for transients and short term flickerings, if any.

2. OBSERVATIONS

The fast photometric observations of AM CVn were conducted during February-March 1985. We employed a photometer using a thermoelectrically cooled RCA C31034 photomultiplier tube attached to the 1 metre telescope at the Kavalur Observatory. The output of the tube was amplified using an ORTEC system consisting of a preamplifier (9301) and an amplifier discriminator (9302). This amplified signal was fed to a photon counter (9315) and the data was fed through a sampling/control unit (9320) to a printer.

The star was observed in U,B,V, filters with the respective sky observations, but the long runs (four hours) were conducted in white light. The comparison stars observed were BD+39 2541 (V=10.11), a comparison star east of AM CVn (RA 1985)≈12^h 35^m Dec (1985)≈+37° 41'; V=10.4) and a check star Landolt 104-306 (V=9.36). The integration time was 1 s and diaphragm used was 24 arc sec for all observations. Fig.1 and 2 show part of the light curves obtained by us.

3. RESULTS AND DISCUSSION

We observed a flare lasting about 3.4min (fig.1) on 24 Feb 1985 at around 20h 17m UT. The flare shows a peak intensity increase of about 37% (≈0.34 mag) in white light above

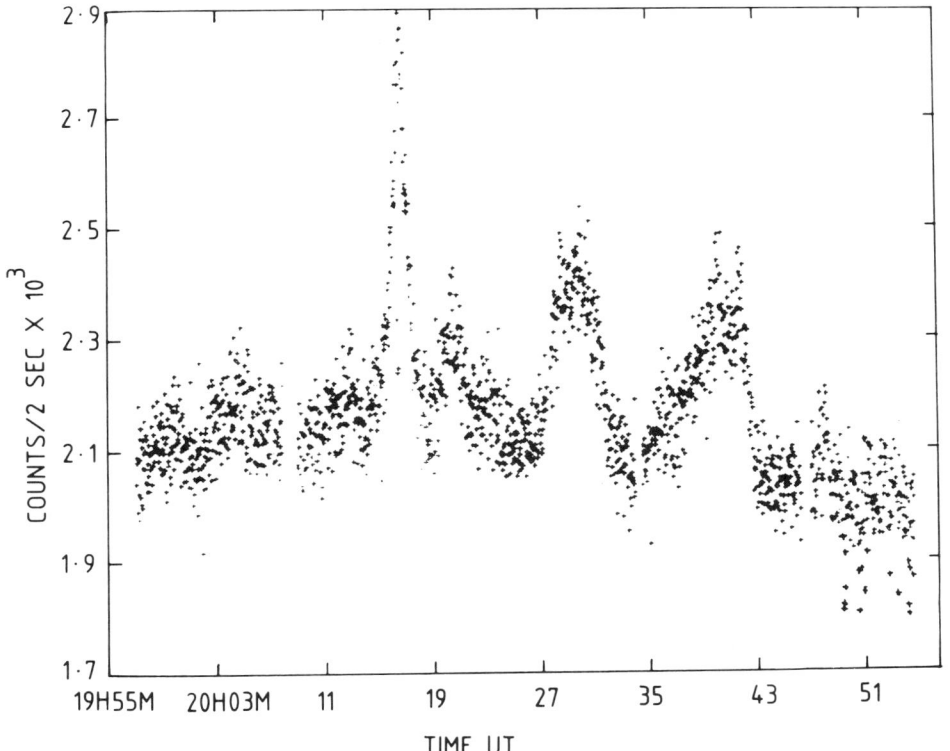

Fig.1. Light curve of AM CVn in white light (Feb.24, 1985)

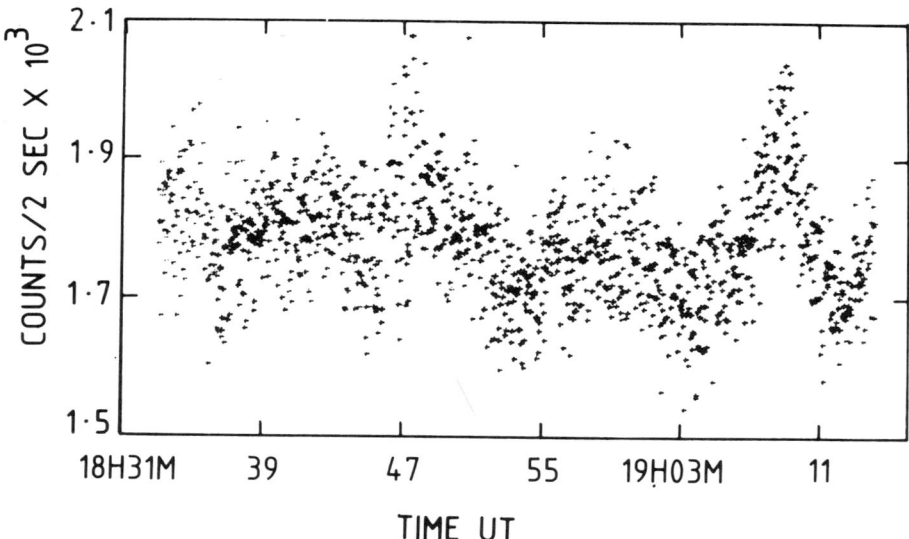

Fig.2. Light curve of AM CVn in white light (March 19, 1985)

the quiescent level. The equivalent duration P of this flare calculated using the relation,

$$P = \int \frac{I_{o+f}(t) - I_o}{I_o} dt$$

is found to be 15 s. Here $I_{o+f}(t)$ is the intensity of the star as a function of time t during the flare, I_o is the quiescent level intensity before the flare and the integration is carried out over the whole duration of the flare. The rise to peak intensity of this outburst is complete in 108s and it decays to ambient level in 96s.

Following this flare we observed a double humped structure, typical of AM CVn light curves, lasting for 1050s which is the dominant period exhibited by AM CVn. The amplitude of this 'M' structure is however much higher than that reported before by other authors. The first hump with a $\Delta m = 0.15$ mag, is nearly symmetric and lasts for about 7.5min, whereas the second hump with a $\Delta m = 0.12$ mag, is skewed to the right and lasts for about 10 min. We point out that the primary minimum (between the two humps) has the lowest intesity level and the duration of these humps are consistent with the phase detected earlier by other observers. Though this double humped structure is observed in other light curves shown in fig.2 (obtained on March 19, 1985) also, the one shown in fig.1 is unique for its enhanced amplitude.

We wish to emphasise that the flare observed by us has

a 37% increase in white light and is clearly seperated from the typical 1050s 'M' structure. This is to be compared with the data presented by Elsworth et al., (1982), where a 30% increase in intensity is reported, but shows a triple humped structure lasting for a total duration of about 50s only.

In addition we note in fig.1 small flickerings typically lasting 1-2 min, which are characteristic of cataclysmic variables. Except for the flicker immediately following the flare all other flickers have about 0.05 mag comparable with the largest amplitude variation of the 1051 s period reported by Smak (1967).

The differential visual magnitudes calculated for AM CVn using BD+39 2541 as the comparison star are given in Table 1.

TABLE 1

Date	Time in UT	Visual Magnitude
24 Feb 1985	23h 00m	14.1
17 Mar 1985	21h 28m	14.2
18 Mar 1985	20h 03m	13.9
18 Mar 1985	21h 25m	14.1
19 Mar 1985	18h 16m	14.0
23 Apr 1985	20h 11m	13.9

In all our observations we detect a 525 s and 1051s period by the folding technique employing synchronous summation. However due to the short length of our data, we are unable to refine the value of the period obtained by earlier observers.

Searches for the detection of coherent periodicities at 26.3s reported by Warner and Robinson (1972) did not yield positive results.

Additional observations are necessary to confirm the type of variation in the 1051s period and its implication in clinching a suitable model for this system.

4. REFERENCES

1. Elsworth.Y., Grimshaw.L. and James.J.F.,1982 M.N.R.A.S., 201, Short communication, 45

2. Faulkner.J., Flannery.B. and Warner.B., 1972. Ap.J.Lett., 175, L79

3. Greenstein.J.L. and Matthews.M.S., 1957, Ap.J., 126, 14

4. Krzeminski.W., 1972, Acta.Astr., 22, 387

5. Malmquist,K.G., 1936, Stockholm Ann., 12 Pt.7

6. Ostriker.J.P. and Hesser.J.E., 1968, Ap.J. Lett., 153, L151

7. Patterson.J., Nather.R.E., Robinson.E.L. and Handler.R.F.,1979, Ap.J., 232, 819

8. Robinson.E.L. and Faulkner.J.,1975, Ap.J.Lett. 200, L23

9. Smak.J., 1967, Acta.Astr.,17, 255

10. Solheim.J.E., Robinson.E.L., Nather.R.E. and Kepler.S.O. ,1984, Astron.Astrophys.135, 1

11. Warner.B. and Robinson.E.L., 1972, M.N.R.A.S.,159, 101

DISCUSSION

LIEBERT: A student at Texas maned Matt Wood, working with Don Winget, is working on a star which is quite similar to AM CVn but shows overall variations in brightness level of up to 4 magnitudes. I think that further monitoring of AM CVn is worthwhile to see, for example, how closely the mean light level really is constant over short to long timescales.

MARAR: Would you call it a dwarf nova with a say 5 mag. brightness change.

LIEBERT: It is analogous to dwarf novae except that both the accretor and accretee to be He rich. You don't know that directly because you have only one spectrum.

DESHPANDE: Did you repeat the observations?

MARAR: Yes it has been followed several nights. The idea has been to see 1051 seconds variations.

KILAMBI: When it brighened up by 0.28 mag, is there any change in the X-ray flux?

KUTTY: No X-ray flux has been detected so far.

MARAR: Einstein has observed AM Can Van, and puts only an upper limit in the energy range 0.1 to 4 Kev. In the hard X-ray range, however, there has been a probable positive detection by the hard X-ray telescope on one of the Ariel satellites which has no high sensitivity.

VIII   I R A S – R E S U L T S

IRAS RESULTS FOR HYDROGEN DEFICIENT STARS

H. J. Walker
Sterrewacht, Huygens Laboratorium,
Postbus 9513,
2300RA Leiden
Holland

ABSTRACT. The Infra-Red Astronomical Satellite, working at 12µm, 25µm, 60µm and 100µm, has observed stars in several subgroups of the hydrogen deficient stars. Observations of most known R CrB stars are reported here, as well as data for several HdC stars and other carbon-rich stars, together with some intermediate helium rich stars and related objects. Also given are LRS spectra of R CrB, RY Sgr, υ Sgr and some Carbon stars. Initial reductions of the additional observations from IRAS show an extended dust shell around R CrB, and probably also around SU Tau.

1. INTRODUCTION

The major task of IRAS was to survey the whole sky at infrared wavelengths, namely 12µm, 25µm, 60µm and 100µm. Between January 1983 and November 1983 it surveyed over 96% of the sky. This occupied about 60% of the total time, and in some of the remaining time, additional observations could be made, usually small raster scans for nominated interesting objects. As can be seen from Figure II.C.9 in the IRAS Explanatory Supplement (1984), the IRAS survey detectors make broad band photometric measurements of the sources they detect, and these values must be corrected for the spectral shape of the object. In addition to the survey detector array there was also a Low Resolution Spectrometer with two channels, covering the wavelength region between about 8µm and 23µm. Kilkenny and Whittet (1984) give UBVRIJHKLMNQ photometry for several R CrB stars, and match the energy distributions with black body contributions from the star and surrounding dust shell. At wavelengths longer than that of the L band

---

The Infra-Red Astronomical Satellite was developed and operated by the Netherlands Agency for Aerospace Programs, the U.S. National Aeronautic and Space Administration and the U.K. Science and Engineering Research Council.

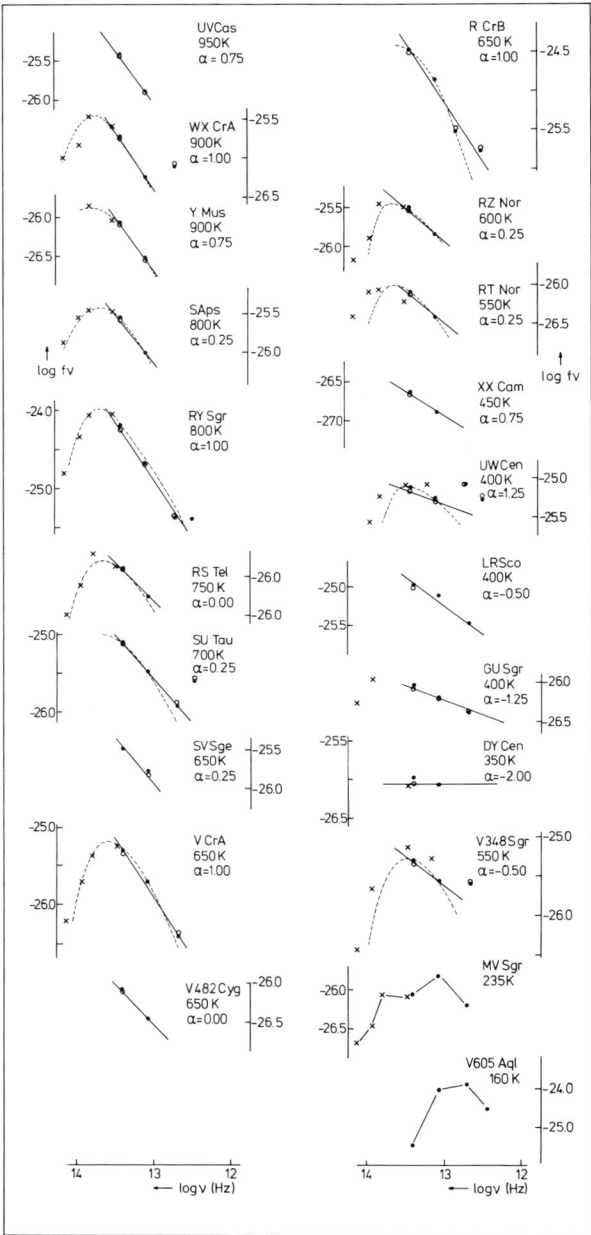

Figure 1   Energy distributions for sources found by IRAS. Filled circles (●) are IRAS colour corrected fluxes; open circles (o) are power law corrected fluxes; crosses (x) show data from Kilkenny & Whittet (1984). The solid line shows the power law fit, and the dashed line comes from the model fit of Walker (1985)

(3.5μm) there is no significant contribution to the flux observed from the stellar energy distribution. This means, for the R CrB stars, IRAS only measures the dust shells around the stars and not the stars themeselves. For some of the brighter helium strong (intermediate helium rich) stars and helium weak stars, it is the stellar flux that IRAS detects.

2. SURVEY RESULTS

Walker (1985) gave fluxes for 19 hydrogen deficient stars, comprising most of the R CrB stars and most of the hydrogen deficient close binaries. Table I includes 54 objects, using much relaxed criteria for flux quality, position and object type. Objects in the groups of the Am/Ap stars, helium weak and helium strong stars are included, since they are believed to be related groups of objects, with the temperature being responsible for the surface abundance anomalies observed (Osmer and Peterson, 1974). Also included are some bright Carbon stars (from the General Catalogue of cool Carbon stars by Stephenson, 1973), an interesting contrast to the hotter HdC stars and R CrB stars. XX Cam was not found in the IRAS survey, but fortunately the object was observed in the additional observation programme, so that fluxes are available, although they are of a lower quality, due to the preliminary nature of the calibration for the additional observations. None of the extreme helium stars were found in the IRAS Point Source Catalog. The fluxes in Table I are corrected assuming a black body energy distribution between 12μm and 100μm. The temperature for the correction factor was derived by taking the ratio of fluxes in two adjacent bands, and a mean correction used for the middle band where the two ratios from three bands gave different values.

Shown in Fig. 1 are the energy distributions for those sources with fluxes in two or more bands (using filled circles for the IRAS data). The energy distributions for the close binaries are shown in Walker (1985). The model fit to the energy distributions shown in Fig. 1, taken from Walker (1985), is sketched as a dashed line, and the temperature of the fit to the model curve is shown in Fig. 1 and given in Table I. Data from Kilkenny and Whittet (1984) are shown in Fig. 1 as crosses. As can be seen from Fig. 1 some energy distributions are poorly represented by a black body curve, and a power law dependence on frequency would be much more appropriate. Schaefer (1986) used this approach for modelling the IRAS data, and his approach (discussed in more detail in 2.1) is followed here. Fig. 1 shows the IRAS fluxes corrected for a power law dependence on frequency, plotted as open circles, with a suitable power law shown as a solid line.

The source MV Sgr is a source of confusion, as well as a confused source as seen by IRAS. Walker (1985) selected an IRAS source different from the one given here, on the basis of the raw data streams. When the position alone (as given in the IRAS Point

Table I Hydrogen deficient stars and related objects with IRAS fluxes

star	IRAS name	IRAS fluxes (colour corr)(Jy) 12μm	25μm	60μm	100μm	temperature(K) 12/25	model	(B-V)	V
R CrB stars									
XX Cam		0.23:	0.14:			450		0.85	7.35
SU Tau	05461+1903	9.50	4.14	1.52	2.78	700	700	1.08	9.77
UW Cen	12404-5415	7.81	5.57	8.46	5.38	400	400		
Y Mus	13025-6514	0.83	0.29	<1.06	<11.59	900			
DY Cen	13224-5359	1.05	0.84	<0.49	<2.21	350		0.31	12.39
S Aps	15043-7152	2.71	1.02	<0.40	<1.00	800	800	1.23	9.79
R CrB	15465+2818	33.83	13.81	3.08	1.72	700	650	0.79	10.24
RT Nor	16200-5913	0.85	0.39	<0.42	<3.04	550	550	1.14	10.72
RZ Nor	16287-5309	3.08	1.77	<5.62	<66.34	550	600		
LR Sco	17243-4348	10.72	7.75	3.41	7.06:	400		0.55	9.72
WX CrA	18054-3720	1.89	0.61	0.78	<2.93	1000	900	1.34	11.0
VZ Sgr	18119-2943	1.13	0.60	<0.40	<30.23				
RS Tel	18151-4634	1.31	0.57	<1.40	<1.48	600	750	1.05	10.0
GU Sgr	18211-2417	0.97	0.66:	0.41	<38.89	400			
V CrA	18441-3812	4.95	2.00	0.39:	<1.27	700	650	0.79	10.24
SV Sge	19059+1732	3.29	1.66	<0.45	<3.81	650		1.89	10.51
RY Sgr	19132-3336	63.88	20.80	4.12	3.99	1000	800	0.65	6.50
V482 Cyg	19577+3351	0.85	0.35	<5.57	<61.30	650			
U Aqr	22006-1652	1.12	<0.51	<0.40	<1.00				
UV Cas	23001+5920	3.81	1.28	<3.35	<48.97	950		1.46	10.65
V348 Sgr	18372-2257	5.05	2.78	2.52	<13.02	500	550	0.30	10.6
MV Sgr	18415-2100	0.86:	1.48	0.64	<3.38	235		0.26	12.7
HdC stars									
HD137613	15248-2459	0.42	<0.39	<0.40	<1.06			1.12	7.54
HD148839	not detected								
HD173409	18433-3123	0.53	<0.35	<0.40	1.89:			0.89	9.54
HD175893	18556-2934	0.44	<0.37	<0.40	<1.71			1.19	9.26
V605 Aql	19158+0141	5.83	30.39	35.70	16.59	160			
HD182040	19204-1048	0.58	<0.32	<0.40	<1.38			1.08	6.96
Close binaries									
KS Per	04453+4311	1.26	0.41	<0.40	<1.31	1000	1000	0.48	7.76
HD37017	not detected								
LSS4300	17346-3521	6.00	2.48	<8.69	<123	650	750	0.82	9.75
β Lyr	18482+3318	4.43	1.95	0.66	<1.00	650	600	0.00	3.45
υ Sgr	19188-1603	136.6	33.69	6.11	2.30	1000	1000	0.10	4.61
He strong stars									
HD37479	05362-0237	5.33	14.89	5.91:	<11.72	200		-0.19	6.65
HD93030	10411-6407	1.04	0.24:	<2.04	<24.01	>10000		-0.22	2.76
He weak stars									
HD5737	00561-2937	0.50	<0.24	<0.40	<1.05			-0.16	4.31
HD19400	03021-7205	0.32	<0.54	<0.40	<1.75			-0.14	5.63
HD120709	13489-3244	0.71	<0.31	<0.40	<1.00			-0.13	4.56
HD143699	16002-3827	0.31	<0.40	<0.52	<15.55			-0.14	4.89
HD175156	18518-1540	0.67	<0.36	<0.73	<4.02			0.17	5.10
Am/Ap stars									
HD18557	02563-0958	0.33	<0.66	<0.40	<1.00			0.22	6.14
HD24712	03529-1214	0.39	<0.25	<0.40	<1.14			0.32	6.00
HD129174	14383+1637	0.68	<0.25	<0.40	<1.00			-0.03	4.94
46 Dra	18416+5529	0.44	<0.25	<0.40	<1.00			-0.09	5.04
HD204411	21250+4837	0.41	<0.30	<0.47	<17.20			0.07	5.31
C stars									
S Aur	05238+3406	119.0	31.51	7.22	10.15:	2500		6.5	10.7
BL Ori	06225+1445	41.23	11.59	3.10	2.56	1200		2.33	6.2
U Ant	10329-3918	125.0	34.16	20.50	18.34	1900		2.95	5.63
U Hya	10350-1307	167.1	57.44	13.16	12.67	900		2.51	4.97
V Aql	19017-0545	110.2	28.96	8.56	5.03	2500		4.19	6.90
U Cyg	20180+4744	89.37	28.17	6.50	<15.78	1100		3.31	8.5
Y Pav	21197-6956	95.21	31.94	7.94	4.50	800		2.82	6.41
TX Psc	23438+0312	118.0	30.12	8.86	5.95	3000		2.60	5.04

Notes for Table I
Sources for V and (B-V)
Blanco et al. (1968) for: LR Sco, HD175893, HD173409
Buscombe (1977) for: HD137613, HD182040
Buscombe (1980) for: XX Cam, WX CrA, RY Sgr, S Aps, RS Tel, SU Tau, V CrA, RT Nor, SV Sge, BL Ori, U Hya, U Ant, S Aur
Fernie et al. (1972) for: R CrB, UV Cas
Hack (1967) for: HD30353
Herbig (1964) for: MV Sgr
Hoffleit + Jaschek (1982) for: β Lyr, υ Sgr, HD37479, HD93030, HD5737, HD19400, HD120709, HD175156, HD143699, HD18557, HD24712, HD129174, 46 Dra, HD204411, V Aql, Y Pav, TX Psc
Houziaux (1968) for: V348 Sgr
Drilling et al. (1984) for: LSS4300
Rao (priv. comm.) for: DY Cen

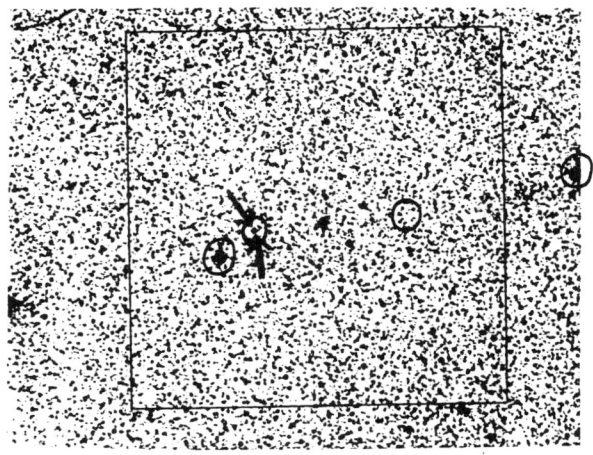

Figure 2   Palomar Sky Survey field around MV Sgr, with MV Sgr arrowed. Circles show the position of IRAS sources. The box is 16 arcmin wide.

Source Catalog) is used as the criterion, the postion agrees very well with the one expected. Lynas-Gray (private communication) provided his finding chart of MV Sgr, shown in Fig. 2, with MV Sgr arrowed. Plotted on top of this, shown as circles (the size of the circle has no meaning) are the IRAS Point Source Catalog sources. The two SAO stars in the field, furthest left and furthest right, agree excellently in position with their IRAS counterparts, as does the source associated with MV Sgr.

2.1 THE POWER LAW DEPENDENCE OF THE ENERGY DISTRIBUTIONS
Following the suggestion by Schaefer (1986) a power law dependence of flux with frequency is fitted to the IRAS data (shown in Fig. 1). As with Schaefer the formula by Rees et al. (1969) is used:

$$f \propto \nu^{\xi} \quad \text{where} \quad \xi = \tfrac{1}{2}[\,2 + \alpha - (4 + \alpha)(2 - \beta)\,]$$

with the grain number density $\rho \propto r^{-\beta}$
and the absorption coefficient $Q \propto \nu^{\alpha}$

Table II   Values for the Power Law Terms

source	if β = 2 α	if α = 1 β	source	if β = 2 α	if α = 1 β
UV Cas	0.75	1.95	R CrB	1.00	2.00
WX CrA	1.00	2.00	RZ Nor	-0.25	1.75
Y Mus	0.75	1.95	RT Nor	-0.25	1.75
S Aps	0.50	1.90	XX Cam	-0.75	1.65
RY Sgr	1.00	2.00	UW Cen	-1.25	1.55
RS Tel	0.00	1.80	LR Sco	-0.50	1.70
SU Tau	0.25	1.85	GU Sgr	-1.25	1.55
SV Sge	0.25	1.85	DY Cen	-2.00	1.40
V CrA	1.00	2.00	V348 Sgr	-0.50	1.70
V482 Cyg	0.00	1.80			

Traditionally, for stars with steady mass loss $\beta = 2$, and for pure graphite $\alpha = 2$. The actual value for α is expected to be lower since the graphite is more likely impure and amorphous.

First, after correcting the IRAS fluxes for a power law dependence, setting $\beta = 2$, I found values for α (shown in Fig. 1 and in Table II), for each of the suitable R CrB star dust shells. The assumption of steady mass loss seems doubtful for the R CrB stars, a suspicion reinforced by the scenario that Feast suggested at this meeting, of mass loss by 'puffs', so I set $\alpha = 1$, a value often currently used, and found which value of β resulted. These values are also shown in Table II. A small change in β has a large effect on ξ, suggesting that the assumption about the dependence of the grain number density on distance from the star is very sensitive. The values derived here are also very sensitive to the flux correction applied. If a power law is fitted to the colour corrected fluxes α is changed by 0.25 in most cases.

2.2 THE (B - V) VS. (V - [12]) DIAGRAM
Following Waters et al. (1986), a colour-colour diagram was plotted for the sources with 12μm fluxes (from which the 12μm magnitude was found) and available V, (B-V), shown in Fig. 3. The values found are shown in Table I, and the source of the values is given in the notes. The parameter (B-V) shows a property of the star, whilst the (V-[12]) colour reflects the contribution of the dust shell, if present. The line followed by the normal stars, as given by Waters et al. is also shown, valid in the interval -0.25 < (B-V) < 1.60. The separation of the sources into several distinct groups is quite obvious. The Carbon stars are reddest in (B-V) and have the greatest excess at (V-[12]).

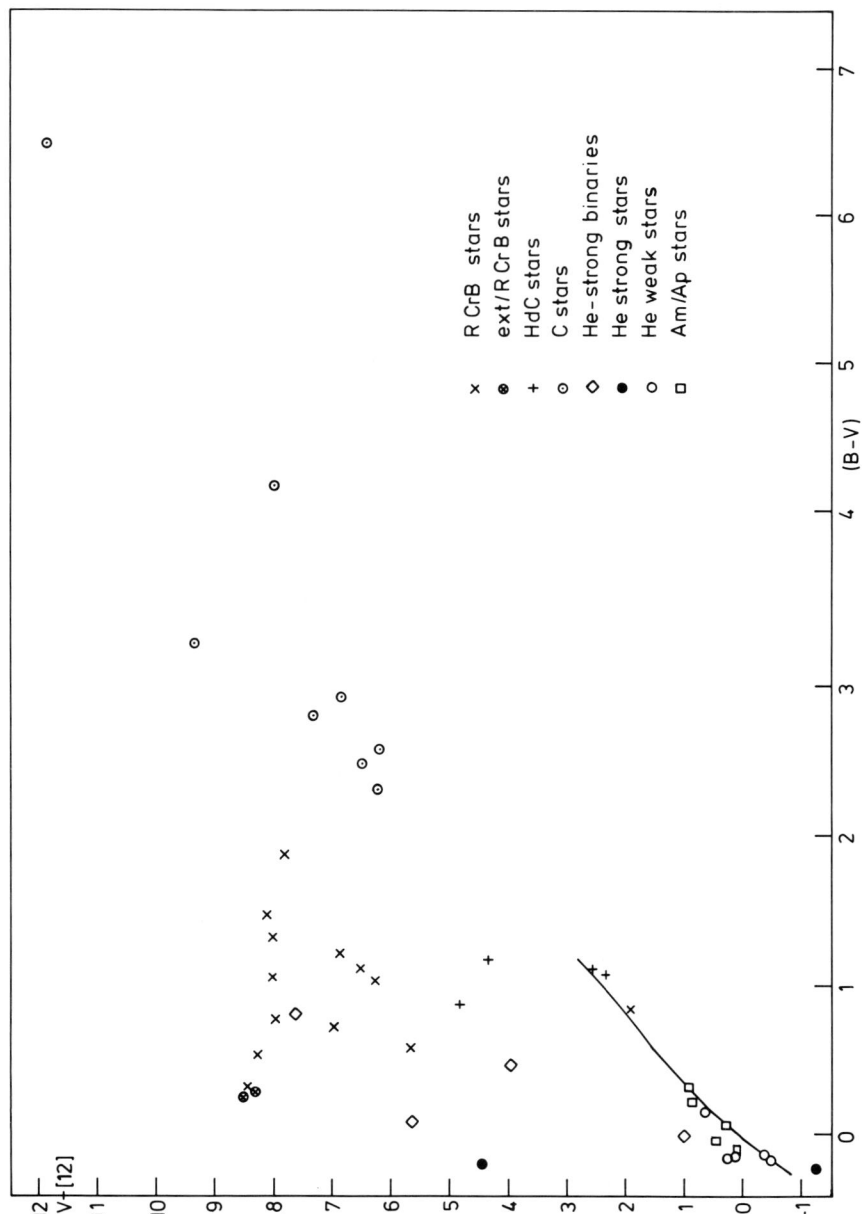

Figure 3 (B-V) vs. (V-[12]) diagram for hydrogen deficient stars and related objects, with the solid line showing the position of normal stars from Waters et al. (1986).

The Am/Ap stars and the helium weak stars follow the line for normal stars, showing that they have no dust shell present. Of the helium strong stars, HD93030 (θ Car) has no evidence of a dust shell, and HD37479 (σ Ori E) shows several magnitudes of excess, confirming the suggestion by Groote and Hunger (1982). They found a shell around HD37479 with a temperature of 270K, which is not very different from the temperature of 200K, found by the 12μm/25μm ratio from IRAS. The HdC stars tend to cluster near the normal stars line, one of the two objects to show an excess is V605 Aql, a rather unique object (Kholopov, 1985). Of the close binaries β Lyr is the star close to the normal stars line, suggesting that the system has little or no dust. This star is the group member with the most nearly normal hydrogen abundance. DY Cen is the R CrB star closest to the two hot R CrB stars, V348 Sgr and MV Sgr, which have the largest (V-[12]) excesses in the group. XX Cam is the R CrB star sitting on the normal stars line. The IRAS observations confirm that it may not be an R CrB star at all, but really an HdC star. This is supported by Rao et al. (1980) who point out that the star has undergone only one minimum between 1898 and 1980.

3. LRS SPECTRA

As reported by Walker (1985) two R CrB stars and one helium rich close binary were bright enough to have LRS spectra available. Fig. 4 shows R CrB and RY Sgr, with the spectra plotted as log (flux) against log (wavelength), together with spectra from some Carbon stars from the General Catalogue of cool Carbon stars by Stephenson (1973). These Carbon stars were retrieved with the help of R. de Grijp (Leiden) and M. de Muizon (Leiden/Paris). The R CrB stars have a very smooth decline in energy with wavelength, to be expected from carbon rich dust. The Carbon stars, characteristically, have a broad emission feature around 11μm, ascribed to SiC at 11.5μm, shown clearly in the spectra of V CrB and V Aql. TX Psc, thought by Goebel and Johnson (1984) to be deficient in hydrogen, looks very similar to the R CrB stars, until the 12μm/25μm ratio is checked, and that reveals that the LRS spectrum is the tail of the stellar energy distribution. U Hya and Y Pav are closest to the R CrB stars with very weak 11μm features. υ Sgr, at the bottom right of Fig. 4, shows an emission feature around 9.5μm, which is attributed to silicates.

4. IRAS ADDITIONAL OBSERVATIONS

Three groups had independently applied to the UK Guest Observer programme for time to observe R CrB stars with IRAS. These groups were led by Evans (Keele), Hill (St. Andrews) and Nandy (ROE), including about twelve people in the proposals, and time was awarded in both observing rounds for the R CrB stars. The group led by Hill was also awarded some observations of the hotter hydrogen deficient

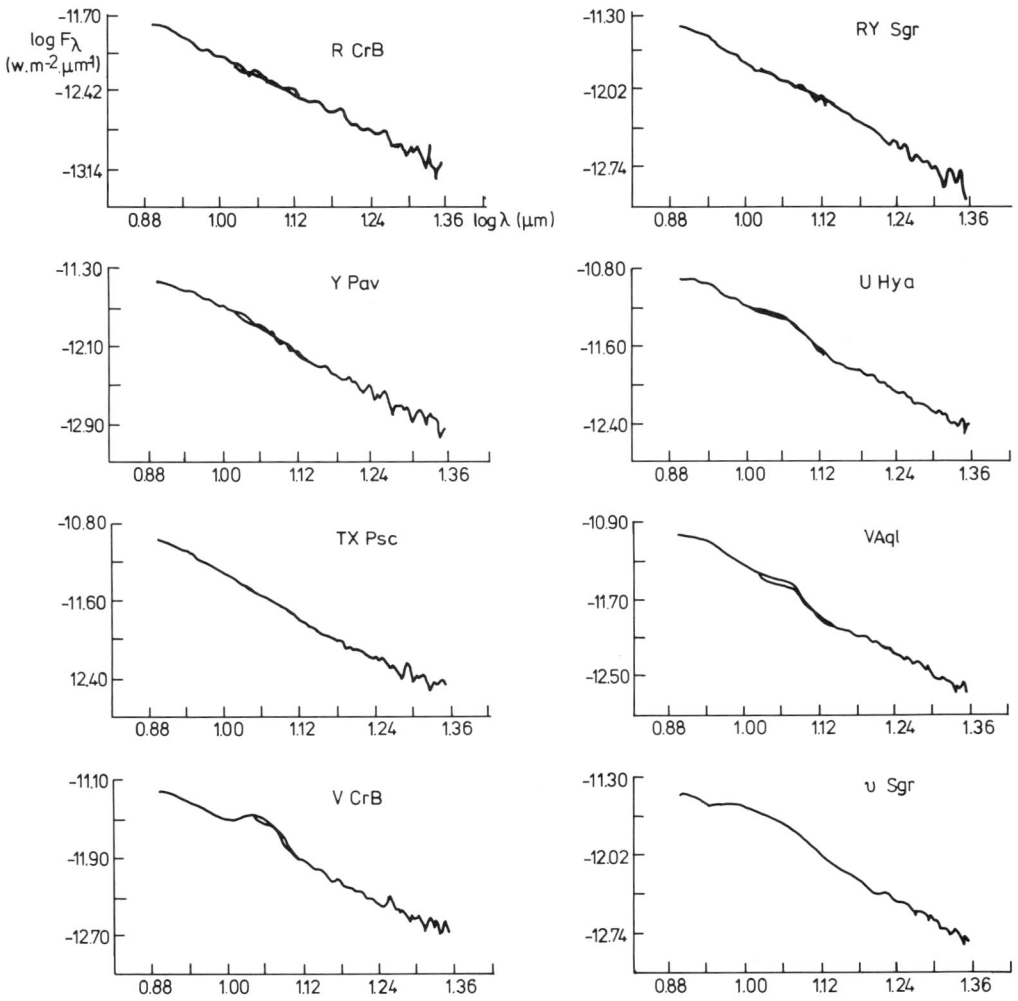

Figure 4 LRS spectra for some hydrogen deficient stars and Carbon stars

stars. As stated earlier, the fluxes derived from the additional observations made with IRAS are not as reliable as the fluxes from the IRAS catalog, so they will not be used here, except in the case of XX Cam where no other fluxes at the IRAS wavelengths are available. The fluxes shown for XX Cam in Table I and Fig. 1 have been corrected using the updated correction factors released in late 1985, but the values may change in the future as the calibration of the additional observations is improved.

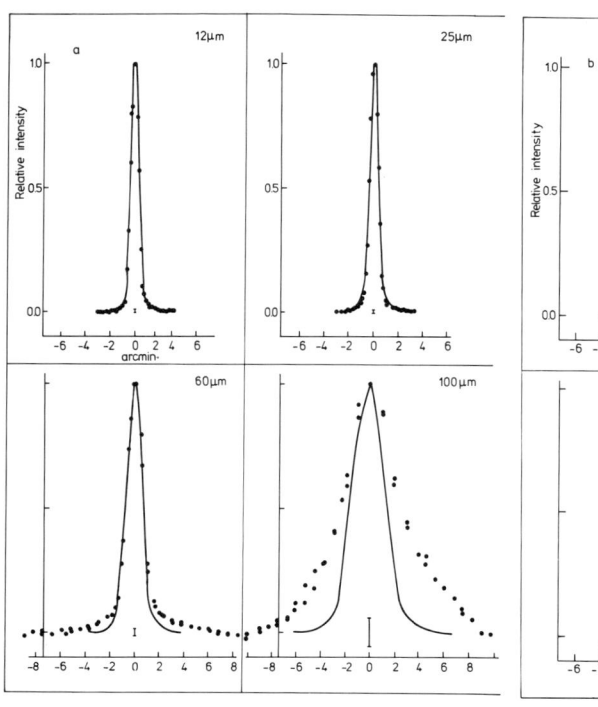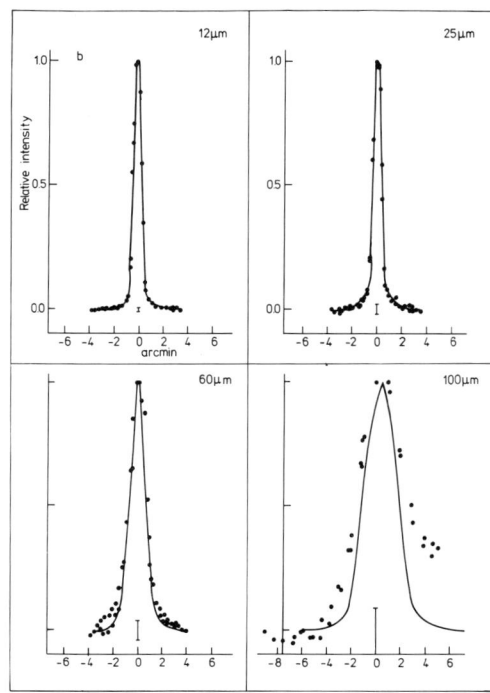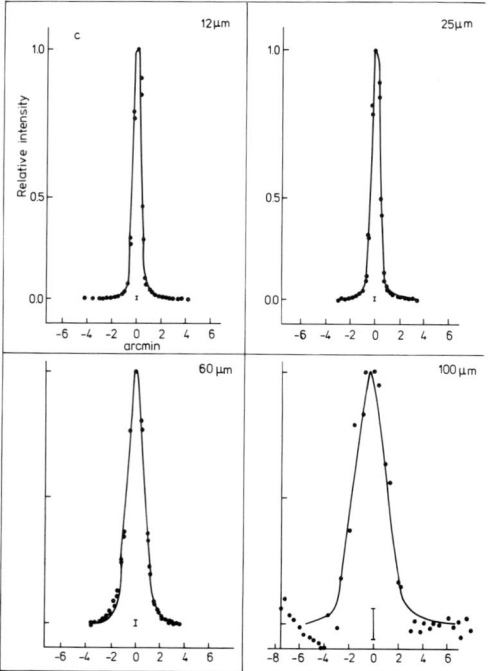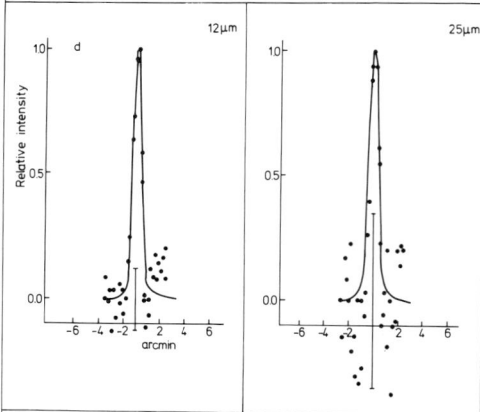

Figure 5 Profiles for some R CrB stars from the IRAS additional observations, with the profile of NGC 6543 shown on each as a solid line.
(a) R CrB   (b) SU Tau
(c) RY Sgr   (d) XX Cam

As someone closely involved in the IRAS mission, having written the command sequences for the additional observations and helped in the scheduling of them, it is a pleasant suprise to find that the scheduling occasionally worked better than expected. The R CrB stars were monitored on the ground during the IRAS mission, in case they went into minimum, and Evans (private communication) recently sent me a graph of the decline in R CrB in 1983, with the IRAS additional observations scheduled when the star was halfway towards minimum light.

Since the additional observations were taken using several different command sequences, involving different satellite speeds, I have reduced the data relative to the calibration source NGC 6543, a planetary nebula at the north ecliptic pole. This source is regarded by IRAS as a point source, although it does sit on a weak plateau of emission at the longer wavelengths. With the aid of P. Schwering (Leiden) I was able to plot the profiles of the sources observed in the additional observation programme, and I measured the full width at half maximum intensity, and also at zero intensity (shown in Table III). Each profile is the result of two independent observations of the source, except for β Lyr and υ Sgr where only one additional observation was taken. Fig. 5 shows the source profiles of R CrB, SU Tau, RY Sgr, and XX Cam, plotted as filled circles, with the profile of the reference point source NGC 6543 drawn in as a solid line. R CrB has a very obviously different profile at longer wavelengths when compared with the reference point source. SU Tau also deviates significantly from the reference point source at 100μm, although the source is weak at the longer wavelengths, and has a close neighbour.

β Lyr is very weak at 100μm. The measurements of the full width at zero intensity obviously have a large error attached, due to the difficulty in judging where the slow decline in flux crosses the background level, but the values derived do show the radical difference in the shape of the profile of R CrB, and to a lesser extent that of SU Tau, which has a width similar to the plateau for NGC 6543. RY Sgr is the better guide to the point source profile at 100μm.

The full width at zero intensity for R CrB is almost 20 arcmin, which implies a very large dust shell. Rao and Nandy (1986) have made an analysis of this large shell and obtained a black body temperature of 30K. They found a dust mass of 4.8×10^{-5} $M_\odot$ (which with a gas to dust ratio of 100 implied a total mass in the shell of about 5×10^{-3} $M_\odot$). There is, however, still the inner extended shell to be studied at 100μm for R CrB and SU Tau. For R CrB, if it is at a distance of 2kpc (as suggested by Schonberner, 1975), the inner shell may be 1pc across, which is a considerable size for a dust shell.

5. CONCLUSIONS

As reported in previous work (for example, Feast and Glass,1973; Kilkenny and Whittet, 1984; Drilling, Landolt and Schonberner, 1984; Walker 1985) the extreme helium stars and HdC stars generally do not

Table III Values for the profile widths from the additional observations

Full-Width-Half-Maximum (arcmin)

source	12μm	25μm	60μm	100μm
NGC 6543	0.85 ± 0.02	0.84 ± 0.02	1.52 ± 0.02	3.03 ± 0.04
RY Sgr	0.81 0.02	0.80 0.03	1.61 0.04	3.11 0.08
R CrB	0.79 0.02	0.78 0.03	1.63 0.04	5.20 0.08
SU Tau	0.84 0.05	0.79 0.08	1.71 0.10	4.31 0.20
XX Cam	0.88 0.08	0.84 0.20	----	----
UW Cen	0.77 0.12	0.77 0.20		
S Aps	0.79 0.12	0.77 0.20		
KS Per	0.78 0.08	0.79 0.08	1.28 0.50	----
υ Sgr	0.81 0.20	0.86 0.20	1.51 0.20	3.37 0.50
β Lyr	0.79 0.20	0.77 0.20	1.42 0.20	1.87 0.60

Full-Width-Zero-Intensity (arcmin)

source	12μm	25μm	60μm	100μm
NGC 6543	5 ± 1	5 ± 1	8 ± 3	11 ± 3
RY Sgr	6	5	6	7
R CrB	5	5	14	18
SU Tau	6	4	7	10
XX Cam	(1)	(1)	–	–
UW Cen	3	4		
S Aps	3	(2)		
KS Per	3	(2)	(2)	–
υ Sgr	4	4	7	9
β Lyr	5	3	3	4

show infrared excesses, and the R CrB stars do show an excess. XX Cam does not have any indication of an infrared excess, and so it is more likely to be an HdC star. MV Sgr has a very cool shell with a temperature of 235K from the 12μm/25μm ratio, in addition to the hotter shell. HD37479 (σ Ori E) has a dust shell with a temperature, from the 12μm/25μm ratio, of 200K. V605 Aql, not an R CrB star but a unique variable, according to Kholopov (1985), has a cool shell of around 160K. R CrB and probably SU Tau appear to have extended dust shells around them, at the longest IRAS wavelengths.

REFERENCES

Blanco, V. M., Demers, S., Douglass, G. G., Fitzgerald, M. P.: 1980, Publ. U.S. Naval Obs. XXI
Buscombe, W.: 1977, 3rd General Catalogue of MK Spectral Classifications, Northwestern University
Buscombe, W.: 1980, 4th General Catalogue of MK Spectral Classifications, Northwestern University
Drilling, J. S., Landolt, A. U., Schonberner, D.: 1984, Astrophys. J. 279, 748
Feast, M. W., Glass, I. S.: 1973, Mon. Not. Roy. Astr. Soc. 161, 293
Fernie, J. D., Sherwood, V., Dupuy, D. L.: 1972, Astrophys. J. 179, 493
Goebel, J. H., Johnson, H. R.: 1984, Astrophys. J. 284, L39
Groote, D., Hunger, K.: 1982, Astron. Astrophys. 116, 64
Hack, M: 1967, Modern Astrophys., publ. Gauthier-Villars, Paris
Herbig, G. H.: 1964, Astrophys. J. 140, 1317
Hoffleit, D., Jaschek, C.: 1982, Bright Star Catalogue (4th rev. ed.), Yale University Obs.
Houziaux, L.: 1968, Bull. Astr. Inst. Czechoslovakia 19, 265
IRAS Explanatory Supplement: 1984, ed. C. A. Beichman, G. Neugebauer, H. J. Habing, P. E. Clegg, T. J. Chester, Jet Propulsion Laboratory
Kholopov, P. N.: 1985, General Catalogue of Variable Stars (4th ed.) Vol I, publ. Nauka, Moscow
Kilkenny, D., Whittet, D. C. B.: 1984, Mon. Not. Roy. Astr. Soc. 208, 25
Osmer, P. S., Peterson, D. M.: 1974, Astrophys. J. 187, 117
Rao, N. K., Nandy, K.: 1986, preprint
Rao, N. K., Ashok, N. M., Kulkarni, P. V.: 1980, J. Astrophys. Astron. 1, 71
Rees, M. J., Silk, J. I., Werner, M. W., Wickramasinghe, N. C.: 1969, Nature 223, 788
Schaefer, B. E.: 1986, preprint
Schonberner, D.:1975, Astron. Astrophys. 44, 383
Stephenson, C. B.: 1973, Publ. Warner and Swasey Obs. 1, no. 4
Walker, H. J.: 1985, Astron. Astrophys. 152, 58

DISCUSSION

HILL: I notice that two of the HdC stars stand above the normal star line in your (V-(12μ))(B-V) relation. Is that significant?

WALKER: I am not sure. One of them comes from Stephenson's Carbon star catalogue with a note in the back that says something about hydrogen deficiency. Again I think those are far above the line which make them incredibly curious and worth following up with near IR photometry.

FEAST: Could I comment on that because at least one of the HdC stars has shown excess; which one it is I don't remember.

HILL: I was wondering how many R CrB stars have been missed.

WALKER: I think there might be some that might be quiescent R CrB stars. XX Cam is one.

FEAST: XX Cam was originally classified as hydrogen deficient and somebody simply went back through the literature and found the light variation.

RAO: Bidelman found it to be HdC and later Yudin showed it to be a variable.

FEAST: It is true that you get quite reasonably good black body fits. If you look at the colors, I think there is more to it than black body fits. In the crude model I have been working on, you would expect the longer wavelength to come from further away and from lower temperatures. From the very few computations I have done for the brighter ones of the IRAS catalogue, for which we got shortward data as well from ground, I would expect λ^{-1} emissivity to fit better than black body. Clearly that warrants a detailed model. It is unrealistic to assume that the soot particles radiate as black bodies. Since the 3.5μ flux of the R CrB shells vary substantially with time, it will be important to put together the IRAS results with near simultaneous ground based data.

WALKER: I started to do some modelling along the lines of Harvey et al., 1979. As you say λ^{-1} is much better for energy distribution variation than r^{-2}, I am using up till now.

POTTASCH: Aren't you afraid in modelling these things, for you have too many free parameters, that you can get a fit with various gradients of temperatures and densities.

WALKER: Yes, it is difficult to do with only four wave length points, when there are 5 free parameters.

WING: A comment: I think I called the temperature as 700 K for the dust shell around R CrB in 1972 in a paper published in PASP. This was based on 1 to 10μ photometry obtained during a visual minimum.

FEAST: I called it 700 K in 1969 (Laughter). This agrees very well with your IRAS result and suggests that the grain temperature may be reasonably constant.

WING: In cases where the IRAS data don't fit the same black body curve as the ground based data - which you suggest may indicate two shells at two different temperatures - couldn't the interpretation equally well be the time variability of the amount of dust in the shell? This was, I believe, the interpretation that Feast gave on Tuesday for the large, slow variations that he has observed in the L magnitude.

WALKER: Yes, It seems as though the variability gets less at the longer wavelengths. You find dramatic variations in the visual, not much in L. I am not sure whether it is reflected in the IRAS domain, but you do not expect here much variability.

RAO: The R Cor Bor dust temperature varies even at maximum light from 600 to 900 K, as shown in Forrest's thesis. But surprisingly, the IR distribution always reflected a black body.

WALKER: You do have to be careful when you make color corrections in the IRAS fluxes, it is based on the shape of the energy distribution. It shouldn't normally change very much. On occasions, you find the temperature is altered quite significantly. I think the same could happen to values of α which is more than the color corrections. This means that the fluxes themselves will get affected.

VARDYA: In the spectra that you showed of R Cor Bor I think there is no feature at 11.3μ; but around 10μ there is a small blip. I do not know whether it is in absorption or emission, what could that be?

WALKER: I think it is just noise.

POTTASCH: There may be as many as 6 individual LRS spectra, so may be that could be checked.

WALKER: I have Mane de Muizon (at Leiden/Paris) with me who is experienced with faint features. She agrees with me: these are not real features – but some spectroscopists never give up!

IX THEORY

PULSATIONS OF HYDROGEN DEFICIENT STARS

Hideyuki SAIO
Department of Astronomy
University of Tokyo
Tokyo 113
Japan

ABSTRACT. A review is given on the properties of the pulsating hydrogen-deficient stars. Since they lie in a large area of the HR diagram, the properties of pulsations are greatly different among them. Pulsations tend to change from radial pulsations to non-radial pulsations as the surface gravity increases. Strong nonadiabaticity affects the dynamical properties of pulsations in high luminosity helium stars. The kappa and gamma mechanisms for helium ionization excite radial pulsations in relatively cool helium stars and nonradial pulsations in DB (helium atmosphere) white dwarfs, while the cyclical ionization of the K-shell electrons of carbon and oxygen seems to be responsible for the excitation of the nonradial pulsations in the GW Vir (PG1159-035) stars (very hot hydrogen-deficient pre-white dwarf stars). The excitation mechanism of the radial pulsations in a unique extreme helium star V652 Her (BD+13°3224) is not known. Period changes have been detected for many pulsating helium stars except the pulsating DB white dwarfs.

1. INTRODUCTION

The first periodic light variation in a hydrogen deficient star was discovered by Jacchia (1933, quoted by Alexander et al. 1972). Jacchia found RY Sgr, an R Coronae Borealis (R CrB) star, to show a semi-regular variation with an average periodicity of 39 days. The number of the presently known pulsating hydrogen deficient stars is about 15 or more. Some of the pulsating hydrogen deficient stars and their pulsation periods are shown in the HR diagram in Figure 1. R CrB and RY Sgr are the best studied pulsating R CrB stars. Pulsations of most of the other stars in this figure were discovered recently. The pulsations of the exteme helium stars, BD+1°4381, BD-1°3438, and BD-9°4395 were discovered by Jeffery and Malaney (1985), Jeffery, Hill, and Morrison (1986), and Jeffery et al. (1985), respectively. Landolt (1975) discovered the light variation of a unique hydrogen deficient star V652 Her (BD+13°3224) which is less luminous than the above stars by about one order of magnitude. Grauer and Bond (1984) found the central star of the planetary nebula Kohoutek 1-16 (k1-16) to be a pulsating variable. The spectrum of K1-16 is similar to that of GW Vir (PG1159-035) whose variability was discovered by McGraw et al. (1979). Winget et al. (1982b) discovered that GD 358, a helium atmosphere (DB) white dwarf, is a pulsating variable star.

These pulsating hydrogen-deficient stars may be classified into the following groups; 1) pulsating high luminosity helium stars (pulsating R CrB stars and extreme helium stars), 2) V652 Her (BD+13°3224), 3) the GW Vir (PG1159-035) stars including the pulsating planetary nebulae nuclei, and 4) pulsating DB white dwarf stars. Each group represents different evolutionary phase. The pulsation properties for each of these groups are reviewed in the following sections.

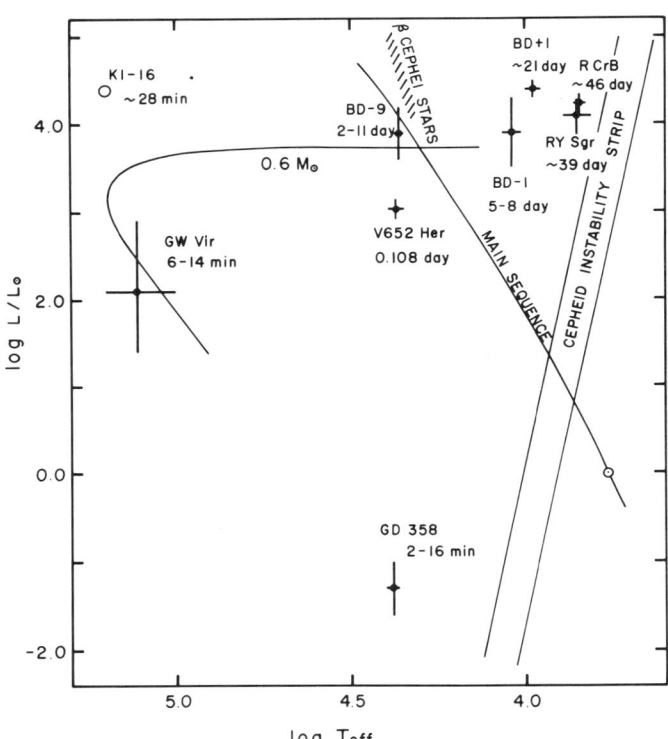

Figure 1. Location of pulsating hydrogen-deficient stars on the HR diagram. The periods of pulsations are shown below the names of the stars (BD+1=BD+1°4381, BD-1=BD-1°3438, BD-9=BD-9°4395). In addition, the evolutionary track of 0.6 $M_\odot$ pre-white dwarf star computed by Schönberner (1979), the Cepheid instability strip, the Beta Cephei instability zone, and the Pop.I main sequence are shown.

2. PULSATIONS OF HIGH LUMINOSITY HELIUM STARS

2.1. Introduction

Many R CrB stars show semi-regular light variations with periods of a few tens of days (Feast 1975, Rao, Ashok, and Kulkarni 1980). An intensive investigation by Alexander et al. (1972) on the variability of RY Sgr revealed that its semi-regular variations are due to stellar pulsations. Most of the extreme helium stars seem to have small amplitude light variations (e.g., Walker and Kilkenny 1980). However, periods are obtained only for three or four stars (other than V652 Her, which is discussed in §3). These stars have luminosities $L/L_\odot \sim 10^4$ similar to those of the R CrB stars, and the latter stars are believed to be the immediate progenitors of the former stars (Heber and Schonberner 1981). The rapid period decrease of RY Sgr (dln P/dt $\sim -10^{-11}$/s) obtained by Kilkenny (1982) and Marraco and Milesi (1982) supports this evolutionary connection. The period changes of other R CrB stars are, however, somewhat confusing: Kilkenny and Flanagan's (1983) analysis suggests that the period of UW Cen might be increasing. Furthermore, Kilkenny (1983) found that the period of pulsations of S Aps had changed from about 120 days to about 40 days and that the ~ 120 day variations exhibited a period decrease while the ~ 40 day period seemed to increase with time.

Some properties of well studied pulsating high luminosity helium stars are listed in Table 1, in which P and Q are, respectively, pulsation period and the pulsation constant ($Q=P(R/R_\odot)^{-1.5}(M/M_\odot)^{0.5}$). The luminosities and masses of R CrB, RY Sgr, and BD+1°4381 were obtained by incorporating the effective temperatures and the pulsation periods with theoretical linear pulsation periods calculated by Saio, Wheeler, and Cox (1984), and a mass luminosity relation

$$\log (L/L_\odot) = 3.2 \log (M/M_\odot) + 4.6, \qquad (1)$$

which is based on Schönberner's (1977) models for post-giant helium stars. Equation (1) is applicable for $\log(L/L_\odot) \gtrsim 4$. Since the mode identifications to the observed periods of BD-1°3438 and BD-9°4395 are uncertain, the luminosities of these stars were obtained from the surface gravities with the effective temperatures and equation (1). Note that the listed ranges of luminosity and mass for RY Sgr and R CrB are considerably narrower than those given in Saio and Wheeler (1983), because by using equation (1), it is assumed that the total mass is nearly equal to the CO core mass.

The theoretical investigation of the pulsations of the luminous helium stars was started by Trimble (1972). A good review of the earlier theoretical works on the pulsations of R CrB stars is given in King (1980).

2.2 Very nonadiabatic radial pulstations

Pulsations in high luminosity helium stars are extremely nonadiabatic (Trimble 1972, Wood 1976, Cox et al. 1980, and Saio, Wheeler, and Cox 1984). Since the thermal time scale is roughly proportional to the mass-luminosity ratio M/L, the nonadiabaticity of pulsations can be significant in a high L/M star (L/M $\sim 10^4 L_\odot/M_\odot$). When the nonadiabaticity is very strong, the dynamical properties of pulsations deviate significantly from what we know for classical pulsators like Cepheids and RR Lyrae variables. The periods of pulsations deviate from the corresponding adiabatic periods. Furthermore, one to one correspondence between adiabatic and nonadiabatic modes is destroyed by the appearance of new nonadiabatic modes (Wood 1976). It should be mentioned here that Shibahashi and Osaki (1981) found that the breakdown of the one to one correspondence occurs for <u>nonradial</u> pulsations, too, in $L/M \sim 10^4 L_\odot/M_\odot$ models.

The nonadiabatic period which is close to the adiabatic period of the k-th overtone in the low effective temperature models tends to approach the adiabatic period of the (k+1)-th overtone in the high effective temperature models. Saio, Wheeler, and Cox (1984) showed that this phenomenon occurs because the nonadiabaticity of a pulsation mode increases as the effective temperature of the model increases. They also showed higher overtone modes to be more nonadiabatic.

Since the thermal timescale is comparable to the dynamical timescale, the growth or damping time can be comparable to the pulsation period. Moreover, the periods of thermal-dynamical damping oscillations which are called "strange modes" enter in the same range of the periods of dynamical (ordinary) pulsations. The properties of the strange modes are discussed in detail in Saio and Wheeler (1982), and Saio, Wheeler, and Cox (1984). When an ordinary pulsation mode has a period close to that of a strange mode, the ordinary mode tends to be stablized. This causes a "stable strip" in the instability region for the first overtone modes in the HR diagram (Saio, Wheeler, and Cox 1984).

Table 1. Properties of pulsating extreme helium stars and GW Vir

name	log L	log T_{eff}	period (day)	$M/M_\odot$	Q (day)	ΔM_V (mag)	ΔV (km/s)	ref.
R CrB	4.3±0.1	3.845±.015	46±5	0.8±.1	.044	.1-.2	∼4	1,2,3
RY Sgr	4.1±0.2	3.850±.037	39	0.8±.1	.046	.5	40	4,5,6
BD+1°4381	4.4±0.1	3.977±.018	22	0.85±.05	.045	.06		7,8,9
BD-1°3438	3.9±0.4	4.037±.024	5-8	(0.6)	(.03-.05)	.07		7,9,10
BD-9°4395	3.9±0.3	4.362±.013	2-11	(0.6)	(.1-.6)	.03	∼10	7,11,12
V652 Her (BD+13°3224)	3.0±0.1	4.370±.024	.108	0.75+.3−.2	.034	.06	70	13,14

1) Cottrell and Lambert (1982)
2) Fernie (1982)
3) Fernie, Sherwood, and DuPuy (1972)
4) Schönberner (1975)
5) Alexander et al. (1972)
6) Lawson (1985)
7) Drilling et al. (1984)
8) Jeffery and Malaney (1985)
9) Jeffery, Hill, Morrison (1986)
10) Schönberner (1978)
11) Jeffery et al. (1985)
12) Kaufmann and Schönberner (1977)
13) Hill et al. (1981)
14) Lynas-Gray et al. (1984)

The strong nonadiabaticity also has a destabilizing effect on pulsations and modifies the instability boundary in the HR diagram significantly. Figure 2 shows the location of the blue edges of the fundamental mode instability region in the HR diagram for helium star models with 90% helium and 10% carbon by mass (Saio, Wheeler, and Cox 1984). (The fundamental mode and a k-th overtone mode are conventionally defined in this paper as the modes which approach the adiabatic fundamental mode and the k-th overtone mode, respectively, in the limit of low effective temperature.) Below a critical luminosity the blue edge of the instability region due to the helium ionization is very similar to that for the Cepheids but becomes nearly horizontal in the HR diagram above the critical luminosity. The critical luminosity is higher for a larger mass, because the effect of nonadiabaticity is stronger in a higher L/M star. This extension of the pulsationally unstable region is interpreted as follows: The vibrational stability is roughly determined by the balance between the driving in the helium ionization zone and the radiative damping in the region interior to the ionization zone. When the luminosity is sufficiently high, the effect of the radiative damping decreases because the local thermal timescale becomes very short there. (The luminosity perturbations are "frozen in" in space [Cox 1974].) Then the excitation due to the ionization zone exceeds the radiative damping even in the models hotter than the bule edge of the classical instability strip (Saio, Wheeler, and Cox 1984). The decrease of the effect of the radiative damping occurs also in high luminosity models with hydrogen rich envelopes (Fadeyev and Fokin 1985, Zalewski 1985).

Several hydrogen deficient stars are also plotted in Figure 2, where the symbols ●, ∆, and o are for (semi-)periodic variables, suspected variables, and non-variables, respectively. [It should be noted here that only BD+10°2179 has been confirmed to have no light variation (Hill Lynas-Gray, and Kilkenny 1984, Grauer, Drilling and Schönberner 1984).] The blueward excursion of the blue edge of the instability region accounts for the existence of pulsations in R CrB stars and BD+1°4381, which are fundamental mode pulsators. BD-1°3438 seems to reside outside the instability region for the fundamental mode, but inside the instability region for the first overtone mode. The instability region for the first overtone mode extends into less luminous region than the fundamental mode instability boundary (Saio, Wheeler, and Cox 1984). Actually, the period of BD-1°3438 seems to be consistent with the period of the first overtone mode. Nonradial g-mode pulsations seem to be excited in BD-9°4395 (Jeffery et al. 1985) and V2076 Oph (HD160641, Lynas-Gray et al. 1986). [The latter star (T_{eff}=31900±1500; Drilling et al. 1984) is not shown in Fig. 2 because its surface gravity is not available.] No theoretical investigation of the nonradial pulsations in such stars has not, to the authers' knowledge, been published. We need extensive theoretical investigations of radial and nonradial pulsations in hot (10^4K $\lesssim T_{eff} \lesssim 3 \times 10^4$K) hydrogen deficient stars.

2.3 Nonlinear pulsations

The nonlinear hydrodynamic calculations for the luminous helium stars had posed a serious difficulty. Pulsations of 1 $M_\odot$ models with T_{eff}

∼6000 K and $L \sim 10^4 L_\odot$ are so violent that the amplitude seems to grow until the expansion velocity attains the escape velocity (Trimble 1972, Wood 1976, King et al. 1980). Pulsations of lower L/M models are less violent. King et al.(1980) showed that for models with T_{eff}=6300 K, $L=1.13 \times 10^4 L_\odot$ and $M \geq 1.6 M_\odot$ pulsations attained stable limit cycles, of which light and velocity curves were quite regular. According to the result, they suggested that the mass of a pulsating R CrB star should be larger than ∼1.4 $M_\odot$. However, this conclusion contradicts the result of Schönberner (1977) who obtained the typical mass of R CrB stars to be about 0.7 $M_\odot$ by using the spectroscopically determined surface gravities and evolutionary models with a CO core and a thin helium envelope.

The mass discrepancy of the pulsating R CrB stars was solved by Saio and Wheeler (1985) who showed that the nonlinear pulsation properties are sensitive to the effective temperature of the models. In the previous nonlinear calculations models with $T_{eff} \sim 6000$ K are adopted, while the spectroscopic analyses by Schönberner (1975) and Cottrell and Lambert (1982) show the effective temperatures of R CrB and RY Sgr to be

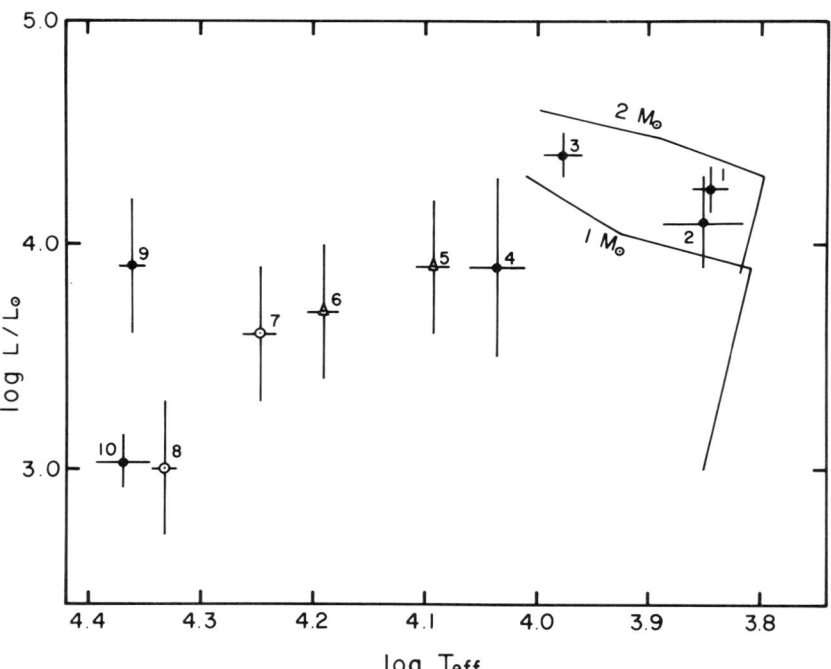

Figure 2. The blue edges of the fundamental mode instability region as a function of mass for helium star models with 90% helium and 10% carbon by mass, and location of extreme helium stars on the HR diagram. The symbols, ●, △, and o indicate (semi-)periodic variables, suspected variables and non-variables, respectively. The names of the plotted stars are 1=R CrB, 2=RY Sgr, 3=BD+1°4381, 4=BD-1°3438, 5=PV Tel (HD168476), 6=HD124448, 7=BD+10°2179, 8=HD144941, 9=BD-9°4395, and 10=V652 Her (BD+1°3224).

about 7000 K. Saio and Wheeler (1985) showed that if $T_{eff} \sim$ 7000 K was adopted, the amplitude of pulsations remained bounded even in models with $M < 1\ M_\odot$. The light and the velocity curves are irregular for low mass models ($M \lesssim 1.4\ M_\odot$), while pulsations in the 2 $M_\odot$ model are quite regular. Thus, pulsations in low mass models are consistent with the observed irregular light and velocity curves of pulsating R CrB stars. The growth of pulsation amplitude in a low mass model with $T_{eff} \sim 7000$ K is restrained by a strong shock wave which dissipates the kinetic energy of pulsations. Such a strong shock wave does not appear in the models with $T_{eff} \sim 6000$ K.

Figure 3 shows the light and velocity curves obtained by Lawson (1985) for RY Sgr and the theoretical ones for the model with $L=1.7\times10^4 L_\odot$, $T_{eff}=7200$ K, $M=0.9\ M_\odot$, and a chemical composition of 90% helium and 10% carbon by mass (Saio and Wheeler 1985). The observed velocity amplitude of RY Sgr is similar to the theoretical one. Although the bolometric correction is not applied in this figure, the observed amplitude of the light variations seems to be considerably smaller than the theoretical one, and the theoretical phase relation between light and velocity curves seems not to agree with the observed one.

Wood (1976) calculated the nonlinear <u>first-overtone</u> pulsations for a model with $L=10^4 L_\odot$, $M=1\ M_\odot$, and $\log T_{eff}=4.0$. The period of the pulsations is 5 days, which is comparable to the period of BD-1°3438. He obtained the amplitudes, $\Delta M_{bol} \sim 1$ mag and $\Delta V \sim 40$ km/s. Again, the theoretical amplitude in luminosity seems too large compared to the observed

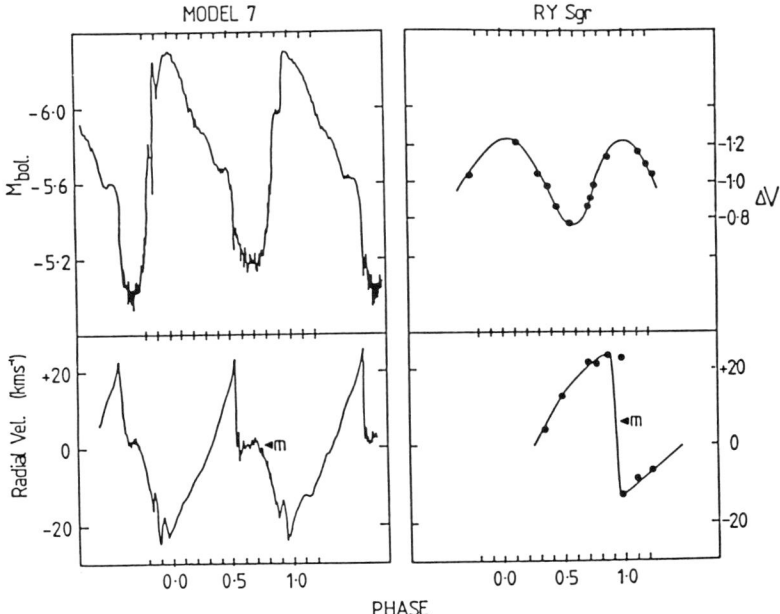

Figure 3. The light and velocity curves of Model 7 ($M=0.9 M_\odot$, $T_{eff}=7200$ K, $L=1.7\times10^4 L_\odot$) of Saio and Wheeler (1985) and observed ones by Lawson (1985). From Figure 7 of Lawson (1985).

amplitude of BD-1°3438. Accurate bolometric correction as a function of the pulsation phase is needed for detailed comparison between theoretical and observational light curves. Moreover, it is also needed to improve theoretical nonlinear pulsation models. For example, the effect of mass loss may affect the amplitude of pulsations.

3. PULSATIONS OF V652 HER (BD+13°3224)

The hot extreme helium star V652 Her is different from other extreme helium stars in many respects. The hydrogen abundance, $n_H=0.01\ n_{He}$ (Hill et al. 1981), and the surface gravity, $\log g = 3.7$ (Lynas-Gray et al. 1984) of V652 Her are higher than those of the extreme helium stars discussed in §2. Besides, its luminosity ($L/L_\odot \sim 10^3$: Lynas-Gray et al. 1984) is lower than the extreme helium stars by about one order of magnitude. Lynas-Gray et al. (1984) indicated that V652 Her is comparable with HD144941 which has $n_H = 0.07\ n_{He}$ and $\log g = 3.5$ (Hunger and Kaufmann 1973). Figure 2 shows that both stars are closely located in the HR diagram, although no periodic variation has been detected for HD144941. V652 Her shows regular variations in light and radial velocity (e.g., Lynas-Gray et al. 1984, Jeffery and Hill 1986) with a period of 0.108 day. The amplitudes in visual magnitude and radial velocity are listed in Table 1 with other quantities. The radial velocity curve is very similar to those of the classical Cepheids (see e.g. Cox (1974) for a review of the pulsations of the classical Cepheids). Therefore, we can safely say that the regular variations of V652 Her are due to radial pulsations.

The observed luminosity and effective temperature of V652 Her can be reproduced by a "helium horizontal branch" model (Jeffery 1984), which consists of a core with a mass $\sim 0.45\ M_\odot$, hydrogen burning shell, and an envelope with the observed hydrogen abundance. Jeffery (1984) also showed that the rapid evolution of the model is consistent with the rate of period change (dln P/dt $\sim -10^{-11}$/s) obtained by Kilkenny and Lynas-Gray (1982, 1984). The mass of the models which can reproduce the observed luminosity and the effective temperature is in the range of 0.6 - 0.7 $M_\odot$, which agrees with $0.7^{+0.4}_{-0.3} M_\odot$ obtained by Lynas-Gray et al. (1984) from the surface gravity.

Another way to estimate the mass of this star is using the pulsation period. Linear nonadiabatic pulsation periods were obtained for the helium horizontal branch models and some envelope models with excess carbon abundance ($Y=0.9$, $X_C=0.1$; X_C=mass fraction of carbon). (All the models considered turned out stable to radial pulsations. The stability will be discussed later.) The results are plotted in the log Q - log (M/R) plane in Figure 4. The two sequences are for the fundamental (F) and the first overtone (1H) modes. Since the luminosity of the star is about $10^3 L_\odot$, the effect of nonadiabaticity on the dynamical properties of pulsations is small and the pulsation periods are very close to the corresponding adiabatic ones. The period and the mean radius of V652 Her give a relation between Q and M/R, which is shown by the solid curve in Figure 4. From this figure we obtain a mass of $0.75^{+0.3}_{-0.2} M_\odot$ if the fundamental mode is assumed (about 0.5 $M_\odot$ if the first overtone is assumed).

The obtained mass (assuming the fundamental mode) is consistent with the results of other estimations (see above), which confirms the validity of the radius and luminosity obtained by Lynas-Gray et al. (1984).

Although the pulsation period is thus consistent with other theoretical and observational results, no radial pulsation is excited in the models with luminosities and effective temperatures similar to those of V652 Her. The effective temperature is too high for the helium-ionization driving mechanism to be effective. The epsilon mechanism due to the hydrogen burning shell is negligible because the fractional radius at the burning shell (r/R ∼ 0.1) is too small. The chemical composition could change from the observed surface composition to a carbon/oxygen-rich mixture at a layer slightly deeper than the photosphere as the models considered by Starrfield, et al. (1984) for the GW Vir variables (§4). In this case we can expect some driving effect for pulsations due to the high order ionization of carbon and oxygen. However, such a model seems unlikely because there is no indication of carbon enrichment in the atmosphere of V652 Her (Lynas-Gray et al. 1984).

Another possibility is the excitation due to Stellingwerf's (1978) helium opacity bump which exists around $T=1.5 \times 10^5$K (Osaki 1982, Cox 1985). This bump is confined in a small range of temperature so that it

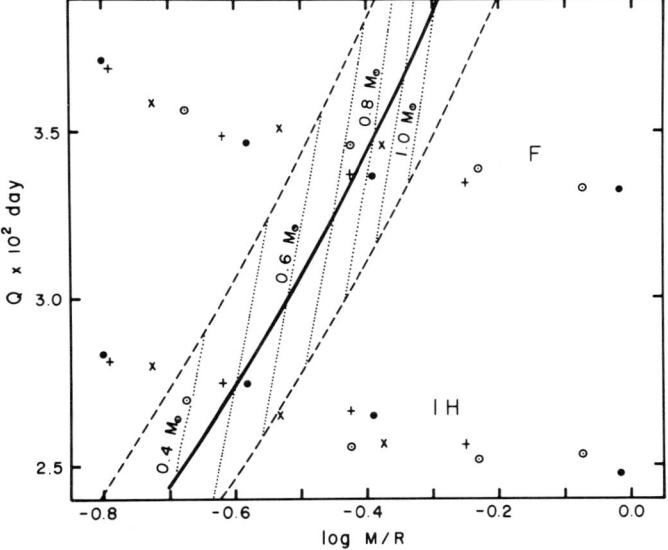

Figure 4. The pulsation constant Q for the fundamental mode (F) and the first overtone mode (1H) versus log M/R (M and R are in the solar units). The symbols ● and + are for the 'helium horizontal branch' models ($M_c=0.475$ $M_\odot$) with X=0.005 and X=0.002, respectively, and o and x are for the envelope models (L=$10^3 L_\odot$, 1.5×10^4K $\leq T_{eff} \leq 3 \times 10^4$K; Y=0.9, X_C=0.1) with 1 $M_\odot$ and 0.5 $M_\odot$, respectively. The solid curve shows the relation which is satisfied by the period and the mean radius of V652 Her. The dashed curves indicate allowable region due to probable error in the mean radius. The dotted lines indicate the loci of constant masses.

is hardly apparent in a coase grid opacity table. Therefore, the effect of the opacity bump was negligible in the models presented in the previous paragraphs. Stellingwerf (1978) showed that the opacity bump is most effective (but still unable to excite pulsations) in models with effective temperatures similar to those of the Beta Cephei stars. The effect of the opacity bump in models of V652 Her is worthy of studying in detail, because the effective temperature of V652 Her is similar to those of the Beta Cephei stars and it has a helium-rich envelope.

Very accurate light and velocity curves of V652 Her have been obtained by Lynas-Gray et al. (1984) and Jeffery and Hill (1986). In particular, Jeffery and Hill have detected high frequency structure in the radial velocity curve. It is unfortunate for us not to have a theoretical pulsation model which can be compared with these very interesting data.

4. GW VIR (PG1159-035) STARS

Since the discovery of light variations in GW Vir (PG1159-035) by McGraw et al. (1979), a few more GW Vir variables (very hot hydrogen deficient pre-white dwarf stars including the nuclei of planetary nebulae) have been discovered (Grauer and Bond 1984, Bond et al. 1984). The known GW Vir variables are listed by Van Horn (1984). The atmospheric parameters lie in the ranges of $8 \times 10^4 K < T_{eff} < 1.6 \times 10^5 K$ and $\log g \gtrsim 7$ for GW Vir (Wesemael, Green, and Liebert 1985), and $T_{eff} > 8 \times 10^4$ K and $\log L/L_\odot > 2.5$ for the central star of the planetary nebula K1-16 (Kaler 1983). If we assume $M=0.6 M_\odot$, $T_{eff}=(1.3 \pm 0.3) \times 10^5 K$ and $\log g = 7.5 \pm 0.5$ for GW Vir, we obtain $\log L/L_\odot = 2.1 \pm 0.8$. The dominant period of the pulsations of K1-16 is 28.3 minutes (Grauer and Bond 1984), which is considerably longer than the periods of other GW Vir variables (6 - 14 minutes). This is consistent with a preliminary luminosity estimate of $L=2.5 \times 10^4 L_\odot$ obtained by Kaler and Feibelman (1984, quoted by Starrfield et al. 1985) for K1-16. In the HR diagram these stars are located around the "knee" of the evolutionary tracks of the pre-white dwarf phase (e.g., Shönberner 1979, and Iben 1984).

Since the periods of these stars are much longer than their free-fall times (10-100 s), the observed pulsations must be higher order nonradial g-mode pulsations (for reviews of nonradial pulsations see e.g., Unno et al. 1979, Cox 1980). The stability of nonradial g-mode pulsations was examined for the GW Vir stars by Starrfield et al. (1984, 1985). The luminosity of the model for a given effective temperature was adopted from the evolutionary track of $0.6 M_\odot$ model computed by Schönberner(1979). They found that if the matter in the temperature range from $2 \times 10^5 K$ to $3 \times 10^6 K$ consists of carbon and oxygen, say half carbon and half oxygen by mass, kappa and gamma mechanisms due to the cyclical ionization of the K-shell electrons of carbon and oxygen can excite high order g-mode pulsations ($k \sim 10 - 50$; k indicates the number of radial nodes of eigenfunctions). The periods of the unstable modes are $\sim 2- 20$ min for models with $5.6 \leq L/L_\odot \leq 410$ and $\sim 15- 70$ min for models with $1.3 \times 10^3 \leq L/L_\odot \leq 2.8 \times 10^3$. Although these period ranges of unstable g-modes are consistent with the observed periods, the number of the detected periods for a star (Winget et al. [1985] obtained eight periods for GW Vir.) is much smaller than the number of the unstable modes.

Moreover, Kawaler et al. (1986) showed that the helium-burning shell excite some g-mode oscillations whose periods range from 50s to 214s in evolutionary models of hydrogen-deficient planetary nebula nuclei ($2.0 \leq \log L/L_\odot \leq 3.2$). Such a short period variation has not been detected in K1-16. The possible interpretations for this discrepancy are; 1) there exists a mechanism which prevents the amplitudes of such unstable modes from growing large enough to be observable, 2) the basic characteristics of the evolutionary models are incorrect and the He-burning shell is extinguished in the real planetary nebula nuclei. Further investigation is needed to find the answer for the problem.

The possibility of detecting period change of the pulsations of GW Vir was first indicated in the discovery paper McGraw et al. (1979). Later Winget, Hansen, Van Horn (1983) have shown that the period change should be measurable on a time scale of 1-3 yr. Actually, Winget et al. (1985) measured a period change of $d\ln P/dt = (-2.3 \pm 0.2) \times 10^{-14}$ s^{-1} for the 516 s period of GW Vir. The magnitude of the measured rate of period change is consistent with the theoretical estimation for high order g-modes in cooling pre-white dwarf models, but has the opposite sign (Kawalar, Hansen, and Winget 1985). This discrepancy prompted Kawalar, Winget, and Hansen (1985) to consider the effect of rotational spin-up, produced by gravitaional contraction, on the rate of period change. For a star rotating with a rotation frequency Ω, an observed pulsation frequency σ_{obs} is given by

$$\sigma_{obs} = \sigma_0 - m\Omega(1-C), \qquad (2)$$

where σ_0 is the pulsation frequency obtained assuming the star not to rotate, and m is the azimuthal index of the spherical harmonic $Y_\ell^m(\theta,\phi)$, which represents the angular behavior of the pulsation eigenfunction. C is a positive number (< 1) which depends on the structure of the star and the pulsation eigenfunction. For a pulsation mode with m < 0, rotational spin-up has a tendency to decrease the rate of period change. Therefore, if m has a sufficiently large negative value, or the rotation speed is fast enough, the effect of rotational spin-up can exceed the rate of the period change due to evolutionary change of the structure. Kawalar, Winget, and Hansen (1985) found that the rate of period change for a pulsation mode with $m \lesssim -2$ can be consistent with the measured rate of period change of the 516 s period of GW Vir if the star rotates with a period of a few thousand seconds.

5. PULSATING DB WHITE DWARFS

Winget et al. (1982a) predicted that g-mode pulsations should be excited by the He-ionization mechanism in DB white dwarfs. Subsequently, Winget et al. (1982b) discovered the light variation of DB white dwarf GD 358. Three more variable DB white dwarfs have been found. The known variable DB white dwarfs are listed by Van Horn (1984). Winget et al. (1982b) found 28 pulsation modes in the variations of GD 358 with periods ranging from 142.3 s to 952 s. The period range is similar to that of the ZZ Ceti stars (e.g., Winget and Fontaine 1982, Van Horn 1984). Many of the pulsation frequencies of GD 358 fall into groups of 4 or 5 modes that are equally spaced. If the equal spacing is due to the rotational mode splitting (m-splitting; see eq. 2), the observed spacing between two adjacent frequencies implies a rotation period of about 90 min (Winget et al. 1982b), which corresponds to a equatorial rotational velocity of ∿10 km/s. Another variable DB white dwarf PG 1654+160 seems to have some equally spaced frequencies. The frequency spacing gives a rotation period of ∿150 min (Winget, et al. 1984). Thus, determining the periods of pulsations in DB white dwarfs can be a powerfull means to obtain their rotation rates.

Winget et al. (1983) showed that the effective temperature of the models at the blue edge of the instability region is extremely sensitive to the assumed efficiency of convection in the evolutionary models. In a star hotter than the blue edge, the helium ionization zone is located so close to the stellar surface that its driving effect is no longer strong enough to excite pulsations (e.g., Cox 1974). A larger efficiency of convection reduces super-adiabatic temperature gradient in the convective zone, and hence makes temperature variation less steep. So, the helium ionization zone in the convection zone is pushed to a deeper zone. Therefore, the effective temperature at the instability boundary in the HR diagram is increased by an increase of the efficiency of convection. Winget et al. (1983) found that in order for the theoretical instability blue edge to be consistent with the effective tempearature of GD 358 (24000 ± 1000 K; Koester et al. 1985), the ratio of the mixing length to the pressure scale hight must be greater than 1.0. However, it should be noted that in the stability analysis the coupling between pulsation and convection is neglected, which may affect stability of nonradial g-modes significantly.

Conversations with C.S. Jeffery, Y. Osaki, H. Shibahashi, and D.E. Winget have been most helpful. I would like to thank C.S. Jeffery and S.D. Kawaler for sending preprints.

REFERENCES

Alexander, J.B., Andrews, P.J., Catchpole, R.M., Feast, M.W., Lloyd Evans, T. Menzies, J.W., Wisse, P.N.J., and Wisse, M. 1972, Mon. Not. R. astr. Soc. **158**, 305.
Bond, H.E., Grauer, A.D., Green, R.F., and Liebert, J.W. 1984, Astrophys. J. **279**, 751.
Cottrell, P.L. and Lambert, D.L. 1982, Astrophys. J. **261**, 595.
Cox, J.P. 1974, Rep. Prog. Phys. **37**, 563.
Cox, J.P. 1980, 'Theory of Stellar Pulsation' (Princeton: Princeton University Press).
Cox, J.P. 1985, in IAU Colloq. #82, 'Cepheids: Observation and Theory', ed. B.F. Madore (Cambridge: Cambridge University Press), p. 126.
Cox, J.P., King, D.S., Cox, A.N., Wheeler, J.C., Hansen, C.J., and Hodson, S.W. 1980, Space Sci. Rev. **27**, 529.

Drilling, J.S., Schönberner, D., Heber, U., and Lynas-Gray, A.E. 1984, Astrophys. J. **278**, 224.
Fadeyev, Yu.A. and Fokin, A.B. 1985, Astrophys. Space Sci. **111**, 355.
Feast, M.W. 1975, in IAU Symp. #67, 'Variable Stars and Stellar Evolution', eds. V.E.Sherwood and L.Plaut (Dordrecht: Reidel), p.293.
Fernie, J.D. 1982, Pub. Astr. Soc. Pacific **94**, 172.
Fernie, J.D., Sherwood, V., and DuPuy, D.L. 1972, Astrophys. J. **172**, 383.
Grauer, A.D. and Bond, H.E. 1984, Astropys. J. **277**, 211.
Grauer, A.D., Drilling, J.S., and Schönberner, D. 1984, Astr. Astrophys. **133**, 285.
Heber, U. and Schönberner, D. 1981, Astr. Astrophys. **102**, 73.
Hill, P.W., Kilkenny, D., Schönberner, D., and Walker, H.J. 1981, Mon. Not. R. astr. Soc. **197**, 81.
Hill, P.W., Lynas-Gray, A.E., and Kilkenny, D. 1984, Mon. Not. R. astr. Soc. **207**, 823.
Hunger, K. and Kaufmann, J.P. 1973, Astr. Astrophys. **25**, 261.
Iben, I.Jr. 1984, Astrophys. J. **277**, 333.
Jeffery, C.S. 1984, Mon. Not. R. astr. Soc. **210**, 731.
Jeffery, C.S. and Hill, P.W. 1986, 'The radial velocity curve of V652 Her (BD+13°3224)', preprint.
Jeffery, C.S., Hill, P.W., and Morrison, K. 1986, 'The period of the extreme helium star BD+1°4381', preprint.
Jeffery, C.S. and Malaney, R.A. 1985, Mon. Not. R. astr. Soc. **213**, 61p.
Jeffery, C.S., Skillen, I., Hill, P.W., Kilkenny, D., Malaney, R.A., and Morrison, K. 1985, Mon. Not. R. astr. Soc. **217**, 701.
Kaler, J.B. 1983, Astrophys. J. **271**, 188.
Kaufmann, J.P. and Schönberner, D. 1977, Astr. Astrophys. **57**, 169.
Kawaler, S.D., Hansen, C.J., and Winget, D.E. 1985, Astrophys. J. **295**, 547.
Kawaler. S.D., Winget, D.E., Hansen, C.J. 1985, Astrophys. J. **298**, 752.
Kawaler, S.D., Winget, D.E., Hansen, C.J., and Iben, I.Jr. 1986, 'The Helium Shell Game: Nonradial G-mode Instabilities In Hydrogen Deficient Planetary Nebula Nuclei', preprint.
Kilkenny, D. 1982, Mon. Not. R. astr. Soc. **200**, 1019.
Kilkenny, D. 1983, Mon. Not. R. astr. Soc. **205**, 907.
Kilkenny, D. and Flanagan, C. 1983, Mon. Not. R. astr. Soc. **203**, 19.
Kilkenny, D. and Lynas-Gray, A.E. 1982, Mon. Not. R. astr. Soc. **198**, 873.
Kilkenny, D. and Lynas-Gray, A.E. 1984, Mon. Not. R. astr. Soc. **208**, 673.
King, D.S. 1980, Space Sci. Rev. **27**, 519.
King, D.S., Wheeler, J.C., Cox, J.P., Cox, A.N., and Hodson, S.W. 1980, in 'Nonradial and Nonlinear Stellar Pulsation', eds. H.A. Hill and W.A. Dziembowski (Berlin: Springer-Verlag), p. 161.
Koester, D., Vauclair, G., Dolez, N., Oke, J.B., Greenstein, J.L., and Weidemann, V. 1985, Astr. Astrophys. **149**, 423.
Landolt, A.U. 1975, Astrophys. J. **196**, 789.
Lawson, W.A. 1985, MS thesis, University of Canterbury.
Lynas-Gray, A.E., Kilkenny, D., Skillen, I., and Jeffery, C.S. 1986, in IAU Colloq. #87, 'Hydrogen Deficient Stars and Related Objects', eds. K.Hunger,N.K.Rao,and D.Schonberner.
Lynas-Gray, A.E., Schönberner, D., Hill, P.W., and Heber, U. 1984, Mon. Not. R. astr. Soc. **209**, 387.

Marraco, H.G. and Milesi, G.E. 1982, Astr. J. **87**, 1775.
McGraw, J.T., Starrfield, S.G., Liebert, J., and Green, R.F. 1979, in IAU Colloq. #53, 'White Dwarfs and Variable Degenerate Stars', eds. H.M. Van Horn and V. Weidemann (Rochester: University of Rochester), p. 377.
Osaki, Y. 1982, in 'Pulsations in Classical and Cataclysmic Variable Stars', eds. J.P.Cox and C.J.Hansen (Boulder: JILA), p.303.
Rao, N.K., Ashok, N.M., and Kulkarni, P.V. 1980, J. Astrphys. Astr. **1**, 71.
Saio, H. and Wheeler, J.C. 1982, in 'Pulsations in Classical and Cataclysmic Variable Stars', eds. J.P.Cox and C.J.Hansen, (Boulder: JILA), p. 327.
Saio, H. and Wheeler, J.C. 1983, Astrophys. J. **272**, L25.
Saio, H. and Wheeler, J.C. 1985, Astrophys. J. **295**, 38.
Saio, H., Wheeler, J.C., and Cox, J.P. 1984, Astrophys. J. **281**, 318.
Schönberner, D. 1975, Astr. Astrophys. **44**, 383.
Schönberner, D. 1977, Astr. Astrophys. **57**, 437.
Schönberner, D. 1978, Mitt. Astr. Ges. **43**, 266.
Schönberner, D. 1979, Astr. Astrophys. **79**, 108.
Shibahashi, H. and Osaki, Y. 1981, Publ. Astr. Soc. Japan **33**, 427.
Starrfield, S.G., Cox, A.N., Kidman, R.B., and Pesnell, W.D. 1984, Astrophys. J. **281**, 800.
Starrfield, S.G., Cox, A.N., Kidman, R.B., and Pesnell, W.D. 1985, Astrophys. J. **293**, L23.
Stellingwerf, R.F. 1978, Astr. J. **83**, 1184.
Trimble, V. 1972, Mon. Not. R. astr. Soc. **156**, 411.
Unno, W., Osaki, Y., Ando, H., and Shibahashi, H. 1979, 'Nonradial Oscillations of Stars' (Tokyo: University of Tokyo Press).
Van Horn, H.M. 1984, in 25th Liege Astrophysical Colloq. 'Theoretical Problems in Stellar Stability and Oscillations' (Liege: University of Liege), p.307.
Walker, H.J. and Kilkenny, D. 1980, Mon. Not. R. astr. Soc. **190**, 299.
Wesemael, F., Green, R.F., and Liebert, J. 1985, Astrophys. J. Supple. **58**, 379.
Winget, D.E. and Fontaine, G. 1982, in 'Pulsations in Classical and Cataclysmic Variable Stars', eds. J.P. Cox and C.J. Hansen (Boulder: JILA), p.46.
Winget, D.E., Hansen, C.J., Van Horn, H.M. 1983, Nature **303**, 781.
Winget, D.E., Kepler, S.O., Robinson, E.L., Nather, R.E., and O'Donoghue, D. 1985, Astrophys. J. **292**, 606.
Winget, D.E., Robinson, E.L., Nather, R.E., and Fontaine, G. 1982b, Astrophys. J. **262**, L11.
Winget, D.E., Robinson, E.L., Nather, R.E., and Balachandran, S. 1984, Astrophys. J. **279**, L15.
Winget, D.E., Van Horn, H.M., Tassoul, M., Hansen, C.J., and Fontaine, G. 1983, Astrophys. J. **268**, L33.
Winget, D.E., Van Horn, H.M., Tassoul, M., Hansen, C.J., Fontaine, G., and Carroll, B.W. 1982a, Astrophys. J. **252**, L65.
Wood, P.R. 1976, Mon. Not. R. astr. Soc. **174**, 531.
Zalewski, J. 1985, Acta Astr. **35**, 51.

DISCUSSION

HUNGER: I must say we are really very happy that you have found these small masses for the R CrB stars. For ten years or more Craig Wheeler was telling us that our analyses of helium stars were wrong because they led to masses of the order of one solar mass, where as he claimed 2 $M_\odot$.

FEAST: If the atmospheres of R CrB stars are very deep, one might expect the spectroscopically determined temperatures to be lower than the T_{eff} which is relevant to the pulsation calculations.

SAIO: Yes.

FEAST: The other point is related. If the atmospheres of R CRB stars are not in equilibrium, but are slowly expanding, what will be the effect on pulsation?

SAIO: During pulsation, a moving envelope is produced by mass loss. I guess the expanding atmosphere has a damping effect on pulsations.

FEAST: One should perhaps be cautious in adopting spectroscopically determined masses for R CrB stars since it is not entirely clear that these stars have atmospheres in hydrostatic equilibrium (the log g may not give the mass).

HUNGER: Do I understand you right? You think, if one has to include an inertial term, such as times r, in the hydrostatic equation then the effective gravity one derives may be different. However, the density is so low in the atmosphere that the inertial term is small compared to the other terms, which means that effective gravity and mass gravity will hardly differ.

LIEBERT: Concerning the PG 1159 stars, I wanted to make it clear that although models including carbon and oxygen have not yet been calculated, the application of helium/hydrogen models (Wesemael) and the analogy with cooler (80000 K) white dwarfs suggests that the atmospheric composition of the "PG 1159" star is helium-dominated and not likely to be 50% carbon/oxygen.

JEFFERY: What I really wanted to know is whether the blue edge of the instability strip, as it bends over, is a clear-cut thing or whether it is a fuzzy blue edge. Because we see that as we go to hotter temperatures the periods seem to be less regular and then we get non-radial pulsation at the very hottest temperatures.

SAIO: The irregularity is due to non-radial pulsations or maybe the appearance of shock waves in the atmosphere of the star. So far, the stability is determined by linear analysis. I think the stability boundary for radial fundamental pulsation is clear-cut and not fuzzy. The blue edge I showed is for radial fundamental mode. Therefore, it may be possible that other radial or non-radial models grow in the region bluer than the blue edge for radial fundamental pulsations.

JEFFERY: In your models for BD+13°3224 you used $Y = 0.9$, $C = 0.1$ composition. The observations suggest photospheric abundances which are extremely N-rich. Supposing the envelope layers to be homogeneous, how would the use of N-rich mixture affect the stability analysis?

SAIO: I think that a N-rich mixture has an effect on stability similar to that a C-rich mixture has. If the ionization of carbon or nitrogen is responsible for the excitation of this pulsation, the abundance is most important in regions of 10^6 K or so. We do not see such a region, so we have a great choice as to any abundance in such a region.

MENDEZ: K1-16 sometimes pulsates and sometimes does not. Grauer and Bond have presented at least one case in which the star suddenly started pulsating at essentially full amplitude. To what extent is this a problem for the proposed pulsation mechanism?

SAIO: I think that phenomenon is beating. I mean, if there are two pulsation frequencies which are spaced very closely, then you get very long beating periods. So I think the quiet part is the duration of just a small amplitude part and I don't think the pulsation is stabilized in this period.

MENDEZ: When it starts the pulsation, it does it suddenly. It is not gradual.

LIEBERT: No, I don't think so. I think every time we looked at it, we did so for 10 months or more. There is considerable complication in the power spectrum of that object. There is a possibility of fundamental changes in modes between the time of the discovery observations by McGraw and the last year or two, although it is probably going to turn out to be some complicated mixing of the same modes. It would be premature to say which modes we have, but certainly the mode structure is very complicated.

THEORY OF DUST FORMATION IN R CORONAE BOREALIS STARS

Yu. A. Fadeyev
Astronomical Council
USSR Academy of Sciences
Pyatnitskaya Str. 48
Moscow 109017, USSR

ABSTRACT. Application of the homogeneous nucleation theory to the problem of R CrB stars shows that the radial distance of the inner boundary of the carbon supersaturation region is at about 12 photospheric radii for a stellar effective temperature T_e = 6000 K. Formation of an optically thick dust shell becomes possible at mass loss rates $\dot{M} \gtrsim 10^{-6}$ $M_\odot$/yr. However, the upper limit of this mass loss rate cannot considerably exceed 10^{-5} $M_\odot$/yr since, at higher $\dot{M}$, the theoretically predicted rate of the visual brightness decline is larger than that derived from observations. Comparison of the theoretically predicted radii of dust grains with those observed in R CrB and RY Sgr shows that the mass loss rate in these stars should be in the range of 1×10^{-7} $M_\odot$/yr to 3×10^{-6} $M_\odot$/yr.

1. INTRODUCTION

The main observational property of R CrB stars is a sudden drop in their visual light with the amplitude ranging from 1 to 9 mag. It is well known that this is ascribed to the formation of optically thick dust shells. Any theoretical model of grain formation must also be able to explain further properties of R CrB stars, the most important of which are the following:
1. Light minimum occurrence is unpredictable.
2. There is a correlation between the rate of the light decline and the intrinsic colour index at normal state (Pugach, 1977). Typical rates of the light decline are 0.2 mag/day at spectral type F and 0.05 mag/day at spectral type R.
3. While the visual light is decreasing, the ratio $R = \Delta V/\Delta(B-V)$ becomes very large. For instance, according to Tempesti and De Santis (1975), this ratio is $R \approx 25$. However, upon recovery of light the ratio is $R \approx 3.5$, which is typical for the interstellar medium.
4. All R CrB stars also show a cyclic light variation with an amplitude of about a few tenths of mag (Alexander et al., 1972; Fernie, 1982; Kilkenny and Flanagan, 1983), which is due to radial

pulsations (King, 1980).
5. During the visual light minimum, excesses of infrared radiation can be observed (Feast and Glass, 1973). This is a direct indication of the formation of optically thick dust shells. IR observations by Kilkenny and Whittet (1984) have shown that during the normal state the spectra of R CrB stars can be considered as a superposition of a photospheric and a dust shell contribution. This indicates the existence of permanent dust shells around R CrB stars.
6. Polarimetric observations by Orlov and Rodriguez (1972) have shown that the polarization increases with the decrease of the visual light. After recovery to the normal state, the polarization is different from that before the minimum. This fact strengthens the evidence of optically thin dust shells during the normal state of R CrB stars.
7. The spectral types of R CrB stars are in the range from R to F. It is suspected that also a few stars from spectral type A and B belong to the group of R CrB stars. Thus, any theoretical grain formation model must be able to explain the existence of circumstellar dust around stars with an effective temperature ranging from 2500 K to 6000 K or to even as much as 20000 K.
8. Previous spectral analyses (Searle, 1961; Schönberner, 1975; Orlov and Rodriguez, 1974; Cottrell and Lambert, 1982) revealed a deficiency of hydrogen and an overabundance of helium and carbon. This means that the formation of the circumstellar dust shell is caused by the phase transition of carbon.
9. While the visual light is decreasing, the spectra of all R CrB stars undergo drastic and complex changes. Firstly, a continuum emission at wavelengths shorter than 4000 Å appears (Alexander et al., 1972). In the next phase a great number of sharp emission lines become visible, the intensity of which decays over the time scale of about 20 days (Payne-Gaposhkin, 1963; Alexander et al., 1972). These lines are displaced bluewards by 10 kms^{-1}. They are also assumed to be present during the normal state (Rao, 1981). Finally, broad emission lines of He I, Ca II and Na I appear, indicating a gas outflow with a velocity as great as 200 km/s (Alexander et al., 1972; Rao, 1981).
10. The R CrB stars are supergiants with a bolometric magnitude ranging from -4 mag to -6 mag (Warner, 1967).

2. MODELS OF R CORONAE BOREALIS STARS

The first explanation of the R CrB phenomenon was given by Loreta (1934) and O'Keefe (1939). They presumed that the fading of the visual light is caused by ejection of matter followed by condensation of carbon particles. This hypothesis also qualitatively explains the increase of infrared radiation during the minimum of visual light.

It was preliminary assumed that the ejection of matter occurs in spherically symmetric shells. However, later observations showed that a change in the visual brightness is not always accompanied by a corresponding change of the infrared flux (Forrest and Gillett, 1971,

1972). These observations favour the idea of local matter ejection (Feast and Glass, 1973). Thus, obscuration of the stellar disc occurs when matter is ejected towards the observer.

It should be noted that the model of local ejection also allows the large values of R during the decline of the visual light to be explained, as well as the dependence of the rate of the light decline on the effective temperature. Following Pugach (1984), let us designate by X the fraction of the stellar disc obscured by the dust cloud. Then, the ratio of the total extinction to the selective extinction can be determined from the following expression:

$$R^{-1} = \frac{\log(1 - X + X \exp(-\tau_B))}{\log(1 - X + X \exp(-\tau_V))} - 1,$$

where τ_V and τ_B are the optical depths of the dust cloud in the V and the B band, respectively. During the initial phase of light decrease, $X < 1$, $\tau_V > 1$ and $\tau_B > 1$ so that R can reach high values. On the other hand, during the recovery phase, $X = 1$, which means that the increase of the visual brightness is caused by the decrease of τ. For the typical relation $\tau_V = 0.8 \tau_B$, we obtain $R \approx 4$. Thus, the rate of the visual brightness increase is determined by the dissipation of the dust cloud and, hence, by the decrease of its optical depth. Pugach (1984) has also shown that the theoretical dependence of the rate of the light decline on the intrinsic colour index $(B-V)_o$ is in good agreement with the empirical relation, if $v_t = 0.8 v_e$ (v_t = tangential expansion velocity of the cloud, v_e = escape velocity).

To explain the time behaviour of the sharp emission line spectrum, Payne-Gaposhkin (1963) proposed that grain formation takes place near the photosphere, whereas the sharp emission lines are formed in the chromosphere. In the frame of this hypothesis, the dust shell blocks the radiative flux and causes the decay of the emission lines. The main difficulty with this model is that there is no way of explaining grain formation close to the photosphere, where the gas temperature can exceed 6000 K.

On the other hand, Fadeyev (1983) proposed that the sharp emission lines are formed in a shock front. This idea is strengthened by the fact that these emission lines are shifted to the blue by about 10 kms^{-1}. Moreover, a simple analysis shows that in pulsating stars with a luminosity to mass ratio of $L/M > 10\ L_\odot/M_\odot$ the frequency of the fundamental mode exceeds the critical frequency for standing waves (Aikawa, 1984). Thus, above the photosphere of such stars, stellar oscillations may occur only in the form of running waves. The decay of the intensity of the emission lines can then be caused by a decrease of the mechanical energy flux in the shock while it propagates through the extended stellar atmosphere.

3. DUST FORMATION IN R CORONAE BOREALIS STARS

So far published models of R CrB stars do not consider grain formation; even the location of the condensation region is still a matter of debate. Simple estimates show that the gas density in a hydrostatic equilibrium atmosphere decreases so rapidly with radius that, in the layers where the gas temperature drops below the evaporation temperature, carbon cannot become supersaturated. The enhancement of the gas density in the outer layers, due to stellar pulsations, seems to be the most probable mechanism for grain formation.

In a first approximation, we assume a spherically symmetric and time-independent outflow of matter, $\dot{M}$. Thus, the gas density ρ at a radial distance R can be estimated from the equation of continuity:

$$\rho = \dot{M}/(4\pi R^2 v) \qquad (1)$$

where v is the velocity of the gas. Let us also assume that the optical thickness of the matter inbetween the photosphere and the considered radius R is negligible. Then the gas temperature T can be estimated from the following relation:

$$T_g = T_e W^{1/4}, \qquad (2)$$

where T_e is the effective temperature and

$$W = 1/2 \ (1 - (1 - (R_{ph}/R)^2)^{1/2})$$

the dilution factor.

The main quantity determining the phase transition is the supersaturation ratio $S = P_c/P_s(T)$, where P_c is the partial pressure of carbon, and

$$P_s(T) = -A/T + B \qquad (3)$$

is the equilibrium vapour pressure. The quantities A and B are constants for a given material and are determined in laboratory measurements. If the supersaturation ratio exceeds unity, vapour condenses to grains whereas, for S < 1, the inverse process, namely evaporation, takes place.

The problem, however, is complicated due to the fact that the phase transition takes place in the stellar radiation field. It is hence necessary to take the energy equilibrium into account, i.e. the radiation absorbed and re-emitted by the dust grains. If we neglect the energy exchange between grains and molecules, the expression for the energy equilibrium can be written in the following form:

$$4\pi r^2 \sigma T_e^4 WQ(T_e, r) = 4\pi r^2 \sigma T_d^4 Q(T_d, r) \qquad (4)$$

where r and T_d are the radius and temperature of a dust particle and $Q(T,r)$ the Planck-averaged absorption coefficient.

According to Lefevre (1979), the supersaturation ratio at the surface of the dust particle, with the temperature T_d, is determined by the relation

$$S = \frac{P_c}{P_s(T)} \left(\frac{T_d}{T_g}\right)^{1/2} \qquad (5)$$

where the equilibrium vapour pressure P_s corresponds to $T = T_d$.

Firstly, it is necessary to find the radial distance of the layer where the supersaturation ratio is $S = 1$. This layer is the inner boundary of the condensation region since, at smaller radii, the supersaturation ratio is $S < 1$. Unfortunately, a direct estimate of this radial distance is impossible since the temperature of the dust particles is determined by their mean radius which is an unknown quantity. The results given below were obtained for the grain radius $r = 10$ Å which is typical for the condensation region.

The radial distance of the saturation level increases rapidly with effective temperature (Fig. 1). This is caused by the stellar radiation which heats the dust particles. The radial distance of the saturation layer also increases when the rate of mass loss decreases since, at the resulting smaller gas densities, saturation of carbon can be reached only when the temperatures of gas and dust particles are also lower. As can be seen from Figure 2, an increased radial distance of the saturation layer leads, furthermore, to an increase of the temperature differences between gas and dust.

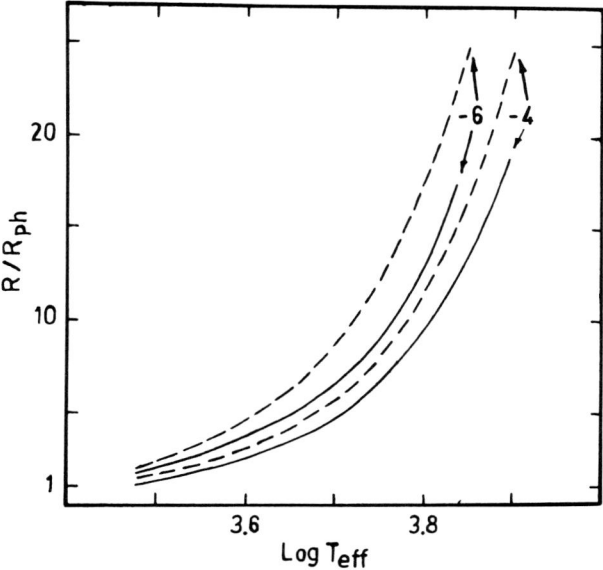

Fig. 1: Radial distances of the saturation layer (fully drawn) and the layer of maximum nucleation (dashed) as a function of the effective temperature for $\log (\dot{M}/M_\odot/\text{yr}) = -6, -4$.

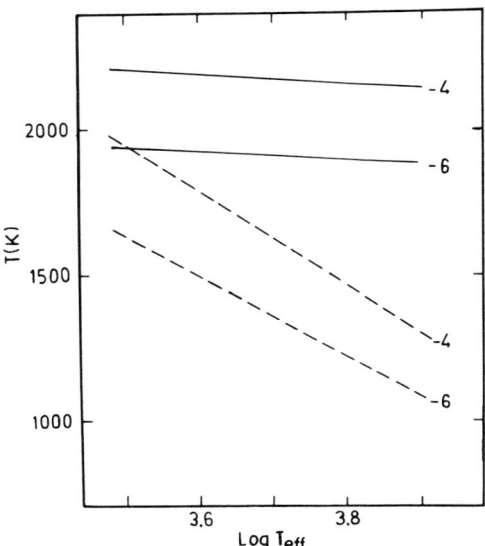

Fig. 2: Temperature of dust particles (fully drawn) and temperature of gas (dashed) at the saturation layer as a function of effective temperature for log $(\dot{M}/M_\odot/yr) = -6, -4$.

$S = 1$ for condensation is only a necessary condition. In reality, vapour has to be supersaturated before a perceptible part of carbon can be transformed into a solids. In supersaturated vapour, the formation of cluster becomes possible due to the agglomeration of monomer molecules. The equilibrium distribution of the resulting clusters of n molecules is determined by the following relation:

$$N_n = N_1 \exp(-\Delta G_n/KT), \qquad (6)$$

where N_1 is the concentration of monomers and ΔG_n the free energy from the cluster formation.

Of decisive importance is the formation of critical-sized clusters. In these sufficiently large clusters or nuclei, the further accretion of monomers is more probable than evaporation. According to Draine and Salpeter (1977), the number of monomers contained in such a nucleus is governed by the following relation:

$$n_* = 1 + (\theta/T_d \ \ln S)^3 \qquad (7)$$

where

$$\theta = 2(4\pi/3)^{1/3} \ \Omega^{2/3} \ \sigma/K \qquad (8)$$

is the bulk volume of the molecule in its solid form and σ the surface tension. Using relations (7) and (8), one can easily obtain the expression for the free energy of nucleus formation:

$$\Delta G_*/KT = 1/2\ (\theta/T_d)^3\ /\ (\ln S)^2 \tag{9}$$

It should be noted that the theory of homogeneous nucleation can be used only for rather large values of n since, for very small clusters, the conception of the surface tension energy becomes meaningless.

The rate of nucleation

$$J = Z\ \omega N_1\ \exp(-\Delta G_*/KT) \tag{10}$$

gives the number of nuclei formed per unit time and per unit volume. Here, Z is the Zeldovich factor, taking into account the fact that the real cluster distribution differs from the equilibrium distribution. According to Draine (1981),

$$Z = (n_*-1)^{-2/3}\ ((1/6\ \pi)\ (\theta/T_d))^{1/2} \tag{11}$$

The second factor in (10) gives the accretion rate of monomers colliding with the nucleus:

$$\omega = \alpha_s\ (4\ \pi r_*^2)\ P_c\ /\ (2\ \pi\ mKT_g)^{1/2} \tag{12}$$

where α_s is the sticking coefficient, r_* the radius of the nucleus and m the mass of the monomer.

The rate of nucleation increases almost exponentially with increasing supersaturation. Simultaneously, the growth of the particles reduces the density of the vapour molecules. Further condensation, hence, is controlled by the growth of the already condensed particles and by the mean frequency of monomer collisions. The time-dependence of the main quantities describing the condensation process in a R CrB star model with T_e = 6000 K, M_b = -5 mag and $\dot{M} = 10^{-6}\ M_\odot$/yr is shown in Figure 3.

The growth of the mean size of the particles after the nucleation rate maximum has been reached is accompanied by the growth of the optical depth of the dust shell and by the increase of the radiative pressure acting on the dust. As a result, dust grains and gas molecules are accelerated, the latter due to momentum transfer from the dust grains so that, within the time interval of about one month, the velocity of the gas exceeds 100 kms^{-1}. The drift velocity of the dust grains through the gas is in the range 10 kms^{-1} to 20 kms^{-1}.

The deviation of the phase transition from the dynamical equilibrium increases both with increasing effective temperature and decreasing mass loss rate (see also Fig. 2). Figure 4 shows the radial dependences of the supersaturation ratio S and the nucleation rate J, calculated for the mass loss rates $\dot{M} = 10^{-6}$, 3.16×10^{-6} and 10^{-7} $M_\odot$/yr. As we have seen previously, decrease of M corresponds to a decrease of the gas density. This leads, hence, to a decrease of the collision frequency. In this case, the exponential growth of the nucleation rate is confined to higher supersaturation ratios which lead not only to an increase of the number density of dust grains but also to a decrease of their finite mean radius.

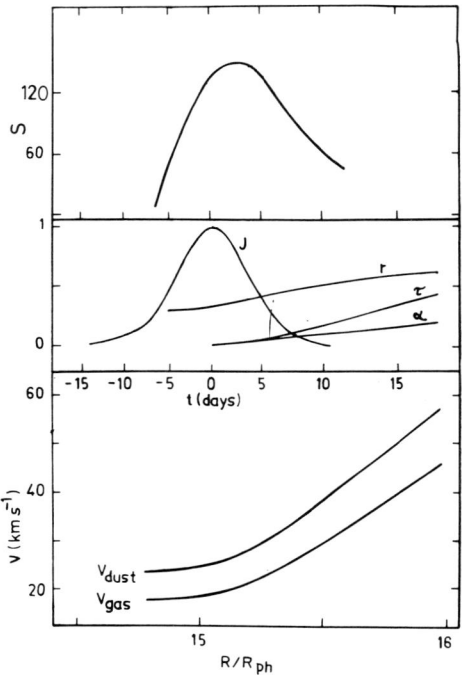

Fig. 3: Supersaturation ratio S, nucleation rate J, mean grain radius r, optical thickness τ, condensation degree α, dust velocity v_{dust} and gas velocity v_{gas} as a function of time in the model with T_e = 6000 K and $\dot{M} = 10^{-6}$ $M_\odot$/yr.

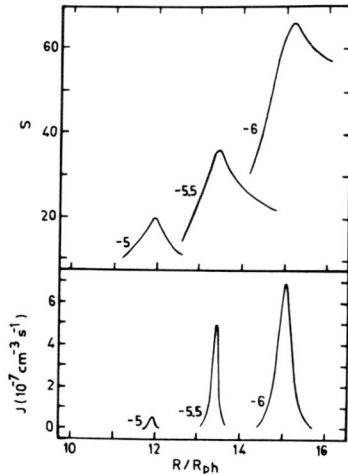

Fig. 4: Supersaturation ratio S and nucleation rate J as a function of radial distance in the model with T_e = 6000 K. The mass loss rates are as follows: log $(\dot{M}/M_\odot/yr)$ = -5, -5.5, -6.

In the limit of very low mass loss, the mean collision time becomes comparable with the expansion time of the outer layers. In this case, carbon does not condense, even if $S > 1$. Unfortunately, the current theory of homogeneous nucleation does not allow this process to be studied in detail, since the number of molecules contained in the critical-sized cluster reduces to the order of unity.

At the effective temperature $T_e \approx 6000$ K, the final mean radius of the dust grains r_∞, depends on the mass loss rate as follows:

$$\text{Log } r_\infty = -0.67 + 0.81 \text{ Log } \dot{M}, \qquad (13)$$

where r_∞ and $\dot{M}$ are expressed in units of cm and $M_\odot$/yr, respectively. This relation can be used to estimate the mass loss rate, provided that the radii of the dust grains are known from observations. For instance, recent IUE observations of R CrB and RY Sgr (Holm et al., 1982; Hecht et al., 1984) have shown the existence of carbon particles with radii ranging from 50 A to 600 A. These radii then correspond to mass loss rates from 1.2×10^{-7} $M_\odot$/yr to 2.5×10^{-6} $M_\odot$/yr.

Of great importance is the comparison of the observed light curves of R CrB stars with predictions. Let us assume that the dust grains completely re-radiate the stellar energy. Then the decline in visual

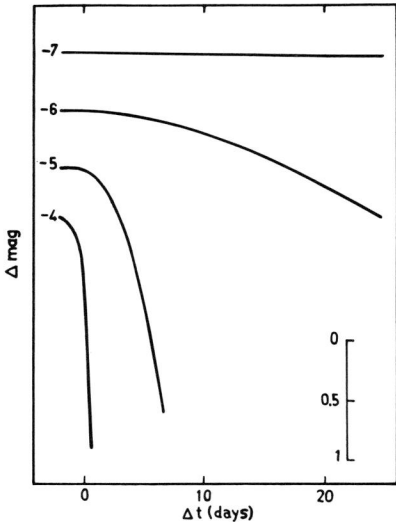

Fig. 5: Temporal behaviour of the visual brightness for the model with $T_e = 6000$ K. The mass loss rates are log $(\dot{M}/M_\odot/\text{yr}) = -7, -6, -4$.

brightness Δm is related to the optical thicknes of the dust cloud as $\Delta m = 1.086 \tau$. In Figure 5, for models with $T_e = 6000$ K and mass loss rates of $\dot{M} = 10^{-7}$, 10^{-6}, 10^{-5} and 10^{-4} $M_\odot$/yr, the variation of Δm with time is shown. Accordingly, for $\dot{M} > 10^{-5}$ $M_\odot$/yr, the rate of brightness decline is faster than observed, and we conclude that dust formation in R CrB stars occurs in outflowing matter with rates $\dot{M} < 10^{-5}$ $M_\odot$/yr. Figure 6 shows the predicted rate of the visual brightness decline, dm/dt, as a function of the stellar effective temperature. We see that the spherically symmetric model of dust shell formation predicts a decrease of dm/dt with increasing T_e, for a given mass loss rate $\dot{M}$. This behaviour comes about because at higher effective temperatures the condensation process takes place at lower gas densities which results in a decrease of the growth rate of dust particles with T_e. In Figure 6 also the observed values of T_e and dm/dt for R CrB and RY Sgr, respectively, are reproduced (error bars). From this diagram follows that the mass loss rate in these stars is of the order of 10^{-5} $M_\odot$/yr to 10^{-6} $M_\odot$/yr.

If we insert in Figure 6, the typical rate brightness decline for the (cool) R type R CrB stars, namely dm/dt = 0.05 mag/day, then we read off a mass loss rate of only $M \approx 10^{-7}$ $M_\odot$/yr, which is below what has been observed and far below the rate obtained for the hotter stars, R CrB and RY Sgr. If M varies with T_e (at constant luminosity), then one should expect a decrease with T_e rather than an increase. Hence, we conclude that the spherically symmetric model is not adequate to describe dust around R CrB stars (see also the review given by Feast, these proceedings).

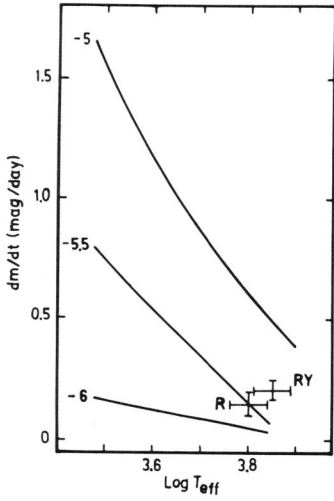

Fig. 6: The rate of the visual brightness decrease as a function of effective temperature. the mass loss rates are log ($\dot{M}/M_\odot$/yr) = -6, -5.5, -5. The observational estimates of T_e and dm/dt for R CrB and RY Sgr are shown as R and RY, respectively.

4. CONCLUSIONS

Application of the homogeneous nucleation theory to the problem of R CrB stars allows the conclusion that the grain formation process can occur only at larger distances from the stellar surface. At the effective temperature T_e = 6000 K, the radius of the inner boundary of the dust shell is in the range of 10 to 15 photospheric radii. In order to supersaturate carbon vapour one has to postulate another mechanism for the origin mass loss. At present we only know of one mechanism, namely the radial stellar oscillations with accompanying shock waves.

If the matter were leaving the star in a spherically symmetric way, the formation of an optically thick dust shell would happen for $\dot{M} \geq 10^{-6}$ $M_\odot$/yr at T_e = 6000 K. However, observations indicate that the geometry of the dust clouds formed in R CrB stars is non-spherical. Thus, the above estimates of M are an upper limit. The actual mass loss rate depends on what fraction of the surface area is covered by the dust cloud.

REFERENCES

Aikawa, T.: 1984, Astrophys. Space Sci. **104**, 405
Alexander, J.B., Andrews, P.J., Catchpole, R.M., Feast, M.W., Lloyd Evans, T., Menzies, J.W., Wisse, P.N.J., Wisse, M.: 1972, Monthly Not. Roy. Astron. Soc. **158**, 305
Cottrell, P.L., Lambert, D.L.: 1982, Astrophys. J. **261**, 595
Draine, B.T.: 1981, in Physical Processes in Red Giants, ed. I. Iben and A. Renzini, D. Reidel, Dordrecht, p. 317
Draine, B.T., Salpeter, E.E.: 1977, J. Chem. Phys. **67**, 2230
Fadeyev, Yu. A.: 1983, Astrophys. Space Sci. **95**, 357
Feast, M.W., Glass, I.S.: 1973, Monthly Not. Roy. Astron. Soc. **161**, 293
Fernie, J.D.: 1982, Publ. Astron. Soc. Pacific **94**, 172
Fernie, J.D., Sherwood, V., DuPuy, D.L.: 1972, Astrophys. J. **172**, 383
Forrest, W.J., Gillett, F.C.: 1971, Astrophys. J. **170**, L21
Forrest, W.J., Gillett, F.C.: 1972, Astrophys. J. **178**, L129
Hecht, J.H., Holm, A.V., Donn, B., Chi-Chao, Wu.: 1984, Astrophys. J. **280**, 228
Holm, A.V., Chi-Chao, Wu, Doherty, L.R.: 1982, Publ. Astron. Soc. Pacific **94**, 548
Kilkenny, D., Flanagan, C.: 1983, Monthly Not. Roy. Astron. Soc. **203**, 19
Kilkenny, D., Whittet, D.C.B.: 1984, Monthly Not. Roy. Astron. Soc. **208**, 25
King, D.S.: 1980, Space Sci. Rev. **27**, 519
Lefevre, J.: 1979, Astron. Astrophys. **72**, 61
Loreta, E.: 1934, Astron. Nachr. **254**, 151
O'Keefe, J.A.: 1939, Astrophys. J. **90**, 294
Orlov, M.Ya., Rodriguez, M.H.: 1972, Inform. Bull. Var. Stars No. **742**, 1
Orlov, M.Ya., Rodriguez, M.H.: 1974, Astron. Astrophys. **31**, 203
Payne-Gaposhkin, C.: 1963, Astrophys. J. **138**, 320
Pugach, A.F.: 1977, Perem. Zvezdi **20**, 391
Pugach, A.F.: 1984, Astron. J. (Soviet) **61**, 491
Rao, N.K.: 1981, in Effects of Mass Loss in Stellar Evolution, ed. C. Chiosi and R. Stalio, D. Reidel, Dordrecht, p. 461
Schönberner, D.: 1975, Astron. Astrophys. **44**, 383
Searle, L.: 1961, Astrophys. J. **133**, 531
Tempesti, P., De Santis, R.: 1975, Mem. Soc. Astron. Ital. **46**, 443
Warner, B.: 1967, Monthly Not. Roy. Astron. Soc. **137**, 119

DIFFUSION AND HE OVERABUNDANCES: HYDRODYNAMICAL IMPLICATIONS

G. Michaud
Département de Physique
Université de Montréal
C.P. 6128, Succ. A, Montréal
CANADA H3C 3J7

ABSTRACT. In the absence of mass loss, diffusion leads to *underabundances* of He in main sequence stars. Because of a very strong observational link with Ap and He weak stars, it has however been suggested that diffusion is the explanation for the He rich stars of the upper main sequence. This requires a mass loss rate of 10^{-12} $M_\odot$ yr^{-1} or slightly lower. The mass loss rate must decrease as T_{eff} increases. Magnetic fields must apparently be involved to *reduce* the mass loss rate. Since this model predicts that the CNO abundances should be normal in the cooler He rich stars, it leads to a clear observational test. Detailed calculations should be made to confirm the importance of this test. The effects of separation in the wind, the atmosphere and the envelope are discussed to conclude that separation in the atmosphere is likely to be most important. The importance of diffusion for He rich white dwarfs and horizontal branch stars are briefly discussed.

1. THE OBSERVATIONAL LINK

The difficulty to find a model based on nucleosynthesis to explain the abundance anomalies of the AmFm, HgMn and magnetic Ap stars has been a major factor in the acceptance of the model based on chemical separation for those objects. It is not easy to find a nuclear process that will build large overabundances of, say, Mn and Hg while not modifying Fe or O. However, most of the He rich stars are believed not to have large anomalies of other elements. Furthermore, it is well known that stellar evolution produces He. At first sight the He rich stars are *not* a likely product of diffusion.

In their original paper, however, Osmer and Peterson (1974) already suggested that diffusion was the likely process responsible for the He rich stars and this suggestion is now generally accepted even if it has been shown that diffusion, by itself, always leads to He underabundances. One of the main reasons invoked by Osmer and Peterson was the surface gravity of the He rich stars. It is the same as that of young main sequence stars of the same effective temperature suggesting that the anomalies are only superficial. Their argument was supported by the magnetic field measurements of Borra and Landstreet (1979). Their results suggested that more than 50% of the He rich stars (6 out of 9 observed) had measurable magnetic fields, linking them to the magnetic Ap stars where diffusion is believed to be responsible for the anomalies. Since they furthermore constitute a temperature sequence with the Ap and He weak stars (Osmer and Peterson 1974), the observational link now seems overwhelming.

After showing that the parameter free model always leads to He underabundances (§ 2), except perhaps for some white dwarfs (§ 3), we will study the effect that mass loss (§ 4) and magnetic fields (§ 5) may have on helium abundances and show that, through the interaction of hydrodynamical processes and diffusion (§ 6), abundances are powerful indicators of stellar hydrodynamics. Only through that interaction are overabundances of helium produced by chemical separation.

2. THE PARAMETER FREE MODEL

Diffusion is a basic physical process and plays a role everywhere a more efficient transport process does not wipe out its effects. If one assumes that a star arrives on the main sequence with the convection zones as given by standard evolutionary models and one allows the chemical separation to go on unimpeded, one obtains a parameter free model. It is a simply defined stellar model that, as we shall see, leads to underabundances of He but to large overabundances of many heavy elements. It is hardly in agreement with the observations of He rich stars! It contains the hydrodynamics that can currently be described without arbitrary parameters.

As the star arrives on the main sequence, diffusion starts occuring below the He II convection zone. The diffusion velocity gives the direction that migration takes:

$$v_D = -D_{12} \left\{ \frac{\partial \ln c}{\partial r} + [g(A - \frac{Z}{2} - 1) - A \, g_R] \frac{m_p}{kT} - k_T \frac{\partial \ln T}{\partial r} \right\}. \quad (1)$$

Since the star is presumably formed homogeneous, the derivative of c (c = N(A)/[N(H) + N(A)]) is originally zero. The thermal diffusion term is never very large according to Michaud et al. (1979) and Paquette et al. (1985) have recently shown that the thermal diffusion coefficient is even smaller than that used by Michaud et al. (1979). It goes in the same direction as the gravitational settling term which is much larger. To stop gravitational settling, the radiative acceleration must then nearly equal gravity. Michaud et al. have however shown that it was never the case on the main sequence for normal helium abundances. The diffusion of helium then always starts downwards and its abundance decreases. Equation (1) assumes trace abundance of an element diffusing in H. This is not the case for He because of its relatively large abundance, but Montmerle and Michaud (1976) have shown that this did not change substantially the diffusion equation. I refer the reader to Pelletier et al. (1985) for a simple diffusion equation that is accurate even for elements that are not trace. It is shown there that the new diffusion coefficients of Paquette et al. (1985) increase the He abundance that can be supported by the radiative acceleration by a factor of about 3 compared to the results of Michaud et al. (1979) even though underabundances of He by a factor of 30 are still predicted.

The time scales for the appearance of underabundances are short, typically of the order of 10^5 years in main sequence stars of T_{eff} = 20 000 K (Martel 1979).

One may question one aspect of the calculations I just referred to. They were all made in the diffusion approximation for the radiative transfer. This is valid below the atmosphere but could it happen that in the atmosphere, where the radiative transfer is more complicated, the radiative accelerations be actually much larger? It seems highly unlikely that the increase could be large enough to explain the He rich stars since the radiative accelerations would need to be multiplied by a factor of ten to be able to support the observed He abundance. Still, it should, some day, be investigated.

As can be expected the degree of He underabundance that diffusion leads to depends on both T_{eff} and log g. The cooler the star, for a given gravity, the smaller the radiative flux and so the smaller the radiative acceleration. This can be seen by comparing Figures 2 and 3 of Michaud et al. (1979). At T_{eff} = 10 000 K the radiative acceleration can support some ten times less helium than at T_{eff} = 20 000 K. The difference is even more striking between main sequence stars, horizontal branch stars and white dwarfs. For the radiative acceleration to support the same He mass fraction as in a 20 000 K main sequence star, a subdwarf star must be 40 000 K while a white dwarf must be 90 000 K (Vennes 1985). In all cases only underabundances can be supported.

3. HYDROGEN BURNING IN WHITE DWARFS

The gravitational settling of He concentrates hydrogen in the superficial layers of white dwarfs, leading in particular to the H rich white dwarfs (Schatzman 1958). Diffusion does not however lead to a complete disappearance of H from the interior. At equilibrium, the hydrogen abundance decreases inwards exponentially as given by:

$$v_D = D_{12} \left(- \frac{\partial \ln c}{\partial r} - \frac{5}{4} \frac{\partial \ln p}{\partial r} \right) \qquad (2)$$

for v_D = 0. Equation (2) gives the diffusion velocity for trace hydrogen diffusing in helium. In words, diffusion leads to the hydrostatic equilibrium gradient for hydrogen, it stops once this is achieved and starts in the opposite direction if it is exceeded. In white dwarfs with regions warm enough for hydrogen to burn (T > 10^7 K), the nuclear reactions will drive down the hydrogen abundance where it can burn, the equilibrium gradient will be exceeded and diffusion will start transporting hydrogen downwards (Michaud, Fontaine and Charland 1984, Michaud and Fontaine 1984). The most important nuclear reaction for this process is with carbon since the pp chain shuts itself off at low hydrogen abundances. The carbon is being constantly replenished by diffusion from the interior since there is a carbon layer starting at less than 1% of the stellar mass. See Pelletier et al. (1985a) for a detailed discussion of the uppard diffusion of C from the interior and how it pollutes the surface.

How important the process is for white dwarfs of a given effective temperature depends sensitively on the internal temperature. A temperature change by a factor of 1.6, changes the nuclear burning timescale by 4 orders of magnitude, so making the process either very efficient or very inefficient (see Table 1 of Michaud, Fontaine and Charland 1984).

This process has potential effects for white dwarfs but not for main sequence stars. For it to be efficient, the distances that hydrogen has to travel to get to the burning region must be small, otherwise, the time scales become too long. It could play a role both in maintaining at a low level the surface abundance of hydrogen in DB white dwarfs and in reducing the depth of the hydrogen layer on DA white dwarfs, even conceivably transforming DA into DB white dwarfs.

Some white dwarfs are known to be extremely He rich and to have traces of metals (Liebert 1980). Because of the high efficiency of gravitational settling in white dwarfs, it appears that the metals can only come from recent accretion episodes. But then how could the star have accreted the metals efficiently while not accreting the hydrogen? Even a very small amount of hydrogen would show strong H lines once concentrated by diffusion to the surface. The process just described can reduce the hydrogen abundance below the observational upper limit of $N(H)/N(He) = 10^{-4}$ so long as the accretion rate remains below 10^{-19} $M_\odot$ yr^{-1} (Michaud, Fontaine and Charland 1984).

In DA white dwarfs, the superficial hydrogen layer can also be reduced by hydrogen burning. Evolutionary models typically leave a surface layer of 10^{-4} $M_\odot$ of hydrogen. During the white dwarf cooling, Michaud and Fontaine (1984) obtained that the hydrogen layer could be reduced to 10^{-9} $M_\odot$. The exact factor by which the hydrogen abundance is reduced is however very sensitive on the detailed internal structure of white dwarfs and can only be determined by stellar evolution models. In their evolutionary models, Iben and MacDonald (1985) obtained that the process just described consumed a significant fraction of the hydrogen remaining at the end of the shell burning phase though it either left a mass fraction of hydrogen of order 10^{-4} $M_\odot$ or lead to an hydrogen flash as the star was approaching the white dwarf sequence. This depended mainly on the exact depth of the He layer buffer between the hydrogen surface and the carbon core. Furthermore given the great sensitivity of the burning rate on temperature, an increase of the internal temperature by a factor of 1.5 would considerably lengthen the time during which the process is effective and so increase the effect on the hydrogen layer. For the destruction of H to be as efficient as proposed by Michaud and Fontaine (1984), the internal temperature of white dwarfs would need to be about 50% larger than in the models of Iben and Macdonald (1985).

It should be pointed out that any small turbulence would strongly enhance the efficiency of the process to the point that even such a small turbulence implied by $D_T = D_{12}$ nearly completely eliminates DA white dwarfs. The mere existence of DA white dwarfs then puts a strong upper limit on the amount of turbulence in white dwarf interiors (see Fig. 2 of Michaud and Fontaine 1984).

This process is however negligible for main sequence or horizontal branch stars because of the larger distances that have to be covered by the diffusing elements and so the much longer time scales involved. Then other hydrodynamical processes must be included in the model to explain helium overabundances.

4. MASS LOSS AND HELIUM OVERABUNDANCES

Vauclair (1975) has suggested that, if the mass loss rate is appropriate, it could lead to overabundances of helium in the atmospheres of the relatively hot stars observed to have such anomalies (Osmer and Peterson 1974). If the mass loss rate is appropriate, hydrogen will drag helium along and He will accumulate where the dragging is least effective that is where He is most in the form of neutral helium, since the diffusion coefficient is then some two orders of magnitude larger than when it is ionized. In the stars where it is observed to be overabundant, helium is least ionized in the atmosphere, so that is where it accumulates.

The simplest model based on mass loss assumes that the star is losing mass at a constant rate in a spherically symmetrical way. The chemical separation can be calculated using equation (1) along with the mass conservation equation of the dominant specie in the presence of mass loss:

$$\frac{dM}{dt} = -4\pi R^2 N_H m_p v_w \quad . \qquad (3)$$

In order to simplify the argumentation, we neglect, in this discussion, the fact that He is not really a test element. The diffusing element must also satisfy a conservation equation:

$$\nabla (c \, N_H (v_w + v_D)) = 0 \quad . \qquad (4)$$

To calculate the chemical separation, it is convenient to separate the star in three zones: the coronal-wind region, the photospheric region and the envelope region. The frontier of each of these zones is somewhat arbitrary. For the corona, it can be fixed where the hydrostatic solution stops to be accurate; for this reason I also call it the wind region even though the outward movement caused by the wind pervades the whole photosphere and envelope but with negligible dynamic effects. This definition applies even to stars that would have a cool wind and so no proper corona. For the envelope we choose the bottom of the He II convection zone for a normal He abundance as a convenient boundary (see Figure 1).

4.1 The Wind

To what extent does the wind cause the abundance anomalies in the atmosphere? Does the wind leave with the abundances of the top of the atmos-

T_{eff} = 25 000 K

HYDRODYNAMIC REGION (WIND)

———————————————————— $\tau = 10^{-4}$

PHOTOSPHERE (OVERSHOOTING?)

———————————————————— $\tau = 3$

HELIUM II CZ

———————————————————— $\tau = 30$

ENVELOPE

Figure 1. Outer structure of a 25000 K main sequence star. The chemical separation could take place in three regions, the envelope, the atmosphere and the dynamical or wind region. Two of them are separated by a convection zone.

phere or does additional separation occur in it? This depends to some extent on the unknown hydrodynamic structure of the wind but we investigate these questions under the assumption of the simplest wind structure: that of a corona with a constant temperature. We investigate the uncertainty of such a model by considering the effect of varying Z, the degree of ionization, on the separation.

The equations governing the element separation in the wind region are actually a little different from those given above because of the importance of the dynamical terms. Details of the calculations will be given elsewhere (Michaud et al. in preparation). It turns out that the solution is dominated by a comparison of the gravitational settling and wind velocities. Depending on which is largest, a term changes sign in the differential equation and the nature of the solution changes completely from one where element A is completely dragged to one where it is essentially left behind (Fakir 1985). Using equations (1) and (3) and equating the wind velocity to the gravitational settling velocity, one obtains:

$$v_W = \frac{-dM/dt}{4\pi R^2 N_H m_p} = D_{12} \frac{A g m_p}{kT}, \qquad (5)$$

where it has been assumed that only gravitational settling is important in the diffusion equation. After replacing the physical constants, one obtains:

$$\frac{dM}{dt} = \frac{2.4 \; 10^{-15} M \; A \; T_5^{1.5}}{Z^2} \qquad (6)$$

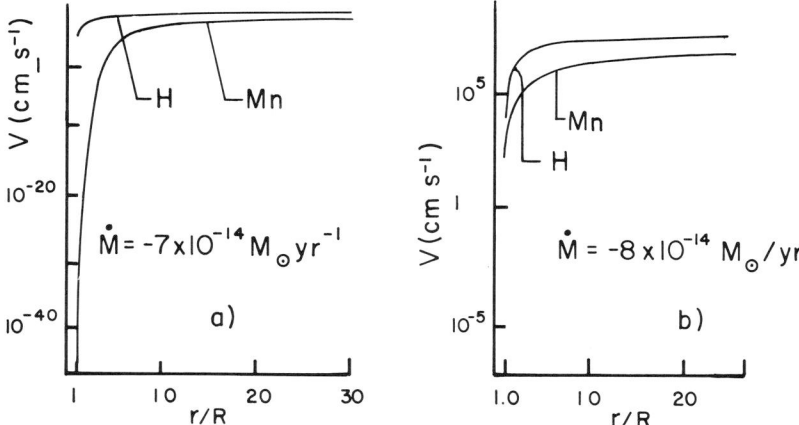

Figure 2. Transport velocity in the wind as a function of the distance to the star center in units of the stellar radius. Hydrogen is assumed to be the dominant specie. Because of chemical separation the velocity of Mn is smaller than that of hydrogen. For a mass loss rate of 10^{-12} $M_\odot$ yr^{-1} the curve for Mn cannot be distinguished from that for H. Note the difference in scale for parts a and b.

where the logarithmic Coulomb term has been replaced by an average value. This causes the mass loss rate to be overestimated by a factor of 1.5 in the wind but underestimated by a factor of up to 3 in deep stellar interiors. The quantity T_5 is the temperature in units of 10^5 Kelvin. Detailed solutions of the chemical separation in presence of a wind are currently being calculated and I will use equation (6) to discuss the main results.

In Figure 2 is shown the particle transport velocity of Mn as a function of the radius in constant temperature coronas for a 3 $M_\odot$ star. The flux is conserved and is proportional to the velocity, so that if the velocity of Mn at the bottom of the corona is smaller than that of H by 47 orders of magnitude, so is the flux of Mn and no Mn then leaves the star. First, notice the high sensitivity on the exact value of the mass loss rate. A few percent change in the mass loss rate changes completely the solution if one is close to the limiting value given by equation (6). The limiting mass loss rate is close to 10^{-13} $M_\odot$ yr^{-1} in this case. Below that value no Mn is dragged by the wind while above that value Mn leaves with the abundance at the top of the atmosphere. There is a transition region but as can be seen from the figure it is approximately a factor of three in mass loss rate.

Since the important parameter is the expression appearing in equation (6), changing the atomic mass has an equivalent effect to changing the mass loss rate. A lighter element such as He would be dragged for a wind smaller by a factor of 14, if it were not for the fact that He is less ionized than Mn. Since He is 4.5 times less ionized than Mn, the critical wind is actually 1.5 times as large for He as it is for Mn. Note that the figure does not correspond to exactly the mass loss rate of equation (6) because of the approximations made in deriving equation (6).

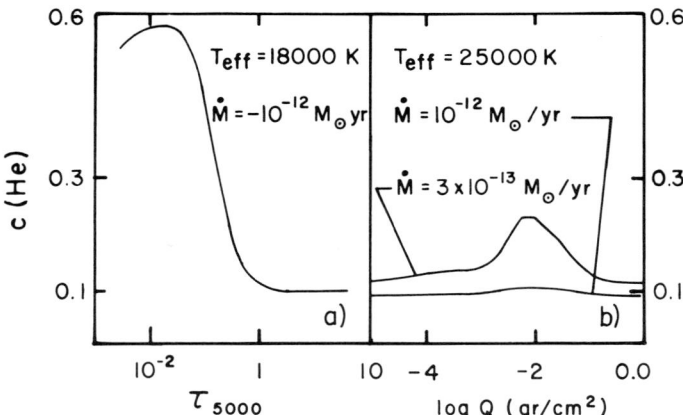

Figure 3. Abundance of He in the atmospheres of an 18000 K and of a 25000 K star. Because of higher ionization, the mass loss must be smaller in the hotter star for overabundances to materialize.

The critical mass loss rate for helium is then about 1.5 10^{-13} $M_\odot$ yr^{-1} in 3 $M_\odot$ stars while it is about 3 10^{-13} $M_\odot$ yr^{-1} in 6 $M_\odot$ stars, as is more appropriate for He rich stars.

The model used here assumes a hot wind with a corona. Even if we have not calculated such models in detail, it is possible (using equation 6) to evaluate the effect of reducing the temperature of the wind on the critical parameter. If the charge remained the same, the critical wind would be 192 times smaller if the temperature were 30000 instead of 10^6 K. However the ionization of Mn is then reduced from Z = 9 to Z = 2 and that of He from 2 to 1. The diffusion coefficient of once ionized He is 4 times that of twice ionized helium. The critical mass loss rate is then, for helium, at 30000 K decreased by a factor of 50. It is equal to about 10^{-14} $M_\odot$ yr^{-1}.

Only in stars with T_{eff} < 20000 K does He become mainly neutral in the wind so that the separation can increase (to about 10^{-12} $M_\odot$ yr^{-1}) because of the much larger diffusion coefficient of neutral helium (Michaud et al. 1978). From equation (6) the same separation would be expected in a cool wind as in the atmosphere, since the temperature is about the same, however for helium in stars of T_{eff} = 20000 K, the ionization is larger in the wind. The separation should then be larger in the atmosphere than in a cool wind of a given star.

4.2 The Atmosphere

The separation in the atmosphere was first studied by Vauclair (1975). She considered the time dependant solution of the He abundance in the presence of a wind. She showed, for a mass loss rate of 10^{-12} $M_\odot$ yr^{-1} in a star with T_{eff} = 20000 K, that He would accumulate in the photosphere. Her result still stands.

I will discuss this model in a little more detail. The whole process is due to He being hardly ionized in the atmosphere at 20000 K and so

T_{eff}	N(He I)/N(He)
17 500	1.0
20 000	0.8
22 500	0.4
25 000	0.1
27 500	0.014
30 000	0.0013

Table 1. Maximum of N(He I)/N(He) from Mihalas (1972).

having a much larger diffusion flux than in the envelope, where it is ionized. Generally an element is least ionized in the atmosphere because of the smaller temperature there.

Since her paper appeared, new evaluations of the neutral helium diffusion coefficient have become available (Michaud et al. 1978). They are more accurate than her hard sphere approximation. They use a more realistic polarization potential to represent the interaction. They lead to diffusion coefficients that are some 4 to 6 times smaller than the hard sphere approximation. It reduces the critical flux by a factor of 4 to 6.

To understand in a little more detail the physics of the outgoing wind, it is interesting to consider an equilibrium solution in the atmosphere with as much He leaving the atmosphere as entering from the bottom. The time evolution of the flux arriving from the envelope is taken into account. The flux conservation (equation [4]) implies that, where the wind and diffusion velocities nearly cancel each other, the He abundance must increase so that the flux remains constant. When this happens in the atmosphere, the He abundance increases in the atmosphere. This occurs for wind velocities just slightly larger than the critical value given by equation (6). Such a solution is shown in Figure 3. This He distribution occurs for about a mass loss rate of 10^{-12} $M_\odot$ yr^{-1} at T_{eff} = 18000 K. At T_{eff} = 25000 K, there is hardly any separation for this mass loss rate. The critical mass loss rate decreases as the effective temperature increases. This can easily be understood from Table 1 where is shown the maximum fraction of unionized He in the atmosphere (from Mihalas 1972). As the temperature increases, the ionization increases and the large diffusion coefficient of neutral helium plays a smaller and smaller role. The maximum overabundance that can be supported by this process also decreases as T_{eff} is increased.

Larger mass loss rates lead to smaller overabundances. Smaller mass loss rates lead to a discontinuity when the wind and diffusion velocities are equal. Some of our hypotheses have broken down: it is not possible to assume flux conservation any more. Below that critical value of the flux, He accumulates in the atmosphere and, for yet smaller mass loss rates, settles gravitationnally.

4.3 The Envelope

In the absence of mass loss, gravitational settling through the bottom

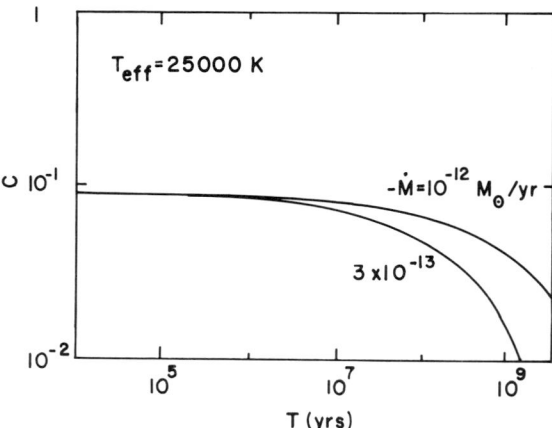

Figure 4. Time dependance of the He abundance in the convection zone of a 25000 K star. If the mass loss rate is 3 10⁻¹³ M☉ yr⁻¹, the gravitational settling does not have time to materialize in the life time of the star and overabundances can materialize in the atmosphere.

of the He II convection zone reduces the helium abundance in the atmosphere by a factor of 3 in about 10^5 years (Martel 1979). For overabundances to be produced, the wind velocity (equation [3]) below the He II convection zone must be large enough to make gravitational settling inefficient. It must then be significantly larger than the diffusion velocity. Indeed the wind model is possible for He rich stars because, when He is neutral in the atmosphere, its diffusion velocity is substantially larger there than below the convection zone, where it is necessarily twice ionized. There comes however a point in the envelope where the two velocities are equal since, as the temperature increases, the ratio of the wind velocity to the diffusion velocity decreases as $T^{1.5}$, due to the $T^{2.5}$ dependence of the diffusion coefficient. As time proceeds and mass loss goes on, the matter that was originally where the combined diffusion and mass loss velocity was downwards finally arrives in the atmosphere, and the He abundance starts decreasing in the atmosphere. To evaluate how long this takes as a function of the mass loss rate, one must calculate the time evolution of the He abundance in the convection zone of a main sequence star with T_{eff} = 25 000 K. To separate the effects of the separation in the wind region and in the envelope, it is assumed that no separation occurs in the wind. The time evolution is shown in Figure 4 for two values of the mass loss rate. Since the stars of interest have main sequence life times of only a few times 10^7 yr (Iben 1966), it is clear from the figure that the envelope always supplies the convection zone with a normal abundance of He. The He abundance starts to decrease in the atmosphere only after the main sequence life is over. In Table 2 is shown the time it takes for the He abundance to be reduced by a factor of 3 in the convection zones as a function of the mass loss rate. Since this result is not very dependent on the mass of the star, and since mass loss rates of 10^{-13} to 10^{-12} M☉ yr⁻¹ always lead to some He overabundances, one can say that the He overabundances are maintained for most of the main sequence life time only for stars that have lifetimes shorter than 10^8 yr.

dM/dt ($M_\odot$ yr^{-1})	τ (yr)
10^{-14}	2.0×10^6
3×10^{-14}	10^7
3×10^{-13}	3.0×10^8
10^{-12}	1.5×10^9

Table 2. Timescale for the helium abundance to decrease by a factor of 3 in the convection zone (T_{eff} = 18 000 K).

5. MAGNETIC FIELDS, MASS LOSS AND CHEMICAL SEPARATION

Many of the He rich stars have been observed to be magnetic (Borra and Landstreet 1979) so that it is generally believed that the magnetic field plays an important role in the appearance of the He rich phenomenon in at last some of the stars. Just as for the Ap stars, the magnetic field is probably essential in stabilizing the atmosphere and so to allow the separation to go on. However here we will only discuss its effect on the chemical separation and the mass loss rate. Both of these imply diffusion through the magnetic field lines when they are horizontal.

5.1 Diffusion Across Magnetic Field Lines

Magnetic fields affect diffusion across magnetic field lines according to the classical formulas that can be found for instance in Chapman and Cowling (1970). The diffusion velocity across magnetic field lines is multiplied by a factor:

$$\frac{1}{1 + \omega_i^2 \tau_c^2}$$

where:

$$\omega_i \tau_c = 1.7 \times 10^4 H \, T^{1.5}/(N_p Z) .$$

All quantities are in the CGS electromagnetic system. For magnetic fields typical of those observed on Ap or He rich stars, diffusion is affected by the magnetic field only outside of the photosphere. Michaud et al. (1981) have shown that it started guiding elements only for τ_{5000} smaller than 10^{-2}. At that depth, the diffusion velocity perpendicular to the magnetic field is reduced by a factor of 2. It is only at an optical depth 10 times smaller that the diffusion velocity is reduced by a factor of 100. Deeper than this in the star, the proton density is too large and the diffusion velocity is hardly affected by the magnetic field. Note however that the diffusion flux is very rapidly reduced as one goes further outside the star since the reduction factor goes as the square of the proton density. Where the magnetic field lines are horizontal this leads to a nearly complete illimination of the wind if the density structure is not modified by the combined presence of the wind and the magnetic field.

This correction factor has however been found not to be an accurate description of the flux diffusing across magnetic field lines in fusion devices. The flux has been observed to be much larger than predicted. It has lead to the development of the Bohm formula which is discussed for instance by Laing (1981). It is an empirical formula which does explain the measured flux in a number of fusion devices but not in all. In some cases the classical formula appears closer to reality, in particular in the thetatron where the gas is confined to a straight line. Laing suggests that the Bohm formula can be understood from the classical formula if proper account is taken of the detailed geometry of the field lines and of the instabilities that can develop in the magnetic field.

We cannot be sure that similar effects are important in stars since the dimensions are so different. However it remains a possibility and while we feel that it is more appropriate to use the classical formula, it remains that it may overestimate the correction to the diffusion velocity. The effective reduction factor may possibly go as $1/H$ as Bohm suggests instead of as $1/H^2$ as the classical formula says.

This could be important when one considers the amplification of the effective radiative acceleration caused by horizontal magnetic fields as discussed by Alecian and Vauclair (1981), Mégessier (1984), Michaud, Mégessier and Charland (1981) and Vauclair, Hardorp and Peterson (1979). It is there argued that where the magnetic field lines are horizontal, the neutral state can diffuse accross the magnetic field lines but not the ionized states. If an element is efficiently pushed upwards by the radiative acceleration in the neutral state, but settles gravitationaly in the ionized state, the efficiency of the radiative acceleration is considerably increased. Even if, in non magnetic He rich stars, the radiative acceleration on He were to be smaller than gravity (Michaud et al. 1979), this effect may make it larger than gravity where the magnetic field lines are horizontal in magnetic stars. This however requires the magnetic field to be larger than is in practice observed (at least 10000 Gauss where the field lines are horizontal) and further requires it to be horizontal to a high accuracy (Michaud, Mégessier and Charland 1981).

5.2 Models of Magnetic He Rich Stars.

There are two papers where aspects of the interaction of magnetic fields with He overabundances are discussed in some detail.

Havnes and Goertz (1984) have studied the structure of magnetospheres of chemically peculiar stars. Their study applies in particular to He rich stars. They study what happens in the magnetosphere for a given mass loss. They do not assume the mass loss to be due to the radiation pressure (Abbott 1982) but rather look for the structure of the magnetosphere under the combined effects of gravity and rotation. This leads to an abundance gap close to the star that might not be there if the driving mechanism of the wind were included in the model. By comparing the particle energy density to the magnetic field energy density, they conclude that mass loss rates of the order of magnitude of those considered here lead to instabilities in the far magnetosphere. It does not appear possible, according to their result, for the wind to diffuse quietly across magnetic field lines. The various diffusion mechanisms they consider are not efficient enough. They conclude that the wind accumulates where the lines are horizontal until the magnetic field lines break. They do not study the problem of how the flux is reduced when the magnetic field is parallel to the surface in the photosphere.

This is however central to the separation process. According to the Shore and Bolton model as described by Bolton (1984) the mass loss is proportional to how horizontal the magnetic field lines are and so varies over the surface. Using the formula of Abbott (1982) for the mass loss rate, he argues it is essential that the mass loss rate be reduced for the separation to be effective. Without a magnetic field it would be too large. Where the magnetic field is horizontal, the mass loss is smallest. Various combinations of mass loss rates and magnetic geometries can then lead to a large variety in the distribution of anomalies over the surface. The basic idea of the importance of the magnetic fiel in reducing the mass loss rate and making the separation possible is probably right. Whether it occurs as they have calculated is impossible to tell since there are no details of the calculations in the account given by Bolton.

6. HYDRODYNAMICS AND ABUNDANCE ANOMALIES

Using the empirical formula of Abbott (1982), it is easy to calculate that the mass loss rate to be expected in main sequence stars with T_{eff} = 20000 K is about $6 \cdot 10^{-11}$ $M_\odot$ yr^{-1}. It should increase as T^7_{eff} because of the dependence of Abbott's formula with temperature. This is more than an order of magnitude larger than the mass loss leading to He overabundances. There is no region in the star where He can separate if the mass loss rate is that large. For chemical separation to occur there must be a mechanism to reduce the mass loss rate as concluded by Bolton (1984). Since, as T_{eff} is increased, the mass loss rates leading to He overabundances go down (see §4.2), the mechanism must become more effective in reducing mass loss as the temperature is increased. Presumably that must be an horizontal magnetic field. If one accepts the empirical values of the mass loss as determined by Abbott (1982), there must be a magnetic field to reduce mass loss in all He rich stars. According to Abbott (1979), the mass loss formula should be reasonably accurate for the higher temperature He rich stars but may be an overestimate around T_{eff} = 20000 K.

In their outer regions, He rich stars have an He convection zone. It starts at an optical depth of 2 or 3 and ends at an optical depth of 30. It is due to He II ionization. It cannot disappear by He settling at its bottom because the mass loss rate is too large (see §4.3) to allow He settling. Some atmosphere models also have convection zones due to He I ionization at optical depths that are smaller than 1 (see e.g. Mihalas 1965). Given the high velocity of random motions in convection zones, they are nearly certainly homogeneous (Schatzman 1969). The separation cannot take place in convection zones. It is also believed that convection zones lead to some overshooting (Latour, Toomre and Zahn 1981). In the presence of such convection zones, can separation take place in the atmosphere of He rich stars? The He separation can only start where overshooting is stabilized and the atmosphere is stable. Perhaps the magnetic field could eliminate the convection or at least the overshooting in parts of the surface. If, because of overshooting, the atmosphere were mixed, the separation could only take place in the wind. Whether the separation takes place in the wind or the atmosphere leads itself to an observational test.

The model that relies on the separation of neutral He in the atmosphere *requires* a relatively large mass loss rate of 10^{-12} $M_\odot$ yr^{-1}. From equation (6) this mass loss rate allows the separation of no other element. All other elements should be normal in He rich stars if this model is right. This applies to the CNO elements in particular. It does not appear possible to evade this consequence of the model though it should be checked by more detailed calculations. Furthermore I know of no other model to explain He overabundances that makes such a prediction. Surely, if He overabundances were to be explained by H burning in 5 $M_\odot$ stars, the relative abundances of the CNO isotopes would be modified.

Similarly, if the separation were to take place in the wind, other anomalies should be present. The mass loss rate is then smaller than that needed to cause overabundance from separation in the atmosphere and additional anomalies are expected from separation in the wind itself or the envelope. The search for other anomalies in He rich stars becomes a precise test of the models that have been proposed for these objects. It should be conducted in the cooler He rich stars since only there is the mass loss rate that allows He separation clearly larger than that allowing the separation of heavier elements.

The decrease in the fraction of neutral He as the T_{eff} increases puts the model based on the separation in the atmosphere in difficulty in the hotter He rich stars. According to Osmer and Peterson (1974) some of the He rich stars have T_{eff} = 29000 K. In such stars, the effect of neutral He is clearly negligible and an alternate model seems necessary. It could be that the separation actually occurs in the wind and not in the atmosphere in the hotter He rich stars. It is also possible that the effective temperature of some He rich stars has been overestimated by some 2000 K. It would be important to determine more precisely the effective temperature range of the phenomenom.

On the other hand the observation of abundance anomalies in various stars of the Herzprung Russel diagram allows the determination of constraints on the mass loss rates that are allowed. On the main sequence, the rates go from 10^{-12} to 10^{-15} $M_\odot$ yr^{-1} depending on the effective temperature. There are similarly constraining limits on the horizontal branch and its continuation. Heber (1985) has discussed constraints coming from the diffusion of He in the atmosphere. If one uses the upper limit determined by Michaud et al. (1985) to evaluate the amount of mass that can have been lost in the sdOB state one finds approximately:

$$\Delta = -\frac{dM}{dt}\tau = 10^{-14}\tau \quad M_\odot \text{ yr}^{-1} \tag{7}$$

where τ is the life time in the sdOB stage. Clearly, if the sdOB stage lasts only 10^6 yr, only a very small mass (10^{-8} $M_\odot$) may have been lost and this has important consequences for the models that attempt to explain the He richness in the sdOs as a consequence of the loss of the H rich envelope during the sdOB stage. They would need to have become a sdOB with only 10^{-8} M$_\odot$ of hydrogen.

An alternative would try to explain the He richness of the sdOs as due to a wind in the sdO stage, just as is preferred for the main sequence. It must be noticed that similar constraints would apply. The implied mass loss rate would then be by 10^{-13} to 10^{-12} M$_\odot$ yr^{-1}. It is a very unlikely model since horizontal branch stars are known to be He rich objects in their interior.

REFERENCES

Abbott, D. C. 1979, in **Mass Loss and the Evolution of O type Stars** Eds: P.S. Conti and C. W. H. de Loore.
Abbott, D. C. 1982, Ap. J., **259**, 282.
Alecian, G., and Vauclair, S. 1981, Astr. Ap., 101, 16.
Bolton, C. T. 1984, in **Proceedings of the HVAR Workshop on Rapid Variability of Early-Type Stars**,
Borra, E. F., and Landstreet, J. D. 1979, Ap. J., **228**, 809.
Chapman, S., and Cowling, T. G. 1970, **The Mathematical Theory of non-Uniform Gases** (3d ed.; Cambridge University Press).
Dupuis, J. 1985, Internal Report, Université de Montréal.
Fakir, R. 1985, Internal Report, Université de Montréal.
Havnes, O., and Goertz, C., K. 1984, Astr. Ap., **138**, 421.
Heber, U. 1985, Astr. Ap., **290**, 000.
Iben, I. 1966, Ap. J., **143**, 505.
Iben, I, and MacDonald, J. 1985, preprint.
Laing, E. W. 1981, in **Plasma Physics and Nuclear Fusion Research**, Ed. R. D. Gill (London: Academic Press).
Latour, J., Toomre, J., and Zahn, J.-P. 1981, Ap. J., **248**, 1081.
Liebert, J. 1980, Ann. Rev. Astr. Ap., 18, 363.
Martel, A. 1979, Internal Report, Université de Montréal.
Mégessier, C. 1984, Astr. Ap., **138**, 267.
Michaud, G., Bergeron, P., Wesemael, F., and Fontaine, G. 1985, Ap. J., **299**, 000.
Michaud, G., Fontaine, G. 1984, Ap. J., **283**, 787.
Michaud, G., Fontaine, G., and Charland, Y. 1984, Ap. J., **280**, 247.
Michaud, G., Martel, A., and Ratel, A. 1978, Ap. J., **226**, 483.
Michaud, G., Mégessier, C., and Charland, Y. 1981, Astr. Ap., **103**, 244.
Michaud, G., Montmerle, T., Cox, A.N., Magee N.H., Hodson, S.W., and Martel, A. 1979, Ap. J., **234**, 206.
Mihalas, D. 1965, Ap. J. Suppl., **9**, 321.
Mihalas, D. 1972, **NCAR Technical Note NCAR-TN/STR-76**, (Boulder, National Center for Atmospheric Research).
Montmerle, T., and Michaud, G. 1976, Ap. J. Suppl., 31, 489.
Osmer, P. S., and Peterson. D. 1974, Ap. J., **187**, 117.
Paquette, C., Pelletier, C., Fontaine, G., and Michaud, G. 1985, Ap. J. Supl., in press.
Pelletier, C., Paquette, C., Michaud, G., and Fontaine, G. 1985, Submitted for publication.

Schatzman, E. 1958, White Dwarf, (Amsterdam: North Holland).
Schatzman, E. 1969, Astr. Ap., 3, 331.
Vauclair, S. 1975, Astr. Ap., 45, 233.
Vauclair, S. 1981, A. J., 86, 513.
Vauclair, S., Hardorp, J., and Peterson, D. A. 1979, Ap. J., 227, 526.

DISCUSSION

LYNAS-GRAY: The upper limits to mass-loss rates for helium depletion that you suggest are consistent with observation since no large mass-loss rates have been detected for such objects.

MICHAUD: I mainly wanted to insist on the strict constraint on mass loss rates that is implied by observations of He underabundances. There is no alternative to gravitational settling to explain the underabundance of He. The latter is possible only if the mass-loss rate is very small. This may have implications for some models of sdO's.

HEBER: What is the evolutionary time scale you adopted for the estimates of mass-loss rates for helium poor sdO stars.

MICHAUD: The most important time scale is probably for the disappearance of the He convection zone. For a mass loss rate of 10^{-14}, you have a factor of 3 underabundance of He after 2.10^6 years. If you want to have a factor of 10 underabundance of He, then I think you run into problems with the sdO's. If you take 10^{-13}, then the time scale for diffusion would be about 10^8 years, maybe a bit less.

HEBER: Since we believe that these stars are extended horizontal branch stars and, therefore, should have rather large evolutionary time scales, around 10^8 years, the long time scale would pose no problems.

MICHAUD: Hence the mass-loss rate should be 10^{-13} solar masses per year, which is generally accepted.

EVOLUTIONARY STATUS AND ORIGIN OF EXTREMELY HYDROGEN-DEFICIENT STARS

D. Schönberner
Institut für Theoretische Physik und Sternwarte
der Universität Kiel
Olshausenstr. 40, 2300 Kiel, F.R.G.

ABSTRACT. Our present knowledge about the evolutionary status of extremely hydrogen-deficient stars is reviewed. Possible schemes for their creation are discussed, with special emphasis on a recently proposed binary scenario, according to which they are the result of the merging of two white dwarfs. Finally, the possible generic links between the different groups of extremely hydrogen-deficient stars are briefly discussed.

1. INTRODUCTION

The main question that we have been interested in since the detection of the first extremely hydrogen-deficient (EHd) star by Popper in 1942 is the following: how can a low mass helium star come into being from a low mass precursor? Unfortunately, we are not yet in a position to answer this question definitively, although several scenarios have been put forward. However, we are able to say something about the internal stucture of EHd-stars and, in a few cases, also about their evolution. I will firstly, therefore, concentrate on our present knowledge of the internal structure and observed evolution of EHd-stars before I discuss the various schemes for their origin.
    In the following, the term "EHd-star" means an object in which the atmospheric hydrogen content is reduced by at least a factor of 10^3 (i.e. $H/He \leq 10^{-3}$, by number fractions). Consequently, the Extreme Helium (EHe) stars, the R CrB-stars, and the Hydrogen Deficient Carbon (HdC) stars are also EHd-stars. All intermediate helium stars are thereby automatically excluded, as well as the somewhat "peculiar" helium star BD +13°3224, also known as V652 Her. I have also excluded the hydrogen deficient binaries since their origin and evolution is understood within the frame work of binary evolution (Schönberner and Drilling 1983; see also the review articles of M. Plavec and A. Tutukov in these proceedings). Furthermore, all helium rich underluminous O-stars are excluded, despite the fact that some of them are carbon-rich and might be further evolved EHe-stars (see the contribution by Husfeld, Heber and Drilling, these proceedings).

2. INTERNAL STRUCTURE OF EHd-STARS

Spectroscopic observations and their interpretations by model atmospheres yield the data necessary for the determination of the internal structure of EHd-stars: effective temperature T_{eff}, gravity g, and chemical composition. With effective temperature and gravity known, the objects can be placed in a (g, T_{eff})-plane (equivalent to the conventional H.R.-diagram) and compared with the loci as predicted by stellar model calculations. Since the results of various fine analyses of EHd-stars have been compiled and discussed in the reviews by Heber and Lambert (these proceedings), I will therefore only briefly summarize the main results:

i) Besides their hydrogen deficiency, EHd-stars are carbon-rich with C/N > 1, as opposed to the hydrogen-deficient binaries which have C/N < 1 (cf. Schönberner and Drilling, 1984);

ii) they occupy a narrow strip in the (g, T_{eff})-diagram (cf. Fig. 1), ranging from the (normal) main sequence towards the giant region;

iii) the luminosity to mass ratio is about constant, and is given by log(L/M) = 4.1 ± 0.5 (solar units).

It should be noted that these results are only based on the analyses of a very limited number of objects, namely on 4 EHe- and 3 R CrB-stars. Furthermore, recent observations seem to indicate that the spread in L/M is intrinsic, i.e. not only caused by observational errors.

From their position in the (g, T_{eff})-diagram above and far to the right of the helium main sequence, we infer that EHd-stars are inhomogeneous helium stars evolved far beyond the core helium burning phase. The first comprehensive study on how low mass helium stars evolve after the core helium burning phase has finished was that of Paczynski (1971). Other calculations of relevance are those of Rose (1969), Biermann and Kippenhahn (1971), Dinger (1972), Trimble and Paczynski (1973), Savonije and Takens (1976), Schönberner (1977), Law (1982) and Habets (1985). It emerges from all these calculations that the observed loci of EHd-stars can be accounted for by inhomogeneous models within the mass range $0.7 \lesssim M/M_\odot \lesssim 2.7$ and luminosity range $3.8 \lesssim \log(L/L_\odot) \lesssim 4.7$. According to Dinger (1972), a small admixture of carbon does not change the results of these evolutionary calculations as long as C/He << 1 (number fractions). For instance, a 1 $M_\odot$ model with C/He = 0.44 does not become a giant at all after core helium burning has finished. Fortunately, the observed carbon to helium number ratio is ≈ 0.01, too small for any noticeable effects on the internal evolution of helium stars (Dinger, 1972). A small extra carbon content may, however, influence to some extent the redward excursion of the giant models. A comparison between the observations of EHd-stars and evolutionary calculations of Paczynski (1971) is shown in Figure 1. These calculations simulate the evolution of helium stars with X = 0, Z = 0.03, i.e. with no extra carbon, from their "main sequence" till the onset of carbon burning or the pre-white dwarf stage. If we neglect for the moment the problem of how to create helium main-sequence stars, we find a very gratifying agreement between stellar evolution theory and the observations.

The details of the helium star evolution are as follows: models with $M > 1.1\ M_\odot$ start carbon burning at, or on their way to, the Hayashi limit. Paczynski (1971) found a mild off-center carbon ignition in his 1.5 and 2.0 $M_\odot$ models but did not compute any further. Rose (1969) argued that such off-center ignition of carbon in only weakly electron-degenerated regions does not seriously alter the structure of the model. Instead, his 1.45 $M_\odot$ model was evolved further to larger radii until carbon burning started in the highly degenerate center of the CO-core ($M_{CO} \approx 1.4\ M_\odot$). More massive helium stars ($M \gtrsim 2\ M_\odot$) do not develop degenerate CO-cores at all, but for $M \lesssim 2.7\ M_\odot$, they also expand to the Hayashi limit during carbon burning (Savonije and Takens, 1976; Habets, 1985). More massive helium stars do not evolve into the cool region of the H.R.-diagram. If mass loss is neglected, the ultimate fate of all helium stars with $M \gtrsim 1.4\ M_\odot$ will be a supernova explosion (Wheeler, 1978).

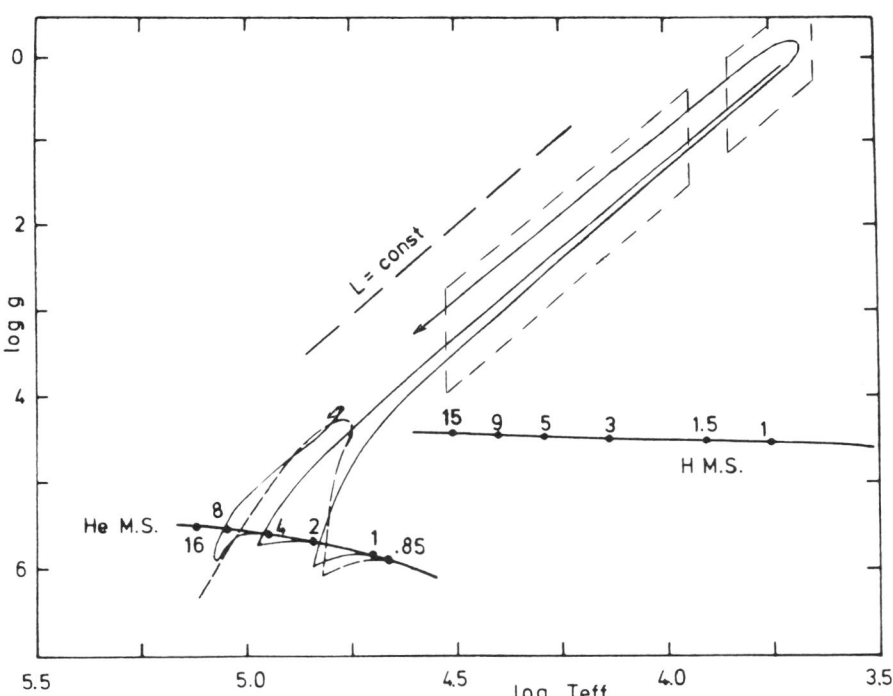

Fig. 1 Observed loci of EHd-stars in the (g, T_{eff})-plane and evolutionary tracks of helium stars with different masses according to Paczynski (1971). EHe-stars occur for $4.5 \gtrsim \log T_{eff} \gtrsim 3.95$, whereas the HdC- and most of the R CrB-stars are confined to $\log T_{eff} \lesssim 3.85$. The numbers along the main sequences give stellar masses in solar units.

We conclude that helium giants can be modelled in two ways:
i) by low mass models ($0.7 \lesssim M/M_\odot \lesssim 2$) which burn helium in a shell on top of a degenerate, inert CO-core;
ii) by more massive models ($2 \lesssim M/M_\odot \lesssim 2.7$) which burn carbon in the center or in a shell.

Helium giants which are able to avoid a supernova explosion, either by mass loss or because they started their evolution with $M < 1.4\ M_\odot$, will finally contract again when their envelope mass falls below a certain value ($\approx 10^{-2}\ M_\odot$), thereby crossing the observed loci of EHd-stars for a second time before they become white dwarfs. Thus, a definite assignment between the evolutionary tracks and the observations is not possible. This is a serious problem because it is crucial for our understanding of the origin and evolution of EHd-stars to know whether they are born as helium main sequence stars which evolve to the Hayashi limit, or as red giants which then further contract from the Hayashi limit to the white dwarf region.

Detailed calculations for models contracting from the Hayashi limit have only been performed by Schönberner (1977). It turned out that the times for crossing the B-type spectral region (i.e. the region where the EHe-stars are found) are very dependent on the direction of evolution: the predicted lifetimes as B-type stars are about ten times larger for evolution "to" the Hayashi limit than those for evolution "from" the Hayashi limit (Table I). Since the latter models, owing to their more advanced evolution, also have larger luminosities (cf. the 1 $M_\odot$ track in Fig. 1), comparision between theory and observation also leads to different mass assignments. The details are given in Table I. It is clear that birthrate estimates are practically impossible as long as we are unable to discriminate between both directions of evolution.

TABLE I

	M_*	M_{CO}	t(B-star)/yr	$(L/M)/(L_\odot/M_\odot)$
"TO" :	(1-2) $M_\odot$	(0.6-0.7) M_*	$2 \cdot 10^4$	$10^{3.9}$
"FROM":	$\approx 0.7\ M_\odot$	$\approx M_*$	$2 \cdot 10^3$	$10^{4.1}$

3. THE OBSERVED EVOLUTION OF EHd-STARS

Concerning the present evolution of EHd-stars, some progress has been made recently by analyzing the the radial pulsations of some R CrB-stars. Pugach (1977) was the first to notice that the period of RY Sgr (P = 39 days) had decreased by about 1 day within the last 40 years. More detailed analyses performed by Kilkenny (1982) and Marraco and Milesi (1982) yielded a period decrease of P = - 0.011 and - 0.015 days/yr, respectively, which corresponds to a pulsational lifetime of about 3000 years. Other R CrB-stars seem also to exhibit period variations (Kilkenny and Flanagan 1983), but the analyses of these are as yet inconclusive. At least one EHe-star, BD+1°4381, also pulsates radially, with a period of about 21 days (Jeffery and Malaney, 1985), but these observations are still too preliminary to be able to say something about period changes. Thus, our information on the present evolution of EHd-stars is based, strictly speaking, on only one single object, namely on RY Sgr.

The consequences, however, are very important. First of all, RY Sgr is obviously evolving to hotter effective temperatures, i.e. it is shrinking and will become a white dwarf in the future. This implies that its mass (more precisely, the mass of its CO-core) is definitely below the SN-limit of 1.4 $M_\odot$, and probably also below 1.1 $M_\odot$, the carbon burning mass limit. Moreover, Kilkenny (1982) has shown its observed evolutionary speed to be consistent with that of post-red giant helium star models of 0.7 to 1 $M_\odot$ as computed by Schönberner (1977).

Recent theoretical calculations of nonlinear radial pulsations of helium-carbon envelopes (Saio and Wheeler, 1985; Saio, these proceedings) have also given interesting results. They show that stellar models with T_{eff} = 7000 K pulsate with a period of 40 days (as is observed for RY Sgr), and that this period is stable even for M = 0.7 $M_\odot$. At lower effective temperatures ($\approx$ 6000 K), the pulsation amplitudes grow without bound if the stellar mass is below 1.6 $M_\odot$. Since the effective temperature of RY Sgr is about 7000 K (Schönberner, 1975; Cottrell and Lambert, 1982), these new calculations do not contradict a mass estimate for RY Sgr of less than 1 $M_\odot$. In the past, when RY Sgr was cooler, it may have experienced severe mass loss owing to these unstable pulsations. Such mass loss leads to an accelerated blueward evolution, thereby reducing the lifetime of R CrB-stars with $T_{eff} \lesssim$ 6000 K as compared to the hotter ones and leading to a deficit of such cool R CrB-stars. It would be very important to check this prediction by determining the effective temperatures of as many R CrB-stars as possible. Pulsationally induced mass loss may also prevent more massive R CrB-stars from exploding as Supernovae.

Heber and Schönberner (1981) found that the temperature distribution of EHe-stars could be explained by Schönberner's (1977) models of contracting helium giants. The evidence is, however, weak because of the poor statistics and further investigations with a larger sample are necessary.

4. ORIGIN OF EHd-STARS

Each scenario for the origin of EHd-stars that we can think of must meet two observational constraints:
i) EHd-stars are evolved single objects of low mass, belonging to a rather old population;
ii) their surface abundances indicate that their envelope consists of a mixture between the original matter (trace of hydrogen), CN-processed matter (enhanced nitrogen content) and 3α-processed matter (enhanced carbon content).

Owing to the latter constraint, a simple mass loss scenario, according to which the hydrogen rich envelope is removed, is completely ruled out. Also, the exposed helium remnant would, in the case of an AGB-progenitor, be very compact ($R \approx 2R_{WD}$) and very hot ($T_{eff} \approx 10^5$ K). One could, of course, imagine more massive progenitors. For instance, a star with $M = 5\ M_\odot$ develops a helium core of $\approx 1\ M_\odot$ during the M.S.-phase. But how could the star get rid of $\approx 4\ M_\odot$ if not by binary evolution? Furthermore, a $5\ M_\odot$ has too short a life for its remnant to be considered as an old star. A violent core helium flash is also ruled out. A stripped-off helium core is also rather compact and has only $0.5\ M_\odot$. Helium giants with such a small total mass could not be constructed numerically. Instead, the high luminosity to mass ratio and the internal structure of EHd-stars suggest a relationship to asymptotic giant branch (AGB) stars (Schönberner, 1975, 1977).

Paczynski (1971) was the first to propose that deep envelope mixing on the AGB could be responsible for the formation of a helium giant. The first detailed modelling of such a mixing process was presented by Sackmann et al. (1974). In their so-called "Eruptive Model" they artificially extended the envelope convection in a low-mass AGB-star somewhat beneath the hydrogen-helium discontinuity. Owing to the large and very rapid energy production, triggered by the injection of fresh fuel, part of the envelope is rapidly blown off whereas the rest is processed. However, when hydrogen becomes sufficiently depleted (H/He $\lesssim 10^{-2}$), a carbon isotopic ratio $^{12}C/^{13}C \approx 6$ is established. This is in strong contradiction to the observations in R CrB-stars where $^{12}C/^{13}C \approx 40$ is found (Cottrell and Lambert 1982). Thus, this model cannot explain the observed surface abundances in R CrB-stars. It is also unclear if, and how the remnant is able to settle down into a quiet nuclear burning state after this explosive event. This is an important point since, as was mentioned in the previous chapter, the observed evolutionary time scale of RY Sgr is in agreement with helium star models in thermal equilibrium evolving to hotter temperatures and smaller radii.

Another mechanism proposed as leading to hydrogen-deficient giants is the so-called "Hot Bottom Burning" of Scalo et al. (1975). Their computations showed that the base of the convective envelope in AGB-stars of at least $1.5\ M_\odot$ and $10^4\ L_\odot$ is sufficiently hot as to allow nuclear processing via the CNO-bicycle. This can also lead to a hydrogen depletion of the whole envelope, provided the available time is sufficiently large. If we demand a hydrogen depletion by a factor of 100 during the typical lifetime of a luminous AGB-star of about 10^6 yrs, a minimum base temperature of $70 \cdot 10^6$ K, corresponding to $M \geq 2\ M_\odot$, follows from the computations of Scalo et al. Such high progenitor masses are also difficult to reconcile with the observed population characteristic of EHd-stars. Moreover, any larger hydrogen reduction will obviously lead to structural changes of the envelope in such a way that the results of Scalo et al. are no longer applicable. We also expect, owing to the CNO-process, a nitrogen to carbon ratio $N/C > 1$ to be established very rapidly, even if fresh carbon is dredged up in the aftermath of a thermal pulse. In summary, we must state that this model also fails to explain the observed surface chemistry of EHd-stars.

Yet another scheme was proposed by Iben et al. (1983), and may be called the "Last Helium Shell Flash Scenario". As is well known from evolutionary calculations, post-AGB models may experience a last thermal pulse immediately before they embark onto the white dwarf track (Schönberner, 1979; Iben 1982). The energy output of this last shell flash leads to large-scale mixing and a brief expansion of the envelope to giant dimensions. Only very approximate calculations are as yet available, but it is clear that the brief excursion to the vicinity of the Hayashi limit is governed by the thermal time scale of the envelope. Because of the high luminosity ($\approx 10^4$ $L_\odot$) and the small envelope mass ($\approx 10^{-4}$ $M_\odot$), this time scale is very short: the model lifetime in the R CrB-star domain is only about 10^2 yrs (Iben et al. 1983), much too short to account for the observations. In particular, the lifetime of RY Sgr, as indicated by its pulsation properties (cf. Chapter 3) is about thirty times larger. We conclude that this scenario also seems to fail in explaining the origin of EHd-stars.

So far, all the schemes that have tried to trace the origin of EHd-stars back to the AGB have severe drawbacks and are thus not convincing. A scenario proposed by Webbink (1984) is completely different and has no relationship to the AGB-evolution of single stars. It is essentially based on two merging white dwarfs which are the product of binary evolution. The details are as follows (cf. Webbink, 1984): the initial systems in question are fairly massive ($2 \lesssim M_p/M_\odot \lesssim 3.5$), with mass ratios close to unity, and begin their evolution with primordial periods between 1 and 10 d. Conservative case B mass transfer transforms the primary into a helium degenerate of about 0.3 to 0.4 $M_\odot$. In the meantime the secondary, which is now much more massive ($M_s/M_p \approx 10$), is able to evolve to the giant branch and to develop a degenerate CO-core before it overflows its Roche lobe. Owing to the large mass ratio, this second mass transfer is very unstable and will obviously lead to a common envelope stage, the result of which being a substantial systemic loss of mass and angular momentum. The secondary is thereby transformed into a CO-white dwarf, and the final periods are now between 1 and 10 h. The system now consists of two close white dwarfs (a He-degenerate and a CO-degenerate) which emit gravitational radiation and thereby lose angular momentum. After a time span of between 10^8 and 10^{10} yr, depending on their final separation, the lighter of these two white dwarfs, which is the He-degenerate, is forced to overflow its Roche lobe and transfer to the companion mass at a rate of about 10^{-4} $M_\odot$ yr^{-1}. It is thought that accretion of helium at such a rate very soon leads to quiet helium burning at the base of the helium envelope which is built up on top of the CO-degenerate. This growing helium envelope is expected to increase in radius and eventually to engulf the remains of the He-degenerate within only about 10^3 yr. If, in this final mass exchange, no matter leaves the system, we end up with a single helium red giant (i.e. a R CrB- or HdC-star) with a total mass between 0.8 and 1.4 $M_\odot$, consisting of a CO-core of about 0.5 to 1.0 $M_\odot$ (originating from the CO-degenerate), and a helium envelope of 0.3 to 0.4 $M_\odot$ (originating from the He-degenerate).

At a first glance, although this scenario leads to helium giants in the appropriate mass range, it seems to violate the population constraint posed above. The observations can surely not be reconciled with progenitor masses of 2 or 3 $M_\odot$! The solution comes from the fact that the gravitational radiation emitted from the white dwarf binary leads to only a very slow orbital decay, which may last up to the order of 10^{10} yrs. Thus, even if a helium giant is presently created by the merging of two white dwarfs which originated, according to Webbink's scenario, from rather massive progenitors, it may still belong to an old stellar population. Webbink's scenario also has other advantages:

i) The trace of hydrogen that is observed in the surfaces of EHd-stars stems from the thin layer of unburned hydrogen at the surface of the He-degenerate after case B mass exchange. For instance, a He-degenerate of 0.38 $M_\odot$ still contains 2.10^{-4} $M_\odot$ of unburned hydrogen (Iben and Tutukov, 1985). This hydrogen would be mixed with helium during the mass transfer process, leading to a hydrogen to helium ratio of 5.10^{-4} in this particular case.

ii) Furthermore, the nitrogen is also provided by the helium degenerate, where all the primordial carbon and oxygen have been converted to nitrogen.

iii) The high atmospheric carbon content of EHd-stars can be accounted for by assuming that the accretion onto the CO-degenerate occurs through a heavy disk (Iben and Tutukov, 1985). Shear mixing may then dredge up some carbon (^{12}C!) from the surface of the white dwarf into the accreted helium envelope. A total amount of only 5.10^{-3} $M_\odot$ would be sufficient! Of course, some oxygen also has to be dredged up.

According to this binary scenario, EHd-stars are born as cool giants due to the above described merging process and, subsequently, evolve to higher effective temperatures. Their lifetime in the vicinity of the Hayashi-limit can be estimated by $M_{env}/\dot{M}_{co} \approx (0.3-0.4)/1.10^{-6} = (3-4)\,10^5$ yr, which is only an upper limit since mass loss is neglected. Mass loss rates in excess of 10^{-6} $M_\odot$ yr^{-1} (i.e. $|\dot{M}| > \dot{M}_{co}$) would shorten the lifetime considerably.

Despite the apparent success of Webbink's proposal for the origin of EHd-Stars, a cautionary remark seems to be in order: we urgently need its confirmation by detailed evolutionary calculations! Assuming for the moment that Webbink's scenario is correct, we can go even further since he also estimated a birthrate for He-Co white dwarf close binary systems, i.e. for the immediate progenitors of EHd-stars: 2.10^{-11} pc^{-2} yr^{-1}, or 0.015 yr^{-1} for the whole galaxis. With this birthrate, and assuming that each merging He-Co system leads to a shell burning helium giant, we estimate a galactic total of $\approx 0.015 \times 2.10^3 = 30$ EHe-stars. A significant fraction of our galaxis has been surveyed for EHe-stars (see Drilling, these proceedings) and 17 objects have been detected so far. It is difficult to estimate the fraction of objects hidden behind the galactic dust layer, but it seems very likely that the total galactic population of EHe-stars is not an order of magnitude different from the figure estimated above.

The situation is unclear in the case of the cool EHd-stars, i.e. the R CrB- and HdC-stars. Assuming a lifetime of 2.10^5 yr, we estimate in the same way as above a total of about 3.10^3 objects. If cool R CrB-stars suffer severe mass loss ($> 2.10^{-6}$ $M_\odot$ yr^{-1}), their lifetime is reduced accordingly and thus also their space density. For instance, in order to end up with a total of 50 R CrB-stars in our galaxis, their lifetimes must be reduced to only 3.10^3 yr, i.e. to values comparative to those of EHe-stars. Mass loss rates of the order of 10^{-4} $M_\odot$ yr^{-1} are necessary to speed up the evolution to this extent. It would be very important in this context to determine the actual mas loss rates of cool R CrB-stars. It should also be noted that, according to Feast (these proceedings), the mass loss rates of the hotter R CrB-Stars are only of the order of 10^{-6} $M_\odot$ yr^{-1}, and therefore of minor importance for the evolutionary time scale. The mass loss rates of EHe-stars are even smaller: $< 10^{-7}$ $M_\odot$ yr^{-2} (Hamann et al. 1982).

5. ON THE RELATIONSHIP BETWEEN THE DIFFERENT CLASSES OF EHd-STARS

In closing this review, I would like to speculate about the evolutionary connections between the different classes of EHd-stars. As outlined in Chapters 2 and 3, the EHe-, R CrB- and HdC-stars are linked together by evolutionary tracks, the main difference between them being the effective temperatures or the envelope masses, respectively. They also seem to have the same, or at least a very similar, surface chemistry (cf. the reviews of Heber and Lambert, these proceedings). But differences must certainly exist, an important one being the existence of dust shells around cool and hot R CrB-stars, which are most likely made up of carbon grains. The EHe- and HdC-stars, on the other hand, have no observed IR-excess, hence no dust shells (Feast and Glass, 1973; Drilling et al., 1984; Walker, 1985). Maybe they lost their dust shells during an earlier evolutionary phase when they were close to the Hayashi-limit. It might also be that the HdC-stars are the ancestors of the EHe-stars and that they did not develop dust shells at all. This would be consistent with their lack of pulsations since it is believed that these lead to heavy mass loss (see Chapter 3).

According to theory, however, HdC-stars should pulsate (Saio and Wheeler, 1985); either the HdC-stars are hotter than believed and on the stability side of the instability line, or their luminosity to mass ratio is smaller than that of R CrB-stars in such a manner that they pulsate only with a very small, not easily detectable amplitude. The luminosity to mass ratios of post-red giant helium stars depend very much on their masses: for instance, a model with $M = 0.6$ $M_\odot$ has $L/M = 10^{3.8}$ $L_\odot/M_\odot$, that of 1 $M_\odot$ has $L/M = 10^{4.5}$ $L_\odot/M_\odot$.

In concluding, it must be stated that we are still far from clearly understanding the origin and evolution of EHd-stars. Also, the connection between the different subclasses is not well-understood. Obviously, more observational and theoretical studies are needed.

ACKNOWLEDGMENTS. The author gratefully acknowledges the travel grant from the Deutsche Forschungsgemeinschaft

REFERENCES

Biermann, P., Kippenhahn, R.: 1971, Astron. Astrophys. **14,** 32
Cottrell, P.L., Lambert, D.L.: 1982, Astrophys. J. **261,** 595
Dinger, A.St.C.: 1972, Mon. Not. R. Astron. Soc. **158,** 383
Drilling, J.S., Landolt, A.U., Schönberner, D.: 1984,
    Astrophys. J. **279,** 748
Feast, M.W., Glass, I.S.: 1973, Mon. Not. Roy. Astron. Soc. **161,** 293
Habets, G.M.H.J.: 1985, Ph.D.-Thesis, University of Amsterdam
Hamann, W.-R., Schönberner, D., Heber, U.: 1982,
    Astron. Astrophys. **116,** 273
Heber, U., Schönberner, D.: 1981, Astron. Astrophys. **102,** 73
Iben, I.Jr.: 1982, Astrophys. J. **260,** 821
Iben, I.Jr., Kaler, J.B., Truran, J.W., Renzini, A.: 1983, Astrophys.
    J. **264,** 605
Iben, I.Jr., Tutukov, A.V.: 1985, Astrophys. J. Suppl. **58,** 661
Jeffery, C.S., Malaney, R.A.: 1985, Mon. Not. R. Astron. Soc. **213,** 61p
Kilkenny, D.: 1982, Mon. Not. R. Astron. Soc. **200,** 1019
Kilkenny, D., Flanagan, C.: 1983, Mon. Not. R. Astr. Soc. **203,** 19
Law, W.-Y.: 1982, Astron. Astrophys. **108,** 118
Marraco, H.G., Milesi, G.E.: 1982, Astron. J. **87,** 1775
Paczynski, B.: 1971, Acta Astron. **21,** 1
Pugach, A.F.: 1977, Inf. Bull. Var. Stars No. 1277
Rose, W.K.: 1969, Astrophys. J. **155,** 491
Sackmann, I.-J., Smith, R.L., Despain, K.H.: 1974, Astrophys. J.
    187, 555
Saio, H., Wheeler, J.C.: 1985, Astrophys. J. **295,** 38
Savonije, G.J., Takens, R.J.: 1976, Astron. Astrophys. **47,** 231
Scalo, J.M., Despain, K.H., Ulrich, R.K.: 1975, Astrophys. J. **196,** 805
Schönberner, D.: 1975, Astron. Astrophys. **44,** 383
Schönberner, D.: 1977, Astron. Astrophys. **57,** 437
Schönberner, D.: 1979, Astron. Astrophys. **103,** 119
Schönberner, D., Drilling, J.S.: 1983, Astrophys. J. **268,** 225
Schönberner, D., Drilling, J.S.: 1984, Astrophys. J. **276,** 229
Trimble, V., Paczynski, B.: 1973, Astron. Astrophys. **22,** 9
Walker, H.: 1985, Astron. Astrophys. **152,** 58
Webbink, R.F.: 1984, Astrophys. J. **277,** 355
Wheeler, J.C.: 1978, Astrophys. J. **225,** 212

DISCUSSION

HILL: If these are 10^{10} years old where do the solar metal abundances come from?

SCHÖNBERNER: I don't know. We know that some of the extreme helium stars probably have lower than solar metallicity. Please remember that this scenario has to be worked out in detail and that the metallicity of most of the extreme He stars is also still unknown.

FEAST: I would think it is fairly risky to divide the hydrogen-deficient carbon stars from the R CrB stars on the basis that they don't pulsate. The evidence that hydrogen-deficient carbon stars do not pulsate is very thin.

SCHÖNBERNER: I only tried to find a reason why hydrogen-deficient carbon stars do not pulsate. Careful photometric observations should be done in the future to settle this question.

RAMADURAI: In the case of the core He flash you mentioned, I would like to bring to your attention the work of Eggleton and Schramm on the role of meridional circulation during the first ascent on the giant branch.

SCHÖNBERNER: I doubt whether a meridional circulation of this sort would be able to mix hydrogen down and bring helium up to such an extent as that observed in the extremely hydrogen-deficient stars.

FEAST: I am puzzled about the discussions on changing periods of RY Sgr and S Aps. I really don't understand, if your are invoking mass loss to explain S Aps, why it is of no importance for RY Sgr. It seems to me that largely the mass loss is comparable in the two cases.

SCHÖNBERNER: I don't know of any reliable mass loss determinations for R CrB stars. Thus, I cannot comment on your last sentence. You advocated a rate of about 10^{-6} $M_\odot$/yr, if I remember correctly. Such a rate is comparable to the growth rate of the helium-exhausted core and hence does not speed up the blueward evolution very much. My arguments were mainly based on the results of pulsational calculations, which predict a strong increase of the mass loss rate for effective temperatures below 6000 K.

FEAST: Do I understand that you think you would rule out the Iben's "born again" planetary business on the basis of abundances?

SCHÖNBERNER: I would rule it out because the predicted lifetimes in the R CrB-domain are too short.

FEAST: Too short for what?

SCHÖNBERNER: Too short for the observed pulsational lifetimes of R CrB and RY Sgr which are about several thousand years.

FEAST: Can we really say that? I mean, for instance, for R CrB stars we do not have a complete range from the hot to the cool end. If you want to connect the two together you have to go very rapidly between these two phases.

SCHÖNBERNER: A 0.7 $M_\odot$ contracting helium giant needs about 5000 yr to evolve from 7000 K to, say, 20000 K. I think that is a quite rapid evolution.

N.K. RAO: We know that R CrB has not changed much in the last two hundred years.

LIEBERT: What do you think of the relevance of Iben et al. scenario for the purpose for which it was proposed, i.e. for the formation of DB white dwarfs in an old population, presumably from the helium-rich progenitors, which we identified as SdOs.

SCHÖNBERNER: I think that is a reasonable explanation.

N.K. RAO: Is there any way to delay the evolution in the red giant phase and increase the time scale there?

SCHÖNBERNER: The timescale of the redward excursion is solely determined by the small envelope masses ($\approx 10^{-4}$ M_o) and the large luminosities ($\approx 10^{+4}$ L_o) involved.

ON THE ORIGIN OF HELIUM RICH STARS

A. Tutukov
Astronomical Council of the USSR Academy of Sciences
Pyatnitskaya St. 48
Moscow, USSR

ABSTRACT: *The modern theory of stellar evolution leads to several possible scenaria for the formation of many types of hydrogen deficient (usually helium rich) stars. Some of these scenaria are briefly discussed in this paper to provide possible explanations for origins of R Cr B, hot helium and Wolf-Rayet stars; hot sub dwarfs, non-DA degenerate dwarfs and helium stars in binaries like KS Per. The merger of two degenerate stars in close binaries due to the radiation of gravitational waves can lead to several types of helium and carbon rich stars as well as to supernova explosions.*

Stars with hydrogen-deficient (and usually helium rich) envelopes have been found in many regions of the Hertzsprung-Russel diagram (Fig. 1). The brightest of all helium rich stars are massive Wolf-Rayet stars. Most of them are products of the evolution of close binaries with initial component masses above ~ 20 $M_\odot$. About one third of all classical WR stars can be products of the evolution of single massive stars with initial masses exceeding ~ 40 $M_\odot$. The theoretical frequency of their formation together with their lifetimes leads to a reasonable estimate of $\sim 10^3$ for the total number in our Galaxy. Masses of WR stars, as a rule, exceed ~ 10 $M_\odot$. The natural extension of helium burning helium rich stars in binaries to lower luminosities in the H-R diagram is observationally less known. The theoretical estimate of the total number of such stars in our galaxy, most of which have masses in the range 0.3-1.0 $M_\odot$ is about 10^7 (Iben and Tutukov, 1985). The observational estimate of the number of sdB and sdO stars, which are possible counterparts of these helium stars in the core helium burning stage now exceeds $\sim 10^6$ (Heber, 1985). Additional studies of these stars will be helpful in understanding the reasons behind this persistent discrepancy. Some fraction of hot subdwarfs can also be explained as limiting cases of horizontal branch stars which have lost their hydrogen rich envelopes during the core helium flash (Greenstein and Sargent, 1977). Their possible position on the line marked helium stars in binaries is shown by a black square in Figure 1. The differences in their luminosity functions can help in distinguishing these two populations. Possible products of the merger of the two helium degenerates, due to the radiation of gravitational waves, will have masses in the range 0.25 $M_\odot$- 0.95 $M_\odot$ and can be hot subdwarfs.

About 30 stars have been discovered, thus far, that are giants with hydrogen deficient envelopes and which display from time to time large decreases in their visual brightness (Warner, 1967). These stars are known as R Cr B stars. The temporary dimming is usually attributed to the formation of an optically thick dust shell above the helium and carbon-rich envelopes. The hydrogen abundance in their atmospheres by mass is less than 10^{-4}. Members of another class, the non-variable hydrogen deficient stars have usually higher surface temperatures but approximately the same luminosities (Fig. 1). The total number of helium rich giants (both variable and non-variable) in our Galaxy consists of several thousands. The z-distribution of helium rich giants is very wide, spread over 400-1500 parsecs (Warner, 1967, Iben and Tutukov, 1985). Several mechanisms have been proposed till now to explain the origin of helium rich giants. Evolution of close binaries with components of initial masses 5-10 $M_\odot$ which undergo two mass exchange phases produces giants with helium rich envelopes. KS Per and Upsilon Sagitarii are examples of such giants. The estimated number of their total population in our Galaxy, agrees well with the observed number, but the z-distribution and especially the chemistry are still unsolved problems for this scenario. The merging of two degenerate dwarfs, the lighter of which is composed of helium and the other of carbon + oxygen or oxygen + neon has been proposed as an alternative possibility to explain the helium giant formation (Tutukov and Yungelson, 1979; Iben and Tutukov, 1985). The lifetime of a star with a helium rich envelope is $\sim 10^4 - 10^5$ yr. A more detailed study of this scenario is required now. The details of the process of merger are still not clear, e.g., the result of the rather frequent (0.1 yr^{-1}) merging of a close pair consisting of two degenerate helium dwarfs. If the product of merger is a non-degenerate helium star a fraction of hot subdwarfs could be such stars which will evolve finally into non DA dwarfs. Non-degenerate products of the merger of CO dwarfs with masses between 1 - 2.8 $M_\odot$ will populate the short sequence of CO stars (Fig. 1). Models of these stars have been computed by Beaudet and Salpeter (1969). Their lifetime is $\sim 10^5$ yr because most of their energy is carried away by neutrinos. The total number of carbon burning stars in our Galaxy can thus be estimated to be about several thousand.

The natural evolution of a helium rich giant with a CO-core will produce a CO degenerate dwarf after the loss of a part of the helium rich envelope. Evidently this is a way to form non-DA white dwarfs, if the mass of such a star doesn't exceed ~ 1.4 $M_\odot$. But the production rate of this scenario, is possibly not enough to explain all non DA dwarfs. Additional possibilities can be attributed to stellar wind from cooling degenerate dwarfs (Iben and Tutukov, 1986). This way can be, of course, applied to double helium dwarfs that are products of close binary evolution with components of initial masses below ~ 2.5 $M_\odot$. The track of the cooling 0.3 $M_\odot$ helium dwarf is shown in Fig. 1 according to Iben and Tutukov (1986).

If the mass of the carbon-oxygen degenerate core of an R Cr B star can exceed 1.4 $M_\odot$ the supernova explosion becomes inevitable (Fig. 1).

ON THE ORIGIN OF HELIUM RICH STARS

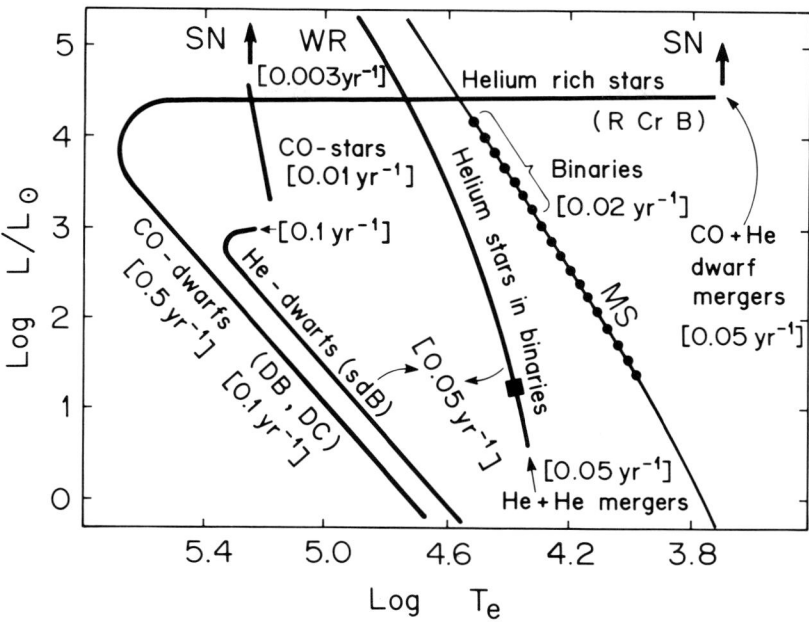

Fig. 1: The Hertzsprung-Russel diagram for Hydrogen deficient stars. The line MS showing the position of main sequence stars some of which can be helium enriched (dots). The position of helium star remnants from close binary evolution products is also shown. The position of a 0.5 $M_\odot$ helium star with a low mass hydrogen envelope is marked by the black square. The evolutionary track of a star consisting of a degenerate carbon-oxygen or oxygen-neon core and a helium rich envelope with the total mass below ~ 1.4 $M_\odot$ is marked with positions of R Cr B and helium rich stars, and non-DA degenerate dwarfs. Also shown is the evolutionary track for the cooling of a 0.3 $M_\odot$ degenerate helium dwarf and the position of CO stars in the core carbon burning stage. Numbers in brackets are rough estimates for formation rates of the corresponding stars.

One more type of helium rich stars can exist as a prdocut of the evolution of close binaries with initial component masses of 5 - 10 $M_\odot$. The two-time mass exchange (BB-case) will evidently considerably enrich with helium the envelope of the initial secondary component of the close binary. This helium rich matter will mix with deeper hydrogen-rich shells. As the time scale of such mixing is not known yet, one may assume that some fraction (less than 0.02) of binary MS stars can be enriched by helium. Masses of these secondaries can be well outside the above range of initial masses because of, e.g., a low initial mass ratio. This possibility also needs to be studied for the explanation of the possible presence of helium rich stars on the main-sequence (Fig. 1).

Secondary components of classical Algols are frequently helium enriched stars. Their evolution has been understood as due to an extensive mass exchange between components of close binaries. As the evidence for the mass exchange is very reliable in this case it would be of interest to check whether the chemical composition of the atmospheres of both the components are similar in algol-type binaries.

In conclusion, I can say that the modern theory of stellar evolution can now lead to several promising possibilities for the origin and evolution of helium rich stars. Many of these scenarios at present involve binary stars. Thus it is now time to study all these possibilities in more detail to compare predictions of different scenaria with observations.

ACKNOWLEDGEMENTS: These short remarks were written during my visit to the Raman Research Institute. It is a great pleasure to thank the Raman Research Institute staff and Prof. V. Radhakrishnan for kind hospitality. I also thank Professors I. Iben Jr., K. Hunger, S. Pottasch and Drs. J. Drilling, J. Libert, U. Heber, K. Rao and D. Schonbernen for many useful discussions on the problem of helium stars.

REFERENCES

Beaudet, G., Salpeter, E.E. 1969, Ap.J. 155, 203.
Greenstein, J.L., Sargent, A.I. 1974, Ap.J. Suppl. 28, 157.
Heber, U. 1985, Astron. Astrophys. (in press).
Iben, I. Jr., Tutukov, A.V. 1985, Ap.J. Suppl. 58, 661.
Iben, I. Jr., Tutukov, A.V. 1986, Ap.J. (in press).
Tutukov, A.V., Yungelson, L.R. 1979, Acta. Astr. 29, 665.
Warner, B. 1967, Mon. Not. R. astr. Soc. 137, 113.

DISCUSSION

SCHÖNBERNER: Your binary scenario for R CrB stars predicts nitrogen-rich and carbon-poor atmospheres, just the opposite to what is observed.

TUTUKOV: Yes, I know. I said that BB mass exchange works for stars like KS Persei. For the single extreme helium stars, the idea of merger was proposed by Iben, Webbink and myself. Then we have some part of the carbon-rich matter mixed with pure helium. Two or three years ago, nobody believed that merging of two degenerate stars would really be possible.

HUNGER: If you start with a specific star on the helium branch, then how would the evolution proceed?

TUTUKOV: It would depend on the mass of the star. If the mass of the helium star is above 4 solar masses then, as has been shown, for instance, by Paczynski and ourselves, it will develop a core of 2.5 solar mass. After the core helium-burning phase, it will never evolve to low effective temperatures.

HUNGER: What is your guess concerning the extreme helium stars: do they belong to the R CrB track or to the generalized main sequence?

TUTUKOV: It is very easy to make a guess; it is very difficult to prove something. So everything is consistent in the diagram.

SUMMARY

P.W.Hill
University Observatory
Buchanan Gardens
St Andrews
Fife KY16 9LZ
Scotland

1. INTRODUCTION

This has been the first international conference devoted to hydrogen-deficient stars and related objects. As Professor Hunger stated in his introduction, one aim of a meeting such as this is to review our achievements and find out what conclusions we can reach about the nature and origin of the objects we study. However, it is even more important that we identify the questions that we must ask and particularly which questions it may be within our means to answer. A newspaper reporter is said to have asked the Chairman of our Local Organizing Committee what we proposed doing to correct this hydrogen deficiency!

Our meeting started with Drilling's review of the galactic distributions and velocity dispersions of the various groups of hydrogen-deficient stars. They cover a wide range from the Population I rapidly rotating intermediate helium stars to the extreme helium stars and cool hydrogen-deficient stars, including the R Coronae Borealis (RCB) variables, which show characteristics of Intermediate Population II. We certainly need to be very clear how each of these classes is defined, particularly as there are a few objects, very definitely hydrogen-deficient, which fail to fit neatly into any of our categories. One star which has been the subject of three contributed papers and mentioned in others has too much hydrogen and too much gravity to be an extreme helium star, too little hydrogen to be an intermediate and too little gravity to be a subdwarf. I refer to BD+13°3224, otherwise known as V652 Her. As many of these stars are discovered to be variable they will be allotted variable star names. To avoid confusion it seems best to mention all known names.

I will briefly summarize what we know about the various groups. Most are shown in the log g - log T_{eff} diagram (Fig.1).

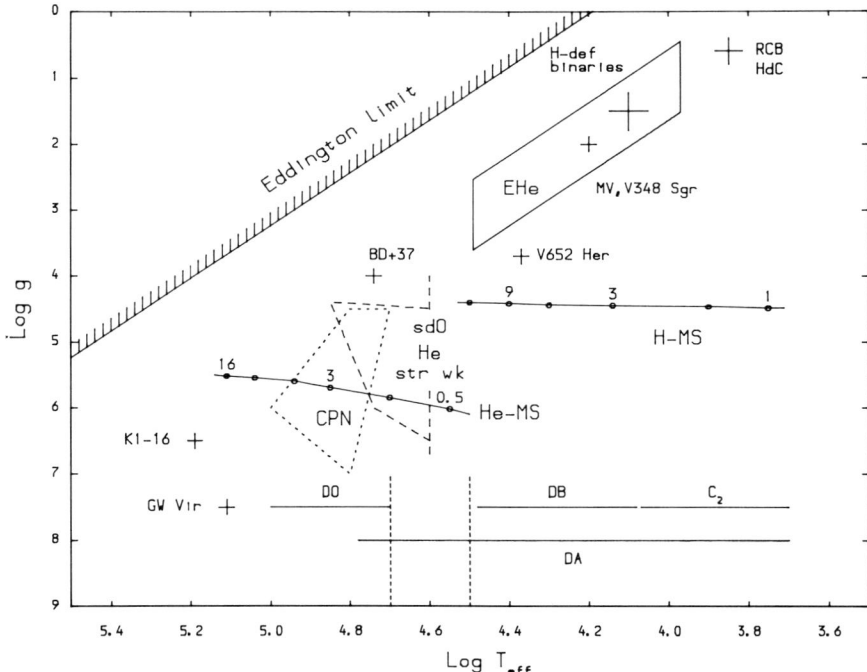

Figure 1. The log g - log T_{eff} diagram for low-mass hydrogen-deficient stars. The objects and regions shown are indicative rather than definitive.

2. HOT HYDROGEN-DEFICIENT STARS

2.1. Extreme helium stars

I shall consider first the extreme hydrogen-deficient stars. Heber defined the class of hot extreme helium (EHe) stars with $n_H/n_{He} < 10^{-3}$ and overabundant carbon ranging in temperature from spectral type A to late O. There are now 17 of these known, Drilling having been responsible for the discovery of a very large proportion of them. Spectroscopic analysis, particularly from Kiel, has been used to determine their surface abundances and place them in the log g - log T_{eff} diagram (Fig.1). Heber demonstrated how they occupy a small strip in the diagram but so far only 4 have a complete fine analysis and even these four are not homogenous in chemical composition for nitrogen and heavier elements. IUE and the new echelle spectrographs, particularly CASPEC, have provided a lot of improved data, with better equivalent widths at higher resolutions, on which to base future analyses. These will perhaps help to solve the evolutionary problem, discussed by Schönberner, of the origin of the range of masses and luminosities seen in preliminary analyses of these stars. We already find differences for one which has had a fine analysis. Are there more real differences?

For those stars bright enough for high resolution IUE spectra to have been obtained the mass loss rate is of the same order of magnitude as for normal stars of the same luminosity. Variable emission lines are seen in the visible spectrum of BD-9°4395 (Jeffery), one of the hotter members of this group which appears to be a non-radial pulsator. V2076 Oph (HD 160641) is another non-radial pulsator (Lynas-Gray et al) and the hottest of the EHe stars which has had any sort of analysis. The cooler EHe stars vary in perhaps a more regular way, maybe radial pulsation (Jeffery et al), and would seem to fit much better the model for RCB pulsation discussed by Saio. In fact, as Landolt's review of the photometric properties demonstrated, most of these stars are variable in light. They are classified as PV Tel stars in the new (4th) edition of the General Catalogue of Variable Stars after the first (HD 168476) for which variability was established. One of the important needs for the future is to really ascertain the nature and timescales of these variations. The amplitudes are small and known timescales vary from hours to weeks. This work needs a lot of careful observing, good results not being possible just by occasional photometric sampling although relatively small telescopes can be used. We saw how important pulsation is to try and tie down parameters such as the mass.

2.2. Hot peculiar RCB stars

Rao discussed two peculiar stars usually classified as hot RCB stars. We don't know very much about them and understand less. Although there are considerable differences between them MV Sgr and V348 Sgr each have the appearance of a hot extreme helium star in a shell and have the IR excesses of the dust clouds. Their long-term light curves resemble those of the RCB stars. V348 Sgr was demonstrated by Schönberner to be in the same region of the log g - log T_{eff} diagram (Fig. 1) as the extreme helium stars of similar effective temperature. We clearly identify this as an area needing more investigation.

2.3. V652 Herculis

The radially pulsating extreme helium star V652 Her is probably an unrelated object. It is carbon-poor, nitrogen and silicon-rich (Jeffery et al). The period decrease rate was reviewed by Lynas-Gray and Kilkenny and a new radial velocity curve presented (Hill & Jeffery) showing an increase in the effective gravity by a factor of 4 at the phase of minimum radius. It is either a unique object or in a very rapid phase of evolution. Saio pointed out that it poses a major problem for pulsation theory as the excitation mechanism is not known and Jeffery's evolutionary model, which fits everything else, appears stable against pulsation.

2.4. Hydrogen-deficient binaries

The hydrogen-deficient binaries were reviewed by Plavec and their evolution discussed by Tutukov. They are nitrogen-rich and carbon-poor and may be a different type of object with different distribution, their hydrogen deficiency being caused by mass exchange. They tend to be discovered along with the extreme helium stars and observed in much the same way in the first place. Four of these are now known, the binarity and pulsation of CPD-58°2721 (LSS 1922) being demonstrated by Jeffery on behalf of a collaboration between Kiel, LSU and St Andrews. A paper from Bangalore showed variable polarization at H-alpha in KS Per (HD 30353) on a timescale of maybe a day. LSS 1922 also has variable Balmer line emission which may also change on the timescale of around a day. The AM CVn stars are hydrogen-deficient binaries of another sort for which flare activity was reported by Kutty. The known members of this group form an inhomogenous group of three. If they are related they may be the hydrogen-deficient analogues of the cataclysmic variables.

3. COOL HYDROGEN-DEFICIENT STARS

3.1. R Coronae Borealis stars

Feast described the so-called "consortium of puffs" model of the RCB minimum which seems to be well-established and in reasonable agreement with the crude description of the observations. Approximately every 40 days at the maximum of the pulsation light curve, where Lawson demonstrated line-splitting at minimum radius in RY Sgr and therefore some correlation with a shock in the pulsation, a puff of carbon particles is emitted which appears in our direction on average every 1000 days or so. The resultant dust shell produces the IR excess which is confirmed by the IRAS results (Walker). The correctness of this model seems confirmed by the relative constancy of fluxes at wavelengths of the L band and longer which arise from the dust cloud and the minima shown at J and shorter wavelengths arising from the eclipse of the star by a puff in our direction. On the other hand L varies with small amplitude in a cycle of about 1000 days. An interesting test is proposed by Menzies who predicts a deep minimum for RY Sgr in the middle of 1986 on the basis of the last two happening about a year after the 1000 day maximum in L. There are many details to be filled in for this model and many problems. How and where do the puffs originate? What is the origin of the 1000 day variation in the dust cloud? Do all the RCB stars pulsate? Do they all have the same effective temperature around 7000 K and if so how do we explain the strengthened carbon bands in some and the apparent historical variations in the carbon bands in others? Lambert showed us how to analyse the spectra of RCB stars but with only two or three analysed in a dozen papers on their composition in the last 50 years there is a clear need for more studies.

Saio demonstrated that the pulsation models are stable at an effective temperature of 7000 K. Indeed, the results of non-linear pulsation analysis fit the observations extremely well, including the period and amplitude irregularities although the calculated amplitudes are just a little large. Another confirmation is that the masses

derived from pulsation agree very well with those determined from
spectroscopic parameters and evolutionary models.

3.2. Hydrogen-deficient carbon stars

The hydrogen-deficient carbon (HdC) stars occupy the same region of the
log g - log T_{eff} diagram (Fig. 1) as the RCB stars, but are not known
to be variable. There are only 4 or 5 known and it seems that these do
not have the dust shells seen around the RCB stars. There is no recent
spectroscopic analysis of HdC stars, the last being a curve of growth
analysis in 1967. The RCB star XX Cam does not have the IR excess
characteristic of the dust shells surrounding the RCB stars. This
raises the possibility that it may be an HdC star and that variations
in other HdC stars have not been detected.

4. ORIGIN AND EVOLUTION OF HYDROGEN-DEFICIENT STARS

The evolution of the extreme hydrogen-deficient group (RCB, HdC and EHe
stars) was discussed by Schönberner who left us with the conclusion
that these stars evolve blueward from the Hayashi limit in the Red
Giant region rather than leaping up from the helium main sequence. The
evidence from pulsation and evolutionary models and spectroscopic
analyses suggested that as being more likely. The RCB stars which have
been analysed have compositions like the extreme helium stars but as
the latter have no IR excess there may not be a direct evolutionary
link. Schönberner asked if the dust cloud could dissipate or whether
the extreme helium stars originate from the hydrogen-deficient carbon
stars which also don't show the dust clouds.

What about their origin? The IRAS results for R CrB itself and
possibly one other show very extended wings at 60 and 100 microns,
suggested by Rao to be evidence for a very cool "fossil" shell. On the
other hand RY Sgr looks like a point source to IRAS. Rao also reported
observations of nebular emission during a minimum of R CrB. Taken
together these were proposed as evidence for the "Born Again" planetary
nebula theory of Iben. Many theories of the origin of extreme hydrogen-
deficient stars were considered by Schönberner but most fail to predict
the observed abundances or lifetimes. Most favoured, in the sense of
raising the fewest objections, is the binary white dwarf coalescence
hypothesis first proposed by Webbink, but the situation is still very
confused.

5. HYDROGEN-POOR SUBLUMINOUS STARS

These occupy the lower part of the log g - log T_{eff} diagram (Fig. 1).
Why do we call them "related objects" when many are as deficient in
hydrogen as the RCB and EHe stars? Because they tend to be classified
as something else first and only some are found to be hydrogen-
deficient. The subdwarf O stars and Central Stars of Planetary Nebulae

(CPN), reviewed by Mendez, had tended to occupy distinct regions in the log g - log T_{eff} diagram until 3 new helium-rich sdO stars found by Drilling, and with no sign of any nebulosity, came into the CPN region as the result of an NLTE analysis described by Heber. These may connect with the extreme helium star region via the two BD+37 stars which are variously described as either sdO or EHe stars. About 20-25 per cent of all subdwarf O stars, including some of the extremes discussed by Heber where hydrogen is completely depleted, appear to be helium-rich. Subdwarf O stars with temperatures less than 40000 K are helium-weak, the boundary being due to gravitational settling rather than a definite abundance effect. The connection with the extreme helium stars is not clear because only about two-thirds of the sdO stars analysed (including two of the three new ones discussed above) are carbon-rich and the effect is not temperature dependent.

Mendez divided the Central Stars of Planetary Nebulae into hydrogen-rich and hydrogen-deficient groups on the basis of a revised classification scheme with about one-third of them apparently being hydrogen-deficient including those in the strange planetary nebulae of the A78 type. Pottasch demonstrated for us with new spectra how abundances in these nebulae change with distance from the centre. Could these be the precursors of the RCB stars?

Further down the diagram (Fig. 1) between most of the planetary nebulae nuclei and the white dwarfs we find the nucleus of the planetary nebula K1-16. This is known to pulsate and may be intermediate between the CPN and the very hot pulsating hydrogen-deficient pre-white dwarfs of the GW Vir (PG1159-035) type described by Liebert. An interesting similarity that has arisen between the hydrogen-deficient CPN and the DO/DB white dwarfs is the apparent absence in both groups of any stars in the effective temperature range from about 30000-50000 K. Another interesting comparison is that there is little evidence for differences in mass between the hydrogen-deficient and hydrogen-rich CPN, as is also the case for DA and DB white dwarfs as shown in Liebert's careful analysis. Liebert demonstrated the two sequences of hydrogen-rich DAs, on the one hand, and hydrogen-deficient DOs and DBs, on the other, with arrows joining them together showing the stars popping up and down between the two sequences as one possible consequence of evolution.

Finally, a careful study by Bues of polarization spectrum variability in the carbon bands of cool DB white dwarfs, the C_2 type, suggests that these could evolve from intermediate helium stars with magnetic fields.

6. INTERMEDIATE HELIUM STARS

The intermediate helium (IHe) stars were reviewed by Hunger. To avoid confusion they have not been included in Fig. 1. A detailed analysis is needed in the first place to identify them because they have abundance ratios N_H/N_{He} from about 3 to 0.1. These appear to split into a low mass non-rotating group for which the helium enrichment could be evolutionary, and a high mass rotating group of probably main sequence stars with magnetic fields which appear to form a continuous temperature sequence with the helium-weak and Ap stars, as described by Barker. The apparent helium enrichment is due to the interaction of diffusion with the stellar wind and magnetic field which Michaud showed should apply to the limited effective temperature range of the intermediate helium stars. Sigma Ori E is the best studied of the IHe stars. It has a shell and radio emission suggesting interaction in a magnetosphere between the stellar wind and the magnetic field. But there are a number of problems with the observations, problems with the equivalent widths which will not resolve the problem of whether these stars have main sequence masses until new high resolution data from linear detectors has been analysed. From the point of view of looking at the end points of stellar evolution through hydrogen deficiency it seems that this particular group of stars perhaps belongs properly with the Ap stars in the chemically-peculiar upper main sequence star stable rather than in the stable of true hydrogen deficiency.

7. CONCLUSIONS

What conclusions can we draw? Our understanding of the origin of the hydrogen deficiency and the evolution of many of these stars is still extremely uncertain. We need many more observations, a lot of good observations with the best detectors and of course we must find the best way to use the new devices becoming available. By combining results we can see how the various groups of stars fit together in the log g - log T_{eff} diagram (Fig. 1). I think this is of extreme value. People like me who have been working very much in one limited region of the diagram have learnt a great deal of what is happening elsewhere and how this will interact with our work on the more luminous extreme helium stars. I am sure that this will also be true for those more experienced with other groups of hydrogen-deficient stars.

ACKNOWLEDGMENTS

I am grateful to Simon Jeffery for assistance in preparing the diagram and for lending me his notes of those sessions I unavoidably missed. Kameswara Rao initiated the idea of this meeting during the IAU General Assembly at Patras in 1982 and we must thank him and his indefatigable team of helpers from Bangalore for the hospitality and impeccable organisation which have made our meeting such a pleasant one. To Kurt Hunger we are indebted for shouldering the scientific organisation and for inspiring so many of us. I would only conclude by thanking everybody for their contributions and I think I can say that we have had a really good, stimulating and exciting scientific meeting.

X APPENDIX

APPENDIX A: A CATALOG OF HYDROGEN-DEFICIENT STARS

J. S. Drilling
Department of Physics and Astronomy
Louisiana State University
Baton Rouge, USA

P. W. Hill
University Observatory
St. Andrews, Scotland

The following four tables were originally presented as part of the paper entitled 'Basic Data on Hydrogen-Deficient Stars' by J. S. Drilling, which appears earlier in this volume. A number of corrections and additions have been made by the participants, mostly by P. W. Hill using the SIMBAD data base. A much improved version of the catalog therefore follows. Helium-rich central stars of planetary nebulae, helium-rich white dwarfs, and Wolf-Rayet stars are not included. A complete list of helium-rich central stars is given by Mendez et al. elsewhere in this volume.

TABLE A1. EXTREME HELIUM STARS

Name	α	(2000)	δ	ℓ	b	V	B-V	r.v. (km/s)	Ref.	Remarks
BD+37°442	01 58 35.4	+38 34 08		137	-22	9.99	-0.29	-156	1,50	sdO?
KS Per	04 48 53.8	+43 16 32		162	-1	7.85	0.49	+5	3,50	HD 30353; binary
LSS 99	06 54 46.3	-10 48 41		223	-4	12.29	0.70	+109	6,50,68	
BD+37°1977	09 24 23.9	+36 42 54		187	46	10.21		-59	7,22,41	sdO?
BD+10°2179	10 38 55.2	+10 03 48		235	54	9.95	-0.19	+158	33,50,68	
CPD-58°2721	10 47 56.8	-59 08 37		288	0	10.50	0.72	-12	9,10,50	LSS 1922; binary
DY Cen	13 25 34.0	-54 14 47		308	8	12.52*	0.35*		12,19	RCB
LSS 3184	14 01 36.5	-66 10 02		310	-4	12.60	0.03	-89	6,68	
HD 124448	14 14 58.6	-46 17 19		318	14	9.98	-0.10	-65	34,50,68	
CoD-48°10153	15 38 59.4	-48 35 57		329	6	11.48	0.44	-4	14,50,68	LSS 3378
BD-9°4395	16 28 35.2	-09 19 34		6	26	10.54	0.06	-58	16,50,68	
V652 Her	16 48 04.7	+13 15 41		31	33	10.51*	-0.18*	+3	42,23,58	BD+13°3224; pec.
HDE 320156	17 37 58.5	-35 23 05		354	-2	9.78	0.84	+7	9,10,50	LSS 4300; binary
V2076 Oph	17 41 50.2	-17 54 08		9	6	9.83	0.14	+70	35,50,17	HD 160641
CoD-46°11775	17 42 33.7	-46 58 46		344	-9	11.22	0.06	-91	6,68	LSE 78
LSS 4357	17 44 25.4	-19 38 03		8	5	12.62	0.41	-99	6,50,68	
LSIV-1°2	17 51 26.7	-01 43 15		24	13	11.01	0.38		9,50	
BD-1°3438	18 03 55.3	-01 00 13		27	10	10.33	0.46		16,50	LSIV-1°3
LSIV+6°2	18 06 55.3	+06 21 46		34	13	12.17	-0.07		6	
PV Tel	18 23 14.7	-56 37 43		338	-19	9.27	-0.01	-171	36,50,68	HD 168476
V348 Sgr	18 40 19.8	-22 54 29		11	-8	11.83*	0.30*	+174	19,24	RCB?
LSS 5121	18 43 16.4	-18 31 47		15	-7	13.25	0.32	-62	6,50,68	
MV Sgr	18 44 32.1	-20 57 16		13	-8	12.70*	0.26*	-68	25	RCB
LSIV-14°109	18 59 39.4	-14 26 11		21	-8	11.15	0.33		27,50	
υ Sgr	19 21 43.6	-15 57 18		22	-14	4.61	0.10	+12	31,50,59	binary
HDE 225642	19 45 17.0	+33 58 25		69	5	10.31	0.16	-88	32,50,68	LSII+33°5
BD+1°4381	20 51 21.4	+02 18 47		50	-25	9.56	0.19	+12	27,50,68	LSIV+2°13

*At or near maximum light.

APPENDIX A: A CATALOG OF HYDROGEN-DEFICIENT STARS

TABLE A2. COOL HYDROGEN-DEFICIENT STARS

Name	α (2000) δ	ℓ	b	V*	B-V*	r.v. (km/s)	Ref.
XX Cam	04 08 38.7 +53 21 39	150	1	7.30	0.87	+16	11,19,28
HV 5637	05 11 32 -67 56 00	LMC		15.79	1.37		4,48
W Men	05 26 24 -71 11 18	LMC		13.86	0.42	+264	4,8,48
HV 12842	05 45 03 -64 24 24	LMC		13.65	0.51		4,48
SU Tau	05 49 06 +19 04 00	189	-4	9.70	1.10	+37	11,19,54
UW Cen	12 43 17.1 -54 31 41	302	8	9.11	0.67		12,19
Y Mus	13 05 48.4 -65 30 48	304	-3	10.37	0.97		12,19
S Aps	15 09 24.6 -72 03 45	313	-12	9.88	1.28		12,19
HD 137613	15 27 48.3 -25 10 11	342	26	7.50	1.19	+55	12,19,28
R CrB	15 48 34.4 +28 09 24	45	51	5.83	0.59	+21	11,19,28
RT Nor	16 24 19.0 -59 20 42	327	-7	10.24	1.12		12,19
HD 148839	16 35 45.9 -67 07 37	322	-13	8.31	0.93	-31	12,13
RZ Nor	16 32 41.6 -53 17 09	332	-4	11.00	1.33		12,19
LR Sco	17 27 54 -43 50 54	345	-5	9.72	0.55		15,19
WX CrA	18 08 50.4 -37 19 46	355	-8	10.43	1.26		12,19
V3795 Sgr	18 13 24 -25 47 24	6	-4	10.97	1.02		12,80
VZ Sgr	18 15 09 -29 42 24	3	-6	10.15	0.73		12,19
RS Tel	18 18 51.3 -46 32 54	348	-14	9.77	0.81		12,19
GU Sgr	18 24 15.5 -24 15 29	8	-5	10.11	1.17		12,19
HD 173409	18 46 26.5 -31 20 34	4	-13	9.54	0.89	-65	19,20,28
V CrA	18 47 32.2 -38 09 31	358	-16	10.24	0.79		15,19
HD 175893	18 58 47.4 -29 30 17	7	-14	9.30	1.15	+42	12,19,28
SV Sge	19 08 12 +17 37 42	51	4	10.39	1.86	+4	11,19,28
RY Sgr	19 16 32.8 -33 31 18	4	-19	6.18	0.62	-10	5,19,29
V605 Aql	19 18 20.4 +01 46 51	38	-5	11.0p			19,26
HD 182040	19 23 10.1 -10 42 10	27	-12	6.98	1.05	-47	12,19,28
V482 Cyg	19 59 44 +33 58 30	70	2	12.1p			19,26
U Aqr	22 03 20.0 -16 37 40	39	-50	11.17	0.99	+103	12,19,21
UV Cas	23 02 13 +59 36 42	110	0	10.60	1.38	-27	2,19,54

*At or near maximum light.

TABLE A3. INTERMEDIATE HELIUM STARS

Name	α (2000) δ	ℓ	b	V	B-V	r.v. (km/s)	Ref.	Remarks
δ Ori C	05 32 00.5 -00 17 04	204	-18	6.85	-0.15	+12	59,67,70	HD 36485
HD 37017	05 35 21.8 -04 29 36	208	-19	6.54	-0.14	+29	59,67,70	
σ Ori E	05 38 47.1 -02 35 39	207	-17	6.65*	-0.19*	+29	59,67,70	HD 37479
HD 37776	05 40 56.3 -01 30 26	206	-16	6.98	-0.14	+27	59,67,70	
HDE 260858	06 37 46.7 +12 46 04	200	3	9.14			18,70	
HDE 264111	06 47 53.8 +04 40 01	208	1	9.65	0.04	var?	30,51,70	
CoD-27°3748	07 12 02.3 -27 43 04	240	-8	9.27	-0.19		44,70	CPD-27°1791
HD 58260	07 23 19.7 -36 20 26	249	-10	6.73	-0.14	+36	59,67,70	
HD 60344	07 33 02.2 -23 56 03	239	-2	7.71	-0.17	+30	52,74,77	
HD 64740	07 53 03.7 -49 36 47	263	-11	4.62	-0.23	+8	53,59,70	
HD 66522	08 01 35.1 -50 36 22	265	-11	7.21	0.05	+15	70,74,78	
CoD-46°4639	08 49 39.7 -46 50 51	266	-2	10.0	0.08		49,52,70	CPD-46°3093
HD 96446	11 06 05.7 -59 56 59	290	0	6.68	-0.16	+7	59,67,70	
CPD-62°2124	11 35 37.7 -63 15 54	295	-2	11.04	0.10	-7	55,57,70	LSS 2394
HD 133518	15 06 56.0 -52 01 49	323	5	6.39	-0.10	-2	52,59,70	
HD 144941	16 09 24.6 -27 16 30	348	18	10.11	0.05	-53	52,64,70	
HD 149257	16 35 45.3 -45 37 17	338	1	8.48	-0.04	+6	70,73,74	
CPD-69°2698	17 12 32.8 -70 05 07	322	-18	9.36	-0.11	-65	52,64,70	CoD-69°1618
HD 164769	18 03 51.1 -27 18 15	3	-3	9.25	-0.07		44,70	
HD 168785	18 22 45.3 -30 08 23	3	-8	8.49	-0.04	+5	52,65,70	
HD 184927	19 35 32.0 +31 16 36	66	5	7.46	-0.17	-16	56,59,60	
HD 186205	19 42 37.9 +09 13 40	47	-7	8.53	0.05	-3	59,67,70	
LSII+35°51	20 08 58.2 +35 28 25	73	1	11.1p			70	
LSII+36°37	20 14 08.5 +36 46 58	74	1	11.30	0.33		55,70	

*At or near maximum light.

APPENDIX A: A CATALOG OF HYDROGEN-DEFICIENT STARS

TABLE A4. HELIUM-RICH sdO STARS WITH B < 14.5.

Name	α (2000) δ	ℓ	b	V	B-V	r.v. (km/s)	Ref.	Remarks
SB 21	00 04 24 -24 25	45	-79	13.87			45,72	TON S 137
SB 58	00 10 00 -26 13	36	-81	12.90			45,72	
LB 1566	00 40 16 -55 02	306	-62	13.11	-0.30		45,74	JL 202
SB 705	01 43 12 -38 33	263	-74	13.03			45,72	
HD 49798	06 48 04.8 -44 18 59	254	-19	8.29	-0.24	-18	37,67	
TD1 32705	07 14 30.8 +22 17 17	195	15	11.7			81	UV0711+22
CoD-31°4800	07 36 30.0 -32 12 57	246	-6	10.52	-0.31		38	CPD-31°1701
LSS 630	07 39 41.9 -27 27 48	243	-3	13.56	-0.28	-9	40,45	
BD-3°2179	08 02 14.5 -03 58 16	225	14	10.33	-0.30		71	
BD+75°325	08 10 49.2 +74 57 55	140	31	9.54	-0.37	-19	39,67	
TD1 32708	08 35 20.1 -01 55 45	227	22	11.44	-0.31		6,81	UV0832-01
CoD-34°5246	08 46 53.0 -35 24 11	257	5	12.55	-0.27		6	LSS 1150
TD1 32709	09 07 08.3 -03 06 09	233	28	11.93	-0.31		6,81	UV0904-02
LSS 1274	09 18 55.7 -57 04 38	277	-5	12.91	-0.21	+24	6,68	
BD+48°1777	09 30 39.7 +48 15 43	170	46	10.75	-0.34	-29	7	
LSS 1349	09 46 56.9 -50 12 39	275	3	13.36	0.05		6	
CoD-24°9052	10 25 51.1 -24 53 22	266	27	9.6		+30	47	
HD 113001B	13 00 26.0 +35 45 23	111	81	10.58	-0.25	-9	49,59,67	
HZ 44	13 23 42 +36 07	89	79	11.71	-0.27		49,67	
LSE 153	13 53 06 -46 45	314	15	11.35	-0.26	-17	43,68	
HD 127493	14 32 21.6 -22 39 25	331	35	10.05	-0.24	+13	49,59,61	
HD 128220B	14 35 15.8 +19 12 54	20	65	8.54*	+0.21*	-7	49,59,67	
LSE 259	16 53 54 -56 02	332	-8	12.6		+43	43,68	
BD+39°3226	17 46 31.9 +39 19 09	65	29	10.21	-0.29	-273	46,75	
LSE 263	19 02 12 -51 30	345	-23	11.8		+13	43,68	
JL 9	19 08 18 -72 30	323	-27	13.24	-0.28		46,74	
LSIV+10°9	20 43 02.5 +10 34 10	56	-19	11.99	-0.27		43,71	
BD+25°4655	21 59 42.0 +26 25 57	82	-22	9.69	-0.26	+59	49,75,79	
GS 259-8	22 49 03 +37 54	97	-19	12.5			49	
PHL 540	23 29 12 -10 05	70	-64	13.38			46,76	
TON S 103	23 33 54 -28 51	23	-73	14.64	-0.23		45,74	PHL 561
SB 933	23 59 12 -40 32	338	-73	14.24			45,72	

*AB

REFERENCES

1. Rebeirot, E. 1966, *Publ. Obs. Haute Provence 8*, No. 19.
2. Shenavrin, V. I. 1979, *Soviet Astr.* **23**, 696.
3. Nariai, K. 1972, *PASJ* **24**, 495.
4. Feast, M. W. 1972, *M.N.R.A.S.* **158**, 11P.
5. Feast, M. W., et. al. 1977, *M.N.R.A.S.* **178**, 415.
6. Drilling, J. S. 1986, in preparation.
7. Berger, J., Fringant, A. M., and Rebeirot, E. 1974, *Compt. Rend. Serie B* **278**, 227.
8. Rodgers, A. W. 1970, *Obs.* **90**, 197.
9. Drilling, J. S. 1980, *Ap. J.* **242**, L43.
10. Jeffery, C. S., and Drilling, J. S. 1986, in preparation.
11. Fernie, J. D., Sherwood, V., and DuPuy, D. L. 1972, *Ap. J.* **172**, 383.
12. Kilkenny, D., Coulson, I. M., Laing, J. D., Jones, J. S., and Engelbrecht, C. 1985, *S. African Astr. Obs. Circ. No. 9*, p. 87.
13. Warner, B. 1967, *M.N.R.A.S.* **137**, 119.
14. Drilling, J. S. 1973, *Ap. J.* **179**, L31.
15. Walker, H. J. 1986, private communication.
16. MacConnell, D. J., Frye, R. L., and Bidelman, W. P. 1972, *PASP* **84**, 388.
17. Lynas-Gray, A. E., et al. 1986, this volume.
18. Hill, P. W. 1986, private communication.
19. Bidelman, W. P. 1979, *Mass Loss and Evolution of O-type Stars*, eds. P. S. Conti and C. W. H. de Loore (Dordrecht: D. Reidel), p. 305.
20. Vandervort, G. L. 1958, *AJ* **63**, 477.
21. Bond, H. E., Luck, R. E., and Newman, M. J. 1979, *Ap. J.* **233**, 205.
22. Wolff, S. C., Pilachowski, C. A., and Wolstencroft, R. D. 1974, *Ap. J.*, **194**, L83.
23. Hill, P. W., Kilkenny, D., Schönberner, D., and Walker, H. J. 1981, *M.N.R.A.S.* **197**, 81.
24. Houziaux, L. 1968, *BAC* **19**, 265.
25. Herbig, G. H. 1964, *Ap. J.*, **140**, 1317.
26. General Catalog of Variable Stars.
27. Drilling, J. S. 1979, *Ap. J.* **228**, 491.
28. Bidelman, W. P. 1953, *Ap. J.* **117**, 25.
29. Alexander, J. B. et al. 1972, *M.N.R.A.S.* **158**, 305.
30. Landolt, A. U. 1973, *PASP* **85**, 661.
31. Schönberner, D., and Drilling, J. S. 1983, *Ap. J.* **268**, 225.
32. Drilling, J. S. 1978, *Ap. J.* **223**, L29.
33. Klemola, A. R. 1961, *Ap. J.* **134**, 130.
34. Popper, D. M. 1942, *PASP* **54**, 160.
35. Bidelman, W. P. 1952, *Ap. J.* **116**, 227.
36. Thackeray, A. D., and Wesselink, A. J. 1952, *Obs.* **72**, 248.
37. Jaschek, M., and Jaschek, C. 1963, *PASP* **75**, 365.
38. Garrison, R. F., and Hiltner, W. A. 1973, *Ap. J.* **179**, L117.

39. Gould, N. L., Herbig, G. H., and Morgan, W. W. 1957, *PASP* **69**, 242.
40. Havlen, R. J. 1976, *PASP* **88**, 685.
41. Rossi, L., Viotti, R., Darius, J., and D'Antona, F. 1980, *Proceedings of Second European IUE Conference* (ESA SP-157), p. 323.
42. Berger, J., and Greenstein, J. L. 1963, *PASP*, **75**, 336.
43. Drilling, J. S. 1983, *Ap. J.* **270**, L13.
44. Buscombe, W. 1980, *MK Spectral Classifications: Fourth General Catalogue* (Evanston: Northwestern U.).
45. Hunger, K., Gruschinske, J., Kudritzki, R. P., and Simon, K. P. 1981, *AA* **95**, 244.
46. Heber, U. 1986, private communication.
47. Kilkenny, D., Heber, U., and Hunger, K. 1986, *AA* **155**, 175.
48. Feast, M. W. 1979, *Changing Trends in Variable Star Research*, eds. F. M. Bateson, J. Smak, and I. H. Urch (Hamilton, New Zealand: U. Waikato), p. 246.
49. Hunger, K. 1975, *Problems in Stellar Atmospheres and Envelopes*, eds. B. Baschek, W. H. Kegel, and G. Traving (New York, Heidelberg, Berlin: Springer-Verlag), p. 57.
50. Landolt, A. U. 1986, this volume.
51. Stephenson, C. B. 1967, *Ap. J.* **149**, 35.
52. MacConnell, D. J., Frye, R. L., and Bidelman, W. P. 1970, *PASP* **82**, 730.
53. Hiltner, W. A., Garrison, R. F., and Schildt, R. E., 1969, *Ap. J.* **157**, 313.
54. Rao, N. K. 1986, private communication.
55. Drilling, J. S. 1981, *Ap. J.* **250**, 701.
56. Bond, H. E. 1970, *PASP* **82**, 321.
57. Ardeberg, A., and Maurice, E. 1977, *AA Suppl.* **28**, 153.
58. Landolt, A. U. 1975, *Ap. J.* **196**, 789.
59. Abt, H. A., and Biggs, E. S. 1972, *Bibliography of Stellar Radial Velocities* (Tucson: Kitt Peak National Obs.).
60. Lee, P., and Daigle, P. 1972, *PASP* **84**, 842.
61. Hill, P. W., Kilkenny, D., and van Breda, I. G. 1974, *M.N.R.A.S.* **168**, 451.
62. Dinger, A. C. L. 1969, *Astrophys. Space Sci.* **6**, 118.
63. Wolf, R. E. A. 1973, *AA* **26**, 127.
64. Hunger, K., and Kaufmann, J. P. 1973, *AA* **25**, 261.
65. Kaufman, J. P., Rahe, J., and Schönberner, D. 1974, *AA* **36**, 201.
66. Hack, M. 1967, *Modern Astrophysics*, ed. M. Hack (New York: Gordon and Breach), p. 163.
67. Blanco, V. M., Demers, S., Douglass, G. G., and FitzGerald, M. P. 1968, *Publ. U. S. Naval Obs.* Ser. II, Vol. 21.
68. Drilling, J. S., and Heber, U. 1986, in this volume.
69. Peterson, A. V. 1970, thesis, Calif. Inst. Tech.
70. Walborn, N. R. 1983, *Ap. J.* **268**, 195.
71. Walker, A. R. 1981, *M.N.R.A.S.* **197**, 241.
72. Graham, J. A., and Slettebak, A. 1973, *AJ* **78**, 295.
73. Hron, J., Maitzen, H. M., Moffat, A. F. J., Schmidt-Kaler, Th., and Vogt, N. 1985, *AA Suppl.* **60**, 355.

74. Nicolet, B. 1978, *AA Suppl.* **34**, 1.
75. Dworetsky, M. M., Whitelock, P. A., and Carnochan, D. J. 1982, *M.N.R.A.S.* **201**, 901.
76. Kilkenny, D., Hill, P. W., and Brown, A. 1977, *M.N.R.A.S.* **178**, 123.
77. Kaufmann, J. P., and Hunger, K. 1975, *AA* **38**, 351.
78. Thackeray, A. D., Tritton, S. B., and Walker, E. N. 1973, *Mem.N.R.A.S.* **77**, 199.
79. Greenstein, J. L., and Sargent, A. I. 1974, *Ap. J. Suppl.* **28**, 157.
80. Hoffleit, D. 1972, *IBVS* No. 16.
81. Berger, J. and Fringant, A.-M.: 1980, AA **85**, 367